Chevrolet Camaro & Pontiac Firebird Automotive Repair Manual

by Mike Stubblefield and John H Haynes

Member of the Guild of Motoring Writers

Models covered:

All Chevrolet Camaro and Pontiac Firebird models 1993 through 2002

(24017-5AA22)

Haynes Group Limited
Haynes North America, Inc.

www.haynes.com

About this manual

Its purpose

The purpose of this manual is to provide comprehensive, useful and accessible automotive repair information, to help you get the best value from your vehicle. It can do so in several ways. It can help you decide what work must be done, even if you choose to have it done by a dealer service department or a repair shop; it provides information and procedures for routine maintenance and servicing; and it offers diagnostic and repair procedures to follow when trouble occurs.

We hope you use the manual to tackle the work yourself. For many simpler jobs, doing it yourself may be quicker than arranging an appointment to get the vehicle into a shop and making the trips to leave it and pick it up. More importantly, a lot of money can be saved by avoiding the expense the shop must pass on to you to cover its labor and overhead costs. An added benefit is the sense of satisfaction and accomplishment that you feel after doing the job yourself. However, this manual is not a substitute for a professional certified technician or mechanic. There are risks associated with automotive repairs. The ability to make repairs on a vehicle depends on individual skill, experience and proper tools. Individuals should act with due care and acknowledge and assume the risk of performing automotive repairs.

Using the manual

The manual is divided into Chapters. Each Chapter is divided into numbered Sections, which are headed in bold type between horizontal lines. Each Section consists of consecutively numbered paragraphs.

The reference numbers used in illustration captions pinpoint the pertinent Section and the Step within that Section. That is, illustration 3.2 means the illustration refers to Section 3 and Step (or paragraph) 2 within that Section.

Procedures, once described in the text, are not normally repeated. When it's necessary to refer to another Chapter, the reference will be given as Chapter and Section number. Cross references given without use of the word "Chapter" apply to Sections and/or paragraphs in the same Chapter. For example, "see Section 8" means in the same Chapter. References to the left or right side of the vehicle assume you are sitting in the driver's seat, facing forward.

This repair manual is produced by a third party and is not associated with an individual car manufacturer. If there is any doubt or discrepancy between this manual and the owner's manual or the factory service manual, please refer to factory service manual or seek assistance from a professional certified technician or mechanic. Even though we have prepared this manual with extreme care, neither the publisher nor the author can accept responsibility for any errors in, or omissions from, the information given.

NOTE

A **Note** provides information necessary to properly complete a procedure or information which will make the procedure easier to understand.

CAUTION

A **Caution** provides a special procedure or special steps which must be taken while completing the procedure where the Caution is found. Not heeding a Caution can result in damage to the assembly being worked on.

WARNING

A **Warning** provides a special procedure or special steps which must be taken while completing the procedure where the Warning is found. Not heeding a Warning can result in personal injury.

Acknowledgements

Technical writers who contributed to this project include Rob Maddox, Jay Storer and Larry Warren. Wiring diagrams originated exclusively for Haynes North America, Inc. by Valley Forge Technical Information Services.

A book in the Haynes Automotive Repair Manual Series

ISBN-10: 1-56392-556-7
ISBN-13: 978-1-56392-556-6

Library of Congress Control Number 2004115061

Disclaimer

There are risks associated with automotive repairs. The ability to make repairs depends on individual skill, experience and proper tools. Individuals should act with due care and acknowledge and assume the risk of making automotive repairs. While every attempt is made to ensure that the information in this manual is correct, no liability can be accepted by the authors or publishers for loss, damage or injury caused by any errors in, or omissions from, the information given.

Contents

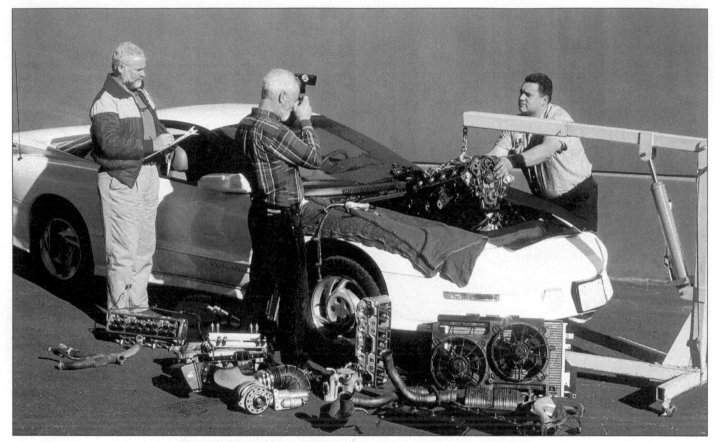

Haynes author, photographer and mechanic with 1994 Pontiac Trans Am

Introduction to the Chevrolet Camaro and Pontiac Firebird

These models are available in two-door liftback body styles only. The body construction is of the space frame type with plastic body panels.

Engines used in these vehicles include the 3.4 liter V6, the 3800 V6, and the 5.7 liter V8. Multi-port fuel injection (MPFI) is used on all models. Later models are equipped with the On Board Diagnostic Second Generation (OBDII) computerized engine management system that controls virtually every aspect of engine operation. OBDII is designed to keep the emissions system operating at the feder-

ally specified level for the life of the vehicle. OBDII monitors emissions system components for signs of degradation and engine operation for any malfunction that could affect emissions, turning on the Service Engine Soon light if any faults are detected.

The engine drives the rear wheels through either a five- or six-speed manual or four-speed automatic transmission via a driveshaft, differential and axles.

The power assisted rack-and-pinion steering is mounted behind the engine. Power assisted steering is standard.

The front suspension is composed of upper and lower control arms, coil spring/shock absorber assemblies and a stabilizer bar. The rear suspension is solid axle with trailing arms, a torque arm, coil springs, shock absorber units, stabilizer bar and a track bar.

The brakes are either four wheel disc or disc on the front and drum on the rear wheels, depending on model, with an Anti-lock Brake System (ABS) standard.

Vehicle identification numbers

Modifications are a continuing and unpublicized part of vehicle manufacturing. Since spare parts manuals and lists are compiled on a numerical basis, the individual vehicle numbers are essential to correctly identify the component required.

Vehicle Identification Number (VIN)

This very important identification number is stamped on a plate attached to the left side of the dashboard and is visible through the driver's side of the windshield (see illustration). The VIN also appears on the Vehicle Certificate of Title and Registration. It contains valuable information such as where and when the vehicle was manufactured, the model year and the body style.

VIN engine and model year codes

Two particularly important pieces of information found in the VIN are the engine code and the model year code. Counting from the left, the engine code letter designation is the 8th digit and the model year code letter designation is the 10th digit.

On the models covered by this manual the engine codes are:

S	3.4L V6
K	3800 V6

P	5.7L V8 (LT1)
	(1997 and earlier)
G	5.7L V8 (LS1)
	(1998 and later)

On the models covered by this manual the model year codes are:

P	1993
R	1994
S	1995
T	1996
V	1997
W	1998
X	1999
Y	2000
1	2001
2	2002

Service parts identification label

This label is located on the glove compartment door (see illustration). It lists the VIN, paint number, options and other information specific to the vehicle it's attached to. Always refer to this label when ordering parts.

Engine identification numbers

On 3.4L engines this number is found on a pad at the front of the block, just above the water pump. Optional locations for the number are on the lower left side of the engine block, below the exhaust manifold and on the left rear side of the block on a casting to the rear of the manifold. On 3800 engines the number is found on a sticker below the coil pack and on a pad on the left rear side of the block on a casting to the rear of the manifold. On 2000 3800 engines, the front sticker is below the water pump, and on 2001 and later models it's below the air-conditioning compressor. On V8 engines the numbers are found on the lower left side of the block, just above the oil pan rail and on the left rear side of the block on a casting to the rear of the manifold.

Automatic transmission number

On the 4L60-E transmission, the ID number is stamped into the casting on the right rear corner of the housing on models through 1999, and on later models on the left side of the case, above the fluid pan.

Manual transmission number

The nameplate/ID number on the both the five- and six-speed transmission can be found on the left side of the case.

Vehicle Emissions Control Information label

This label is found in the engine compartment. See Chapter 6 for more information on this label.

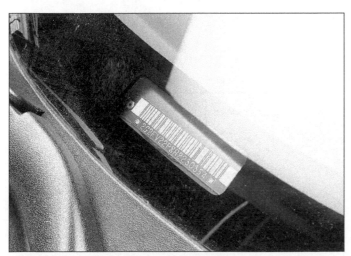

The Vehicle Identification Number (VIN) is visible through the driver's side of the windshield

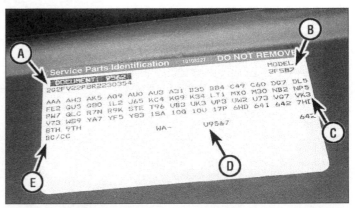

The service parts identification label is located on the glove box door and contains much important information

A	Vehicle Identification Number (VIN)	C	Options
		D	Paint codes
B	Body type and style	E	Paint type

Buying parts

Replacement parts are available from many sources, which generally fall into one of two categories - authorized dealer parts departments and independent retail auto parts stores. Our advice concerning these parts is as follows:

Retail auto parts stores: Good auto parts stores will stock frequently needed components which wear out relatively fast, such as clutch components, exhaust systems, brake parts, tune-up parts, etc. These stores often supply new or reconditioned parts on an exchange basis, which can save a considerable amount of money. Discount auto parts stores are often very good places to buy materials and parts needed for general vehicle maintenance such as oil, grease, filters, spark plugs, belts, touch-up paint, bulbs, etc. They also usually sell tools and general accessories, have convenient hours, charge lower prices and can often be found not far from home.

Authorized dealer parts department: This is the best source for parts which are unique to the vehicle and not generally available elsewhere (such as major engine parts, transmission parts, trim pieces, etc.).

Warranty information: If the vehicle is still covered under warranty, be sure that any replacement parts purchased - regardless of the source - do not invalidate the warranty!

To be sure of obtaining the correct parts, have engine and chassis numbers available and, if possible, take the old parts along for positive identification.

Maintenance techniques, tools and working facilities

Maintenance techniques

There are a number of techniques involved in maintenance and repair that will be referred to throughout this manual. Application of these techniques will enable the home mechanic to be more efficient, better organized and capable of performing the various tasks properly, which will ensure that the repair job is thorough and complete.

Fasteners

Fasteners are nuts, bolts, studs and screws used to hold two or more parts together. There are a few things to keep in mind when working with fasteners. Almost all of them use a locking device of some type, either a lockwasher, locknut, locking tab or thread adhesive. All threaded fasteners should be clean and straight, with undamaged threads and undamaged corners on the hex head where the wrench fits. Develop the habit of replacing all damaged nuts and bolts with new ones. Special locknuts with nylon or fiber inserts can only be used once. If they are removed, they lose their locking ability and must be replaced with new ones.

Rusted nuts and bolts should be treated with a penetrating fluid to ease removal and prevent breakage. Some mechanics use turpentine in a spout-type oil can, which works quite well. After applying the rust penetrant, let it work for a few minutes before trying to loosen the nut or bolt. Badly rusted fasteners may have to be chiseled or sawed off or removed with a special nut breaker, available at tool stores.

If a bolt or stud breaks off in an assembly, it can be drilled and removed with a special tool commonly available for this purpose. Most automotive machine shops can perform this task, as well as other repair procedures, such as the repair of threaded holes that have been stripped out.

Flat washers and lockwashers, when removed from an assembly, should always be replaced exactly as removed. Replace any damaged washers with new ones. Never use a lockwasher on any soft metal surface (such as aluminum), thin sheet metal or plastic.

Grade 1 or 2   Grade 5   Grade 8

Bolt strength marking (standard/SAE/USS; bottom - metric)

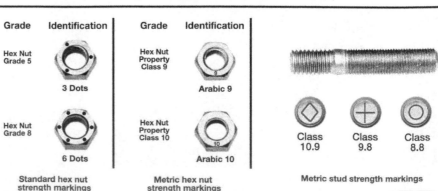

Grade	Identification
Hex Nut Grade 5	3 Dots
Hex Nut Grade 8	6 Dots

Standard hex nut strength markings

Grade	Identification
Hex Nut Property Class 9	Arabic 9
Hex Nut Property Class 10	Arabic 10

Metric hex nut strength markings

Class 10.9   Class 9.8   Class 8.8

Metric stud strength markings

00-1 HAYNES

Fastener sizes

For a number of reasons, automobile manufacturers are making wider and wider use of metric fasteners. Therefore, it is important to be able to tell the difference between standard (sometimes called U.S. or SAE) and metric hardware, since they cannot be interchanged.

All bolts, whether standard or metric, are sized according to diameter, thread pitch and length. For example, a standard 1/2 - 13 x 1 bolt is 1/2 inch in diameter, has 13 threads per inch and is 1 inch long. An M12 - 1.75 x 25 metric bolt is 12 mm in diameter, has a thread pitch of 1.75 mm (the distance between threads) and is 25 mm long. The two bolts are nearly identical, and easily confused, but they are not interchangeable.

In addition to the differences in diameter, thread pitch and length, metric and standard bolts can also be distinguished by examining the bolt heads. To begin with, the distance across the flats on a standard bolt head is measured in inches, while the same dimension on a metric bolt is sized in millimeters (the same is true for nuts). As a result, a standard wrench should not be used on a metric bolt and a metric wrench should not be used on a standard bolt. Also, most standard bolts have slashes radiating out from the center of the head to denote the grade or strength of the bolt, which is an indication of the amount of torque that can be applied to it. The greater the number of slashes, the greater the strength of the bolt. Grades 0 through 5 are commonly used on automobiles. Metric bolts have a property class (grade) number, rather than a slash, molded into their heads to indicate bolt strength. In this case, the higher the number, the stronger the bolt. Property class numbers 8.8, 9.8 and 10.9 are commonly used on automobiles.

Strength markings can also be used to distinguish standard hex nuts from metric hex nuts. Many standard nuts have dots stamped into one side, while metric nuts are marked with a number. The greater the number of dots, or the higher the number, the greater the strength of the nut.

Metric studs are also marked on their ends according to property class (grade). Larger studs are numbered (the same as metric bolts), while smaller studs carry a geometric code to denote grade.

It should be noted that many fasteners, especially Grades 0 through 2, have no distinguishing marks on them. When such is the case, the only way to determine whether it is standard or metric is to measure the thread pitch or compare it to a known fastener of the same size.

Standard fasteners are often referred to as SAE, as opposed to metric. However, it should be noted that SAE technically refers to a non-metric fine thread fastener only. Coarse thread non-metric fasteners are referred to as USS sizes.

Since fasteners of the same size (both standard and metric) may have different

Metric thread sizes	Ft-lbs	Nm
M-6	6 to 9	9 to 12
M-8	14 to 21	19 to 28
M-10	28 to 40	38 to 54
M-12	50 to 71	68 to 96
M-14	80 to 140	109 to 154

Pipe thread sizes		
1/8	5 to 8	7 to 10
1/4	12 to 18	17 to 24
3/8	22 to 33	30 to 44
1/2	25 to 35	34 to 47

U.S. thread sizes		
1/4 - 20	6 to 9	9 to 12
5/16 - 18	12 to 18	17 to 24
5/16 - 24	14 to 20	19 to 27
3/8 - 16	22 to 32	30 to 43
3/8 - 24	27 to 38	37 to 51
7/16 - 14	40 to 55	55 to 74
7/16 - 20	40 to 60	55 to 81
1/2 - 13	55 to 80	75 to 108

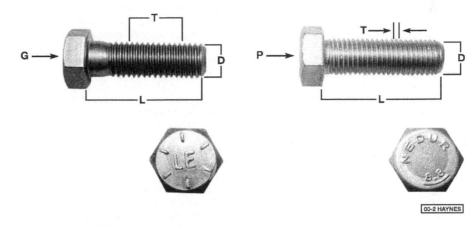

Standard (SAE and USS) bolt dimensions/grade marks

G Grade marks (bolt strength)
L Length (in inches)
T Thread pitch (number of threads per inch)
D Nominal diameter (in inches)

Metric bolt dimensions/grade marks

P Property class (bolt strength)
L Length (in millimeters)
T Thread pitch (distance between threads in millimeters)
D Diameter

strength ratings, be sure to reinstall any bolts, studs or nuts removed from your vehicle in their original locations. Also, when replacing a fastener with a new one, make sure that the new one has a strength rating equal to or greater than the original.

Tightening sequences and procedures

Most threaded fasteners should be tightened to a specific torque value (torque is the twisting force applied to a threaded component such as a nut or bolt). Overtightening the fastener can weaken it and cause it to break, while undertightening can cause it to eventually come loose. Bolts, screws and studs, depending on the material they are made of and their thread diameters, have

specific torque values, many of which are noted in the Specifications at the beginning of each Chapter. Be sure to follow the torque recommendations closely. For fasteners not assigned a specific torque, a general torque value chart is presented here as a guide. These torque values are for dry (unlubricated) fasteners threaded into steel or cast iron (not aluminum). As was previously mentioned, the size and grade of a fastener determine the amount of torque that can safely be applied to it. The figures listed here are approximate for Grade 2 and Grade 3 fasteners. Higher grades can tolerate higher torque values.

Fasteners laid out in a pattern, such as cylinder head bolts, oil pan bolts, differential cover bolts, etc., must be loosened or tightened in sequence to avoid warping the com-

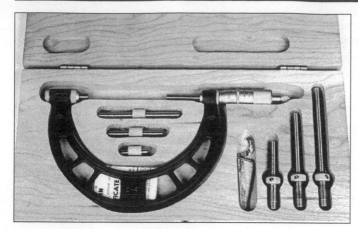

Micrometer set

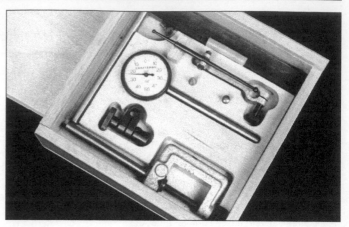

Dial indicator set

ponent. This sequence will normally be shown in the appropriate Chapter. If a specific pattern is not given, the following procedures can be used to prevent warping.

Initially, the bolts or nuts should be assembled finger-tight only. Next, they should be tightened one full turn each, in a crisscross or diagonal pattern. After each one has been tightened one full turn, return to the first one and tighten them all one-half turn, following the same pattern. Finally, tighten each of them one-quarter turn at a time until each fastener has been tightened to the proper torque. To loosen and remove the fasteners, the procedure would be reversed.

Component disassembly

Component disassembly should be done with care and purpose to help ensure that the parts go back together properly. Always keep track of the sequence in which parts are removed. Make note of special characteristics or marks on parts that can be installed more than one way, such as a grooved thrust washer on a shaft. It is a good idea to lay the disassembled parts out on a clean surface in the order that they were removed. It may also be helpful to make sketches or take instant photos of components before removal.

When removing fasteners from a component, keep track of their locations. Sometimes threading a bolt back in a part, or putting the washers and nut back on a stud, can prevent mix-ups later. If nuts and bolts cannot be returned to their original locations, they should be kept in a compartmented box or a series of small boxes. A cupcake or muffin tin is ideal for this purpose, since each cavity can hold the bolts and nuts from a particular area (i.e. oil pan bolts, valve cover bolts, engine mount bolts, etc.). A pan of this type is especially helpful when working on assemblies with very small parts, such as the carburetor, alternator, valve train or interior dash and trim pieces. The cavities can be marked with paint or tape to identify the contents.

Whenever wiring looms, harnesses or connectors are separated, it is a good idea to identify the two halves with numbered pieces of masking tape so they can be easily reconnected.

Gasket sealing surfaces

Throughout any vehicle, gaskets are used to seal the mating surfaces between two parts and keep lubricants, fluids, vacuum or pressure contained in an assembly.

Many times these gaskets are coated with a liquid or paste-type gasket sealing compound before assembly. Age, heat and pressure can sometimes cause the two parts to stick together so tightly that they are very difficult to separate. Often, the assembly can be loosened by striking it with a soft-face hammer near the mating surfaces. A regular hammer can be used if a block of wood is placed between the hammer and the part. Do not hammer on cast parts or parts that could be easily damaged. With any particularly stubborn part, always recheck to make sure that every fastener has been removed.

Avoid using a screwdriver or bar to pry apart an assembly, as they can easily mar the gasket sealing surfaces of the parts, which must remain smooth. If prying is absolutely necessary, use an old broom handle, but keep in mind that extra clean up will be necessary if the wood splinters.

After the parts are separated, the old gasket must be carefully scraped off and the gasket surfaces cleaned. Stubborn gasket material can be soaked with rust penetrant or treated with a special chemical to soften it so it can be easily scraped off. A scraper can be fashioned from a piece of copper tubing by flattening and sharpening one end. Copper is recommended because it is usually softer than the surfaces to be scraped, which reduces the chance of gouging the part. Some gaskets can be removed with a wire brush, but regardless of the method used, the mating surfaces must be left clean and smooth. If for some reason the gasket surface is gouged, then a gasket sealer thick enough to fill scratches will have to be used during reassembly of the components. For most applications, a non-drying (or semi-drying) gasket sealer should be used.

Hose removal tips

Warning: *If the vehicle is equipped with air conditioning, do not disconnect any of the A/C hoses without first having the system depressurized by a dealer service department or a service station.*

Hose removal precautions closely parallel gasket removal precautions. Avoid scratching or gouging the surface that the hose mates against or the connection may leak. This is especially true for radiator hoses. Because of various chemical reactions, the rubber in hoses can bond itself to the metal spigot that the hose fits over. To remove a hose, first loosen the hose clamps that secure it to the spigot. Then, with slip-joint pliers, grab the hose at the clamp and rotate it around the spigot. Work it back and forth until it is completely free, then pull it off. Silicone or other lubricants will ease removal if they can be applied between the hose and the outside of the spigot. Apply the same lubricant to the inside of the hose and the outside of the spigot to simplify installation.

As a last resort (and if the hose is to be replaced with a new one anyway), the rubber can be slit with a knife and the hose peeled from the spigot. If this must be done, be careful that the metal connection is not damaged.

If a hose clamp is broken or damaged, do not reuse it. Wire-type clamps usually weaken with age, so it is a good idea to replace them with screw-type clamps whenever a hose is removed.

Tools

A selection of good tools is a basic requirement for anyone who plans to maintain and repair his or her own vehicle. For the owner who has few tools, the initial investment might seem high, but when compared to the spiraling costs of professional auto maintenance and repair, it is a wise one.

To help the owner decide which tools are needed to perform the tasks detailed in this manual, the following tool lists are offered: *Maintenance and minor repair, Repair/overhaul* and *Special*.

The newcomer to practical mechanics

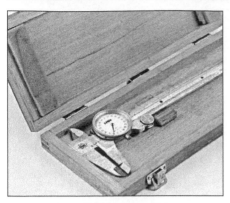

Dial caliper

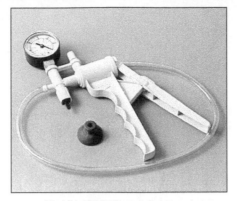

Hand-operated vacuum pump

Timing light

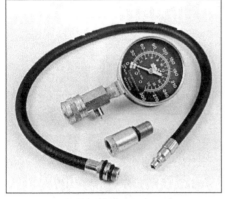

Compression gauge with spark plug hole adapter

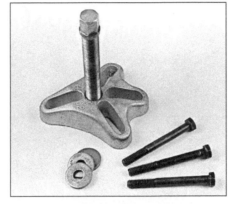

Damper/steering wheel puller

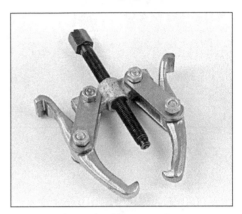

General purpose puller

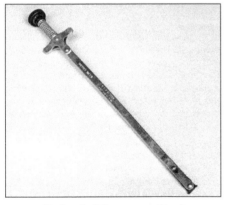

Hydraulic lifter removal tool

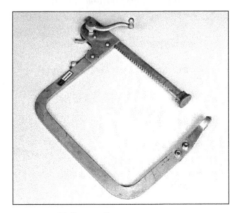

Valve spring compressor

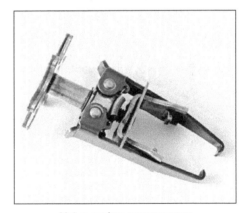

Valve spring compressor

Ridge reamer

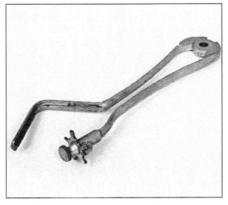

Piston ring groove cleaning tool

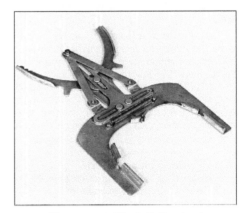

Ring removal/installation tool

Ring compressor

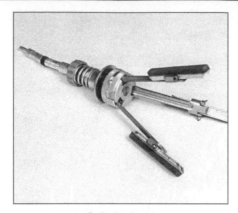

Cylinder hone

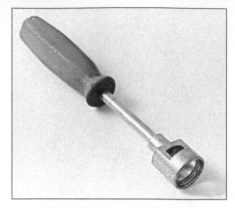

Brake hold-down spring tool

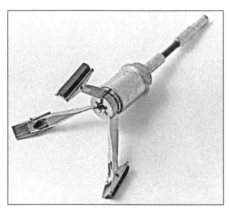

Brake cylinder hone

Clutch plate alignment tool

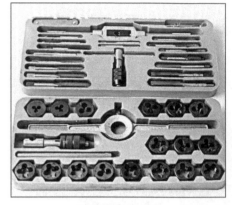

Tap and die set

should start off with the *maintenance and minor repair* tool kit, which is adequate for the simpler jobs performed on a vehicle. Then, as confidence and experience grow, the owner can tackle more difficult tasks, buying additional tools as they are needed. Eventually the basic kit will be expanded into the *repair and overhaul* tool set. Over a period of time, the experienced do-it-yourselfer will assemble a tool set complete enough for most repair and overhaul procedures and will add tools from the special category when it is felt that the expense is justified by the frequency of use.

Maintenance and minor repair tool kit

The tools in this list should be considered the minimum required for performance of routine maintenance, servicing and minor repair work. We recommend the purchase of combination wrenches (box-end and open-end combined in one wrench). While more expensive than open end wrenches, they offer the advantages of both types of wrench.

> *Combination wrench set (1/4-inch to 1 inch or 6 mm to 19 mm)*
> *Adjustable wrench, 8 inch*
> *Spark plug wrench with rubber insert*
> *Spark plug gap adjusting tool*
> *Feeler gauge set*
> *Brake bleeder wrench*

> *Standard screwdriver (5/16-inch x 6 inch)*
> *Phillips screwdriver (No. 2 x 6 inch)*
> *Combination pliers - 6 inch*
> *Hacksaw and assortment of blades*
> *Tire pressure gauge*
> *Grease gun*
> *Oil can*
> *Fine emery cloth*
> *Wire brush*
> *Battery post and cable cleaning tool*
> *Oil filter wrench*
> *Funnel (medium size)*
> *Safety goggles*
> *Jackstands (2)*
> *Drain pan*

Note: *If basic tune-ups are going to be part of routine maintenance, it will be necessary to purchase a good quality stroboscopic timing light and combination tachometer/dwell meter. Although they are included in the list of special tools, it is mentioned here because they are absolutely necessary for tuning most vehicles properly.*

Repair and overhaul tool set

These tools are essential for anyone who plans to perform major repairs and are in addition to those in the maintenance and minor repair tool kit. Included is a comprehensive set of sockets which, though expensive, are invaluable because of their versatil-

ity, especially when various extensions and drives are available. We recommend the 1/2-inch drive over the 3/8-inch drive. Although the larger drive is bulky and more expensive, it has the capacity of accepting a very wide range of large sockets. Ideally, however, the mechanic should have a 3/8-inch drive set and a 1/2-inch drive set.

> *Socket set(s)*
> *Reversible ratchet*
> *Extension - 10 inch*
> *Universal joint*
> *Torque wrench (same size drive as sockets)*
> *Ball peen hammer - 8 ounce*
> *Soft-face hammer (plastic/rubber)*
> *Standard screwdriver (1/4-inch x 6 inch)*
> *Standard screwdriver (stubby - 5/16-inch)*
> *Phillips screwdriver (No. 3 x 8 inch)*
> *Phillips screwdriver (stubby - No. 2)*
> *Pliers - vise grip*
> *Pliers - lineman's*
> *Pliers - needle nose*
> *Pliers - snap-ring (internal and external)*
> *Cold chisel - 1/2-inch*
> *Scribe*
> *Scraper (made from flattened copper tubing)*
> *Centerpunch*
> *Pin punches (1/16, 1/8, 3/16 inch)*
> *Steel rule/straightedge - 12 inch*

*Allen wrench set (1/8 to 3/8-inch or
    4 mm to 10 mm)
A selection of files
Wire brush (large)
Jackstands (second set)
Jack (scissor or hydraulic type)*

Note: *Another tool which is often useful is an
electric drill with a chuck capacity of 3/8-inch
and a set of good quality drill bits.*

Special tools

The tools in this list include those which
are not used regularly, are expensive to buy,
or which need to be used in accordance with
their manufacturer's instructions. Unless
these tools will be used frequently, it is not
very economical to purchase many of them.
A consideration would be to split the cost
and use between yourself and a friend or
friends. In addition, most of these tools can
be obtained from a tool rental shop on a temporary basis.

This list primarily contains only those
tools and instruments widely available to the
public, and not those special tools produced
by the vehicle manufacturer for distribution to
dealer service departments. Occasionally,
references to the manufacturer's special
tools are included in the text of this manual.
Generally, an alternative method of doing the
job without the special tool is offered. However, sometimes there is no alternative to
their use. Where this is the case, and the tool
cannot be purchased or borrowed, the work
should be turned over to the dealer service
department or an automotive repair shop.

*Valve spring compressor
Piston ring groove cleaning tool
Piston ring compressor
Piston ring installation tool
Cylinder compression gauge
Cylinder ridge reamer
Cylinder surfacing hone
Cylinder bore gauge
Micrometers and/or dial calipers
Hydraulic lifter removal tool
Balljoint separator
Universal-type puller
Impact screwdriver
Dial indicator set
Stroboscopic timing light (inductive
    pick-up)
Hand operated vacuum/pressure pump
Tachometer/dwell meter
Universal electrical multimeter
Cable hoist
Brake spring removal and installation
    tools
Floor jack*

Buying tools

For the do-it-yourselfer who is just starting to get involved in vehicle maintenance
and repair, there are a number of options
available when purchasing tools. If maintenance and minor repair is the extent of the
work to be done, the purchase of individual
tools is satisfactory. If, on the other hand,
extensive work is planned, it would be a good
idea to purchase a modest tool set from one

of the large retail chain stores. A set can usually be bought at a substantial savings over
the individual tool prices, and they often
come with a tool box. As additional tools are
needed, add-on sets, individual tools and a
larger tool box can be purchased to expand
the tool selection. Building a tool set gradually allows the cost of the tools to be spread
over a longer period of time and gives the
mechanic the freedom to choose only those
tools that will actually be used.

Tool stores will often be the only source
of some of the special tools that are needed,
but regardless of where tools are bought, try
to avoid cheap ones, especially when buying
screwdrivers and sockets, because they
won't last very long. The expense involved in
replacing cheap tools will eventually be
greater than the initial cost of quality tools.

Care and maintenance of tools

Good tools are expensive, so it makes
sense to treat them with respect. Keep them
clean and in usable condition and store them
properly when not in use. Always wipe off any
dirt, grease or metal chips before putting
them away. Never leave tools lying around in
the work area. Upon completion of a job,
always check closely under the hood for tools
that may have been left there so they won't
get lost during a test drive.

Some tools, such as screwdrivers, pliers, wrenches and sockets, can be hung on a
panel mounted on the garage or workshop
wall, while others should be kept in a tool box
or tray. Measuring instruments, gauges,
meters, etc. must be carefully stored where
they cannot be damaged by weather or
impact from other tools.

When tools are used with care and
stored properly, they will last a very long
time. Even with the best of care, though,
tools will wear out if used frequently. When a
tool is damaged or worn out, replace it. Subsequent jobs will be safer and more enjoyable
if you do.

How to repair damaged threads

Sometimes, the internal threads of a nut
or bolt hole can become stripped, usually
from overtightening. Stripping threads is an
all-too-common occurrence, especially when
working with aluminum parts, because aluminum is so soft that it easily strips out.

Usually, external or internal threads are
only partially stripped. After they've been
cleaned up with a tap or die, they'll still work.
Sometimes, however, threads are badly damaged. When this happens, you've got three
choices:

1) *Drill and tap the hole to the next suitable
oversize and install a larger diameter
bolt, screw or stud.*
2) *Drill and tap the hole to accept a
threaded plug, then drill and tap the plug
to the original screw size. You can also
buy a plug already threaded to the original size. Then you simply drill a hole to*

*the specified size, then run the threaded
plug into the hole with a bolt and jam
nut. Once the plug is fully seated, remove
the jam nut and bolt.*
3) *The third method uses a patented
thread repair kit like Heli-Coil or Slimsert.
These easy-to-use kits are designed to
repair damaged threads in straight-
through holes and blind holes. Both are
available as kits which can handle a variety of sizes and thread patterns. Drill the
hole, then tap it with the special
included tap. Install the Heli-Coil and the
hole is back to its original diameter and
thread pitch.*

Regardless of which method you use,
be sure to proceed calmly and carefully. A little impatience or carelessness during one of
these relatively simple procedures can ruin
your whole day's work and cost you a bundle
if you wreck an expensive part.

Working facilities

Not to be overlooked when discussing
tools is the workshop. If anything more than
routine maintenance is to be carried out,
some sort of suitable work area is essential.

It is understood, and appreciated, that
many home mechanics do not have a good
workshop or garage available, and end up
removing an engine or doing major repairs
outside. It is recommended, however, that
the overhaul or repair be completed under
the cover of a roof.

A clean, flat workbench or table of com-
fortable working height is an absolute neces-
sity. The workbench should be equipped with
a vise that has a jaw opening of at least four
inches.

As mentioned previously, some clean,
dry storage space is also required for tools,
as well as the lubricants, fluids, cleaning solvents, etc. which soon become necessary.

Sometimes waste oil and fluids, drained
from the engine or cooling system during normal maintenance or repairs, present a disposal problem. To avoid pouring them on the
ground or into a sewage system, pour the
used fluids into large containers, seal them
with caps and take them to an authorized
disposal site or recycling center. Plastic jugs,
such as old antifreeze containers, are ideal
for this purpose.

Always keep a supply of old newspapers and clean rags available. Old towels are
excellent for mopping up spills. Many
mechanics use rolls of paper towels for most
work because they are readily available and
disposable. To help keep the area under the
vehicle clean, a large cardboard box can be
cut open and flattened to protect the garage
or shop floor.

Whenever working over a painted surface, such as when leaning over a fender to
service something under the hood, always
cover it with an old blanket or bedspread to
protect the finish. Vinyl covered pads, made
especially for this purpose, are available at
auto parts stores.

Anti-theft audio system

General information

1   Some of these models are equipped with the Delco Loc II (early models) or THEFT-LOCK (later models) audio systems, which include an anti-theft feature that will render the stereo inoperative if stolen. If the power source to the stereo is cut with the anti-theft feature activated, the stereo will be inoperative. Even if the power source is immediately re-connected, the stereo will not function.

2   If your vehicle is equipped with this anti-theft system, do not disconnect the battery, remove the stereo or disconnect related components unless you have either turned off the feature or have the individual ID (code) number for the stereo.

Delco-Loc II

Disabling the anti-theft feature

3   Press the stereo's PREV and FWD buttons at the same time for five seconds with the ignition on and the radio power off. The display will show SEC, indicating the unit is in the secure mode (anti-theft feature enabled).

4   Press the SET button. The display will show "000"

5   Press the SEEK button to make the first number appear.

6   Rotate the TUNE knob right or left until the second and third digit of your code appear. The numbers will be displayed as entered.

7   Press the lower BAND knob. "000" will be displayed.

8   Press the SEEK button to make the fourth number appear.

9   Rotate the lower knob until the fifth and sixth digits of the code are displayed.

10   Press the lower BAND knob. If the display shows "_ _," you have successfully disabled the anti-theft feature. If SEC is displayed, the code you entered was incorrect and the anti-theft feature is still enabled.

Unlocking the stereo after a power loss

11   When power is restored to the stereo, the stereo won't turn on and LOC will appear on the display. Enter your ID code as follows; pause no more than 15 seconds between Steps.

12   Turn the ignition switch to ON, but leave the stereo off.

13   Press the SET button. "000" should display.

14   Press the SEEK button to make the first number appear, then release it.

15   Rotate the TUNE knob right or left to make sure the second and third numbers agree with your code.

16   Repeat Steps 8 and 9 for the last three digits of your code.

17   Press the lower BAND knob. If SEC appears, the numbers you entered were correct and the stereo will work. If LOC appears, the numbers you entered were not correct and the stereo is still inoperative

THEFTLOCK

Disabling the anti-theft feature

18   Press the stereo's 1 and 4 buttons at the same time for five seconds with the ignition on and the radio power off. The display will show SEC, indicating the unit is in the secure mode (anti-theft feature enabled).

19   Press the MIN button. The display will show "000."

20   Press the MIN button to make the last two numbers appear.

21   Press HR to display the first one or two numbers of your code. The numbers will be displayed as entered.

22   Press AM/FM. If the display shows "_ _," you have successfully disabled the anti-theft feature. If SEC is displayed, the code you entered was incorrect and the anti-theft feature is still enabled.

Unlocking the stereo after a power loss

23   When power is restored to the stereo, the stereo won't turn on and LOC will appear on the display. Enter your ID code as follows; pause no more than 15 seconds between Steps.

24   Turn the ignition switch to ON, but leave the stereo off.

25   Press the MIN button. "000" should display.

26   Press the HR button to make the last two numbers appear, then release the button.

27   Press the HR button until the first one or two numbers appear.

28   Press AM/FM. SEC should appear, indicating the stereo is unlocked. If LOC appears, the numbers you entered were not correct and the stereo is still inoperative.

Jacking and towing

Jacking

Warning: *The jack supplied with the vehicle should only be used for raising the vehicle when changing a tire or placing jackstands under the frame. Never work under the vehicle or start the engine while the jack is being used as the only means of support.*

The vehicle must be on a level surface with the wheels blocked and the transmission in Park. Apply the parking brake if the front of the vehicle must be raised. Make sure no one is in the vehicle as it's being raised with the jack.

Remove the jack and lug nut wrench and spare tire from the compartment in the right rear corner of the rear passenger compartment.

To replace the tire, use the tapered end of the lug wrench to pry loose the wheel cover. **Note:** *If the vehicle is equipped with aluminum wheels, it may be necessary to pry out the special lug nut covers. Also, aluminum wheels normally have anti-theft lug nuts (one per wheel) which require using a special "key" between the lug wrench and lug nut. The key is usually in the glove compartment. Loosen the lug nuts one half turn, but leave them in place until the tire is raised off the ground.*

Position the jack under the side of the

vehicle at the indicated jacking points. There's a front and rear jacking point on each side of the vehicle **(see illustration)**.

Turn the jack handle clockwise (the lug wrench also serves as the jack handle) until the tire clears the ground. Remove the lug nuts and pull the tire off. Clean the mating surfaces of the hub and wheel, then install the spare. Replace the lug nuts with the beveled edges facing in and tighten them snugly. Don't attempt to tighten them completely until the vehicle is lowered or it could slip off the jack.

Turn the jack handle counterclockwise to lower the vehicle. Remove the jack and tighten the lug nuts in a criss-cross pattern. If possible, tighten the nuts with a torque wrench (see Chapter 1 for the torque figures). If you don't have access to a torque wrench, have the nuts checked by a service station or repair shop as soon as possible. **Caution:** *The compact spare included with these vehicles is intended for temporary use only. Have the tire repaired and reinstall it on the vehicle at the earliest opportunity and don't exceed 50 mph with the spare tire on the car.*

Install the wheel cover, then stow the tire, jack and wrench and unblock the wheels.

Towing

We recommend these vehicles be towed

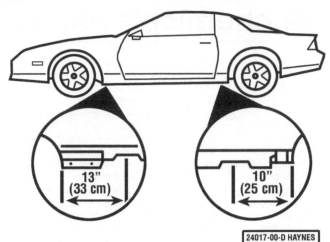

The head of the jack should engage the notch in the rocker panel flange - there's one at the front and one at the rear on each side of the vehicle

24017-00-D HAYNES

from the rear, with the rear wheels off the ground. If it's absolutely necessary, these vehicles can be towed from the front with the front wheels off the ground, provided that speeds don't exceed 35 mph and the distance is less than 50 miles; the transmission can be damaged if these mileage/speed limitations are exceeded.

Equipment specifically designed for towing should be used. It must be attached to the main structural members of the vehicle,

not the bumpers or brackets.

Safety is a major consideration when towing and all applicable state and local laws must be obeyed. A safety chain must be used at all times.

The parking brake must be released and the transmission must be in Neutral. The steering must be unlocked (ignition switch in the Off position). Remember that power steering and power brakes won't work with the engine off.

Booster battery (jump) starting

Observe these precautions when using a booster battery to start a vehicle:

a) *Before connecting the booster battery, make sure the ignition switch is in the Off position.*

b) *Turn off the lights, heater and other electrical loads.*

c) *Your eyes should be shielded. Safety goggles are a good idea.*

d) *Make sure the booster battery is the same voltage as the dead one in the vehicle.*

e) *The two vehicles MUST NOT TOUCH each other!*

f) *Make sure the transaxle is in Neutral (manual) or Park (automatic).*

g) *If the booster battery is not a maintenance-free type, remove the vent caps and lay a cloth over the vent holes.*

Connect the red jumper cable to the positive (+) terminals of each battery **(see illustration)**.

Connect one end of the black jumper cable to the negative (-) terminal of the booster battery. The other end of this cable should be connected to a good ground on the vehicle to be started, such as a bolt or bracket on the body.

Start the engine using the booster battery, then, with the engine running at idle speed, disconnect the jumper cables in the reverse order of connection.

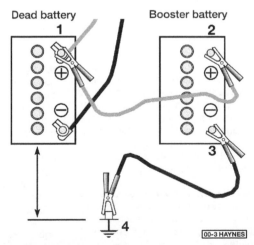

Dead battery   Booster battery

00-3 HAYNES

Make the booster battery cable connections in the numerical order shown (note that the negative cable of the booster battery is NOT attached to the negative terminal of the dead battery)

Automotive chemicals and lubricants

A number of automotive chemicals and lubricants are available for use during vehicle maintenance and repair. They include a wide variety of products ranging from cleaning solvents and degreasers to lubricants and protective sprays for rubber, plastic and vinyl.

Cleaners

Carburetor cleaner and choke cleaner is a strong solvent for gum, varnish and carbon. Most carburetor cleaners leave a dry-type lubricant film which will not harden or gum up. Because of this film it is not recommended for use on electrical components.

Brake system cleaner is used to remove grease and brake fluid from the brake system, where clean surfaces are absolutely necessary. It leaves no residue and often eliminates brake squeal caused by contaminants.

Electrical cleaner removes oxidation, corrosion and carbon deposits from electrical contacts, restoring full current flow. It can also be used to clean spark plugs, carburetor jets, voltage regulators and other parts where an oil-free surface is desired.

Demoisturants remove water and moisture from electrical components such as alternators, voltage regulators, electrical connectors and fuse blocks. They are non-conductive, non-corrosive and non-flammable.

Degreasers are heavy-duty solvents used to remove grease from the outside of the engine and from chassis components. They can be sprayed or brushed on and, depending on the type, are rinsed off either with water or solvent.

Lubricants

Motor oil is the lubricant formulated for use in engines. It normally contains a wide variety of additives to prevent corrosion and reduce foaming and wear. Motor oil comes in various weights (viscosity ratings) from 5 to 80. The recommended weight of the oil depends on the season, temperature and the demands on the engine. Light oil is used in cold climates and under light load conditions. Heavy oil is used in hot climates and where high loads are encountered. Multi-viscosity oils are designed to have characteristics of both light and heavy oils and are available in a number of weights from 5W-20 to 20W-50.

Gear oil is designed to be used in differentials, manual transmissions and other areas where high-temperature lubrication is required.

Chassis and wheel bearing grease is a heavy grease used where increased loads and friction are encountered, such as for wheel bearings, balljoints, tie-rod ends and universal joints.

High-temperature wheel bearing grease is designed to withstand the extreme temperatures encountered by wheel bearings in disc brake equipped vehicles. It usually contains molybdenum disulfide (moly), which is a dry-type lubricant.

White grease is a heavy grease for metal-to-metal applications where water is a problem. White grease stays soft under both low and high temperatures (usually from -100 to +190-degrees F), and will not wash off or dilute in the presence of water.

Assembly lube is a special extreme pressure lubricant, usually containing moly, used to lubricate high-load parts (such as main and rod bearings and cam lobes) for initial start-up of a new engine. The assembly lube lubricates the parts without being squeezed out or washed away until the engine oiling system begins to function.

Silicone lubricants are used to protect rubber, plastic, vinyl and nylon parts.

Graphite lubricants are used where oils cannot be used due to contamination problems, such as in locks. The dry graphite will lubricate metal parts while remaining uncontaminated by dirt, water, oil or acids. It is electrically conductive and will not foul electrical contacts in locks such as the ignition switch.

Moly penetrants loosen and lubricate frozen, rusted and corroded fasteners and prevent future rusting or freezing.

Heat-sink grease is a special electrically non-conductive grease that is used for mounting electronic ignition modules where it is essential that heat is transferred away from the module.

Sealants

RTV sealant is one of the most widely used gasket compounds. Made from silicone, RTV is air curing, it seals, bonds, waterproofs, fills surface irregularities, remains flexible, doesn't shrink, is relatively easy to remove, and is used as a supplementary sealer with almost all low and medium temperature gaskets.

Anaerobic sealant is much like RTV in that it can be used either to seal gaskets or to form gaskets by itself. It remains flexible, is solvent resistant and fills surface imperfections. The difference between an anaerobic sealant and an RTV-type sealant is in the curing. RTV cures when exposed to air, while an anaerobic sealant cures only in the absence of air. This means that an anaerobic sealant cures only after the assembly of parts, sealing them together.

Thread and pipe sealant is used for sealing hydraulic and pneumatic fittings and vacuum lines. It is usually made from a Teflon compound, and comes in a spray, a paint-on liquid and as a wrap-around tape.

Chemicals

Anti-seize compound prevents seizing, galling, cold welding, rust and corrosion in fasteners. High-temperature anti-seize, usually made with copper and graphite lubricants, is used for exhaust system and exhaust manifold bolts.

Anaerobic locking compounds are used to keep fasteners from vibrating or working loose and cure only after installation, in the absence of air. Medium strength locking compound is used for small nuts, bolts and screws that may be removed later. High-strength locking compound is for large nuts, bolts and studs which aren't removed on a regular basis.

Oil additives range from viscosity index improvers to chemical treatments that claim to reduce internal engine friction. It should be noted that most oil manufacturers caution against using additives with their oils.

Gas additives perform several functions, depending on their chemical makeup. They usually contain solvents that help dissolve gum and varnish that build up on carburetor, fuel injection and intake parts. They also serve to break down carbon deposits that form on the inside surfaces of the combustion chambers. Some additives contain upper cylinder lubricants for valves and piston rings, and others contain chemicals to remove condensation from the gas tank.

Miscellaneous

Brake fluid is specially formulated hydraulic fluid that can withstand the heat and pressure encountered in brake systems. Care must be taken so this fluid does not come in contact with painted surfaces or plastics. An opened container should always be resealed to prevent contamination by water or dirt.

Weatherstrip adhesive is used to bond weatherstripping around doors, windows and trunk lids. It is sometimes used to attach trim pieces.

Undercoating is a petroleum-based, tar-like substance that is designed to protect metal surfaces on the underside of the vehicle from corrosion. It also acts as a sound-deadening agent by insulating the bottom of the vehicle.

Waxes and polishes are used to help protect painted and plated surfaces from the weather. Different types of paint may require the use of different types of wax and polish. Some polishes utilize a chemical or abrasive cleaner to help remove the top layer of oxidized (dull) paint on older vehicles. In recent years many non-wax polishes that contain a wide variety of chemicals such as polymers and silicones have been introduced. These non-wax polishes are usually easier to apply and last longer than conventional waxes and polishes.

Conversion factors

Length (distance)

Inches (in)	X	25.4	= Millimetres (mm)	X	0.0394	= Inches (in)
Feet (ft)	X	0.305	= Metres (m)	X	3.281	= Feet (ft)
Miles	X	1.609	= Kilometres (km)	X	0.621	= Miles

Inches (in) X 25.4 = Millimetres (mm) X 0.0394 = Inches (in)
Feet (ft) X 0.305 = Metres (m) X 3.281 = Feet (ft)
Miles X 1.609 = Kilometres (km) X 0.621 = Miles

Volume (capacity)

Cubic inches (cu in; in³) X 16.387 = Cubic centimetres (cc; cm³) X 0.061 = Cubic inches (cu in; in³)
Imperial pints (Imp pt) X 0.568 = Litres (l) X 1.76 = Imperial pints (Imp pt)
Imperial quarts (Imp qt) X 1.137 = Litres (l) X 0.88 = Imperial quarts (Imp qt)
Imperial quarts (Imp qt) X 1.201 = US quarts (US qt) X 0.833 = Imperial quarts (Imp qt)
US quarts (US qt) X 0.946 = Litres (l) X 1.057 = US quarts (US qt)
Imperial gallons (Imp gal) X 4.546 = Litres (l) X 0.22 = Imperial gallons (Imp gal)
Imperial gallons (Imp gal) X 1.201 = US gallons (US gal) X 0.833 = Imperial gallons (Imp gal)
US gallons (US gal) X 3.785 = Litres (l) X 0.264 = US gallons (US gal)

Mass (weight)

Ounces (oz) X 28.35 = Grams (g) X 0.035 = Ounces (oz)
Pounds (lb) X 0.454 = Kilograms (kg) X 2.205 = Pounds (lb)

Force

Ounces-force (ozf; oz) X 0.278 = Newtons (N) X 3.6 = Ounces-force (ozf; oz)
Pounds-force (lbf; lb) X 4.448 = Newtons (N) X 0.225 = Pounds-force (lbf; lb)
Newtons (N) X 0.1 = Kilograms-force (kgf; kg) X 9.81 = Newtons (N)

Pressure

Pounds-force per square inch (psi; lbf/in²; lb/in²) X 0.070 = Kilograms-force per square centimetre (kgf/cm²; kg/cm²) X 14.223 = Pounds-force per square inch (psi; lbf/in²; lb/in²)
Pounds-force per square inch (psi; lbf/in²; lb/in²) X 0.068 = Atmospheres (atm) X 14.696 = Pounds-force per square inch (psi; lbf/in²; lb/in²)
Pounds-force per square inch (psi; lbf/in²; lb/in²) X 0.069 = Bars X 14.5 = Pounds-force per square inch (psi; lbf/in²; lb/in²)
Pounds-force per square inch (psi; lbf/in²; lb/in²) X 6.895 = Kilopascals (kPa) X 0.145 = Pounds-force per square inch (psi; lbf/in²; lb/in²)
Kilopascals (kPa) X 0.01 = Kilograms-force per square centimetre (kgf/cm²; kg/cm²) X 98.1 = Kilopascals (kPa)

Torque (moment of force)

Pounds-force inches (lbf in; lb in) X 1.152 = Kilograms-force centimetre (kgf cm; kg cm) X 0.868 = Pounds-force inches (lbf in; lb in)
Pounds-force inches (lbf in; lb in) X 0.113 = Newton metres (Nm) X 8.85 = Pounds-force inches (lbf in; lb in)
Pounds-force inches (lbf in; lb in) X 0.083 = Pounds-force feet (lbf ft; lb ft) X 12 = Pounds-force inches (lbf in; lb in)
Pounds-force feet (lbf ft; lb ft) X 0.138 = Kilograms-force metres (kgf m; kg m) X 7.233 = Pounds-force feet (lbf ft; lb ft)
Pounds-force feet (lbf ft; lb ft) X 1.356 = Newton metres (Nm) X 0.738 = Pounds-force feet (lbf ft; lb ft)
Newton metres (Nm) X 0.102 = Kilograms-force metres (kgf m; kg m) X 9.804 = Newton metres (Nm)

Vacuum

Inches mercury (in. Hg) X 3.377 = Kilopascals (kPa) X 0.2961 = Inches mercury
Inches mercury (in. Hg) X 25.4 = Millimeters mercury (mm Hg) X 0.0394 = Inches mercury

Power

Horsepower (hp) X 745.7 = Watts (W) X 0.0013 = Horsepower (hp)

Velocity (speed)

Miles per hour (miles/hr; mph) X 1.609 = Kilometres per hour (km/hr; kph) X 0.621 = Miles per hour (miles/hr; mph)

Fuel consumption*

Miles per gallon, Imperial (mpg) X 0.354 = Kilometres per litre (km/l) X 2.825 = Miles per gallon, Imperial (mpg)
Miles per gallon, US (mpg) X 0.425 = Kilometres per litre (km/l) X 2.352 = Miles per gallon, US (mpg)

Temperature

Degrees Fahrenheit = (°C x 1.8) + 32   Degrees Celsius (Degrees Centigrade; °C) = (°F - 32) x 0.56

*It is common practice to convert from miles per gallon (mpg) to litres/100 kilometres (l/100km),
where mpg (Imperial) x l/100 km = 282 and mpg (US) x l/100 km = 235

Safety first!

Regardless of how enthusiastic you may be about getting on with the job at hand, take the time to ensure that your safety is not jeopardized. A moment's lack of attention can result in an accident, as can failure to observe certain simple safety precautions. The possibility of an accident will always exist, and the following points should not be considered a comprehensive list of all dangers. Rather, they are intended to make you aware of the risks and to encourage a safety conscious approach to all work you carry out on your vehicle.

Essential DOs and DON'Ts

DON'T rely on a jack when working under the vehicle. Always use approved jackstands to support the weight of the vehicle and place them under the recommended lift or support points.

DON'T attempt to loosen extremely tight fasteners (i.e. wheel lug nuts) while the vehicle is on a jack - it may fall.

DON'T start the engine without first making sure that the transmission is in Neutral (or Park where applicable) and the parking brake is set.

DON'T remove the radiator cap from a hot cooling system - let it cool or cover it with a cloth and release the pressure gradually.

DON'T attempt to drain the engine oil until you are sure it has cooled to the point that it will not burn you.

DON'T touch any part of the engine or exhaust system until it has cooled sufficiently to avoid burns.

DON'T siphon toxic liquids such as gasoline, antifreeze and brake fluid by mouth, or allow them to remain on your skin.

DON'T inhale brake lining dust - it is potentially hazardous (see *Asbestos* below).

DON'T allow spilled oil or grease to remain on the floor - wipe it up before someone slips on it.

DON'T use loose fitting wrenches or other tools which may slip and cause injury.

DON'T push on wrenches when loosening or tightening nuts or bolts. Always try to pull the wrench toward you. If the situation calls for pushing the wrench away, push with an open hand to avoid scraped knuckles if the wrench should slip.

DON'T attempt to lift a heavy component alone - get someone to help you.

DON'T rush or take unsafe shortcuts to finish a job.

DON'T allow children or animals in or around the vehicle while you are working on it.

DO wear eye protection when using power tools such as a drill, sander, bench grinder, etc. and when working under a vehicle.

DO keep loose clothing and long hair well out of the way of moving parts.

DO make sure that any hoist used has a safe working load rating adequate for the job.

DO get someone to check on you periodically when working alone on a vehicle.

DO carry out work in a logical sequence and make sure that everything is correctly assembled and tightened.

DO keep chemicals and fluids tightly capped and out of the reach of children and pets.

DO remember that your vehicle's safety affects that of yourself and others. If in doubt on any point, get professional advice.

Asbestos

Certain friction, insulating, sealing, and other products - such as brake linings, brake bands, clutch linings, torque converters, gaskets, etc. - may contain asbestos. Extreme care must be taken to avoid inhalation of dust from such products, since it is hazardous to health. If in doubt, assume that they do contain asbestos.

Fire

Remember at all times that gasoline is highly flammable. Never smoke or have any kind of open flame around when working on a vehicle. But the risk does not end there. A spark caused by an electrical short circuit, by two metal surfaces contacting each other, or even by static electricity built up in your body under certain conditions, can ignite gasoline vapors, which in a confined space are highly explosive. Do not, under any circumstances, use gasoline for cleaning parts. Use an approved safety solvent.

Always disconnect the battery ground (-) cable at the battery before working on any part of the fuel system or electrical system. Never risk spilling fuel on a hot engine or exhaust component. It is strongly recommended that a fire extinguisher suitable for use on fuel and electrical fires be kept handy in the garage or workshop at all times. Never try to extinguish a fuel or electrical fire with water.

Fumes

Certain fumes are highly toxic and can quickly cause unconsciousness and even death if inhaled to any extent. Gasoline vapor falls into this category, as do the vapors from some cleaning solvents. Any draining or pouring of such volatile fluids should be done in a well ventilated area.

When using cleaning fluids and solvents, read the instructions on the container carefully. Never use materials from unmarked containers.

Never run the engine in an enclosed space, such as a garage. Exhaust fumes contain carbon monoxide, which is extremely poisonous. If you need to run the engine, always do so in the open air, or at least have the rear of the vehicle outside the work area.

If you are fortunate enough to have the use of an inspection pit, never drain or pour gasoline and never run the engine while the vehicle is over the pit. The fumes, being heavier than air, will concentrate in the pit with possibly lethal results.

The battery

Never create a spark or allow a bare light bulb near a battery. They normally give off a certain amount of hydrogen gas, which is highly explosive.

Always disconnect the battery ground (-) cable at the battery before working on the fuel or electrical systems.

If possible, loosen the filler caps or cover when charging the battery from an external source (this does not apply to sealed or maintenance-free batteries). Do not charge at an excessive rate or the battery may burst.

Take care when adding water to a non maintenance-free battery and when carrying a battery. The electrolyte, even when diluted, is very corrosive and should not be allowed to contact clothing or skin.

Always wear eye protection when cleaning the battery to prevent the caustic deposits from entering your eyes.

Household current

When using an electric power tool, inspection light, etc., which operates on household current, always make sure that the tool is correctly connected to its plug and that, where necessary, it is properly grounded. Do not use such items in damp conditions and, again, do not create a spark or apply excessive heat in the vicinity of fuel or fuel vapor.

Secondary ignition system voltage

A severe electric shock can result from touching certain parts of the ignition system (such as the spark plug wires) when the engine is running or being cranked, particularly if components are damp or the insulation is defective. In the case of an electronic ignition system, the secondary system voltage is much higher and could prove fatal.

Troubleshooting

Contents

This Section provides an easy reference guide to the more common problems that may occur during the operation of your vehicle. Various symptoms and their probable causes are grouped under headings denoting components or systems, such as Engine, Cooling system, etc. They also refer to the Chapter and/or Section that deals with the problem.

Remember that successful trouble-shooting isn't a mysterious 'black art' practiced only by professional mechanics, it's simply the result of knowledge combined with an intelligent, systematic approach to a problem. Always use a process of elimination starting with the simplest solution and working through to the most complex - and never overlook the obvious. Anyone can run the gas tank dry or leave the lights on overnight, so don't assume that you're exempt from such oversights.

Finally, always establish a clear idea why a problem has occurred and take steps to ensure that it doesn't happen again. If the electrical system fails because of a poor connection, check all other connections in the system to make sure they don't fail as well. If a particular fuse continues to blow, find out why - don't just go on replacing fuses. Remember, failure of a small component can often be indicative of potential failure or incorrect functioning of a more important component or system.

Engine and performance

1 Engine will not rotate when attempting to start

1   Battery terminal connections loose or corroded. Check the cable terminals at the battery; tighten cable clamp and/or clean off corrosion as necessary (see Chapter 1).
2   Battery discharged or faulty. If the cable ends are clean and tight on the battery posts, turn the key to the On position and switch on the headlights or windshield wipers. If they won't run, the battery is discharged.
3   Automatic transmission not engaged in park (P) or Neutral (N).
4   Broken, loose or disconnected wires in the starting circuit. Inspect all wires and connectors at the battery, starter solenoid and ignition switch (on steering column).
5   Starter motor pinion jammed in flywheel ring gear. If manual transmission, place transmission in gear and rock the vehicle to manually turn the engine. Remove starter (Chapter 5) and inspect pinion and flywheel (Chapter 2) at earliest convenience.
6   Starter solenoid faulty (Chapter 5).
7   Starter motor faulty (Chapter 5).
8   Ignition switch faulty (Chapter 12).
9   Engine seized. Try to turn the crankshaft with a large socket and breaker bar on the pulley bolt.

2 Engine rotates but will not start

1   Fuel tank empty.
2   Battery discharged (engine rotates slowly). Check the operation of electrical components as described in previous Section.
3   Battery terminal connections loose or corroded. See previous Section.
4   Fuel not reaching fuel injectors. Check for clogged fuel filter or lines and defective fuel pump. Also make sure the tank vent lines aren't clogged (Chapter 4).
5   Faulty distributor (if equipped) compo-

nents. Check the cap (Chapter 1).
6   Low cylinder compression. Check as described in Chapter 2.
7   Water in fuel. Drain tank and fill with new fuel.
8   Dirty or clogged fuel injectors.
9   Wet or damaged ignition components (Chapters 1 and 5).
10   Worn, faulty or incorrectly gapped spark plugs (Chapter 1).
11   Broken, loose or disconnected wires in the ignition circuit.
12   Broken, loose or disconnected wires at the ignition coil(s) or faulty coil(s) (Chapter 5).
13   Timing chain failure or wear affecting valve timing (Chapter 2).

3 Starter motor operates without turning engine

1   Starter pinion sticking. Remove the starter (Chapter 5) and inspect.
2   Starter pinion or flywheel/driveplate teeth worn or broken. Remove the inspection cover on the left side of the engine and inspect.

4 Engine hard to start when cold

1   Battery discharged or low. Check as described in Chapter 1.
2   Fuel not reaching the fuel injectors. Check the fuel filter and lines (Chapters 1 and 4).
3   Defective spark plugs (Chapter 1).
4   Intake manifold vacuum leaks. Make sure all mounting bolts/nuts are tight and all vacuum hoses connected to the manifold are attached properly and in good condition.

5 Engine hard to start when hot

1   Air filter dirty (Chapter 1).
2   Bad engine ground connection.
3   Fuel not reaching the injectors (Chapter 4).
4   Loose connection in the ignition system (Chapter 5).

6 Starter motor noisy or engages roughly

1   Pinion or flywheel/driveplate teeth worn or broken. Remove the inspection cover and inspect.
2   Starter motor mounting bolts loose or missing.

7 Engine starts but stops immediately

1   Loose or damaged wiring in the ignition system.

2   Intake manifold vacuum leaks. Make sure all mounting bolts/nuts are tight and all vacuum hoses connected to the manifold are attached properly and in good condition.

8 Engine 'lopes' while idling or idles erratically

1   Vacuum leaks. Check mounting bolts at the intake manifold or plenum for tightness. Make sure that all vacuum hoses are connected and in good condition. Use a stethoscope or a length of fuel hose held against your ear to listen for vacuum leaks while the engine is running. A hissing sound will be heard. A soapy water solution will also detect leaks. Check the intake manifold or plenum gasket surfaces.
2   Leaking EGR valve or plugged PCV valve (Chapter 6).
3   Air filter clogged (Chapter 1).
4   Leaking head gasket. Perform a cylinder compression check (Chapter 2).
5   Timing chain (Chapter 2).
6   Camshaft lobes worn (Chapter 2).
7   Valves burned or otherwise leaking (Chapter 2).
8   Ignition system not operating properly (Chapters 1 and 5).
9   Dirty or clogged injectors (Chapter 4).
10   Idle speed out of adjustment (Chapter 4).

9 Engine misses at idle speed

1   Spark plugs faulty or not gapped properly (Chapter 1).
2   Faulty spark plug wires (Chapter 1).
3   Wet or damaged distributor components (Chapter 1).
4   Short circuits in ignition, coil(s) or spark plug wires.
5   Sticking or faulty emissions systems (Chapter 6).
6   Clogged fuel filter and/or foreign matter in fuel. Replace the fuel filter (Chapter 1).
7   Vacuum leaks at intake manifold or plenum or hose connections. Check as described in Section 8.
8   Incorrect idle speed (Chapter 4) or idle mixture.
9   Low or uneven cylinder compression. Check as described in Chapter 2.
10   Clogged or dirty fuel injectors (Chapter 4).
11   Leaky EGR valve (Chapter 6).

10 Excessively high idle speed

1   Sticking throttle linkage (Chapter 4).
2   Idle speed incorrect (Chapter 4).

11 Battery will not hold a charge

1   Drivebelt defective or not adjusted prop-

erly (Chapter 1).

2   Battery cables loose or corroded (Chapter 1).

3   Alternator not charging properly (Chapter 5).

4   Loose, broken or faulty wires in the charging circuit (Chapter 5).

5   Short circuit causing a continuous drain on the battery (Chapter 12).

6   Battery defective internally.

7   Faulty regulator (Chapter 5).

12   Alternator light stays on

1   Fault in alternator or charging circuit (Chapter 5).

2   Drivebelt defective or not properly adjusted (Chapter 1).

13   Alternator light fails to come on when key is turned on

1   Faulty bulb (Chapter 12).

2   Defective alternator (Chapter 5).

3   Fault in the printed circuit, dash wiring or bulb holder (Chapter 12).

14   Engine misses throughout driving speed range

1   Fuel filter clogged and/or impurities in the fuel system. Check fuel filter (Chapter 1) or clean system (Chapter 4).

2   Faulty or incorrectly gapped spark plugs (Chapter 1).

3   Incorrect ignition timing (Chapter 5).

4   Cracked distributor cap - if equipped - or disconnected ignition system wires or damaged ignition system components (Chapter 1).

5   Defective spark plug wires (Chapter 1).

6   Emissions system components faulty (Chapter 6).

7   Low or uneven cylinder compression pressures. Check as described in Chapter 2.

8   Weak or faulty ignition coil(s) (Chapter 5).

9   Weak or faulty ignition system (Chapter 5).

10   Vacuum leaks at intake manifold or plenum or vacuum hoses (see Section 8).

11   Dirty or clogged fuel injector (Chapter 4).

15   Hesitation or stumble during acceleration

1   Ignition timing incorrect (Chapter 5).

2   Ignition system not operating properly (Chapter 5).

3   Dirty or clogged fuel injectors (Chapter 4).

4   Low fuel pressure. Check for proper operation of the fuel pump and for restrictions in the fuel filter and lines (Chapter 4).

16   Engine stalls

1   Idle speed incorrect (Chapter 4).

2   Fuel filter clogged and/or water and impurities in the fuel system (Chapter 1).

3   Damaged or wet ignition system wires or components.

4   Emissions system components faulty (Chapter 6).

5   Faulty or incorrectly gapped spark plugs (Chapter 1). Also check the spark plug wires (Chapter 1).

6   Vacuum leak at the intake manifold or plenum or vacuum hoses. Check as described in Section 8.

17   Engine lacks power

1   Incorrect ignition timing (Chapter 5).

2   Check for faulty distributor cap, wires, etc. (Chapter 1).

3   Faulty or incorrectly gapped spark plugs (Chapter 1).

4   Air filter dirty (Chapter 1).

5   Spark timing control system not operating properly (Chapter 5).

6   Faulty ignition coil(s) (Chapter 5).

7   Brakes binding (Chapters 1 and 9).

8   Automatic transmission fluid level incorrect, causing slippage (Chapter 1).

9   Clutch slipping (Chapter 8).

10   Fuel filter clogged and/or impurities in the fuel system (Chapters 1 and 4).

11   EGR system not functioning properly (Chapter 6).

12   Use of sub-standard fuel. Fill tank with proper octane fuel.

13   Low or uneven cylinder compression pressures. Check as described in Chapter 2.

14   Air (vacuum) leak at intake manifold or plenum (check as described in Section 8).

18   Engine backfires

1   EGR system not functioning properly (Chapter 6).

2   Ignition timing incorrect (Chapter 5).

3   Vacuum leak (refer to Section 8).

4   Damaged valve springs or sticking valves (Chapter 2).

5   Intake air (vacuum) leak (see Section 8).

19   Engine surges while holding accelerator steady

1   Intake air (vacuum) leak (see Section 8).

2   Fuel pump not working properly.

20   Pinging or knocking engine sounds when engine is under load

1   Incorrect grade of fuel. Fill tank with fuel

of the proper octane rating.

2   Ignition timing incorrect (Chapter 5) or problem in the ignition system (Chapter 5).

3   Carbon build-up in combustion chambers. Remove cylinder head and clean combustion chambers (Chapter 2).

4   Incorrect spark plugs (Chapter 1).

5   Electronic Spark Control (ESC) system not functioning properly (Chapter 6).

21   Engine diesels (continues to run) after being turned off

1   Idle speed too high (Chapter 4).

2   Ignition timing incorrect (Chapter 5).

3   Incorrect spark plug heat range (Chapter 1).

4   Intake air (vacuum) leak (see Section 8).

5   Carbon build-up in combustion chambers. Remove the cylinder head and clean the combustion chambers (Chapter 2).

6   Valves sticking (Chapter 2).

7   Valve clearance incorrect (Chapter 1).

8   EGR system not operating properly (Chapter 6).

9   Leaking fuel injector(s) (Chapter 4).

10   Check for causes of overheating (Section 27).

22   Low oil pressure

1   Improper grade of oil.

2   Oil pump regulator valve not operating properly (Chapter 2).

3   Oil pump worn or damaged (Chapter 2).

4   Engine overheating (refer to Section 27).

5   Clogged oil filter (Chapter 1).

6   Clogged oil strainer (Chapter 2).

7   Oil pressure gauge not working properly (Chapter 2).

23   Excessive oil consumption

1   Loose oil drain plug.

2   Loose bolts or damaged oil pan gasket (Chapter 2).

3   Loose bolts or damaged front cover gasket (Chapter 2).

4   Front or rear crankshaft oil seal leaking (Chapter 2).

5   Loose bolts or damaged valve cover gasket (Chapter 2).

6   Loose oil filter (Chapter 1).

7   Loose or damaged oil pressure switch (Chapter 2).

8   Pistons and cylinders excessively worn (Chapter 2).

9   Piston rings not installed correctly on pistons (Chapter 2).

10   Worn or damaged piston rings (Chapter 2).

11   Intake and/or exhaust valve oil seals worn or damaged (Chapter 2).

12   Worn valve stems.

13   Worn or damaged valves/guides (Chapter 2).

24 Excessive fuel consumption

1   Dirty or clogged air filter element (Chapter 1).
2   Incorrect ignition timing (Chapter 5).
3   Incorrect idle speed (Chapter 4).
4   Low tire pressure or incorrect tire size (Chapter 10).
5   Fuel leakage. Check all connections, lines and components in the fuel system (Chapter 4).
6   Dirty or clogged fuel injectors (Chapter 4).
7   Problem in the fuel injection system (Chapter 4).

25 Fuel odor

1   Fuel leakage. Check all connections, lines and components in the fuel system (Chapter 4).
2   Fuel tank overfilled. Fill only to automatic shut-off.
3   Charcoal canister filter in Evaporative Emissions Control system clogged (Chapter 6).
4   Vapor leaks from Evaporative Emissions Control system lines (Chapter 6).

26 Miscellaneous engine noises

1   A strong dull noise that becomes more rapid as the engine accelerates indicates worn or damaged crankshaft bearings or an unevenly worn crankshaft. To pinpoint the trouble spot, remove the spark plug wire from one plug at a time and crank the engine over. If the noise stops, the cylinder with the removed plug wire indicates the problem area. Replace the bearing and/or service or replace the crankshaft (Chapter 2).
2   A similar (yet slightly higher pitched) noise to the crankshaft knocking described in the previous paragraph, that becomes more rapid as the engine accelerates, indicates worn or damaged connecting rod bearings (Chapter 2). The procedure for locating the problem cylinder is the same as described in Paragraph 1.
3   An overlapping metallic noise that increases in intensity as the engine speed increases, yet diminishes as the engine warms up indicates abnormal piston and cylinder wear (Chapter 2).To locate the problem cylinder, use the procedure described in Paragraph 1.
4   A rapid clicking noise that becomes faster as the engine accelerates indicates a worn piston pin or piston pin hole. This sound will happen each time the piston hits the highest and lowest points in the stroke (Chapter 2). The procedure for locating the problem piston is described in Paragraph 1.
5   A metallic clicking noise coming from the water pump indicates worn or damaged water pump bearings or pump. Replace the water pump with a new one (Chapter 3).
6   A rapid tapping sound or clicking sound that becomes faster as the engine speed increases indicates "valve tapping" or stuck valve lifters. This can be identified by holding one end of a section of hose to your ear and placing the other end at different spots along the rocker arm cover. The point where the sound is loudest indicates the problem valve. Adjust the valve clearance (Chapter 1).
7   A steady metallic rattling or rapping sound coming from the area of the timing chain cover indicates a worn, damaged or out-of-adjustment timing chain. Service or replace the chain and related components (Chapter 2).

Cooling system

27 Overheating

1   Insufficient coolant in system (Chapter 1).
2   Drivebelt defective or not adjusted properly (Chapter 1).
3   Radiator core blocked or radiator grille dirty and restricted (Chapter 3).
4   Thermostat faulty (Chapter 3).
5   Fan not functioning properly (Chapter 3).
6   Radiator cap not maintaining proper pressure. Have cap pressure tested by gas station or repair shop.
7   Ignition timing incorrect (Chapter 5).
8   Defective water pump (Chapter 3).
9   Improper grade of engine oil.
10   Inaccurate temperature gauge (Chapter 3).

28 Overcooling

1   Thermostat faulty (Chapter 3).
2   Inaccurate temperature gauge (Chapter 3).

29 External coolant leakage

1   Deteriorated or damaged hoses. Loose clamps at hose connections (Chapter 1).
2   Water pump seals defective. If this is the case, water will drip from the weep hole in the water pump body (Chapter 3).
3   Leakage from radiator core or header tank. This will require the radiator to be professionally repaired (see Chapter 3 for removal procedures).
4   Engine drain plugs or water jacket freeze plugs leaking (see Chapters 1 and 2).
5   Leak from coolant temperature switch (Chapter 3).
6   Leak from damaged gaskets or small cracks (Chapter 2).
7   Damaged head gasket. This can be verified by checking the condition of the engine oil as noted in Section 30.

30 Internal coolant leakage

Note: *Internal coolant leaks can usually be detected by examining the oil. Check the dipstick and inside the valve cover for water deposits and an oil consistency like that of a milkshake.*
1   Leaking cylinder head gasket. Have the system pressure tested or remove the cylinder head (Chapter 2) and inspect.
2   Cracked cylinder bore or cylinder head. Dismantle engine and inspect (Chapter 2).
3   Loose cylinder head bolts (tighten as described in Chapter 2).

31 Abnormal coolant loss

1   Overfilling system (Chapter 1).
2   Coolant boiling away due to overheating (see causes in Section 27).
3   Internal or external leakage (see Sections 29 and 30).
4   Faulty radiator cap. Have the cap pressure tested.
5   Cooling system being pressurized by engine compression. This could be due to a cracked head or block or leaking head gasket(s).

32 Poor coolant circulation

1   Inoperative water pump. A quick test is to pinch the top radiator hose closed with your hand while the engine is idling, then release it. You should feel a surge of coolant if the pump is working properly (Chapter 3).
2   Restriction in cooling system. Drain, flush and refill the system (Chapter 1). If necessary, remove the radiator (Chapter 3) and have it reverse flushed or professionally cleaned.
3   Thermostat sticking (Chapter 3).
4   Insufficient coolant (Chapter 1).

33 Corrosion

1   Excessive impurities in the water. Soft, clean water is recommended. Distilled or rainwater is satisfactory.
2   Insufficient antifreeze solution (refer to Chapter 1 for the proper type and ratio of water to antifreeze).
3   Infrequent flushing and draining of system. Regular flushing of the cooling system should be carried out at the specified intervals as described in Chapter 1.

Clutch

Note: *All clutch related service information is located in Chapter 8, unless otherwise noted.*

34 Fails to release (pedal pressed to the floor - shift lever does not move freely in and out of Reverse)

1   Clutch contaminated with oil. Remove clutch plate and inspect.
2   Clutch plate warped, distorted or otherwise damaged.
3   Diaphragm spring fatigued. Remove clutch cover/pressure plate assembly and inspect.
4   Leakage of fluid from clutch hydraulic system. Inspect master cylinder, operating (release) cylinder and connecting lines.
5   Air in clutch hydraulic system. Bleed the system.
6   Insufficient pedal stroke. Check and adjust as necessary.
7   Piston seal in operating (release) cylinder deformed or damaged.
8   Lack of grease on pilot bushing.
9   Damaged transmission input shaft splines.

35 Clutch slips (engine speed increases with no increase in vehicle speed)

1   Worn or oil soaked clutch plate.
2   Clutch plate not broken in. It may take 30 or 40 normal starts for a new clutch to seat.
3   Diaphragm spring weak or damaged. Remove clutch cover/pressure plate assembly and inspect.
4   Flywheel warped or scored (Chapter 2).
5   Debris in master cylinder preventing the piston from returning to its normal position.

36 Grabbing (chattering) as clutch is engaged

1   Oil on clutch plate. Remove and inspect. Repair any leaks.
2   Worn or loose engine or transmission mounts. They may move slightly when clutch is released. Inspect mounts and bolts.
3   Worn splines on transmission input shaft. Remove clutch components and inspect.
4   Warped pressure plate or flywheel. Remove clutch components and inspect.
5   Diaphragm spring fatigued. Remove clutch cover/pressure plate assembly and inspect.
6   Clutch linings hardened or warped.
7   Clutch lining rivets loose.
8   Engine and transmission not in alignment. Check for foreign object between bellhousing and engine block. Check for loose bellhousing bolts.

37 Squeal or rumble with clutch engaged (pedal released)

1   Release bearing binding on transmission shaft. Remove clutch components and check bearing. Remove any burrs or nicks, clean and relubricate before reinstallation.
2   Clutch rivets loose.
3   Clutch plate cracked.
4   Fatigued clutch plate torsion springs. Replace clutch plate.

38 Squeal or rumble with clutch disengaged (pedal depressed)

1   Worn or damaged release bearing.
2   Worn or broken pressure plate diaphragm fingers.
3   Worn or damaged pilot bearing.

39 Clutch pedal stays on floor when disengaged

1   Binding release fork or release bearing. Inspect release fork or remove clutch components as necessary.
2   Faulty clutch master cylinder or release cylinder.

Manual transmission

Note: *All manual transmission service information is located in Chapter 7, unless otherwise noted.*

40 Noisy in Neutral with engine running

1   Input shaft bearing worn.
2   Damaged main drive gear bearing.
3   Insufficient transmission oil (Chapter 1).
4   Transmission lubricant in poor condition. Drain and fill with proper grade oil. Check old lubricant for water and debris (Chapter 1).
5   Noise can be caused by variations in engine torque.

41 Noisy in all gears

1   Any of the above causes, and/or:
2   Worn or damaged output gear bearings or shaft.

42 Noisy in one particular gear

1   Worn, damaged or chipped gear teeth.
2   Worn or damaged synchronizer.

43 Slips out of gear

1   Transmission loose on clutch housing.
2   Stiff shift lever seal.
3   Shift linkage binding.
4   Broken or loose input gear bearing retainer.
5   Dirt between clutch lever and engine housing.
6   Worn linkage.
7   Damaged or worn check balls, fork rod ball grooves or check springs.
8   Worn mainshaft or countershaft bearings.
9   Loose engine mounts (Chapter 2).
10   Excessive gear end play.
11   Worn synchronizers.

44 Oil leaks

1   Excessive amount of lubricant in transmission (see Chapter 1 for correct checking procedures). Drain lubricant as required.
2   Case retaining bolts loose or gaskets damaged.
3   Rear oil seal or speedometer oil seal damaged.
4   To pinpoint a leak, first remove all built-up dirt and grime from the transmission. Degreasing agents and/or steam cleaning will achieve this. With the underside clean, drive the vehicle at low speeds so the air flow will not blow the leak far from its source. Raise the vehicle and determine where the leak is located.

45 Difficulty engaging gears

1   Clutch not releasing completely.
2   Loose or damaged shift linkage. Make a thorough inspection, replacing parts as necessary.
3   Insufficient transmission oil (Chapter 1).
4   Transmission oil in poor condition. Drain and fill with proper grade oil. Check oil for water and debris (Chapter 1).
5   Worn or damaged shift forks.
6   Sticking or jamming gears.

46 Noise occurs while shifting gears

1   Check for proper operation of the clutch (Chapter 8).
2   Faulty synchronizer assemblies. Measure baulk ring-to-gear clearance. Also, check for wear or damage to baulk rings or any parts of the synchromesh assemblies.

Automatic transmission

Note: *Due to the complexity of the automatic transmission, it's difficult for the home mechanic to properly diagnose and service. For problems other than the following, the* *vehicle should be taken to a reputable mechanic.*

47 Fluid leakage

1   Automatic transmission fluid is a deep red color, and fluid leaks should not be confused with engine oil which can easily be blown by air flow to the transmission.
2   To pinpoint a leak, first remove all built-up dirt and grime from the transmission. Degreasing agents and/or steam cleaning will achieve this. With the underside clean, drive the vehicle at low speeds so the air flow will not blow the leak far from its source. Raise the vehicle and determine where the leak is located. Common areas of leakage are:

a) *Fluid pan: tighten mounting bolts and/or replace pan gasket as necessary (Chapter 1).*
b) *Rear extension: tighten bolts and/or replace oil seal as necessary.*
c) *Filler pipe: replace the rubber oil seal where pipe enters transmission case.*
d) *Transmission oil lines: tighten fittings where lines enter transmission case and/or replace lines.*
e) *Vent pipe: transmission overfilled and/or water in fluid (see checking procedures, Chapter 1).*
f) *Speedometer connector: replace the O-ring where speedometer sensor enters transmission case.*

48 General shift mechanism problems

Chapter 7 deals with checking and adjusting the shift linkage on automatic transmissions. Common problems which may be caused by out of adjustment linkage are:

a) *Engine starting in gears other than P (Park) or N (Neutral).*
b) *Indicator pointing to a gear other than the one actually engaged.*
c) *Vehicle moves with transmission in P (Park) position.*

49 Transmission will not downshift with the accelerator pedal pressed to the floor

Chapter 7 deals with adjusting the throttle valve cable installed on automatic transmissions.

50 Engine will start in gears other than Park or Neutral

Chapter 7 deals with adjusting the Neutral start switch installed on automatic transmissions.

51 Transmission slips, shifts rough, is noisy or has no drive in forward or Reverse gears

1   There are many probable causes for the above problems, but the home mechanic should concern himself only with one possibility; fluid level.
2   Before taking the vehicle to a shop, check the fluid level and condition as described in Chapter 1. Add fluid, if necessary, or change the fluid and filter if needed. If problems persist, have a professional diagnose the transmission.

Driveshaft

52 Leaks at front of driveshaft

Defective transmission rear seal. See Chapter 7 for replacement procedure. As this is done, check the splined yoke for burrs or roughness that could damage the new seal. Remove burrs with a fine file or whetstone.

53 Knock or clunk when transmission is under initial load (just after transmission is put into gear)

1   Loose or disconnected rear suspension components. Check all mounting bolts and bushings (Chapters 1 and 11).
2   Loose driveshaft bolts. Inspect all bolts and nuts and tighten them securely.
3   Worn or damaged universal joint bearings. Replace driveshaft (Chapter 8).
4   Worn sleeve yoke and mainshaft spline.

54 Metallic grating sound consistent with vehicle speed

Pronounced wear in the universal joint bearings. Replace U-joints or driveshafts, as necessary.

55 Vibration

1   Install a tachometer inside the vehicle to monitor engine speed as the vehicle is driven. Drive the vehicle and note the engine speed at which the vibration (roughness) is most pronounced. Now shift the transmission to a different gear and bring the engine speed to the same point.
2   If the vibration occurs at the same engine speed (rpm) regardless of which gear the transmission is in, the driveshaft is NOT at fault since the driveshaft speed varies.
3   If the vibration decreases or is elimi-

nated when the transmission is in a different gear at the same engine speed, refer to the following probable causes.
4   Bent or dented driveshaft. Inspect and replace as necessary.
5   Undercoating or built-up dirt, etc. on the driveshaft. Clean the shaft thoroughly.
6   Worn universal joint bearings. Replace the U-joints or driveshaft as necessary.
7   Driveshaft and/or companion flange out of balance. Check for missing weights on the shaft. Remove driveshaft and reinstall 180-degrees from original position, then recheck. Have the driveshaft balanced if problem persists.
8   Loose driveshaft mounting bolts/nuts.
9   Worn transmission rear bushing.

56 Scraping noise

Make sure the dust cover on the sleeve yoke isn't rubbing on the transmission extension housing.

Rear axle and differential

Note: *For differential servicing information, refer to Chapter 8, unless otherwise specified.*

57 Noise - same when in drive as when vehicle is coasting

1   Road noise. No corrective action available.
2   Tire noise. Inspect tires and check tire pressures (Chapter 1).
3   Front wheel bearings worn or damaged (Chapter 10).
4   Insufficient differential oil (Chapter 1).
5   Defective differential.

58 Knocking sound when starting or shifting gears

Defective or incorrectly adjusted differential.

59 Noise when turning

Defective differential.

60 Vibration

See probable causes under Driveshaft. Proceed under the guidelines listed for the driveshaft. If the problem persists, check the rear wheel bearings by raising the rear of the vehicle and spinning the wheels by hand. Listen for evidence of rough (noisy) bearings. Remove and inspect (Chapter 8).

61 Oil leaks

1 Pinion oil seal damaged (Chapter 8).
2 Axleshaft or driveaxle oil seals damaged (Chapter 8).
3 Loose cover bolts or filler plug on differential (Chapter 1).
4 Clogged or damaged breather on differential.

Brakes

Note: *Before assuming a brake problem exists, make sure the tires are in good condition and inflated properly, the front end alignment is correct and the vehicle is not loaded with weight in an unequal manner. All service procedures for the brakes are included in Chapter 9, unless otherwise noted.*

62 Vehicle pulls to one side during braking

1 Defective, damaged or oil contaminated brake pad on one side. Inspect as described in Chapter 1. Refer to Chapter 9 if replacement is required.
2 Excessive wear of brake pad material or disc on one side. Inspect and repair as necessary.
3 Loose or disconnected front suspension components. Inspect and tighten all bolts securely (Chapters 1 and 10).
4 Defective caliper assembly. Remove caliper and inspect for stuck piston or damage.
5 Brake pad-to-disc adjustment needed. Inspect automatic adjusting mechanism for proper operation.
6 Scored or out of round disc.
7 Loose caliper mounting bolts.

63 Noise (high-pitched squeal)

1 Front brake pads worn out. This noise comes from the wear sensor rubbing against the disc. Replace pads with new ones immediately!
2 Glazed or contaminated pads.
3 Dirty or scored disc.
4 Bent support plate.

64 Excessive brake pedal travel

1 Partial brake system failure. Inspect entire system and correct as required.
2 Insufficient fluid in master cylinder. Check (Chapter 1) and add fluid - bleed system if necessary.
3 Air in system. Bleed system.
4 Excessive lateral disc play.
5 Brakes out of adjustment. Check the operation of the automatic adjusters.
6 Defective check valve. Replace valve and bleed system.

65 Brake pedal feels spongy when depressed

1 Air in brake lines. Bleed the brake system.
2 Deteriorated rubber brake hoses. Inspect all system hoses and lines. Replace parts as necessary.
3 Master cylinder mounting nuts loose. Inspect master cylinder bolts (nuts) and tighten them securely.
4 Master cylinder faulty.
5 Incorrect shoe or pad clearance.
6 Defective check valve. Replace valve and bleed system.
7 Clogged reservoir cap vent hole.
8 Deformed rubber brake lines.
9 Soft or swollen caliper seals.
10 Poor quality brake fluid. Bleed entire system and fill with new approved fluid.

66 Excessive effort required to stop vehicle

1 Power brake booster not operating properly.
2 Excessively worn pads. Check and replace if necessary.
3 One or more caliper pistons seized or sticking. Inspect and rebuild as required.
4 Brake pads contaminated with oil or grease. Inspect and replace as required.
5 New pads installed and not yet seated. It'll take a while for the new material to seat against the disc.
6 Worn or damaged master cylinder or caliper assemblies. Check particularly for frozen pistons.
7 Also see causes listed under Section 65.

67 Pedal travels to the floor with little resistance

Little or no fluid in the master cylinder reservoir caused by leaking caliper piston(s) or loose, damaged or disconnected brake lines. Inspect entire system and repair as necessary.

68 Brake pedal pulsates during brake application

Note: *Brake pedal pulsation during operation of the Anti-Lock Brake System (ABS) is normal.*
1 Wheel bearings damaged or worn (Chapter 10).
2 Caliper not sliding properly due to improper installation or obstructions. Remove and inspect.
3 Disc not within specifications. Remove the disc and check for excessive lateral runout and parallelism. Have the discs resur-

faced or replace them with new ones. Also make sure that all discs are the same thickness.

69 Brakes drag (indicated by sluggish engine performance or wheels being very hot after driving)

1 Output rod adjustment incorrect at the brake pedal.
2 Obstructed master cylinder compensator. Disassemble master cylinder and clean.
3 Master cylinder piston seized in bore. Overhaul master cylinder.
4 Caliper assembly in need of overhaul.
5 Brake pads or shoes worn out.
6 Piston cups in master cylinder or caliper assembly deformed. Overhaul master cylinder.
7 Parking brake assembly will not release.
8 Clogged brake lines.
9 Brake pedal height improperly adjusted.

70 Rear brakes lock up under light brake application

1 Tire pressures too high.
2 Tires excessively worn (Chapter 1).

71 Rear brakes lock up under heavy brake application

1 Tire pressures too high.
2 Tires excessively worn (Chapter 1).
3 Front brake pads contaminated with oil, mud or water. Clean or replace the pads.
4 Front brake pads excessively worn.
5 Defective master cylinder or caliper assembly.

Suspension and steering

Note: *All service procedures for the suspension and steering systems are included in Chapter 10, unless otherwise noted.*

72 Vehicle pulls to one side

1 Tire pressures uneven (Chapter 1).
2 Defective tire (Chapter 1).
3 Excessive wear in suspension or steering components (Chapter 10).
4 Front end alignment incorrect.
5 Front brakes dragging. Inspect as described in Section 69.
6 Wheel lug nuts loose.
7 Worn upper or lower link or strut rod bushings.

73 Shimmy, shake or vibration

1   Tire or wheel out of balance or out of round. Have them balanced on the vehicle.
2   Worn or damaged wheel bearings (Chapter 10).
3   Shock absorbers and/or suspension components worn or damaged. Check for worn bushings in the upper and lower links.
4   Wheel lug nuts loose.
5   Incorrect tire pressures.
6   Excessively worn or damaged tire.
7   Loosely mounted steering gear housing.
8   Steering gear improperly adjusted.
9   Loose, worn or damaged steering components.
10   Worn balljoint.

74 Excessive pitching and/or rolling around corners or during braking

1   Defective shock absorbers. Replace as a set.
2   Broken or weak springs and/or suspension components.
3   Worn or damaged stabilizer bar or bushings.
4   Worn or damaged upper or lower links or bushings.

75 Wandering or general instability

1   Improper tire pressures.
2   Worn or damaged bushings.
3   Incorrect front end alignment.
4   Worn or damaged steering linkage.
5   Improperly adjusted steering gear.
6   Out of balance wheels.
7   Loose wheel lug nuts.
8   Worn rear shock absorbers.
9   Fatigued or damaged rear springs.

76 Excessively stiff steering

1   Lack of lubricant in power steering fluid reservoir, where appropriate (Chapter 1).
2   Incorrect tire pressures (Chapter 1).
3   Front end out of alignment.
4   Steering gear out of adjustment or lacking lubrication.
5   Worn or damaged wheel bearings.
6   Worn or damaged steering gear.
7   Interference of steering column with turn signal switch.
8   Low tire pressures.
9   Worn or damaged balljoints.
10   Worn or damaged steering linkage.

77 Excessive play in steering

1   Worn wheel bearings (Chapter 10).
3   Steering gear improperly adjusted.
2   Incorrect front end alignment.
3   Steering gear mounting bolts loose.
4   Worn steering linkage.

78 Lack of power assistance

1   Drivebelt faulty or not adjusted properly (Chapter 1).
2   Fluid level low (Chapter 1).
3   Hoses or pipes restricting the flow. Inspect and replace parts as necessary.
4   Air in power steering system. Bleed system.
5   Defective power steering pump.

79 Steering wheel fails to return to straight-ahead position

1   Incorrect front end alignment.
2   Tire pressures low.
3   Steering gears improperly engaged.
4   Steering column out of alignment.
5   Worn or damaged balljoint.
6   Worn or damaged steering linkage.
7   Insufficient oil in steering gear.
8   Lack of fluid in power steering pump.

80 Steering effort not the same in both directions (power system)

1   Leaks in steering gear.
2   Clogged fluid passage in steering gear.

81 Noisy power steering pump

1   Insufficient oil in pump.
2   Clogged hoses or oil filter in pump.
3   Loose pulley.
4   Worn or improperly adjusted drivebelt (Chapter 1).
5   Defective pump.

82 Miscellaneous noises

1   Improper tire pressures.
2   Insufficiently lubricated balljoint or steering linkage.
3   Loose or worn steering gear, steering linkage or suspension components.
4   Defective shock absorber.
5   Defective wheel bearing.
6   Damaged spring.
7   Loose wheel lug nuts.
8   Worn or damaged rear axleshaft spline.
9   Worn or damaged rear shock absorber mounting bushing.
10   Worn rear axle bearing.
11   See also causes of noises at the rear axle and driveshaft.

83 Excessive tire wear (not specific to one area)

1   Incorrect tire pressures.
2   Tires out of balance. Have them balanced on the vehicle.
3   Wheels damaged. Inspect and replace as necessary.
4   Suspension or steering components worn (Chapter 1).

84 Excessive tire wear on outside edge

1   Incorrect tire pressure
2   Excessive speed in turns.
3   Wheel alignment incorrect.

85 Excessive tire wear on inside edge

1   Incorrect tire pressure.
2   Wheel alignment incorrect.
3   Loose or damaged steering components (Chapter 1).

86 Tire tread worn in one place

1   Tires out of balance. Have them balanced on the vehicle.
2   Damaged or buckled wheel. Inspect and replace if necessary.
3   Defective tire.

Chapter 1
Tune-up and routine maintenance

Contents

Specifications

Recommended lubricants and fluids

Note: *Listed here are manufacturer recommendations at the time this manual was written. Manufacturers occasionally upgrade their fluid and lubricant specifications, so check with your local auto parts store for current recommendations.*

Engine oil	
Type	API grade SG, SG/CC or SG/CD multigrade and fuel-efficient oil
Viscosity	See accompanying chart
Manual transmission lubricant	Dexron II, IIE or III Automatic Transmission Fluid (ATF)
Automatic transmission fluid	Dexron II, IIE or III Automatic Transmission Fluid (ATF)
Engine coolant	50/50 mixture of water and the specified ethylene glycol-based (green color) antifreeze or "DEX-COOL™ silicate-free (orange-color) coolant - DO NOT mix the two types (refer to Sections 4, 10 and 30)
Brake fluid	DOT 3 fluid
Power steering fluid	GM power steering fluid or equivalent

Capacities*

Engine oil (with filter change)	
1993	
3.4L engine	4.3 qts
5.7L engine	4.5 qts
1994	
3.4L engine	4.3 qts
5.7L engine	5.0 qts
1995	
3.4L engine	5.0 qts
3800 engine	5.0 qts
5.7L engine	5.0 qts
1996 through 1997	
3800 engine	4.5 qts
5.7L engine	5.0 qts
1998 and later	
3800 engine	4.5 qts
5.7L engine	5.5 qts
Fuel tank	
1998 and earlier	15.5 gals
1999 and later	16.8 gals

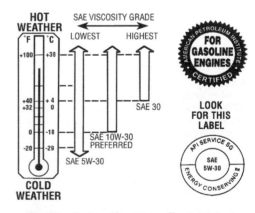

Engine oil viscosity chart - For best fuel economy and cold starting, select the lowest SAE viscosity grade for the expected temperature range

Capacities* (continued)

Cooling system
   3.4L engine... 12 qts
   3800 engine
      through 1999... 12 quarts
      2000 and later
         Automatic.. 11.4 quarts
         Manual.. 11.6 quarts
   5.7L engine
      1997 and earlier .... 15 qts
      1998 .... 15 qts
      1999 and later .... 12 qts
Manual transmission
   Five-speed.. 3.4 qts
   Six-speed .... 4.0 qts
Automatic transmission (fluid and filter replacement) .... 5 qts
All capacities approximate. Add as necessary to bring to appropriate level.

Ignition system

Spark plug type
   1997 and earlier
      3.4L engine .... AC type R43TSK or equivalent
      3800 engine .... AC type R43LTS6 or equivalent
      5.7L engine .... AC type R45LTSP or equivalent
   1998 and 1999
      3800 engine .... AC type 41-921 or equivalent
      5.7L engine .... AC type 41-931 or equivalent
   2000 and later
      3800 engine .... AC type 41-921 or equivalent
      5.7L engine
         2000.. AC type 41-952 or equivalent
         2001 and 2002.................................. AC type 41-921
Spark plug gap
   1997 and earlier
      3.4L engine .... 0.045 inch
      3800 engine .... 0.060 inch
      5.7L engine .... 0.050 inch
   1998 and later
      3800 engine .... 0.060 inch
      5.7L engine .... 0.060 inch
Firing order
   3.4L engine.. 1-2-3-4-5-6
   3800 engine... 1-6-5-4-3-2
   5.7L engine
      1997 and earlier 1-8-4-3-6-5-7-2
      1998 and later 1-8-7-2-6-5-4-3

General

Radiator cap pressure rating .... 15 to 18 psi
Brake pad wear limit.. 1/8 inch

Torque specifications   **Ft-lbs** (unless otherwise indicated)

Spark plugs
   3.4L engine.. 23
   3800 engine... 144 in-lbs
   V8 engines... 132 in-lbs
Engine oil drain plug .... 15 to 20
Automatic transmission pan bolts .... 96 to 120 in-lbs
Wheel lug nuts 100

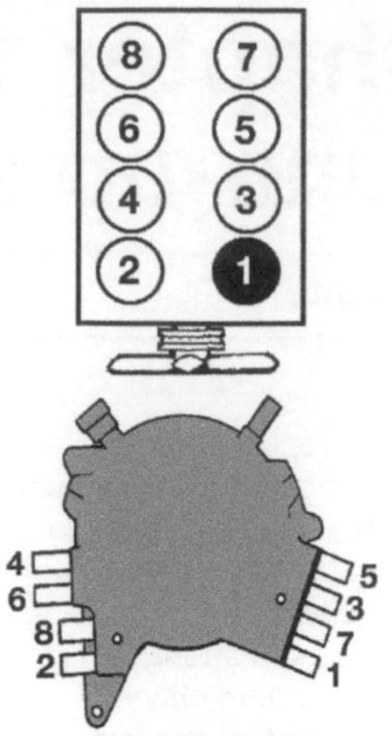

**1997 and earlier
V8 engine**

24017-1-B HAYNES

**1998 and later
V8 engine**

24046-1SPECS.HAYNES

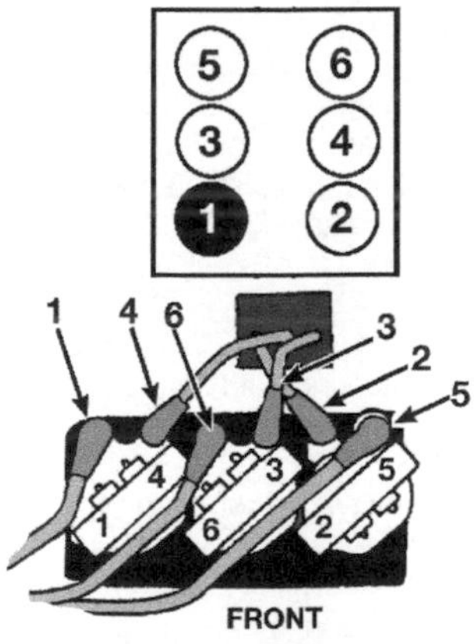

3.4L engine

24017-1-C HAYNES

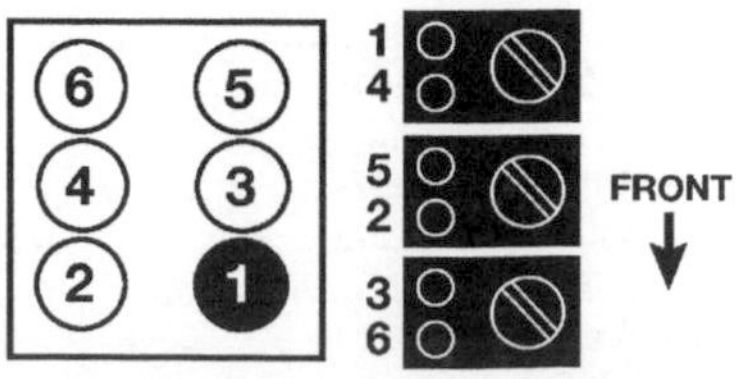

3800 engine

24017-1-D HAYNES

Cylinder location, distributor and
coil terminal identification

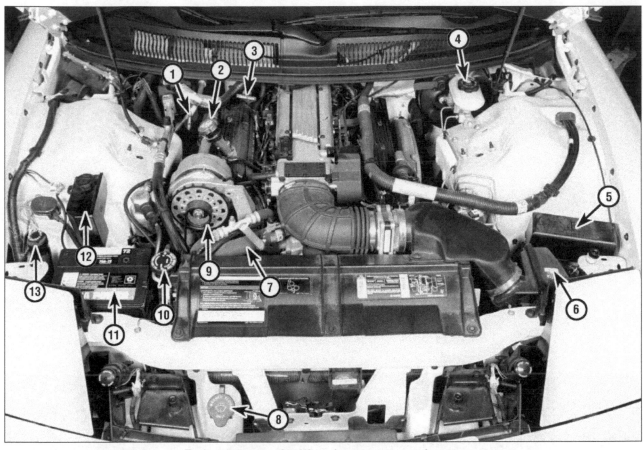

Typical 1st generation V8 engine compartment layout

1	Engine oil dipstick	6	Air filter housing	11	Battery
2	Engine oil filler cap	7	Upper radiator hose	12	Power steering fluid reservoir
3	Automatic transmission dipstick	8	Windshield washer fluid reservoir	13	Engine coolant reservoir
4	Brake master cylinder reservoir	9	Drivebelt		
5	Fuse block	10	Radiator cap		

1   Introduction

This Chapter is designed to help the home mechanic maintain the Chevrolet Camaro and Pontiac Firebird models with the goals of maximum performance, economy, safety and reliability in mind.

Included is a master maintenance schedule (page 1-6), followed by procedures dealing specifically with each item on the schedule. Visual checks, adjustments, component replacement and other helpful items are included. Refer to the accompanying illustrations of the engine compartment and the underside of the vehicle for the locations of various components.

Servicing your vehicle in accordance with the mileage/time maintenance schedule and the step-by-step procedures will result in a planned maintenance program that should produce a long and reliable service life. Keep in mind that it's a comprehensive plan, so maintaining some items but not others at the specified intervals will not produce the same results.

As you service your vehicle, you'll discover that many of the procedures can - and should - be grouped together because of the nature of the particular procedure you're performing or because of the close proximity of two otherwise unrelated components to one another.

For example, if the vehicle is raised, you should inspect the exhaust, suspension, steering and fuel systems while you're under the vehicle. When you're rotating the tires, it makes good sense to check the brakes since the wheels are already removed. Finally, let's suppose you have to borrow or rent a torque wrench. Even if you only need it to tighten the spark plugs, you might as well check the torque of as many critical fasteners as time allows.

The first step in this maintenance program is to prepare yourself before the actual work begins. Read through all the procedures you're planning to do, then gather up all the parts and tools needed. If it looks like you might run into problems during a particular job, seek advice from a mechanic or an experienced do-it-yourselfer.

Caution: *The stereo in your vehicle may be equipped with an anti-theft system. Refer to the information at the front of this manual before performing any procedure which requires disconnecting the battery cable.*

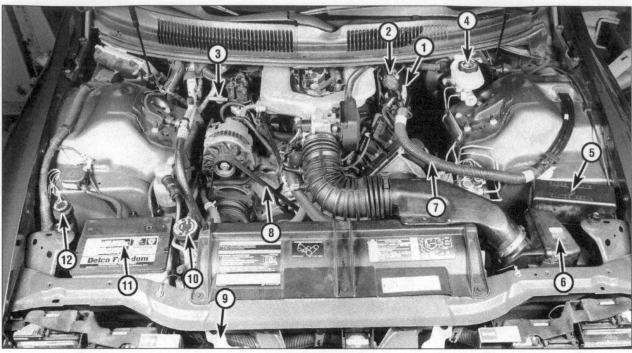

Typical 3.4L V6 engine compartment layout

1	Engine oil dipstick	5	Fuse block	9	Windshield washer fluid reservoir
2	Engine oil filler cap	6	Air filter housing	10	Radiator cap
3	Automatic transmission dipstick	7	Power steering fluid reservoir	11	Battery
4	Brake master cylinder reservoir	8	Drivebelt	12	Engine coolant reservoir

Typical 3800 V6 engine compartment layout

1	Engine oil filler cap	5	Fuse block	9	Windshield washer fluid reservoir
2	Engine oil dipstick	6	Air filter housing	10	Radiator cap
3	Brake master cylinder reservoir	7	Power steering fluid reservoir	11	Battery
4	Clutch master cylinder reservoir	8	Drivebelt	12	Engine coolant reservoir

Typical V8 engine compartment underside components

1	Radiator drain	4	Spring and shock absorber assembly	7	Engine oil drain plug
2	Drivebelt	5	Engine oil filter	8	Exhaust system
3	Front brake caliper	6	Automatic transmission fluid pan	9	Steering gear boot

Typical rear underside components

1	Driveshaft	3	Shock absorber	5	Muffler	7	Rear brake
2	Fuel filter	4	Fuel tank	6	Coil spring		

2 Camaro and Firebird Maintenance schedule

The following maintenance intervals are based on the assumption that the vehicle owner will be doing the maintenance or service work, as opposed to having a dealer service department do the work. Although the time/mileage intervals are loosely based on factory recommendations, most have been shortened to ensure, for example, that such items as lubricants and fluids are checked/changed at intervals that promote maximum engine/driveline service life. Also, subject to the preference of the individual owner interested in keeping his or her vehicle in peak condition at all times, and with the vehicle's ultimate resale in mind, many of the maintenance procedures may be performed more often than recommended in the following schedule. We encourage such owner initiative.

When the vehicle is new it should be serviced initially by a factory authorized dealer service department to protect the factory warranty. In many cases the initial maintenance check is done at no cost to the owner (check with your dealer service department for more information).

Every 250 miles or weekly, whichever comes first

Check the engine oil level (Section 4)
Check the engine coolant level (Section 4)
Check the windshield washer fluid level (Section 4)
Check the brake and clutch fluid levels (Section 4)
Check the tires and tire pressures (Section 5)

Every 3000 miles or 3 months, whichever comes first

All items listed above plus:
Check the automatic transmission fluid level (Section 6)
Check the power steering fluid level (Section 7)
Change the engine oil and filter (Section 8)
Check and service the battery (Section 9)

Every 7500 miles or 6 months, whichever comes first

All items listed above plus:
Check the cooling system (Section 10)
Inspect and replace, if necessary, all underhood hoses (Section 11)
Inspect and replace, if necessary, the windshield wiper blades (Section 12)
Lubricate the chassis (Section 13)
Inspect the suspension and steering components (Section 14)
Inspect the exhaust system (Section 15)
Inspect the clutch hydraulic system (Section 16)
Inspect the seatbelts (Section 17)
Check the starter safety switch (Section 18)

Every 15,000 miles or 12 months, whichever comes first

Check the manual transmission lubricant level (Section 19)
Check the differential lubricant level (Section 20)
Rotate the tires (Section 21)***
Check the brakes (Section 22)*
Inspect the fuel system (Section 23)

Every 30,000 miles or 24 months, whichever comes first

All items listed above plus:
Replace the air filter (Section 24)****
Replace the fuel filter (Section 25)
Check the engine drivebelt (Section 26)
Change the automatic transmission fluid and filter (Section 27)**
Change the manual transmission lubricant (Section 28)
Change the differential lubricant (Section 29)
Service the cooling system (drain, flush and refill) (green-colored ethylene glycol anti-freeze only) (Section 30)
Inspect and replace, if necessary, the PCV valve (Section 31)
Inspect the evaporative emissions control system (Section 32)
Replace the spark plugs (conventional [non-platinum] spark plugs) (Section 33)
Inspect the spark plug wires (Section 34)
Inspect the distributor housing (1997 and earlier V8 engines only) (Section 34)

Every 50,000 miles or 36 months, whichever comes first

Replace the spark plugs (platinum-tipped spark plugs) (Section 33)

Every 100,000 miles or 5 years, whichever comes first

Service the cooling system (drain, flush and refill) (orange-colored "DEX-COOL™ silicate-free coolant only) (Section 30)

* If the vehicle frequently tows a trailer, is operated primarily in stop-and-go conditions or its brakes receive severe usage for any other reason, check the brakes every 3000 miles or three months.
** If operated under one or more of the following conditions, change the automatic transmission fluid every 15,000 miles:
    In heavy city traffic where the outside temperature regularly reaches 90-degrees F (32-degrees C) or higher
    In hilly or mountainous terrain
    Frequent trailer pulling
*** If the vehicle is primarily driven in city traffic, rotate the tires every 7500 miles or 6 months.
**** If the vehicle is frequently operated in dusty conditions, change the air filter every 15,000 miles or 12 months.

4.2 The engine oil dipstick is mounted on the side of the engine

3 Tune-up general information

The term tune-up is used in this manual to represent a combination of individual operations rather than one specific procedure.

If, from the time the vehicle is new, the routine maintenance schedule is followed closely and frequent checks are made of fluid levels and high wear items, as suggested throughout this manual, the engine will be kept in relatively good running condition and the need for additional work will be minimized due to lack of regular maintenance. This is even more likely if a used vehicle, which has not received regular and frequent maintenance checks, is purchased. In such cases, an engine tune-up will be needed outside of the regular routine maintenance intervals.

The first step in any tune-up or diagnostic procedure to help correct a poor running engine is a cylinder compression check. A compression check (see Chapter 2, Part D) will help determine the condition of internal engine components and should be used as a guide for tune-up and repair procedures. If, for instance, a compression check indicates serious internal engine wear, a conventional tune-up won't improve the performance of the engine and would be a waste of time and money. Because of its importance, the compression check should be done by someone with the right equipment and the knowledge to use it properly.

The following procedures are those most often needed to bring a generally poor

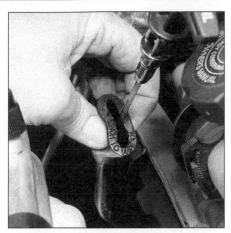

4.4a On early model V8 engines the dipstick tube incorporates a wiper

running engine back into a proper state of tune.

Minor tune-up

Check all engine related fluids (Section 4)
Clean, inspect and test the battery (Section 9)
Check the cooling system (Section 10)
Check all underhood hoses (Section 11)
Check the air filter (Section 24)
Check and adjust the drivebelts (Section 26)
Check the PCV valve (Section 31)
Replace the spark plugs (Section 33)
Inspect the spark plug wires (Section 34)

Major tune-up

All items listed under Minor tune-up plus . . .
Check the fuel system (Section 23)
Replace the air filter (Section 24)
Replace the spark plug wires (Section 33)
Check the charging system (Chapter 5)
Check the EGR system (Chapter 6)

4 Fluid level checks (every 250 miles or weekly)

Note: *The following are fluid level checks to be done on a 250 mile or weekly basis. Additional fluid level checks can be found in specific maintenance procedures which follow. Regardless of intervals, be alert to fluid leaks under the vehicle which would indicate a problem to be corrected immediately.*

1   Fluids are an essential part of the lubrication, cooling, brake and windshield washer systems. Because the fluids gradually become depleted and/or contaminated during normal operation of the vehicle, they must be periodically replenished. See *Recommended lubricants and fluids* at the beginning of this Chapter before adding fluid to any of the following components. **Note:** *The vehicle must be on level ground when fluid levels are checked.*

Engine oil

Refer to illustrations 4.2, 4.4a, 4.4b and 4.6

2   The engine oil level is checked with a dipstick **(see illustration)**. The dipstick extends through a metal tube down into the oil pan.

3   The oil level should be checked before the vehicle has been driven, or about 15 minutes after the engine has been shut off. If the oil is checked immediately after driving the vehicle, some of the oil will remain in the upper part of the engine, resulting in an inaccurate reading on the dipstick.

4   Pull the dipstick from the tube and wipe all the oil from the end with a clean rag or paper towel. This won't be necessary on V8 engines because the dipstick tube incorporates a wiper **(see illustration)**. Insert the clean dipstick all the way back into the tube and pull it out again. Note the oil at the end of the dipstick. Add oil as necessary to keep the level above the ADD mark in the cross hatched area of the dipstick **(see illustration)**.

5   Do not overfill the engine by adding too much oil since this may result in oil fouled spark plugs, oil leaks or oil seal failures.

6   Oil is added to the engine after removing a twist-off cap located on the engine **(see illustration)**. A funnel may help to reduce spills.

7   Checking the oil level is an important preventive maintenance step. A consistently low oil level indicates oil leakage through damaged seals, defective gaskets or past worn rings or valve guides. If the oil looks

4.4b The oil level should be at or near the upper hole or cross-hatched area on the dipstick - if it's below the ADD line, add enough oil to bring the level into the upper hole or cross-hatched area

4.6 The engine oil filler cap is clearly marked and threads into the tube on the valve cover

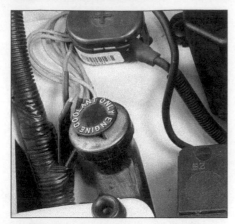

4.9a The coolant reservoir is located in the right (passenger side) front corner of the engine compartment

4.9b Unscrew the reservoir cap and check the coolant level on the attached dipstick

milky in color or has water droplets in it, the cylinder head gasket may be blown or the head or block may be cracked. The engine should be checked immediately. The condition of the oil should also be checked. Whenever you check the oil level, slide your thumb and index finger up the dipstick before wiping off the oil. If you see small dirt or metal particles clinging to the dipstick, the oil should be changed (see Section 8).

Engine coolant

Refer to illustrations 4.9a and 4.9b
Warning: *Do not allow antifreeze to come in contact with your skin or painted surfaces of the vehicle. Flush contaminated areas immediately with plenty of water. Do not store new coolant or leave old coolant lying around where it's accessible to children or pets - they're attracted by its sweet smell. Ingestion of even a small amount of coolant can be fatal! Wipe up garage floor and drip pan coolant spills immediately. Keep antifreeze containers covered and repair leaks in the cooling system immediately.*
Caution: *Never mix green-colored ethylene glycol anti-freeze and orange-colored "DEX-COOL™ silicate-free coolant because doing so will destroy the efficiency of the "DEX-COOL™ coolant which is designed to last for 100,000 miles or five years.*
8   All vehicles covered by this manual are equipped with a pressurized coolant recovery system. A plastic coolant reservoir located at the front of the engine compartment is connected by a hose to the radiator filler neck. As the engine warms up and the coolant expands, it escapes through a valve in the radiator cap and travels through the hose into the reservoir. As the engine cools, the coolant is automatically drawn back into the cooling system to maintain the correct level.
9   The coolant level in the reservoir should be checked regularly. Unscrew the coolant reservoir cap and withdraw the dipstick to check the coolant level **(see illustration)**.
Warning: *Do not remove the radiator cap to check the coolant level when the engine is*

warm. The level in the reservoir varies with the temperature of the engine. When the engine is cold, the coolant level should be at or slightly above the FULL COLD mark on the reservoir **(see illustration)**. Once the engine has warmed up, the level should be at or near the FULL HOT mark. If it isn't, allow the engine to cool, then unscrew the cap from the reservoir and add a 50/50 mixture of ethylene glycol based green-colored antifreeze or orange-colored "DEX-COOL™ silicate-free coolant and water (see Caution above).
10   Drive the vehicle and recheck the coolant level. If only a small amount of coolant is required to bring the system up to the proper level, water can be used. However, repeated additions of water will dilute the antifreeze and water solution. In order to maintain the proper ratio of antifreeze and water, always top up the coolant level with the correct mixture. An empty plastic milk jug or bleach bottle makes an excellent container for mixing coolant. Do not use rust inhibitors or additives.
11   If the coolant level drops consistently, there may be a leak in the system. Inspect the radiator, hoses, filler cap, drain plugs and water pump (see Section 10). If no leaks are noted, have the radiator cap pressure tested by a service station.
12   If you have to remove the radiator cap, wait until the engine has cooled completely, then wrap a thick cloth around the cap and turn it to the first stop. If coolant or steam escapes, let the engine cool down longer, then remove the cap.
13   Check the condition of the coolant as well. It should be relatively clear. If it is brown or rust colored, the system should be drained, flushed and refilled. Even if the coolant appears to be normal, the corrosion inhibitors wear out, so it must be replaced at the specified intervals.

Windshield washer fluid

Refer to illustration 4.14
14   Fluid for the windshield washer system is located in a plastic reservoir at the front of the engine compartment **(see illustration)**. In milder climates, plain water can be used in the reservoir, but it should be kept no more than two-thirds full to allow for expansion if the water freezes. In colder climates, use windshield washer system antifreeze, available at any auto parts store, to lower the freezing point of the fluid. Mix the antifreeze with water in accordance with the manufac-

turer's directions on the container. **Caution:** *Do not use cooling system antifreeze - it will damage the vehicle's paint.*
15   To help prevent icing in cold weather, warm the windshield with the defroster before using the washer.

Battery electrolyte

16   All vehicles covered by this manual are equipped with a battery which is permanently sealed (except for vent holes) and has no filler caps. Water does not have to be added to these batteries at any time; however, if a maintenance-type battery has been installed on the vehicle since it was new, remove all the cell caps on top of the battery (usually there are two caps that cover three cells each). If the electrolyte level is low, add distilled water until the level is above the plates. There is usually a split-ring indicator in each cell to help you judge when enough water has been added. Add water until the electrolyte level is just up to the bottom of the split ring indicator. Do not overfill the battery or it will spew out electrolyte when it is charging.

Brake and clutch fluid

Refer to illustrations 4.17a and 4.17b
17   The brake master cylinder is located on the front of the power booster unit in the engine compartment **(see illustration)**. The clutch master cylinder is mounted on the inner fender to the left of the brake master cylinder **(see illustration)** .
18   The brake reservoir can be checked visually to make sure the level is above the plastic seam. The fluid level in the clutch reservoir should be even with the step mark inside the filler neck.
19   When adding fluid, pour it carefully into the reservoir to avoid spilling it on surrounding painted surfaces. Be sure the specified fluid is used, since mixing different types of brake fluid can cause damage to the system. See *Recommended lubricants and fluids* at the front of this Chapter or your owner's manual. **Warning:** *Brake fluid can harm your eyes and damage painted surfaces, so use extreme caution when handling or pouring it. Do not use brake fluid that has been standing open or is more than one year old. Brake fluid absorbs moisture from the air. Excess moisture can cause a dangerous loss of braking effectiveness.*
20   At this time the fluid and master cylinder can be inspected for contamination. The sys-

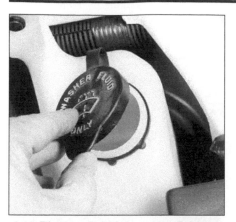

4.14 Flip the windshield washer fluid cap up to add fluid

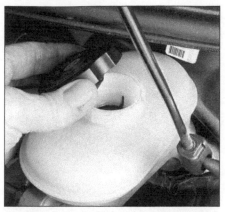

4.17a Make sure the brake fluid level is even with the seam on the reservoir - clean the area around the cap before unscrewing it

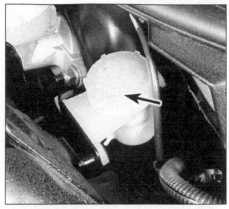

4.17b The clutch fluid reservoir (arrow) is located to the left of the brake reservoir - unscrew the cap to check the level and add fluid

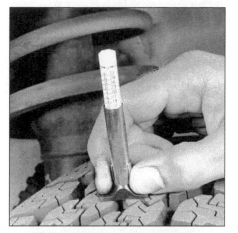

5.2 Use a tire tread depth gauge to monitor tire wear - they are available at auto parts stores and service stations and cost very little

tem should be drained and refilled if deposits, dirt particles or water droplets are seen in the fluid.

21   After filling the reservoir to the proper level, make sure the cap is on tight to prevent fluid leakage.

22   The brake fluid level in the master cylinder will drop slightly as the pads at each wheel wear down during normal operation. If the master cylinder requires repeated replenishing to keep it at the proper level, this is an indication of leakage in the brake system, which should be corrected immediately. Check all brake lines and connections (see Section 22 for more information).

23   If, when checking the master cylinder fluid level, you discover one or both reservoirs empty or nearly empty, the brake system should be bled (see Chapter 9).

5   Tire and tire pressure checks (every 250 miles or weekly)

Refer to illustrations 5.2, 5.3, 5.4a, 5.4b and 5.8

1   Periodic inspection of the tires may spare you the inconvenience of being stranded with a flat tire. It can also provide you with vital information regarding possible problems in the steering and suspension systems before major damage occurs.

2   The original tires on this vehicle are equipped with 1/2-inch wide bands that appear when tread depth reaches 1/16-inch, indicating the tires are worn out. Tread wear can be monitored with a simple, inexpensive device known as a tread depth indicator **(see illustration)**.

3   Note any abnormal tread wear **(see**

UNDERINFLATION

INCORRECT TOE-IN OR EXTREME CAMBER

CUPPING

Cupping may be caused by:

- Underinflation and/or mechanical irregularities such as out-of-balance condition of wheel and/or tire, and bent or damaged wheel.
- Loose or worn steering tie-rod or steering idler arm.
- Loose, damaged or worn front suspension parts.

OVERINFLATION

FEATHERING DUE TO MISALIGNMENT

5.3 This chart will help you determine the condition of the tires, the probable cause(s) of abnormal wear and the corrective action necessary

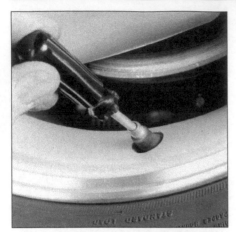

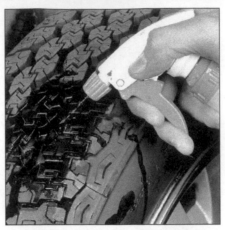

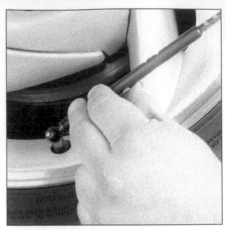

5.4a If a tire looses air on a steady basis, check the valve core first to make sure it's snug (special inexpensive wrenches are commonly available at auto parts stores)

5.4b If the valve core is tight, raise the corner of the vehicle with the low tire and spray a soapy water solution onto the tread as the tire is turned slowly - leaks will cause small bubbles to appear

5.8 To extend the life of the tires, check the air pressure at least once a week with an accurate gauge (don't forget the spare)

illustration). Tread pattern irregularities such as cupping, flat spots and more wear on one side than the other are indications of front end alignment and/or balance problems. If any of these conditions are noted, take the vehicle to a tire shop or service station to correct the problem.

4   Look closely for cuts, punctures and embedded nails or tacks. Sometimes a tire will hold air pressure for short time or leak down very slowly after a nail has embedded itself in the tread. If a slow leak persists, check the valve stem core to make sure it's tight (see illustration). Examine the tread for an object that may have embedded itself in the tire or for a "plug" that may have begun to leak (radial tire punctures are repaired with a plug that's installed in a puncture). If a puncture is suspected, it can be easily verified by spraying a solution of soapy water onto the suspected area (see illustration). The soapy solution will bubble if there's a leak. Unless the puncture is unusually large, a tire shop or service station can usually repair the tire.

5   Carefully inspect the inner sidewall of each tire for evidence of brake fluid. If you see any, inspect the brakes immediately.

6   Correct air pressure adds miles to the lifespan of the tires, improves mileage and enhances overall ride quality. Tire pressure cannot be accurately estimated by looking at a tire, especially if it's a radial. A tire pressure gauge is essential. Keep an accurate gauge in the vehicle. The pressure gauges attached to the nozzles of air hoses at gas stations are often inaccurate.

7   Always check tire pressure when the tires are cold. Cold, in this case, means the vehicle has not been driven over a mile in the three hours preceding a tire pressure check. A pressure rise of four to eight pounds is not uncommon once the tires are warm.

8   Unscrew the valve cap protruding from the wheel or hubcap and push the gauge firmly onto the valve stem (see illustration). Note the reading on the gauge and compare the figure to the recommended tire pressure

6.3 The automatic transmission fluid dipstick (arrow) is located at the rear of the engine compartment

shown on the label attached to the inside of the glove compartment door. Be sure to reinstall the valve cap to keep dirt and moisture out of the valve stem mechanism. Check all four tires and, if necessary, add enough air to bring them up to the recommended pressure.

9   Don't forget to keep the spare tire inflated to the specified pressure (refer to your owner's manual or the tire sidewall).

6   Automatic transmission fluid level check (every 3000 miles or 3 months)

Refer to illustrations 6.3 and 6.6

1   The automatic transmission fluid level should be carefully maintained. Low fluid level can lead to slipping or loss of drive, while overfilling can cause foaming and loss of fluid.

2   With the parking brake set, start the engine, then move the shift lever through all the gear ranges, ending in Park. The fluid level must be checked with the vehicle level and the engine running at idle. **Note:** *Incorrect fluid level readings will result if the vehi-*

6.6 The automatic transmission fluid level must be maintained within the cross-hatched area on the dipstick, between the upper and lower holes

cle has just been driven at high speeds for an extended period, in hot weather in city traffic, or if it has been pulling a trailer. If any of these conditions apply, wait until the fluid has cooled (about 30 minutes).

3   With the transmission at normal operating temperature, remove the dipstick from the filler tube. The dipstick is located at the rear of the engine compartment (see illustration).

4   Carefully touch the fluid at the end of the dipstick to determine if the fluid is cool, warm or hot. Wipe the fluid from the dipstick with a clean rag and push it back into the filler tube until the cap seats.

5   Pull the dipstick out again and note the fluid level.

6   If the fluid felt cool, the level should be within the lower marks on the dipstick (see illustration). If it felt warm or hot, the level should be within the cross-hatched upper areas on the dipstick. If additional fluid is required, pour it directly into the tube using a funnel. It takes about one pint to raise the level from the lower mark to the upper edge of the cross-hatched area with a hot transmission, so add the fluid a little at a time and keep checking the level until it's correct.

7   The condition of the fluid should also be checked along with the level. If the fluid at the

7.2 The power steering fluid reservoir is mounted on the driver's side of the engine compartment on all engines except 1997 and earlier (Gen I) V8's , where it is mounted on the passenger side of the engine compartment - turn the cap clockwise for removal

end of the dipstick is a dark reddish-brown color, or if the fluid has a burned smell, the fluid should be changed. If you're in doubt about the condition of the fluid, purchase some new fluid and compare the two for color and smell.

7 Power steering fluid level check (every 3000 miles or 3 months)

Refer to illustrations 7.2 and 7.6

1   Unlike manual steering, the power steering system relies on fluid which may, over a period of time, require replenishing.

2   The fluid reservoir for the power steering pump is located at the front of the engine on V6 models or on the right (passenger) side of the engine compartment **(see illustration)**. On 2000 and later models, the reservoir on V8 engines is at the front of the engine attached to the power steering pump, while V6 engines have a remote power steering fluid reservoir at the left side of the engine.

3   For the check, the front wheels should be pointed straight ahead and the engine should be off.

4   Use a clean rag to wipe off the reservoir cap and the area around the cap. This will help prevent any foreign matter from entering the reservoir during the check.

5   Twist off the cap and check the temperature of the fluid at the end of the dipstick with your finger.

6   Wipe off the fluid with a clean rag, reinsert it, then withdraw it and read the fluid level. The level should be at the HOT mark if the fluid was hot to the touch **(see illustration)**. It should be at the COLD mark if the fluid was cool to the touch.

7   If additional fluid is required, pour the specified type directly into the reservoir, using a funnel to prevent spills.

8   If the reservoir requires frequent fluid additions, all power steering hoses, hose connections, the power steering pump and

7.6 The marks on the dipstick indicate the safe fluid range

the rack and pinion assembly should be carefully checked for leaks.

8 Engine oil and filter change (every 3000 miles or 3 months)

Refer to illustrations 8.2, 8.7, 8.12 and 8.14

1   Frequent oil changes are the best preventive maintenance the home mechanic can give the engine, because aging oil becomes diluted and contaminated, which leads to premature engine wear.

2   Make sure you have all the necessary tools before you begin this procedure **(see illustration)**. You should also have plenty of rags or newspapers handy for mopping up any spills.

3   Access to the underside of the vehicle is greatly improved if the vehicle can be lifted on a hoist, driven onto ramps or supported by jackstands. **Warning:** *Do not work under a vehicle which is supported only by a hydraulic or scissors-type jack.*

4   If this is your first oil change, get under the vehicle and familiarize yourself with the locations of the oil drain plug and the oil filter. The engine and exhaust components will be warm during the actual work, so try to anticipate any potential problems before the engine and accessories are hot.

5   Park the vehicle on a level spot. Start the engine and allow it to reach its normal operating temperature. Warm oil and sludge will flow out more easily. Turn off the engine when it's warmed up. Remove the filler cap from the valve cover.

6   Raise the vehicle and support it securely on jackstands. **Warning:** *Never get beneath the vehicle when it is supported only by a jack. The jack provided with your vehicle is designed solely for raising the vehicle to remove and replace the wheels. Always use jackstands to support the vehicle when it becomes necessary to place your body underneath the vehicle.*

7   Being careful not to touch the hot exhaust components, place the drain pan under the drain plug in the bottom of the pan and remove the plug **(see illustration)**. You

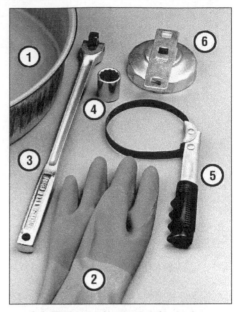

8.2 These tools are required when changing the engine oil and filter

1   **Drain pan** - *It should be fairly shallow in depth, but wide to prevent spills*

2   **Rubber gloves** - *When removing the drain plug and filter, you will get oil on your hands (the gloves will prevent burns)*

3   **Breaker bar** - *Sometimes the oil drain plug is tight, and a long breaker bar is needed to loosen it*

4   **Socket** - *To be used with the breaker bar or a ratchet (must be the correct size to fit the drain plug)*

5   **Filter wrench** - *This is a metal band-type wrench, which requires clearance around the filter to be effective*

6   **Filter wrench** - *This type fits on the bottom of the filter and can be turned with a ratchet or breaker bar (different-size wrenches are available for different types of filters)*

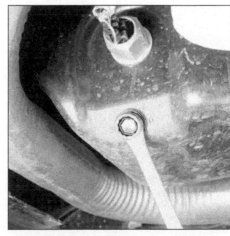

8.7 The engine oil drain plug is located at the rear of the oil pan or on the side of the oil pan - it is usually very tight, so use a box-end wrench to avoid rounding off the hex

8.12 The oil filter is usually on very tight as well and will require a special wrench for removal - DO NOT use the wrench to tighten the new filter!

8.14 Lubricate the oil filter gasket with clean engine oil before installing the filter on the engine

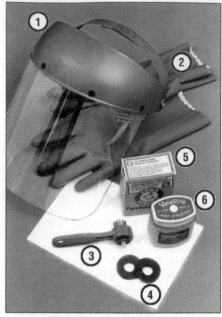

9.1 Tools and materials required for battery maintenance

1   *Face shield/safety goggles - When removing corrosion with a brush, the acidic particles can easily fly up into your eyes*

2   *Rubber gloves - Another safety item to consider when servicing the battery - remember that's acid inside the battery!*

3   *Battery terminal/cable cleaner - This wire brush cleaning tool will remove all traces of corrosion from the battery and cable*

4   *Treated felt washers - Placing one of these on each terminal, directly under the cable end, will help prevent corrosion (be sure to get the correct type for side-terminal batteries)*

5   *Baking soda - A solution of baking soda and water can be used to neutralize corrosion*

6   *Petroleum jelly - A layer of this on the battery terminal bolts will help prevent corrosion*

may want to wear gloves while unscrewing the plug the final few turns if the engine is hot.

8   Allow the old oil to drain into the pan. It may be necessary to move the pan farther under the engine as the oil flow slows to a trickle. Inspect the old oil for the presence of metal shavings and chips.

9   After all the oil has drained, wipe off the drain plug with a clean rag. Even minute metal particles clinging to the plug would immediately contaminate the new oil.

10   Clean the area around the drain plug opening, reinstall the plug and tighten it securely, but do not strip the threads.

11   Move the drain pan into position under the oil filter.

12   Loosen the oil filter **(see illustration)** by turning it counterclockwise with the filter wrench. Any standard filter wrench will work. Sometimes the oil filter is screwed on so tightly that it cannot be loosened. If this situation occurs, punch a metal bar or long screwdriver directly through the side of the canister and use it as a T-bar to turn the filter. Be prepared for oil to spurt out of the canister as it is punctured. Once the filter is loose, use your hands to unscrew it from the block. Just as the filter is detached from the block, immediately tilt the open end up to prevent the oil inside the filter from spilling out. **Warning:** *The exhaust system may still be hot, so be careful.*

13   With a clean rag, wipe off the mounting surface on the block. If a residue of old oil is allowed to remain, it will smoke when the block is heated up. Also make sure that none of the old gasket remains stuck to the mounting surface. It can be removed with a scraper if necessary.

14   Compare the old filter with the new one to make sure they are the same type. Smear some clean engine oil on the rubber gasket of the new filter and screw it into place **(see illustration)**. Because overtightening the filter will damage the gasket, do not use a filter wrench to tighten the filter. Tighten it by hand until the gasket contacts the seating surface. Then seat the filter by giving it an additional 3/4-turn.

15   Remove all tools, rags, etc. from under

the vehicle, being careful not to spill the oil in the drain pan, then lower the vehicle.

16   Add new oil to the engine through the oil filler cap in the valve cover. Use a funnel, if necessary, to prevent oil from spilling onto the top of the engine. Pour three quarts of fresh oil into the engine. Wait a few minutes to allow the oil to drain into the pan, then check the level on the oil dipstick (see Section 4 if necessary). If the oil level is at or near the upper hole on the dipstick, install the filler cap hand tight, start the engine and allow the new oil to circulate.

17   Allow the engine to run for about a minute. While the engine is running, look under the vehicle and check for leaks at the oil pan drain plug and around the oil filter. If either is leaking, stop the engine and tighten the plug or filter.

18   Wait a few minutes to allow the oil to trickle down into the pan, then recheck the level on the dipstick and, if necessary, add enough oil to bring the level to the upper hole.

19   During the first few trips after an oil change, make it a point to check frequently for leaks and proper oil level.

20   The old oil drained from the engine cannot be re-used in its present state and should be discarded. Check with your local refuse disposal company, disposal facility or environmental agency to see whether they will accept the oil for recycling. Don't pour used oil into drains or onto the ground. After the oil has cooled, it can be drained into a suitable container (capped plastic jugs, topped bottles, milk cartons, etc.) for transport to one of these disposal sites.

---

9   Battery check, maintenance and charging (every 7500 miles or 6 months)

Refer to illustrations 9.1, 9.4, 9.5a, 9.5b and 9.5c

Warning: *Hydrogen gas is produced by the battery, so keep open flames and lighted*

tobacco away from it at all times. Always wear eye protection when working around the battery. Rinse off spilled electrolyte immediately with large amounts of water. When removing the battery cables, always detach the negative cable first and hook it up last!

Caution: *If the radio in your vehicle is equipped with an anti-theft system, make sure you have the correct activation code before disconnecting the battery.*

1   Battery maintenance is an important procedure which will help ensure you aren't stranded because of a dead battery. Several tools are required for this procedure **(see illustration)**.

2   A sealed battery is standard equipment on all vehicles covered by this manual. Although this type of battery has many advantages over the older, capped cell type, and never requires the addition of water, it

9.4 Check the tightness of the battery cable terminal bolts

9.5a A tool like this one (available at auto parts stores) is used to clean the side terminal type battery contact area

9.5b Use the brush to finish the cleaning job

9.5c The result should be a clean, shiny terminal area

should still be routinely maintained according to the procedures which follow.

Check

3   The battery is located in the right front corner of the engine compartment. The exterior of the battery should be inspected periodically for damage such as a cracked case or cover.

4   Check the tightness of the battery cable terminals and connections to ensure good electrical connections and check the entire length of each cable for cracks and frayed conductors **(see illustration)**.

5   If corrosion (visible as white, fluffy deposits) is evident, remove the cables from the terminals, clean them with a battery brush and reinstall the cables **(see illustrations)**. Corrosion can be kept to a minimum by using special treated fiber washers available at auto parts stores or by applying a layer of petroleum jelly to the terminals and cables after they are assembled.

6   Make sure that the battery tray is in good condition and the hold-down clamp bolt is tight. If the battery is removed from the tray, make sure no parts remain in the bottom of the tray when the battery is reinstalled. When reinstalling the hold-down clamp bolt, do not overtighten it.

7   Information on removing and installing the battery can be found in Chapter 5. Information on jump starting can be found at the front of this manual. For more detailed battery checking procedures, refer to the Haynes *Automotive Electrical Manual*.

Cleaning

8   Corrosion on the hold-down components, battery case and surrounding areas can be removed with a solution of water and baking soda. Thoroughly rinse all cleaned areas with plain water.

9   Any metal parts of the vehicle damaged by corrosion should be covered with a zinc-based primer, then painted.

Charging

Warning: *When batteries are being charged,*

hydrogen gas, which is very explosive and flammable, is produced. Do not smoke or allow open flames near a charging or a recently charged battery. Wear eye protection when near the battery during charging. Also, make sure the charger is unplugged before connecting or disconnecting the battery from the charger.

10   Slow-rate charging is the best way to restore a battery that's discharged to the point where it will not start the engine. It's also a good way to maintain the battery charge in a vehicle that's only driven a few miles between starts. Maintaining the battery charge is particularly important in the winter when the battery must work harder to start the engine and electrical accessories that drain the battery are in greater use.

11   It's best to use a one or two-amp battery charger (sometimes called a "trickle" charger). They are the safest and put the least strain on the battery. They are also the least expensive. For a faster charge, you can use a higher amperage charger, but don't use one rated more than 1/10th the amp/hour rating of the battery. Rapid boost charges that claim to restore the power of the battery in one to two hours are hardest on the battery and can damage batteries that aren't in good condition. This type of charging should only be used in emergency situations.

12   The average time necessary to charge a battery should be listed in the instructions that come with the charger. As a general rule, a trickle charger will charge a battery in 12 to 16 hours.

13   Remove all of the cell caps (if equipped) and cover the holes with a clean cloth to prevent spattering electrolyte. Disconnect the negative battery cable and hook the battery charger leads to the battery posts (positive to positive, negative to negative), then plug in the charger. Make sure it is set at 12-volts if it has a selector switch.

14   If you're using a charger with a rate higher than two amps, check the battery regularly during charging to make sure it doesn't overheat. If you're using a trickle charger, you can safely let the battery charge overnight

after you've checked it regularly for the first couple of hours.

15   If the battery has removable cell caps, measure the specific gravity with a hydrometer every hour during the last few hours of the charging cycle. Hydrometers are available inexpensively from auto parts stores - follow the instructions that come with the hydrometer. Consider the battery charged when there's no change in the specific gravity reading for two hours and the electrolyte in the cells is gassing (bubbling) freely. The specific gravity reading from each cell should be very close to the others. If not, the battery probably has a bad cell(s).

16   Some batteries with sealed tops have built-in hydrometers on the top that indicate the state of charge by the color displayed in the hydrometer window. Normally, a bright-colored hydrometer indicates a full charge and a dark hydrometer indicates the battery still needs charging. Check the battery manufacturer's instructions to be sure you know what the colors mean.

17   If the battery has a sealed top and no built-in hydrometer, you can hook up a digital voltmeter across the battery terminals to check the charge. A fully charged battery should read 12.5-volts or higher.

10   Cooling system check (every 7500 miles or 6 months)

Refer to illustration 10.4

Caution: *Never mix green-colored ethylene glycol anti-freeze and orange-colored "DEX-COOL™ silicate-free coolant because doing so will destroy the efficiency of the "DEX-COOL™ coolant which is designed to last for 100,000 miles or five years.*

1   Many major engine failures can be attributed to a faulty cooling system. If the vehicle is equipped with an automatic transmission, the cooling system also cools the transmission fluid and plays an important role in prolonging transmission life.

2   The cooling system should be checked with the engine cold. Do this before the vehicle is driven for the day or after the engine

Check for a chafed area that could fail prematurely.

Check for a soft area indicating the hose has deteriorated inside.

Overtightening the clamp on a hardened hose will damage the hose and cause a leak.

Check each hose for swelling and oil-soaked ends. Cracks and breaks can be located by squeezing the hose.

10.4 Hoses, like drivebelts, have a habit of failing at the worst possible time - to prevent the inconvenience of a blown radiator or heater hose, inspect them carefully as shown here

has been shut off for at least three hours.

3   Remove the radiator cap by turning it to the left until it reaches a stop. If you hear any hissing sounds (indicating there is still pressure in the system), wait until it stops. Now press down on the cap with the palm of your hand and continue turning to the left until the cap can be removed. Thoroughly clean the cap, inside and out, with clean water. Also clean the filler neck on the radiator. All traces of corrosion should be removed. The coolant inside the radiator should be relatively transparent. If it is rust colored, the system should be drained and refilled (see Section 30). If the coolant level is not up to the top, add additional antifreeze/coolant mixture (see Section 4).

4   Carefully check the large upper and lower radiator hoses along with any smaller diameter heater hoses which run from the engine to the firewall. Inspect each hose along its entire length, replacing any hose which is cracked, swollen or shows signs of deterioration. Cracks may become more apparent if the hose is squeezed **(see illustration)**.

5   Make sure all hose connections are tight. A leak in the cooling system will usually show up as white or rust colored deposits on the areas adjoining the leak. If wire-type clamps are used at the ends of the hoses, it may be wise to replace them with more secure screw-type clamps.

6   Use compressed air or a soft brush to remove bugs, leaves, etc. from the front of the radiator or air conditioning condenser. Be careful not to damage the delicate cooling fins or cut yourself on them.

7   Every other inspection, or at the first indication of cooling system problems, have the cap and system pressure tested. If you don't have a pressure tester, most gas stations and repair shops will do this for a minimal charge.

11   Underhood hose check and replacement (every 7500 miles or 6 months)

General

1   **Warning:** *Replacement of air conditioning hoses must be left to a dealer service department or air conditioning shop that has the equipment to depressurize the system safely. Never remove air conditioning components or hoses until the system has been depressurized.*

2   High temperatures under the hood can cause the deterioration of the rubber and plastic hoses used for engine, accessory and emission systems operation. Periodic inspection should be made for cracks, loose clamps, material hardening and leaks. Information specific to the cooling system hoses can be found in Section 10.

3   Some, but not all, hoses are secured to the fittings with clamps. Where clamps are

used, check to be sure they haven't lost their tension, allowing the hose to leak. If clamps aren't used, make sure the hose hasn't expanded and/or hardened where it slips over the fitting, allowing it to leak.

Vacuum hoses

4   It's quite common for vacuum hoses, especially those in the emissions system, to be color coded or identified by colored stripes molded into each hose. Various systems require hoses with different wall thicknesses, collapse resistance and temperature resistance. When replacing hoses, be sure the new ones are made of the same material.

5   Often the only effective way to check a hose is to remove it completely from the vehicle. If more than one hose is removed, be sure to label the hoses and fittings to ensure correct installation.

6   When checking vacuum hoses, be sure to include any plastic T-fittings in the check. Inspect the fittings for cracks and the hose where it fits over the fitting for distortion, which could cause leakage.

7   A small piece of vacuum hose (1/4-inch inside diameter) can be used as a stethoscope to detect vacuum leaks. Hold one end of the hose to your ear and probe around vacuum hoses and fittings, listening for the "hissing" sound characteristic of a vacuum leak. **Warning:** *When probing with the vacuum hose stethoscope, be careful not to allow your body or the hose to come into contact with moving engine components such as the drivebelt, cooling fan, etc.*

Fuel hose

Warning: *Gasoline is extremely flammable, so take extra precautions when you work on any part of the fuel system. Don't smoke or allow open flames or bare light bulbs near the work area, and don't work in a garage where a gas-type appliance (such as a water heater or clothes dryer) is present. If you spill any fuel on your skin, rinse it off immediately with soap and water. When you perform any kind of work on the fuel system, wear safety glasses and have a Class B type fire extinguisher on hand. The fuel system is under pressure, so if any lines must be disconnected, the pressure in the system must be relieved first (see Chapter 4 for more information).*

8   Check all rubber fuel lines for deterioration and chafing. Check especially for cracks in areas where the hose bends and just before fittings, such as where a hose attaches to the fuel filter and fuel injection unit.

9   High quality fuel line, usually identified by the word *Fluoroelastomer* printed on the hose, should be used for fuel line replacement. Never, under any circumstances, use unreinforced vacuum line, clear plastic tubing or water hose for fuel lines.

10   Spring-type clamps are commonly used on fuel lines. These clamps often lose their tension over a period of time, and can be

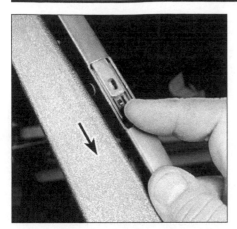

12.5 Depress the release lever, then slide the blade assembly out of the hook in the end of the wiper arm

12.7 Squeeze the end of the wiper element to free it from the bridge claw, then slide the element out

13.1 Materials required for chassis and body lubrication

1   *Engine oil* - *Light engine oil in a can like this can be used for door and hood hinges*
2   *Graphite spray* - *Used to lubricate lock cylinders*
3   *Grease* - *Grease, in a variety of types and weights, is available for use in a grease gun. Check the Specification for your requirements*
4   *Grease gun* - *A common grease gun, shown here with a detachable hose and nozzle, is needed for chassis lubrication. After use, clean it thoroughly!*

"sprung" during the removal process. As a result spring-type clamps should be replaced with screw-type clamps whenever a hose is replaced.

Metal lines

11   Sections of steel tubing often used for fuel line between the fuel pump and fuel injection unit. Check carefully for cracks, kinks and flat spots in the line.
12   If a section of metal fuel line must be replaced, only seamless steel tubing should be used, since copper and aluminum tubing do not have the strength necessary to withstand normal engine vibration.
13   Check the metal brake lines where they enter the master cylinder and brake proportioning unit (if used) for cracks in the lines and loose fittings. Any sign of brake fluid leakage calls for an immediate thorough inspection of the brake system.

12 Wiper blade inspection and replacement (every 7500 miles or 6 months)

Refer to illustrations 12.5 and 12.7
1   The windshield wiper and blade assemblies should be inspected periodically for damage, loose components and cracked or worn blade elements.
2   Road film can build up on the wiper blades and affect their efficiency, so they should be washed regularly with a mild detergent solution.
3   The action of the wiping mechanism can loosen the bolts, nuts and fasteners, so they should be checked and tightened, as necessary, at the same time the wiper blades are checked.
4   If the wiper blade elements (sometimes called inserts) are cracked, worn or warped, they should be replaced with new ones.
5   Remove the wiper blade assembly from the wiper arm by depressing the release lever while pulling on the blade to release it **(see**

illustration).
6   With the blade removed from the vehicle, you can remove the rubber element from the blade.
7   Grasp the end of the wiper bridge securely with one hand and the element with the other. Detach the end of the element from the bridge claw and slide it to free it, then slide the element out **(see illustration)**.
8   Compare the new element with the old for length, design, etc.
9   Slide the new element into the claw into place, notched end last and secure the claw into the notches.
10   Reinstall the blade assembly on the arm, wet the windshield and test for proper operation.

13 Chassis lubrication (every 7500 miles or 6 months)

Refer to illustrations 13.1 and 13.6
1   Refer to *Recommended lubricants and fluids* at the front of this Chapter to obtain the necessary grease, etc. You'll also need a grease gun **(see illustration)**. Occasionally plugs will be installed rather than grease fittings. If so, grease fittings will have to be purchased and installed.
2   Look under the vehicle and see if grease fittings or plugs are installed. If there are plugs, remove them and buy grease fittings, which will thread into the component. A dealer or auto parts store will be able to supply the correct fittings. Straight, as well as angled, fittings are available.
3   For easier access under the vehicle, raise it with a jack and place jackstands under the frame. Make sure it's safely supported by the stands. If the wheels are to be removed at this interval for tire rotation or brake inspection, loosen the lug nuts slightly while the vehicle is still on the ground.
4   Before beginning, force a little grease out of the nozzle to remove any dirt from the end of the gun. Wipe the nozzle clean with a rag.

5   With the grease gun and plenty of clean rags, crawl under the vehicle and begin lubricating the components.
6   Wipe one of the grease fittings clean and push the nozzle firmly over it **(see illustration)**. Pump the gun until the balljoint rubber seal is firm to the touch. Do not pump too much grease into the fitting as it could rupture the seal.
7   Wipe the excess grease from the components and the grease fitting. Repeat the procedure for the remaining fitting.
8   On manual transmission-equipped models, lubricate the clutch linkage pivot points with clean engine oil.

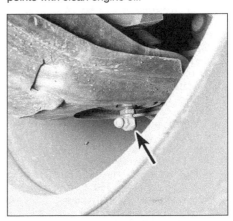

13.6 Location of the lower balljoint grease fitting (arrow)

14.4a Inspect the rack-and-pinion steering boots (there's one on each side) for tears or leaking grease

14.4b Make sure the power steering fluid connections aren't leaking

14.5 Check for oil leaks in this area of the shock absorbers

9   Clean and lubricate the parking brake cable, along with the cable guides and levers. This can be done by smearing some of the chassis grease onto the cable and its related parts with your fingers.

10   Open the hood and smear a little chassis grease on the hood latch mechanism. Have an assistant pull the hood release lever from inside the vehicle as you lubricate the cable at the latch.

11   Lubricate all the hinges (door, hood, etc.) with engine oil to keep them in proper working order.

12   The key lock cylinders can be lubricated with spray graphite or silicone lubricant, which is available at auto parts stores.

13   Lubricate the door weatherstripping with silicone spray. This will reduce chafing and retard wear.

14 Suspension and steering check (every 7500 miles or 6 months)

Refer to illustrations 14.4a, 14.4b and 14.5
Note: *The front hub assembly contains sealed bearings that do not require maintenance. In the event of bearing failure, the hubs must be removed and new bearings pressed in by an automotive machine shop (see Chapter 10).*

1   Indications of a fault in these systems are excessive play in the steering wheel before the front wheels react, excessive sway around corners, body movement over rough roads or binding at some point as the steering wheel is turned.

2   Raise the front of the vehicle periodically and visually check the suspension and steering components for wear. Because of the work to be done, make sure the vehicle cannot fall from the stands.

3   Check the wheel bearings. Do this by spinning the front wheels. Listen for any abnormal noises and watch to make sure the wheel spins true (doesn't wobble). Grab the top and bottom of the tire and pull in-and-out on it. Notice any movement which would indicate a loose wheel bearing assembly. If the

bearings are suspect, refer to Chapter 10 for more information.

4   From under the vehicle check for loose bolts, broken or disconnected parts and deteriorated rubber bushings on all suspension and steering components. Check the steering boots for damage or leaks **(see illustration)**. Check the power steering hoses and connections for leaks **(see illustration)**.

5   Check the shock absorbers or leaking fluid or damage **(see illustration)**.

6   Have an assistant turn the steering wheel from side-to-side and check the steering components for free movement, chafing and binding. If the steering doesn't react with the movement of the steering wheel, try to determine where the slack is located.

15 Exhaust system check (every 7500 miles or 6 months)

Refer to illustration 15.2
1   With the engine cold (at least three hours after the vehicle has been driven), check the complete exhaust system from the engine to the end of the tailpipe. Ideally, the inspection should be done with the vehicle on a hoist to permit unrestricted access. If a hoist is not available, raise the vehicle and support it securely on jackstands.

2   Check the exhaust pipes and connec-

15.2 Check the exhaust system rubber hangers for cracks and damage

tions for evidence of leaks, severe corrosion and damage. Make sure that all brackets and hangers are in good condition and tight **(see illustration)**.

3   At the same time, inspect the underside of the body for holes, corrosion, open seams, etc. which may allow exhaust gases to enter the interior. Seal all body openings with silicone or body putty.

4   Rattles and other noises can often be traced to the exhaust system, especially the mounts and hangers. Try to move the pipes, muffler and catalytic converter. If the components can come in contact with the body or suspension parts, secure the exhaust system with new mounts.

16 Hydraulic clutch check (every 7500 miles or 6 months)

1   Check all hoses for cracks and distortion.

2   Check the clutch master cylinder and slave cylinder for loose mounting screws and leaks.

3   Check for smooth operation of the clutch pedal with no binding, looseness or sponginess.

4   While an assistant depresses the clutch pedal, check the operation of the slave cylinder pushrod. You may need to remove a plug in the clutch housing first. The pushrod should move in and out of the slave cylinder about an inch as the pedal is depressed and released.

5   Replace any damaged or leaking components. Bleed the system if the pedal is spongy, the pushrod travel is not sufficient, or the slave cylinder, master cylinder or any lines were disconnected (see Chapter 8).

17 Seat belt check (every 7500 miles or 6 months)

1   Check the seat belts, buckles, latch plates and guide loops for obvious damage and signs of wear.

2   See if the seat belt reminder light comes on when the key is turned to the Run or Start position. A chime should also sound.

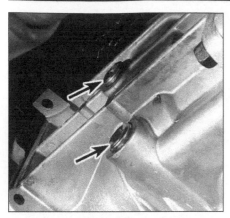

19.1 Locations of the manual transmission check plug (top) and drain plug (bottom)

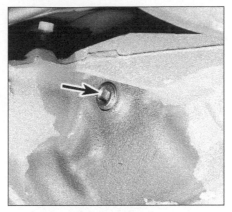

20.2 Differential fill/check plug location (arrow)

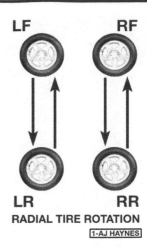

RADIAL TIRE ROTATION

1-AJ HAYNES

21.2 Tire rotation diagram

3   The seat belts are designed to lock up during a sudden stop or impact, yet allow free movement during normal driving. Make sure the retractors return the belt against your chest while driving and rewind the belt fully when the buckle is unlatched.

4   If any of the above checks reveal problems with the seat belt system, replace parts as necessary.

18   Starter safety switch check (every 7500 miles or 6 months)

Warning: *During the following checks there's a chance the vehicle could lunge forward, possibly causing damage or injuries. Allow plenty of room around the vehicle, apply the parking brake and hold down the regular brake pedal during the checks.*

1   Try to start the engine in each gear. The engine should crank only in Park or Neutral.

2   Make sure the steering column lock allows the key to go into the Lock position only when the shift lever is in Park.

3   The ignition key should come out only in the Lock position.

19   Manual transmission lubricant level check (every 15,000 miles or 12 months)

Refer to illustration 19.1

1   The manual transmission has a fill plug which must be removed to check the lubricant level **(see illustration)**. If the vehicle is raised to gain access to the plug, be sure to support it safely on jackstands - DO NOT crawl under a vehicle which is supported only by a jack!

2   Remove the plug from the transmission and use your little finger to reach inside the housing to feel the lubricant level. The level should be at or near the bottom of the plug hole.

3   If it isn't, add the recommended lubricant through the plug hole with a syringe or squeeze bottle.

4   Install and tighten the plug and check for leaks after the first few miles of driving.

20   Differential lubricant level check (every 15,000 miles or 12 months)

Refer to illustration 20.2

1   The differential has a filler plug which must be removed to check the lubricant level. If the vehicle is raised to gain access to the plug, be sure to support it safely on jackstands - DO NOT crawl under the vehicle when it's supported only by the jack.

2   Remove the filler plug from the side of the differential **(see illustration)**.

3   The lubricant level should be at the bottom of the plug opening. If not, use a syringe to add the recommended lubricant until it just starts to run out of the opening. On some models a tag is located in the area of the plug which gives information regarding lubricant type, particularly on models equipped with a limited slip differential.

4   Install the plug and tighten it securely.

21   Tire rotation (every 15,000 miles or 12 months)

Refer to illustration 21.2

1   The tires should be rotated at the specified intervals and whenever uneven wear is noticed.

2   Front wheel drive vehicles require a special tire rotation pattern **(see illustration)**.

3   Refer to the information in *Jacking and towing* at the front of this manual for the proper procedures to follow when raising the vehicle and changing a tire. If the brakes are going to be checked, don't apply the parking brake as stated. Make sure the tires are blocked to prevent the vehicle from rolling as it's raised.

4   The entire vehicle should be raised at the same time. This can be done on a hoist or by jacking up each corner and then lowering the vehicle onto jackstands placed under the

frame rails. Always use four jackstands and make sure the vehicle is safely supported.

5   After rotation, check and adjust the tire pressures as necessary and be sure to check the lug nut tightness.

22   Brake check (every 15,000 miles or 12 months)

Warning: *Brake system dust is hazardous to your health. DO NOT blow it out with compressed air or inhale it. DO NOT use gasoline or solvents to remove the dust. Use brake system cleaner or denatured alcohol only.*
Note 1: *For detailed photographs of the brake system, refer to Chapter 9.*
Note 2: *On 2000 and later models, the rear brakes are a disc-type, not drum. The checking procedure below is applicable to the rear disc brakes also.*

1   In addition to the specified intervals, the brakes should be inspected every time the wheels are removed or whenever a defect is suspected. Raise the vehicle and place it securely on jackstands. Remove the wheels (see *Jacking and towing* at the front of this manual, if necessary).

Disc brakes (front)

Refer to illustrations 22.5

2   Disc brakes can be checked without removing any parts except the wheels. Extensive disc damage can occur if the pads are not replaced when needed.

3   The disc brake pads have built-in wear indicators which make a high-pitched squealing sound when the pads are worn. **Caution:** *Expensive damage to the disc can result if the pads are not replaced soon after the wear indicators start squealing.*

4   The disc brake calipers, which contain the pads, are now visible. There is an outer pad and an inner pad in each caliper. All pads should be inspected.

5   Each caliper has a "window" to inspect

22.5 Look through the opening in the front of the caliper to check the brake pads (arrow) - the pad lining, which rubs against the disc, can also be inspected by looking through each end of the caliper

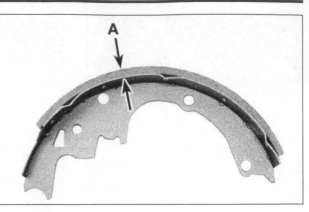

22.14 If the lining is bonded to the brake shoe, measure the lining thickness from the outer surface to the metal shoe, as shown here; if the lining is riveted to the shoe, measure from the lining outer surface to the rivet head

the pads **(see illustration)**. If the pad material has worn to about 1/8-inch thick or less, the pads should be replaced.

6   If you're unsure about the exact thickness of the remaining lining material, remove the pads for further inspection or replacement (refer to Chapter 9).

7   Before installing the wheels, check for leakage and/or damage at the brake hoses and connections. Replace the hose or fittings as necessary, referring to Chapter 9.

8   Check the condition of the brake rotor. Look for score marks, deep scratches and overheated areas (they will appear blue or discolored). If damage or wear is noted, the disc can be removed and resurfaced by an automotive machine shop or replaced with a new one. Refer to Chapter 9 for more detailed inspection and repair procedures.

Drum brakes (rear)

Refer to illustration 22.14

9   Raise the vehicle and support it securely on jackstands. Block the front tires to prevent the vehicle from rolling; however, don't apply the parking brake or it will lock the drums in place.

10   Remove the wheels, referring to *Jacking and towing* at the front of this manual if necessary.

11   Mark the hub so it can be reinstalled in the same position. Use a scribe, chalk, etc. on the drum, hub and backing plate.

12   Remove the brake drum.

13   With the drum removed, carefully clean the brake assembly with brake system cleaner. **Warning:** *Don't blow the dust out with compressed air and don't inhale any of it (it may contain asbestos, which is harmful to your health).*

14   Note the thickness of the lining material on both front and rear brake shoes. If the material has worn away to within 1/8-inch of the recessed rivets or metal backing, the shoes should be replaced **(see illustration)**. The shoes should also be replaced if they're cracked, glazed (shiny areas), or covered with brake fluid.

15   Make sure all the brake assembly springs are connected and in good condition.

16   Check the brake components for signs of fluid leakage. With your finger or a small screwdriver, carefully pry back the rubber cups on the wheel cylinder located at the top of the brake shoes. Any leakage here is an indication that the wheel cylinders should be overhauled immediately (see Chapter 9). Also, check all hoses and connections for signs of leakage.

17   Wipe the inside of the drum with a clean rag and denatured alcohol or brake cleaner. Again, be careful not to breathe the dangerous brake dust.

18   Check the inside of the drum for cracks, score marks, deep scratches and "hard spots" which will appear as small discolored areas. If imperfections cannot be removed with fine emery cloth, the drum must be taken to an automotive machine shop for resurfacing.

19   Repeat the procedure for the remaining wheel. If the inspection reveals that all parts are in good condition, reinstall the brake drums, install the wheels and lower the vehicle to the ground.

Parking brake

20   The parking brake is operated by a hand lever and locks the rear brake system. The easiest, and perhaps most obvious, method of periodically checking the operation of the parking brake assembly is to park the vehicle on a steep hill with the parking brake set and the transmission in Neutral (be sure to stay in the vehicle during this check!). If the parking brake cannot prevent the vehicle from rolling, it needs service (see Chapter 9).

23   Fuel system check (every 15,000 miles or 12 months)

Warning: *Gasoline is extremely flammable, so take extra precautions when you work on any part of the fuel system. Don't smoke or allow open flames or bare light bulbs near the work area, and don't work in a garage where a gas-type appliance (such as a water heater or clothes dryer) is present. Since gasoline is carcinogenic, wear latex gloves when there's a possibility of being exposed to fuel, and, if you spill any fuel on your skin, rinse it off immediately with soap and water. Mop up any spills immediately and do not store fuel-soaked rags*

where they could ignite. The fuel system is under constant pressure, so, if any fuel lines are to be disconnected, the fuel pressure in the system must be relieved first (see Chapter 4 for more information). When you perform any kind of work on the fuel system, wear safety glasses and have a Class B type fire extinguisher on hand.

1   The fuel system is most easily checked with the vehicle raised on a hoist so the components underneath the vehicle are readily visible and accessible.

2   If the smell of gasoline is noticed while driving or after the vehicle has been in the sun, the system should be thoroughly inspected immediately.

3   Remove the gas tank cap and check for damage, corrosion and an unbroken sealing imprint on the gasket. Replace the cap with a new one if necessary.

4   With the vehicle raised, inspect the gas tank and filler neck for cracks and other damage. The connection between the filler neck and tank is especially critical. Sometimes a filler neck will leak due to cracks, problems a home mechanic can't repair. **Warning:** *Do not, under any circumstances, try to repair a fuel tank yourself (except rubber components). A welding torch or any open flame can easily cause the fuel vapors to explode if the proper precautions are not taken.*

5   Carefully check all rubber hoses and metal lines leading away from the fuel tank. Check for loose connections, deteriorated hoses, crimped lines and other damage. Follow the lines to the front of the vehicle, carefully inspecting them all the way. Repair or replace damaged sections as necessary.

24   Air filter replacement (every 30,000 miles or 24 months)

Refer to illustrations 24.4a, 24.4b, 24.5a and 24.5b

1   At the specified intervals, the air filter should be replaced with a new one. The filter should be inspected between changes.

2   The air filter is located inside the air cleaner housing which is mounted in the left front corner of the engine compartment on 1997 and earlier vehicles and on 1998 and later vehicles it is mounted at the front of the engine compartment above the radiator.

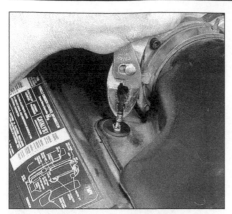

24.4a Use pliers to pull out the plastic pins from the air intake duct

24.4b Pull the air cleaner housing up for access to the clips

24.5a Detach the clips and separate the air cleaner housing halves

3   On 1998 and later vehicles, simply unsnap the cover retaining clips and lift the top cover upward to access the air filter element. Pull the filter element out of the air cleaner housing, while ensuring that the filter frame stays in place in the air cleaner housing. Install a new filter and be sure the rear locating tabs on the top cover are mated with the retainers on the lower section of the air cleaner housing before snapping the top cover back into place.

4   On 1997 and earlier vehicles, use a pair of pliers to remove the plastic pin (some models have two) retaining the air intake duct to the radiator shroud **(see illustration)**. Grasp the air cleaner housing and pull it up for access **(see illustration)**. It may be necessary to remove the clamp and detach the air cleaner duct from the housing.

5   Detach the clips, separate housing halves and lift the filter out **(see illustrations)**.

6   While the filter housing cover is off, be careful not to drop anything down into the air cleaner assembly.

7   Wipe out the inside of the air cleaner housing with a clean rag.

8   Place the new filter in the air cleaner housing. Make sure it seats properly, seat the two halves together and secure them with the clips.

9   The remainder of installation is the reverse of removal.

24.5b Lift the filter element out of the housing

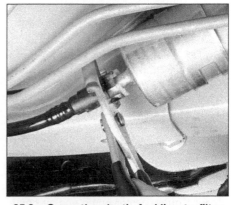

25.3a Grasp the plastic fuel line-to-filter fitting and turn it 1/4-turn to dislodge any dirt

---

25   Fuel filter replacement (every 30,000 miles or 24 months)

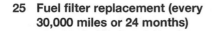

Refer to illustrations 25.3a through 25.3f
Warning: *Gasoline is extremely flammable, so take extra precautions when you work on any part of the fuel system. Don't smoke or allow open flames or bare light bulbs near the work area, and don't work in a garage where a gas-type appliance (such as a water heater or clothes dryer) is present. Since gasoline is carcinogenic, wear latex gloves when there's a possibility of being exposed to fuel, and, if you spill any fuel on your skin, rinse it off immediately with soap and water. Mop up any spills immediately and do not store fuel-*

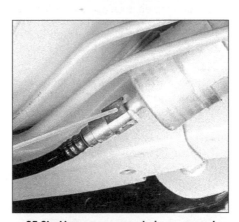

25.3b Use compressed air or aerosol carburetor cleaner to blow or wash the dirt from the fitting

soaked rags where they could ignite. The fuel system is under constant pressure, so, if any fuel lines are to be disconnected, the fuel pressure in the system must be relieved first (see Chapter 4 for more information). When you perform any kind of work on the fuel system, wear safety glasses and have a Class B type fire extinguisher on hand.

1   Relieve the fuel system pressure (see Chapter 4).

2   Raise the vehicle and support it securely on jackstands.

3   Refer to the accompanying illustrations

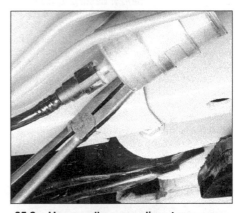

25.3c Use needle-nose pliers to squeeze the filter bracket and detach it from the chassis

and replace the fuel filter **(see illustrations)**.
Note: *On later models there are two types of quick-connect fittings at the fuel-line-to-filter connection. Both the metal-collar and plastic-collar connectors can be disconnected with a simple plastic tool available at auto parts stores.*
Warning: *Quick-connect fittings have an O-ring on the end of the fuel line. When reassembling the connection, always put a drop of engine oil on the O-ring to ease assembly or a swollen O-ring may tear and cause a fuel leak.*

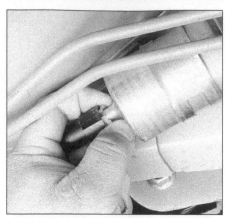

25.3d Depress the white plastic quick-disconnect tabs and detach the fuel lines from the filter - wrap a rag around the fuel line to absorb the fuel that will run out

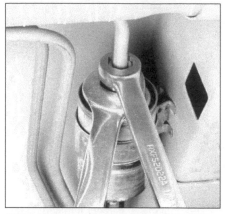

25.3e Use one wrench to steady the filter, then unscrew the fuel line fitting and remove the filter (if available, use a flare nut wrench on the fuel line fitting)

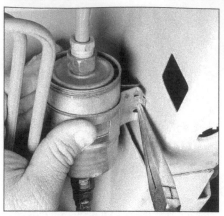

25.3f When installing the new filter, place the bracket in position, then use needle-nose pliers to push the lower tab into the opening in the chassis

26 Drivebelt check and replacement (every 30,000 miles or 24 months)

Refer to illustrations 26.5 and 26.7

1   A single serpentine drivebelt is located at the front of the engine and plays an important role in the overall operation of the engine and its components. Due to its function and material make up, the belt is prone to wear and should be periodically inspected. The serpentine belt drives the alternator, power steering pump, water pump and air conditioning compressor.

2   With the engine off, open the hood and use your fingers (and a flashlight, if necessary), to move along the belt checking for cracks and separation of the belt plies. Also check for fraying and glazing, which gives the belt a shiny appearance. Both sides of the belt should be inspected, which means you will have to twist the belt to check the underside. **Note:** *On 1997 and earlier models, it will be necessary to remove the air intake duct. On 1998 and later models, it will be necessary to remove the air cleaner housing and air intake resonator from the engine compartment to access the drivebelt.*

3   Check the ribs on the underside of the belt. They should all be the same depth, with none of the surface uneven.

4   The tension of the belt is maintained by the tensioner assembly and isn't adjustable. The belt should be replaced at the mileage specified in the maintenance schedule at the front of this chapter, or if it is damaged or worn.

5   To replace the belt, rotate the tensioner clockwise to release belt tension **(see illustration).**

6   Remove the belt from the auxiliary components and slowly release the tensioner.

7   Route the new belt over the various pulleys, again rotating the tensioner to allow the belt to be installed, then release the belt tensioner. **Note:** *These models have a drivebelt routing decal on the radiator shroud to help during drivebelt installation* **(see illustration).**

27 Automatic transmission fluid and filter change (every 30,000 miles or 24 months)

Refer to illustrations 27.6, 27.9 and 27.10

1   At the specified intervals, the transmission fluid should be drained and replaced.

Since the fluid will remain hot long after driving, perform this procedure only after the engine has cooled down completely.

2   Before beginning work, purchase the specified transmission fluid (see *Recommended lubricants and fluids* at the front of this Chapter) and a new filter.

3   Other tools necessary for this job include a floor jack, jackstands to support the vehicle in a raised position, a drain pan capable of holding at least eight pints, newspapers and clean rags.

4   Raise the vehicle and support it securely on jackstands.

5   Place the drain pan underneath the transmission pan. Remove the front and side pan mounting bolts, but only loosen the rear pan bolts approximately four turns. **Note:** *Some models are equipped with a drain plug on the transmission fluid pan. On these models, remove the plug to drain the fluid.*

6   Carefully pry the transmission pan loose with a screwdriver, allowing the fluid to drain **(see illustration).**

7   Remove the remaining bolts, pan and gasket. Carefully clean the gasket surface of the transmission to remove all traces of the old gasket and sealant. **Note:** *On later mod-*

26.5 Rotate the tensioner clockwise to remove or install the belt

26.7 On most models, the serpentine drivebelt routing diagram is located on the radiator support

27.6 With the rear bolts in place but loose, pull the front of the pan down to drain the transmission fluid

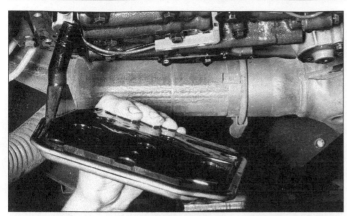

27.9 Rotate the filter out of the retaining clip, then lower it from the transmission

27.10 If necessary, use a screwdriver to remove the seal from the transmission - be careful not to gouge the aluminum housing

29.7a Remove the bolts from the lower edge of the cover . . .

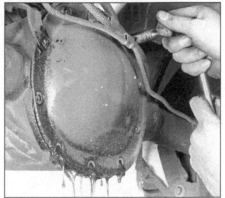

29.7b . . . then loosen the top bolts and let the lubricant drain out

29.7c After the lubricant has drained, remove the bolts and the cover

els you may have to set aside the bracket for the transmission range sensor cable to fully remove the transmission pan.

8   Drain the fluid from the transmission pan, clean it with solvent and dry it with compressed air.

9   Remove the filter from the mount inside the transmission **(see illustration)**.

10   If the seal did not come out with the filter, remove it from the transmission **(see illustration)**. Install a new filter and seal.

11   Make sure the gasket surface on the transmission pan is clean, then install a new gasket on the pan. Put the pan in place against the transmission and, working around the pan, tighten each bolt a little at a time until the final torque figure is reached.

12   Lower the vehicle and add approximately seven pints of the specified type of automatic transmission fluid through the filler tube (Section 6).

13   With the transmission in Park and the parking brake set, run the engine at a fast idle, but don't race it.

14   Move the gear selector through each range and back to Park. Check the fluid level. It will probably be low. Add enough fluid to bring the level up to the COLD FULL range on the dipstick.

15   Check under the vehicle for leaks during the first few trips.

28   Manual transmission lubricant change (every 30,000 miles or 24 months)

1   Raise the vehicle and support it securely on jackstands.

2   Move a drain pan, rags, newspapers and wrenches under the transmission.

3   Remove the transmission drain plug at the bottom of the case and allow the lubricant to drain into the pan **(see illustration 19.1)**.

4   After the lubricant has drained completely, reinstall the plug and tighten it securely.

5   Remove the fill plug from the side of the transmission case. Using a hand pump, syringe or funnel, fill the transmission with the specified lubricant until it begins to leak out through the hole. Reinstall the fill plug and tighten it securely.

6   Lower the vehicle.

7   Drive the vehicle for a short distance, then check the drain and fill plugs for leakage.

29   Differential lubricant change (every 30,000 miles or 24 months)

Refer to illustrations 29.7a, 29.7b, 29.7c and 29.9

1   This procedure should be performed after the vehicle has been driven so the lubri-

cant will be warm and therefore will flow out of the differential more easily.

2   Raise the vehicle and support it securely on jackstands.

3   The easiest way to drain the differential is to remove the lubricant through the filler plug hole with a suction pump. If the differential's bolt-on cover gasket is leaking, it will be necessary to remove the cover to drain the lubricant (which will also allow you to inspect the differential.

Changing the lubricant with a suction pump

4   Remove the fill plug from the differential (see Section 20).

5   Insert the flexible hose. Work the hose down to the bottom of the differential housing and pump the lubricant out.

Changing lubricant by removing the cover

6   Move a drain pan, rags, newspapers and wrenches under the vehicle.

7   Remove the bolts on the lower half of the plate **(see illustration)**. Loosen the bolts on the upper half and use them to keep the cover loosely attached **(see illustration)**. Allow the oil to drain into the pan, then completely remove the cover **(see illustration)**.

8   Using a lint-free rag, clean the inside of

29.9 Carefully scrape the old gasket material off to ensure a leak-free seal

30.6a The drain plug is located at the lower right corner of the radiator - attach a piece of hose so the coolant will drain without splashing

the cover and the accessible areas of the differential housing. As this is done, check for chipped gears and metal particles in the lubricant, indicating that the differential should be more thoroughly inspected and/or repaired.

9   Thoroughly clean the gasket mating surfaces of the differential housing and the cover plate. Use a gasket scraper or putty knife to remove all traces of the old gasket **(see illustration)**.

10   Apply a thin layer of RTV sealant to the cover flange, then press a new gasket into position on the cover. Make sure the bolt holes align properly.

All models

11   Use a hand pump, syringe or funnel to fill the differential housing with the specified lubricant until it's level with the bottom of the plug hole. Install the fill plug.

---

30   Cooling system servicing (draining, flushing and refilling) (see maintenance schedule for service intervals)

Refer to illustrations 30.6a, 30.6b, 30.7a and 30.7b

Caution: *Never mix green-colored ethylene glycol anti-freeze and orange-colored DEX-COOL™ silicate-free coolant because doing so will destroy the efficiency of the DEX-COOL™ coolant which is designed to last for 100,000 miles or five years.*

1   Periodically, the cooling system should be drained, flushed and refilled to replenish the antifreeze mixture and prevent formation of rust and corrosion, which can impair the performance of the cooling system and cause engine damage.

2   At the same time the cooling system is serviced, all hoses and the radiator cap should be inspected and replaced if defective (see Section 10).

3   Since antifreeze is a corrosive and poi-

30.6b Use a screwdriver to open the bleed screws (arrows) - 1997 and earlier (Gen I) V8 shown

sonous solution, be careful not to spill any of the coolant mixture on the vehicle's paint or your skin. If this happens, rinse it off immediately with plenty of clean water. Consult local authorities about the dumping of antifreeze before draining the cooling system. In many areas, reclamation centers have been set up to collect automobile oil and drained antifreeze/water mixtures, rather than allowing them to be added to the sewage system.

4   With the engine cold, remove the radiator cap.

5   Move a large container under the radiator to catch the coolant as it's drained.

6   Drain the radiator by opening the drain plug at the bottom on the left side **(see illustration)**. If the drain plug is corroded and can't be turned easily, or if the radiator isn't equipped with a plug, disconnect the lower radiator hose to allow the coolant to drain. Be careful not to get antifreeze on your skin or in your eyes. On 3800 engines and 1997 and earlier V8 engines, pack rags under the bleed screw(s) so coolant won't run down the side of the engine and use a screwdriver to open the screw(s) two or three turns **(see illustration)**.

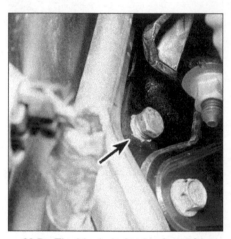

30.7a The block drain plug (arrow) is located below the exhaust manifold

7   After the coolant stops flowing out of the radiator, move the container under the engine block drain plug(s), then remove the plugs, or knock sensors (which double as plugs on some models), on both sides of the block **(see illustrations)**.

30.7b Some models have a knock sensor (arrow) which screws into the block drain hole - unplug the connector and remove the sensor to drain the coolant

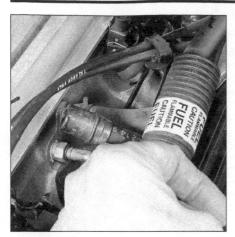

31.1 On 1997 and earlier V8 engines the PCV valve is located in the intake plenum - to check it, pull it out and feel for suction and shake it to make sure it rattles

8   Disconnect the hose from the coolant reservoir and remove the reservoir (see Chapter 3). Flush it out with clean water.
9   Place a garden hose in the radiator filler neck and flush the system until the water runs clear at all drain points.
10   In severe cases of contamination or clogging of the radiator, remove it (see Chapter 3) and reverse flush it. This involves inserting the hose in the bottom radiator outlet to allow the water to run against the normal flow, draining through the top. A radiator repair shop should be consulted if further cleaning or repair is necessary.
11   When the coolant is regularly drained and the system refilled with the correct antifreeze/water mixture, there should be no need to use chemical cleaners or descalers.
12   To refill the system, install the block plugs or knock sensors, reconnect any radiator hoses and install the reservoir and the overflow hose.
13   On later models, make sure to use the proper coolant (see **Caution** above). The manufacturer recommends adding its cooling system sealer any time the coolant is changed. Slowly fill the radiator with the recommended mixture of antifreeze and water to the base of the filler neck. Wait two minutes and recheck the coolant level, adding if necessary, then install the radiator cap. Add more coolant to the reservoir until it reaches the lower mark. On 3800 engines and 1997 and earlier V8 engines, close the bleed screws when the coolant issuing from them is free of bubbles.
14   On 1997 and earlier V8 engines, turn the ignition switch from On to Run then off, otherwise the LOW COOLANT could stay on.
15   Keep a close watch on the coolant level and the cooling system hoses during the first few miles of driving. Tighten the hose clamps and/or add more coolant as necessary. The coolant level should be a little above the HOT mark on the reservoir with the engine at normal operating temperature.

31.9 On 3.4L engines the PCV valve is located near the rear of the valve cover

31   Positive Crankcase Ventilation (PCV) valve check and replacement (every 30,000 miles or 24 months)

Refer to illustrations 31.1 and 31.9

V8 engines

1   On 1997 and earlier V8 engines the PCV valve is located in the intake plenum **(see illustration)**. On 1998 and later V8 engines the PCV valve is located in the front of the passenger side valve cover.
2   With the engine idling at normal operating temperature, pull the valve (with hose attached) out of the rubber grommet. Place your finger over the end of the valve. If there is no vacuum at the valve, check for a plugged hose, manifold port, or the valve itself. Replace any plugged or deteriorated hoses.
3   Turn off the engine and shake the PCV valve, listening for a rattle. If the valve doesn't rattle, replace it with a new one.
4   To replace the valve, pull it out of the end of the hose, noting its installed position and direction.
5   When purchasing a replacement PCV valve, make sure it's for your particular vehicle, model year and engine size. Compare the old valve with the new one to make sure they are the same.
6   Push the valve into the end of the hose until it's seated.
7   Inspect the rubber grommet for damage and replace it with a new one if necessary.
8   Push the PCV valve and hose securely into position in the rubber grommet.

3.4L engine

9   With the engine idling at normal operating temperature, pull the valve (with hose attached) out of the rubber grommet in the valve cover **(see illustration)**.
10   Place your finger over the end of the valve. If there is no vacuum at the valve, check for a plugged hose, manifold port, or the valve itself. Replace any plugged or deteriorated hoses.

11   Turn off the engine and shake the PCV valve, listening for a rattle. If the valve doesn't rattle, replace it with a new one.
12   To replace the valve, pull it out of the end of the hose, noting its installed position and direction.
13   When purchasing a replacement PCV valve, make sure it's for your particular vehicle, model year and engine size. Compare the old valve with the new one to make sure they are the same.
14   Push the valve into the end of the hose until it's seated.
15   Inspect the rubber grommet for damage and replace it with a new one if necessary.
16   Push the PCV valve and hose securely into position in the valve cover.

3800 engine

17   The PCV valve on these models is located in the intake plenum under a cover. Remove the two bolts and detach the cover.
18   Use needle-nose pliers to withdraw the PCV valve and spring from the manifold. Be careful not to lose the O-ring at the end of the valve.
19   Shake the PCV valve, listening for a rattle. If the valve doesn't rattle, replace it with a new one. Make sure the new PCV valve is for your particular vehicle, model year and engine size. Compare the old valve with the new one to make sure they are the same.
20   Insert the new PCV valve, O-ring and spring into position and install the cover and bolts.

32   Evaporative emissions control system check (every 30,000 miles or 12 months)

Refer to illustration 32.2

1   The function of the evaporative emissions control system is to draw fuel vapors from the gas tank and fuel system, store them in a charcoal canister and then burn them during normal engine operation.
2   The most common symptom of a fault in the evaporative emissions system is a strong fuel odor in the engine compartment. If a fuel

32.2 Check the evaporative emissions control canister for damage and the hose connections for cracks and damage

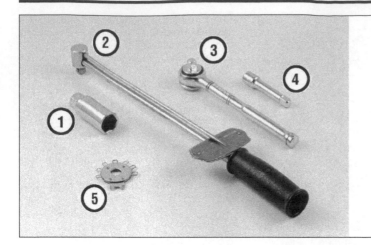

33.2 Tools required for changing spark plugs

1 **Spark plug socket** - This will have special padding inside to protect the spark plug's porcelain insulator
2 **Torque wrench** - Although not mandatory, using this tool is the best way to ensure the plugs are tightened properly
3 **Ratchet** - Standard hand tool to fit the spark plug socket
4 **Extension** - Depending on model and accessories, you may need special extensions and universal joints to reach one or more of the plugs
5 **Spark plug gap gauge** - This gauge for checking the gap comes in a variety of styles. Make sure the gap for your engine is included

odor is detected, inspect the charcoal canister, located in the left rear fender, behind a cover. Check the canister and all hoses for damage and deterioration **(see illustration)**.

3   The evaporative emissions control system is explained in more detail in Chapter 6.

33   Spark plug replacement (see maintenance schedule for service intervals)

Refer to illustrations 33.2, 33.5a, 33.5b, 33.6a, 33.6b, 33.8, 33.9 and 33.10

1   The spark plugs are located at the sides of the engine. The front spark plugs can be reached from the engine compartment. **Note:** *Replace the rear spark plugs from underneath. Raise the vehicle and support it securely on jackstands.*

2   In most cases, the tools necessary for spark plug replacement include a spark plug socket which fits onto a ratchet (spark plug sockets are padded inside to prevent damage to the porcelain insulators on the new plugs), various extensions and a gap gauge to check and adjust the gaps on the new plugs **(see illustration)**. A special plug wire removal tool is available for separating the wire boots from the spark plugs, and is a good idea on these models because the boots fit very tightly. A torque wrench should be used to tighten the new plugs. It is a good idea to allow the engine to cool before removing or installing the spark plugs.

3   The best approach when replacing the spark plugs is to purchase the new ones in advance, adjust them to the proper gap and replace the plugs one at a time. When buying the new spark plugs, be sure to obtain the correct plug type for your particular engine. The plug type can be found in the Specifications at the front of this Chapter and on the Emission Control Information label located under the hood. If these two sources list different plug types, consider the emission control label correct.

4   Allow the engine to cool completely before attempting to remove any of the plugs. While you are waiting for the engine to

33.5a Spark plug manufacturers recommend using a wire type gauge when checking the gap - if the wire does not slide between the electrodes with a slight drag, adjustment is required

33.5b To change the gap, bend the *side* electrode only, as indicated by the arrows, and be very careful not to crack or chip the porcelain insulator surrounding the center electrode

cool, check the new plugs for defects and adjust the gaps.

5   Check the gap by inserting the proper thickness gauge between the electrodes at the tip of the plug **(see illustration)**. The gap between the electrodes should be the same as the one specified on the Emissions Control Information label. The wire should slide between the electrodes with a slight amount of drag. If the gap is incorrect, use the adjuster on the gauge body to bend the curved side electrode slightly until the proper gap is obtained **(see illustration)**. If the side

electrode is not exactly over the center electrode, bend it with the adjuster until it is. Check for cracks in the porcelain insulator (if any are found, the plug should not be used).

6   With the engine cool, remove the spark plug wire from one spark plug. Pull only on the boot at the end of the wire - do not pull on the wire. A plug wire removal tool should be used if available **(see illustration)**. Some boots have a metal heat shield that must be removed before pulling the boot off **(see illustration)**. Be sure to replace the shield after reinstalling the boot.

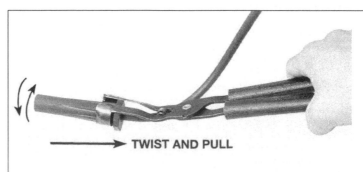

→ TWIST AND PULL

33.6a When removing the spark plug wires, pull only on the boot and use a twisting, pulling motion

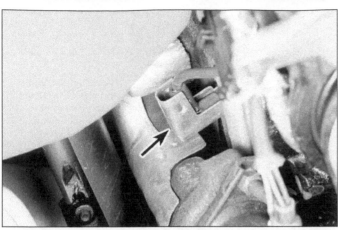

33.6b Some models have metal heat shields to protect the boots (arrow) - these simply pull off (be sure to reinstall them after spark plug replacement)

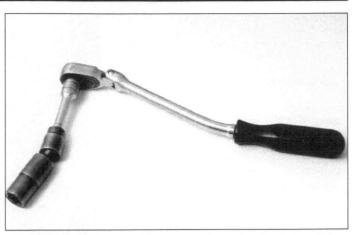

33.8 A socket and extension with a universal swivel joint will be required to remove the spark plugs on these models

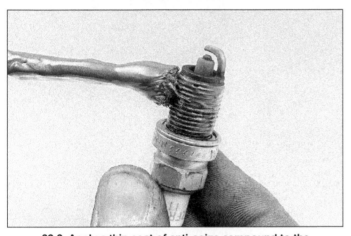

33.9 Apply a thin coat of anti-seize compound to the spark plug threads

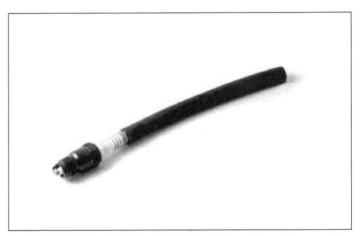

33.10 A length of 3/8-inch ID rubber hose will save time and prevent damaged threads when installing the spark plugs

7   If compressed air is available, use it to blow any dirt or foreign material away from the spark plug hole. A common bicycle pump will also work. The idea here is to eliminate the possibility of debris falling into the cylinder as the spark plug is removed.

8   The spark plugs on these models are, for the most part, difficult to reach so a spark plug socket incorporating a universal joint will be necessary **(see illustration)**. Place the spark plug socket over the plug and remove it from the engine by turning it in a counterclockwise direction. On some models where the plugs are unusually hard to get to, it may be easier to raise the vehicle, support it securely on jackstands and access the plugs from underneath.

9   Compare the spark plug with the chart shown on the inside back cover of this manual to get an indication of the general running condition of the engine. Before installing the new plugs, it is a good idea to apply a thin coat of anti-seize compound to the threads **(see illustration)**.

10   Thread one of the new plugs into the hole until you can no longer turn it with your fingers, then tighten it with a torque wrench (if available) or the ratchet. It's a good idea to

slip a short length of rubber hose over the end of the plug to use as a tool to thread it into place **(see illustration)**. The hose will grip the plug well enough to turn it, but will start to slip if the plug begins to cross-thread in the hole - this will prevent damaged threads and the accompanying repair costs.

11   Before pushing the spark plug wire onto the end of the plug, inspect it following the procedures outlined in Section 34.

12   Attach the plug wire to the new spark plug, again using a twisting motion on the boot until it's seated on the spark plug.

13   Repeat the procedure for the remaining spark plugs, replacing them one at a time to prevent mixing up the spark plug wires.

---

34   Spark plug wire and distributor housing check and replacement (every 30,000 miles or 24 months)

Refer to illustration 34.9
Note: *All V6 engines and 1998 and later V8 engines are equipped with distributorless ignition systems. The spark plug wires are connected directly to the ignition coils. The 1997*

and earlier V8 engine uses a distributor cap housing mounted at the front of the engine.

1   The spark plug wires should be checked at the recommended intervals and whenever new spark plugs are installed in the engine.

2   The wires should be inspected one at a time to prevent mixing up the order, which is essential for proper engine operation.

3   Disconnect the plug wire from the spark plug. To do this, grab the rubber boot, twist slightly and pull the wire off. Do not pull on the wire itself, only on the rubber boot.

4   Check inside the boot for corrosion, which will look like a white crusty powder. Push the wire and boot back onto the end of the spark plug. It should be a tight fit on the plug. If it isn't, remove the wire and use pliers to carefully crimp the metal connector inside the boot until it fits securely on the end of the spark plug.

5   Using a clean rag, wipe the entire length of the wire to remove any built-up dirt and grease. Once the wire is clean, check for burns, cracks and other damage. Do not bend the wire excessively or pull the wire lengthwise - the conductor inside might break.

6   On V6 engines and the 1998 and later (Gen II) V8 engines, disconnect the wire from

the coil pack. Again, pull only on the rubber boot. On 1997 and earlier V8 engines, disconnect the wire from the distributor at the front of the engine. Check for corrosion and a tight fit in the same manner as the spark plug end. Replace the wire at the coil pack or coil.

7   Check the remaining spark plug wires one at a time, making sure they are securely fastened at the ignition coil and the spark plug when the check is complete.

8   If new spark plug wires are required, purchase a set for your specific engine model. Wire sets are available pre-cut, with the rubber boots already installed. Remove and replace the wires one at a time to avoid mix-ups in the firing order.

9   On 1997 and earlier V8 engines, inspect the distributor housing for cracks or other damage **(see illustration)**. If cracks and damage are noted, it will be necessary to remove and replace the distributor, refer to Chapter 5.

34.9 The distributor and coil on 1997 and earlier V8 engines are accessible from below

Chapter 2 Part A
3.4L V6 engine

Contents

Specifications

General

Cylinder numbers (front-to-rear)	
Left (driver's side) bank	2-4-6
Right bank	1-3-5
Firing order	1-2-3-4-5-6

Torque specifications

Ft-lbs (unless otherwise indicated)

Air intake plenum bolts	216 in-lbs
Crankshaft balancer bolt	58
Crankshaft pulley bolts	37
Cylinder head bolts	
First step	41
Second step	Rotate an additional 1/4-turn (90-degrees)
Exhaust manifold-to-cylinder head bolts	216 in-lbs
Flywheel/driveplate-to-crankshaft bolts	61
Intake manifold-to-cylinder head bolts/nuts	22
Oil pan bolts/nuts	
Front corner	24
Rear corner	18
Side bolts	89 in-lbs
Studs	53 in-lbs
Oil pump mounting bolt	30
Timing chain cover bolts	
Small bolts	180 in-lbs
Large bolts	35
Timing chain sprocket-to-camshaft bolt(s)	216 in-lbs
Valve cover nuts	89 in-lbs

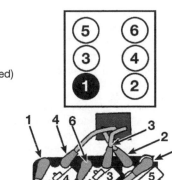

FRONT

24017-1-C HAYNES

3.4L engine cylinder and coil terminal locations

1   General information

Warning: The models covered by this manual are equipped with airbags. Impact sensors for the airbag system on 1993 through 1995 models are located in the dash and just in front of the radiator. On 1996 models a single impact sensor is located under the center console. The airbag(s) could accidentally deploy if these sensors are disturbed, so be extremely careful when working in these areas. Airbag system components are also located in the steering wheel, steering column, the base of the steering column and the passenger side of the dash, so be extremely careful when working in these areas and don't disturb any airbag system components or wiring. You could be injured if an airbag accidentally deploys, and the airbag might not deploy correctly in a collision if any components or wiring in the system have been disturbed.

Note: *On models equipped with a Delco Loc II or Theftlock audio system, be sure the lockout feature is turned off before performing any procedure which requires disconnecting the battery.*

This Part of Chapter 2 is devoted to in-vehicle repair procedures for the 3.4L V6 engine. For repair procedures for the 3800 V6, refer to Part B. These engines utilize cast-iron blocks with six cylinders arranged in a "V" shape at a 60-degree angle between the two banks. The overhead valve cast-iron cylinder heads are equipped with replaceable valve guides and seats. Hydraulic lifters actuate the valves through tubular pushrods.

All information concerning engine removal and installation and engine block and cylinder head overhaul can be found in Part E of this Chapter. The following repair procedures are based on the assumption the engine is installed in the vehicle. If the engine has been removed from the vehicle and mounted on a stand, many of the steps outlined in this Part of Chapter 2 will not apply.

The Specifications included in this Part of Chapter 2 apply only to the procedures contained in this Part. Part E of Chapter 2 contains the Specifications necessary for cylinder head and engine block rebuilding.

2   Repair operations possible with the engine in the vehicle

Many major repair operations can be accomplished without removing the engine from the vehicle.

Clean the engine compartment and the exterior of the engine with some type of degreaser before any work is done. It'll make the job easier and help keep dirt out of the internal areas of the engine.

Depending on the components involved, it may be helpful to remove the hood to improve access to the engine as repairs are performed (refer to Chapter 11 if necessary). Cover the fenders to prevent damage to the paint. Special pads are available, but an old bedspread or blanket will also work.

If vacuum, exhaust, oil or coolant leaks develop, indicating a need for gasket or seal replacement, the repairs can generally be done with the engine in the vehicle. The intake and exhaust manifold gaskets, timing chain cover gasket, oil pan gasket, crankshaft oil seals and cylinder head gaskets are all accessible with the engine in place, although many procedures involving the rear half of the engine may be more difficult than on previous models, because the engine is further back under the cowl.

Exterior engine components, such as the intake and exhaust manifolds, the oil pan (and the oil pump), the water pump, the

starter motor, the alternator and the fuel system components can be removed for repair with the engine in place.

Since the cylinder heads can be removed without pulling the engine, valve component servicing can also be accomplished with the engine in the vehicle. Replacement of the timing chain and sprockets is also possible with the engine in the vehicle.

In extreme cases caused by a lack of necessary equipment, repair or replacement of piston rings, pistons, connecting rods and rod bearings is possible with the engine in the vehicle. However, this practice is not recommended because of the cleaning and preparation work that must be done to the components involved.

3   Valve covers - removal and installation

Removal

Note: *The air intake plenum must be removed for access to either valve cover. Refer to Chapter 4 for the plenum removal procedure.*

1   Disconnect the cable from the negative terminal of the battery. **Note:** *On models equipped with a Delco Loc II or Theftlock audio system, be sure the lockout feature is turned off before performing any procedure which requires disconnecting the battery.*

2   Detach the spark plug wires from the clips at the valve covers.

3   If you're removing the right valve cover, unbolt the EGR valve adapter/EGR valve from the exhaust manifold (see Chapter 6). Also remove the drivebelt (see Chapter 1) and the alternator rear brace, then move the alternator towards the fenderwell.

4   Remove the valve cover mounting nuts.

5   Detach the valve cover. **Note:** *If the cover sticks to the cylinder head, use a block of wood and a hammer to dislodge it. If the cover still won't come loose, pry on it carefully, but don't distort the sealing flange.*

Installation

6   The mating surfaces of each cylinder head and valve cover must be perfectly clean when the covers are installed. Use a gasket scraper to remove all traces of sealant or old gasket material, then clean the mating surfaces with lacquer thinner or acetone (if there's sealant or oil on the mating surfaces when the cover is installed, oil leaks may develop). The valve covers are made of aluminum, so be extra careful not to nick or gouge the mating surfaces with the scraper.

7   Place the valve cover and new gasket in position, then install the bolts. Tighten the nuts in several steps to the torque listed in this Chapter's Specifications. Do not overtighten.

8   Complete the installation by reversing the removal procedure. Start the engine and check carefully for oil leaks at the valve cover-to-cylinder head joints.

4.2 Remove the rocker arm nuts, pivot balls, and rocker arms

4   Rocker arms and pushrods - removal, inspection, installation and adjustment

Refer to illustrations 4.2 and 4.3

Removal

1   Refer to Section 3 and remove the valve covers.

2   Beginning at the front end of one cylinder head, remove the rocker arm mounting nuts one at a time and detach the rocker arms, nuts, and pivot balls **(see illustration)**. Store each set of rocker arm components separately in a marked plastic bag to ensure they're reinstalled in their original locations. **Note:** *If you only need to remove the pushrods, loosen the rocker arm nuts and turn the rocker arms to allow room for pushrod removal.*

3   Remove the pushrods and store them separately to make sure they don't get mixed up during installation **(see illustration)**.

Inspection

4   Inspect each rocker arm for wear, cracks and other damage, especially where the pushrods and valve stems make contact.

5   Check the pivot seat in each rocker arm and the pivot ball faces. Look for galling, stress cracks and unusual wear patterns. If the rocker arms are worn or damaged, replace them with new ones and install new pivot balls as well.

6   Make sure the hole at the pushrod end of each rocker arm is open.

7   Inspect the pushrods for cracks and excessive wear at the ends. Roll each pushrod across a piece of plate glass to see if it's bent (if it wobbles, it's bent).

Installation

8   Lubricate the lower end of each pushrod with clean engine oil or moly-base grease and install them in their original locations. Make sure each pushrod seats completely in the lifter socket.

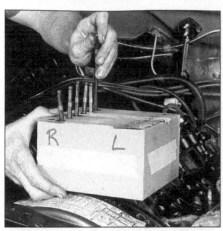

4.3 A perforated cardboard box can be used to store the pushrods to ensure they are reinstalled in their original locations - mark the box to indicate the front, or which side is right or left

9   Apply moly-base grease to the ends of the valve stems and the upper ends of the pushrods.
10   Apply moly-based grease to the pivot balls to prevent damage to the mating surfaces before engine oil pressure builds up. Install the rocker arms, pivot balls and nuts. As the nuts are tightened, make sure the pushrods engage properly in the rocker arms.

Adjustment

Refer to illustration 4.13
11   Turn the engine by hand with a breaker bar and socket on the crankshaft balancer bolt (vehicle must be in Neutral or Park) until the arrow on the balancer points straight up (12 o'clock). Watch the rocker arms on cylinder number 1 while doing this. As the arrow approaches 12 o'clock, if one of the rocker arms moves, you are approaching firing position for cylinder number 4. Turn the engine one more complete revolution for number 1 firing position.
12   In this position, valve adjustment can be performed on exhaust valves 1, 2, and 3, and intake valves 1, 5, and 6.
13   Adjust each valve by backing off the rocker arm nut while feeling for play in the pushrod. When movement is felt, start tightening the rocker arm nut until all lash (play) is eliminated. This can be determined by wig-

4.13 When play is eliminated at the pushrod, tighten the nut an additional 3/4 to 1-1/2 turn

gling the pushrod up-and-down; when you can no longer feel movement, the lash is taken up. At this point, tighten the rocker arm nut an additional 3/4 to 1-1/2 turn.
14   Rotate the engine one full turn of the crankshaft and adjust exhaust valves 4, 5, and 6, and intake valves 2, 3, and 4.
15   Install the valve covers.
16   The remainder of the assembly is the reverse of removal. Start the engine and check for leaks at the valve covers. **Note:** *The lifters may sound noisy at first. Run the engine at idle speed for five minutes to allow the lifter to completely fill with oil and any lifter noise should stop.*

5   Valve springs, retainers and seals - replacement

Refer to illustrations 5.4, 5.7, 5.8 and 5.16
Note: *Broken valve springs and defective valve stem seals can be replaced without removing the cylinder heads. Two special tools and a compressed air source are normally required to perform this operation, so read through this Section carefully and rent or buy the tools before beginning the job.*
1   Remove the valve cover from the affected cylinder head. If all of the valve stem seals are being replaced, remove both valve covers (see Section 3).
2   Remove the spark plug from the cylinder

which has the defective component. If all of the valve stem seals are being replaced, all of the spark plugs should be removed.
3   Turn the crankshaft until the piston in the affected cylinder is at Top Dead Center on the compression stroke. If you're replacing all of the valve stem seals, begin with cylinder number one and work on the valves for one cylinder at a time. Move from cylinder-to-cylinder following the firing order sequence (see this Chapter's Specifications). Each cylinder's TDC is 120-degrees of crankshaft rotation from the previous cylinder in the firing order, i.e. cylinder 2 is 120-degrees after number 1. Make two chalk marks on the crankshaft balancer that, along with the cast-in arrow, divide the balancer into three equal sections with the marks 120-degrees apart.
4   Thread an adapter into the spark plug hole and connect an air hose from a compressed air source to it. Most auto parts stores can supply the air hose adapter **(see illustration)**. **Note:** *Many cylinder compression gauges utilize a screw-in fitting that may work with your air hose quick-disconnect fitting.*
5   Remove the bolt, pivot and rocker arm for the valve with the defective part and pull out the pushrod (see Section 4). If all of the valve stem seals are being replaced, all of the rocker arms and pushrods should be removed.
6   Apply compressed air to the cylinder. The valves should be held in place by the air pressure. If the valve faces or seats are in poor condition, leaks may prevent the air pressure from retaining the valves. **Warning:** *If the cylinder isn't exactly at TDC, air pressure may cause the engine to rotate - do not leave a socket or wrench on the balancer bolt, or you may be injured by the tool.*
7   Stuff shop rags into the cylinder head holes above and below the valves to prevent parts and tools from falling into the engine, then use a valve spring compressor to compress the spring/balancer assembly. Remove the keepers with a pair of small needle-nose pliers or a magnet **(see illustration)**. **Note:** *because the rear cylinders are under the*

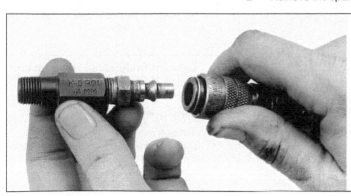

5.4 This is the air hose adapter that threads into the spark plug hole - they're commonly available from auto parts stores

5.7 Compress the valve spring and remove the keepers with a magnet

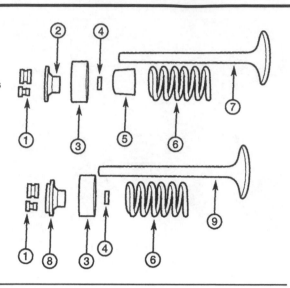

5.8 Typical valve components

1 Keepers
2 Retainer
3 Oil shield
4 O-ring oil seal
5 Umbrella seal
6 Spring and damper
7 Intake valve
8 Retainer/rotator
9 Exhaust valve

5.16 Apply a small dab of grease to each keeper as shown here before installation - it'll hold them in place on the valve stem as the spring is released

cowl, a lever-type spring compressor will not work, the clamp-type must be used.

8   Remove the spring retainer and valve spring, then remove the guide seal **(see illustration)**. **Note:** *If air pressure fails to hold the valve in the closed position during this operation, the valve face or seat is probably damaged. If so, the cylinder head will have to be removed for additional repair operations.*

9   Wrap a rubber band or tape around the top of the valve stem so the valve won't fall into the combustion chamber, then release the air pressure.

10   Inspect the valve stem for damage. Rotate the valve in the guide and check the end for eccentric movement, which would indicate that the valve is bent.

11   Move the valve up-and-down in the guide and make sure it doesn't bind. If the valve stem binds, either the valve is bent or the guide is damaged. In either case, the head will have to be removed for repair.

12   Apply air pressure to the cylinder to retain the valve in the closed position, then remove the tape or rubber band from the valve stem.

13   Lubricate the valve stem with clean engine oil. Install a new guide seal.

14   Install the spring in position over the valve.

15   Install the valve spring retainer and compress the valve spring.

16   Position the keepers in the upper groove. Apply a small dab of grease to the inside of each keeper to hold it in place if necessary **(see illustration)**. Remove the pressure from the spring tool and make sure the keepers are seated.

17   Disconnect the air hose and remove the adapter from the spark plug hole.

18   Install the rocker arm(s) and pushrod(s).

19   Refer to Section 3 and install the valve cover(s).

20   Install the spark plug(s) and connect the wire(s).

21   Start and run the engine, then check for oil leaks and unusual sounds coming from the valve cover area.

6   Intake manifold - removal and installation

Removal

1   Relieve the fuel system pressure (see Chapter 4).

2   Disconnect the cable from the negative terminal of the battery. **Note:** *On models equipped with a Delco Loc II or Theftlock audio system, be sure the lockout feature is turned off before performing any procedure which requires disconnecting the battery.*

3   Remove the plenum, fuel rail and injectors (see Chapter 4). When disconnecting fuel line fittings, be prepared to catch some fuel with a rag, then cap the fittings to prevent contamination.

4   Remove the serpentine drivebelt and drain the cooling system (see Chapter 1).

5   Disconnect the upper radiator hose and heater hose from the intake manifold.

6   Label and disconnect any remaining wires from the manifold.

7   Remove the valve covers (see Section 3).

8   Remove the manifold mounting bolts/nuts and separate the manifold from the engine. Don't pry between the manifold and heads, as damage to the soft aluminum gasket sealing surfaces may result. If you're installing a new manifold, transfer all fittings and sensors to the new manifold. The four corner fasteners are studs with nuts.

Installation

Note: *The mating surfaces of the cylinder heads, block and manifold must be perfectly clean when the manifold is installed. Gasket removal solvents in aerosol cans are available at most auto parts stores and may be helpful when removing old gasket material that's stuck to the heads and manifold (since the manifold is made of aluminum, aggressive scraping can cause damage). Be sure to follow the directions printed on the container.*

9   Lift or scrape the old gaskets off. Use a gasket scraper to remove all traces of sealant and old gasket material, then clean the mating surfaces with lacquer thinner or acetone. If there's old sealant or oil on the mating surfaces when the manifold is installed, oil or vacuum leaks may develop. Use a vacuum cleaner to remove any gasket material that falls into the intake ports or the lifter valley.

10   Use a tap of the correct size to chase the threads in the bolt holes, if necessary, then use compressed air (if available) to remove the debris from the holes. **Warning:** *Wear safety glasses or a face shield to protect your eyes when using compressed air!*

11   Apply a 3/16-inch (5 mm) bead of RTV sealant or equivalent to the front and rear ridges of the engine block between the heads. **Note:** *Some gasket sets include end seals.*

12   Install the intake gaskets to the cylinder heads, noting any markings for "Front" or "Up". **Note:** *Some gasket sets are marked with portions to "cut out". Remove those portions of the gasket before installation.*

13   Install the pushrods and rocker arms (see Section 4).

14   Carefully lower the manifold into place and install the mounting bolts/nuts finger tight.

15   Tighten the mounting bolts/nuts in three steps following the sequence, until they're all at the torque listed in this Chapter's Specifications.

16   Install the remaining components in the reverse order of removal.

17   Change the oil and filter and refill the cooling system (see Chapter 1). Start the engine and check for leaks.

7   Exhaust manifolds - removal and installation

Left manifold

Refer to illustration 7.2

1   Disconnect the cable from the negative terminal of the battery. **Note:** *On models equipped with the Delco Loc II or Theftlock audio system, be sure the lockout feature is*

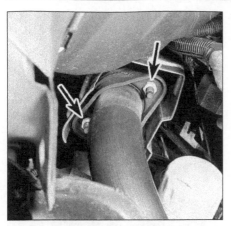

7.2 From underneath, remove the two bolts at the exhaust pipe-to-manifold flanges (left side shown)

8.3a Remove the four crankshaft pulley bolts

8.3b Wedge a large screwdriver in the ring gear teeth to prevent the crankshaft from turning (shown here is a manual transmission - on automatics, you must remove the converter shield)

turned off before performing any procedure which requires disconnecting the battery.

2 Unbolt the exhaust pipe(s) from the manifold(s) **(see illustration)**.

3 Disconnect the electrical connector from the oxygen sensor.

4 Unbolt and move the secondary air injection pipe (manual transmissions only).

5 Remove the mounting bolts/nuts and detach the heat shield, manifold and gasket from the cylinder head.

6 Clean the mating surfaces to remove all traces of old gasket material, then inspect the manifold for distortion and cracks. Warpage can be checked with a precision straightedge held against the mating flange. If a feeler gauge thicker than 0.030-inch can be inserted between the straight-edge and flange surface, take the manifold to an automotive machine shop for resurfacing.

7 Place the manifold in position with a new gasket and the heat shield and install the mounting bolts finger tight.

8 Starting in the middle and working out toward the ends, tighten the mounting bolts a little at a time until all of them are at the torque listed in this Chapter's Specifications.

9 Install the remaining components in the reverse order of removal.

10 Start the engine and check for exhaust leaks between the manifold and cylinder head and between the manifold and exhaust pipe.

Right manifold

11 Disconnect the cable from the negative terminal of the battery. **Note:** *On models equipped with the Delco Loc II or Theftlock audio system, be sure the lockout feature is turned off before performing any procedure which requires disconnecting the battery.*

12 Remove the serpentine drivebelt (see Chapter 1).

13 Allow the engine to cool completely, support the vehicle and unbolt the exhaust crossover pipe (see Chapter 4). You may have to apply penetrating oil to the fastener threads - they're usually corroded.

14 While still working from underneath,

unbolt the automatic transmission filler tube (automatic transmissions), the last two bolts on the exhaust manifold and the rear bolt of the air conditioning compressor mount.

15 Lower the vehicle, and unbolt and set aside the air conditioning compressor and brackets (see Chapter 3).

16 Disconnect the electrical connector from the oxygen sensor.

17 Remove the alternator and rear alternator brace **(see** Chapter 5**)**.

18 Unbolt and move the secondary air injection pipe (standard transmissions only).

19 Unbolt the digital EGR valve assembly from the manifold and pull it up away from the manifold (see Chapter 6).

20 Remove the front and middle exhaust manifold bolts and remove the heat shields, manifold and gasket.

21 Clean the mating surfaces to remove all traces of old sealant, then check for warpage and cracks. Warpage can be checked with a precision straightedge held against the mating flange. If a feeler gauge thicker than 0.030-inch can be inserted between the straightedge and flange surface, take the manifold to an automotive machine shop for resurfacing.

22 Place the manifold in position with the heat shields and a new gasket and install the bolts finger tight.

23 Starting in the middle and working out toward the ends, tighten the mounting bolts a little at a time until all of them are at the torque listed in this Chapter's Specifications.

24 Install the remaining components in the reverse order of removal.

25 Start the engine and check for exhaust leaks between the manifold and cylinder head and between the manifold and exhaust pipe.

8 Crankshaft front oil seal - removal and installation

Refer to illustrations 8.3a, 8.3b, 8.4, 8.5 and 8.6

1 Disconnect the cable from the negative

terminal of the battery. **Note:** *On models equipped with the Delco Loc II or Theftlock audio system, be sure the lockout feature is turned off before performing any procedure which requires disconnecting the battery.*

2 Remove the throttle body air duct (see Chapter 4) and the serpentine drive belt (see Chapter 1).

3 Unbolt the crankshaft pulley and remove the crankshaft balancer bolt. On automatic transmission-equipped models, remove the driveplate cover and position a large screwdriver in the ring gear teeth to keep the crankshaft from turning **(see illustrations)**. On manual transmission-equipped models, engage high gear and apply the brakes while an assistant removes the bolts.

4 Remove the crankshaft balancer-to-crankshaft bolt. The bolt is normally very tight, so use a large breaker bar and a six-point socket. Use the same technique to immobilize the crankshaft as you did in the previous Step. Pull the balancer off the crankshaft with a bolt-type puller **(see illustration)**. **Warning:** *Don't use a jaw-type puller, as it will ruin the balancer.* Leave the

8.4 You'll need a bolt-type puller to remove the crankshaft balancer

8.5 Carefully pry the old seal out of the timing chain cover

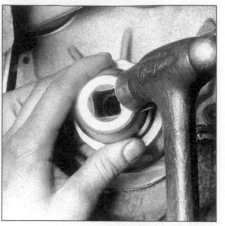

8.6 Drive the new seal in with a large socket and hammer

9.14 At TDC, the timing marks on the camshaft and crankshaft gears align with marks on the chain dampener (shown) or the block under the dampener

Woodruff key in place in the end of the crankshaft.

5   Note how the seal is installed - the new one must be installed to the same depth and facing the same way. Carefully pry the oil seal out of the cover with a seal puller or a large screwdriver **(see illustration)**. Be very careful not to distort the cover or scratch the crankshaft! Wrap electricians tape around the tip of the screwdriver to avoid damage to the crankshaft.

6   Apply clean engine oil or multi-purpose grease to the outer edge of the new seal, then install it in the cover with the lip (spring side) facing IN. Drive the seal into place **(see illustration)** with a large socket and a hammer (if a large socket isn't available, a piece of pipe will also work). Make sure the seal enters the bore squarely and stop when the front face is at the proper depth.

7   Installation is the reverse of removal. Be sure to apply moly-base grease to the seal contact surface of the balancer hub (if it isn't lubricated, the seal lip could be damaged and oil leakage would result). Align the damper hub keyway with the Woodruff key.

8   Tighten the crankshaft balancer-to-crankshaft bolt to the torque listed in this Chapter's Specifications.

9   Reinstall the remaining parts in the reverse order of removal.

9   Timing chain and sprockets - removal and installation

Refer to illustrations 9.14 and 9.15

Front cover removal

1   Disconnect the cable from the negative terminal of the battery. **Note:** *On models equipped with the Delco Loc II or Theftlock audio system, be sure the lockout feature is turned off before performing any procedure which requires disconnecting the battery.*

2   Allow the engine to cool completely and drain the cooling system and engine oil (see Chapter 1).

3   Remove the throttle body air duct (see Chapter 4).

4   Loosen the water pump pulley bolts, then remove the serpentine drivebelt (see Chapter 1).

5   Remove the water pump pulley and water pump (see Chapter 3).

6   Remove the crankshaft balancer (see Section 8).

7   Remove the oil pan (see Section 12).

8   Unbolt the power steering pump (if equipped) and tie it aside (see Chapter 10). Leave the hoses connected.

9   Detach the lower radiator hose from the front cover.

10   Remove the front cover-to-engine block bolts.

11   Separate the cover from the engine. If it's stuck, tap it with a soft-face hammer, but don't try to pry it off.

12   Use a gasket scraper to remove all traces of old gasket material and sealant from the cover and engine block. The cover is made of aluminum, so be careful not to nick or gouge it. Clean the gasket sealing surfaces with lacquer thinner or acetone.

Chain inspection

13   The timing chain should be replaced with a new one if the engine has high mileage, the chain has visible damage, or total freeplay midway between the sprockets exceeds one-inch. Failure to replace a worn timing chain may result in erratic engine performance, loss of power and decreased fuel mileage. Loose chains can "jump" timing. In the worst case, chain "jumping" or breakage will result in severe engine damage.

Chain and sprocket removal

14   Temporarily install the crankshaft balancer bolt and turn the crankshaft with the bolt to align the timing marks on the crankshaft and camshaft sprockets with their respective marks on the engine **(see illustration)**.

15   Remove the camshaft sprocket bolts **(see illustration)**. Do not turn the camshaft in the process (if you do, realign the timing

marks before the bolts are removed).

16   Use two large screwdrivers to carefully pry the camshaft sprocket off the camshaft dowel pin. Slip the timing chain and camshaft sprocket off the engine.

17   Timing chains and sprockets should be replaced in sets. If you intend to install a new timing chain, remove the crankshaft sprocket with a puller and install a new one. Be sure to align the key in the crankshaft with the keyway in the sprocket during installation.

18   Inspect the timing chain dampener (guide) for cracks and wear and replace it if necessary. The dampener is held to the engine block by two bolts.

19   Clean the timing chain and sprockets with solvent and dry them with compressed air (if available). **Warning:** *Wear eye protection when using compressed air.*

20   Inspect the components for wear and damage. Look for teeth that are deformed, chipped, pitted and cracked.

9.15 Use a screwdriver to hold the camshaft sprocket in place while you loosen the mounting bolts

10.9 Don't damage the gasket sealing surface when breaking the cylinder head loose - pry on a casting protrusion

10.12 Remove the old gasket (use a putty knife if necessary to separate it) and carefully scrape off all the old gasket material and sealant

can be removed by hand - work from bolt-to-bolt in a pattern that's the reverse of the tightening sequence **(see illustration 10.18)**. Store the bolts in the cardboard holder as they're removed - this will ensure they are reinstalled in their original locations, which is absolutely essential.

9 Lift the head(s) off the engine. If resistance is felt, only pry on a casting protrusion **(see illustration)** - don't pry on a gasket surface. Recheck for head bolts that may have been overlooked, then use a hammer and block of wood to tap up on the head and break the gasket seal. Be careful because there are locating dowels in the block which position each head. After removal, place the head on blocks of wood to prevent damage to the gasket surfaces.

10 Refer to Chapter 2, Part E, for cylinder head disassembly, inspection and valve service procedures.

Installation

Installation

21 Mesh the timing chain with the camshaft sprocket, then slip it over and engage it with the crankshaft sprocket. The timing marks should be aligned as shown in **illustration 9.14.**

22 Install the camshaft sprocket bolts and tighten them to the torque listed in this Chapter's Specifications.

23 Lubricate the chain and sprocket with clean engine oil. Replace the crankshaft front oil seal (see Section 8) and install the timing chain cover.

24 Apply a thin layer of sealant to both sides of the new gasket, then position the gasket on the engine block (the dowel pins should keep it in place). Attach the cover to the engine and install the bolts.

25 Apply sealant to the bottom of the gasket. Follow a criss-cross pattern when tightening the fasteners and work up to the torque listed in this Chapter's Specifications in three steps.

26 The remainder of installation is the reverse of removal.

27 Add oil and coolant, start the engine and check for leaks.

28 The remaining installation steps are the reverse of removal.

10 Cylinder heads - removal and installation

Refer to illustrations 10.9, 10.12, 10.15a, 10.15b, 10.17 and 10.18
Note: *On engines with high mileage and during an overhaul, camshaft lobe height should be checked prior to cylinder head removal (see Chapter 2, Part E for instructions).*
Caution: *Allow the engine to cool completely before loosening the cylinder head bolts.*

Removal

1 Disconnect the cable from the negative terminal of the battery. **Note:** *On models*

equipped with the Delco Loc II or Theftlock audio system, be sure the lockout feature is turned off before performing any procedure which requires disconnecting the battery.

2 Remove the valve covers as described in Section 3, the intake manifold as described in Section 6, and the exhaust manifolds as described in Section 7.

3 If you're removing the left cylinder head, remove the oil dipstick tube mounting bolt and the dipstick and tube.

4 Disconnect all wires from the cylinder head(s). Be sure to label them to simplify reinstallation.

5 Disconnect the spark plug wires and remove the spark plugs (see Chapter 1). Be sure the plug wires are labeled to simplify reinstallation.

6 Unbolt the power steering pump and lay it aside, keeping it upright (see Chapter 10).

7 Loosen the rocker arms enough to swing them aside to remove the pushrods (see Section 4).

8 Using the new head gasket, outline the cylinders and bolt pattern on a piece of cardboard. Be sure to indicate the front of the engine for reference. Punch holes at the bolt locations. Loosen each of the cylinder head mounting bolts 1/4-turn at a time until they

Installation

11 The mating surfaces of each cylinder head and block must be perfectly clean when the head is installed.

12 Use a gasket scraper to remove all traces of carbon and old gasket material **(see illustration)**, then clean the mating surfaces with lacquer thinner or acetone. If there's oil on the mating surfaces when the head is installed, the gasket may not seal correctly and leaks may develop. When working on the block, it's a good idea to cover the lifter valley with shop rags to keep debris out of the engine. Use a shop rag or vacuum cleaner to remove any debris that falls into the cylinders.

13 Check the block and head mating surfaces for nicks, deep scratches and other damage. If damage is slight, it can be removed with a file; if it's excessive, machining may be the only alternative.

14 Use a tap of the correct size to chase the threads in the head bolt holes. Dirt, corrosion, sealant and damaged threads will affect torque readings.

15 Position the new gasket over the dowel pins in the block. Some gaskets are marked TOP or THIS SIDE UP to ensure correct installation **(see illustrations)**.

10.15a Position the new gasket over the dowel pins (arrows) . . .

10.15b . . . with the correct side facing up

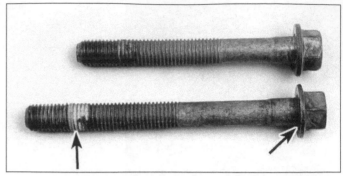

10.17 Two different length cylinder head bolts are used - apply sealant to the threads and undersides of the bolt heads (arrows)

16   Carefully position the head on the block without disturbing the gasket.

17   Apply sealant to the threads and the undersides of the bolt heads. Install the bolts in the correct locations - remember that the lower-front fastener on the left head is a stud. Here's where the cardboard holder comes in handy - there are two bolt lengths **(see illustration)**.

18   Tighten the bolts to 41 ft-lbs in the recommended sequence **(see illustration)**. Tighten each bolt an additional 90-degrees (1/4-turn) following the same sequence.

19   The remaining installation steps are the reverse of removal.

20   Change the oil and filter (see Chapter 1).

11   Camshaft and lifters - removal, inspection and installation

Camshaft lobe lift check

Refer to illustration 11.3

1   To determine the extent of cam lobe wear, the lobe lift should be checked prior to camshaft removal. Refer to Section 3 and remove the valve covers.

2   Position the number one piston at TDC on the compression stroke (see Section 5).

3   Beginning with the number one cylinder, mount a dial indicator on the engine and position the plunger against the top surface of the first rocker arm. The plunger should be directly above and in-line with the pushrod **(see illustration)**.

4   Zero the dial indicator, then very slowly turn the crankshaft in the normal direction of rotation until the indicator needle stops and begins to move in the opposite direction. The point at which it stops indicates maximum cam lobe lift.

5   Record this figure for future reference, then reposition the piston at TDC on the compression stroke.

6   Move the dial indicator to the other number one cylinder rocker arm and repeat the check. Be sure to record the results for each valve.

7   Repeat the check for the remaining valves. Since each piston must be at TDC on the compression stroke for this procedure, work from cylinder-to-cylinder following the firing order sequence.

8   After the check is complete, compare the results to the Chapter 2E Specifications. If camshaft lobe lift is less than specified, cam lobe wear has occurred and a new camshaft should be installed.

Lifter Noise

9   A noisy valve lifter can be isolated when the engine is idling. Hold a mechanic's stethoscope or a length of hose near the location of each valve while listening at the other end.

10   The most likely causes of noisy valve lifters are dirt trapped inside the lifter and lack of oil flow, viscosity or pressure. Before condemning the lifters, check the oil for fuel contamination, correct level, cleanliness and correct viscosity.

Camshaft/lifter removal

Refer to illustrations 11.13, 11.14, 11.17 and 11.18

11   Remove the intake manifold and valve cover(s) as described in Sections 3 and 4.

12   Loosen the rocker arms enough to swing them aside so the pushrods can be removed (see Section 4).

13   There are several ways to extract the lifters from the bores. A special tool designed to grip and remove lifters is manufactured by many tool companies and is widely available, but it may not be required in every case. On newer engines without a lot of varnish buildup, the lifters can often be removed with a small magnet or even with your fingers. A machinist's scribe with a bent end can be used to pull the lifters out by positioning the point under the retainer ring in the top of each lifter **(see illustration)**. **Caution:** *Don't use*

10.18 Cylinder head bolt TIGHTENING sequence (reverse the sequence to loosen the bolts)

11.3 Position a dial indicator over a rocker arm, with the plunger in line with the pushrod

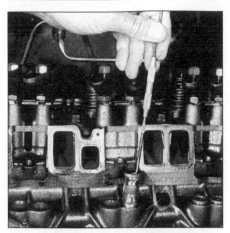

11.13 A scribe can be used to remove the lifters, or a strong magnet

11.14 Store the lifters in a box like this to ensure installation in their original locations

11.17 Remove the oil pump drive at the back of the block - be sure to replace the O-ring (arrow) when reinstalling

11.18 Thread long bolts into the cam sprocket bolt holes to use as handles when extracting the cam

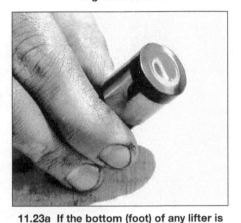

11.23a If the bottom (foot) of any lifter is worn concave (shown here) or rough, replace the camshaft and lifters

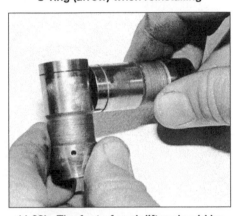

11.23b The foot of each lifter should be slightly convex - the side of another lifter can be used as a straightedge to check it - if it appears flat, it is worn out

pliers to remove the lifters unless you intend to replace them with new ones (along with the camshaft). The pliers may damage the precision machined and hardened lifters, rendering them useless.

14   Before removing the lifters, arrange to store them in a clearly labeled box to ensure they're reinstalled in their original locations. Remove the lifters and store them where they won't get dirty (see illustration). Note: *Some engines may have both standard and 0.25 mm oversize lifters installed at the factory. They are marked on the block "0.25 OS".*

15   To allow room for camshaft removal, the radiator and air conditioning condenser must be removed (see Chapter 3). Warning: *The air conditioning system must first be evacuated and the refrigerant recovered by a dealer service department or service station. Do not disconnect any air conditioning lines until the system has been properly depressurized.*

16   Remove the front cover, the timing chain and camshaft sprocket (see Section 9).

17   At the rear of the engine, remove the bolt, clamp and oil pump drive assembly (see illustration).

18   Thread three 6-inch long 5/16-18 bolts

into the camshaft sprocket bolt holes to use as a handle when removing the camshaft from the block (see illustration).

19   Carefully pull the camshaft out. Support the cam near the block so the lobes do not nick or gouge the bearings as it is withdrawn. Caution: *All the camshaft journals are the same size, so the camshaft must be withdrawn straight out to prevent damage to the bearings or journals.*

Inspection and installation

Refer to illustrations 11.23a, 11.23b, 11.23c, 11.25 and 11.27

20   Parts for valve lifters are not available separately. The work required to remove them from the engine again if cleaning is unsuccessful outweighs any potential savings from repairing them. If the lifters are worn, they must be replaced with new ones and the camshaft must be replaced as well - never install used conventional (non-roller) lifters with a new camshaft or new conventional lifters with a used camshaft.

21   When reinstalling used lifters, make sure they're replaced in their original bores. Soak new lifters in oil to remove trapped air. Coat all lifters with moly-base grease or engine assembly lube prior to installation.

22   Clean the lifters with solvent and dry them thoroughly without mixing them up.

23   Check each lifter wall, pushrod seat and foot for scuffing, score marks and uneven wear. Each lifter foot (the surface that rides on the cam lobe) must be slightly convex, although this can be difficult to determine by eye. If the base of the lifter is concave (see illustrations), the lifters and camshaft must be replaced.

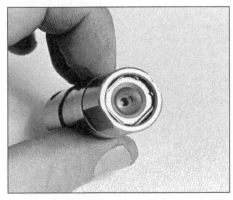

11.23c Check the pushrod seat (arrow) in the top of each lifter for wear

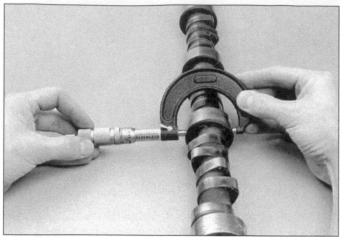

11.25 Measure the camshaft lobes and journals
with a micrometer

11.27 Apply camshaft installation lubricant to the camshaft
journals and lobes before installing it into the block

12.5 At the left side of the pan, unhook the clip holding the wiring
harness (upper arrow), then disconnect the oil level sensor
connector and remove the oil level sensor (lower arrow)

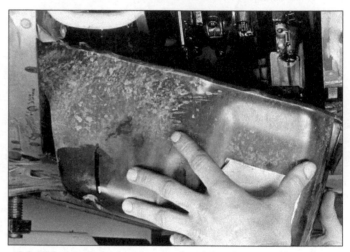

12.8 With the engine safely raised and the oil pan bolts removed,
break the pan loose and pull it down and back

24   After the camshaft has been removed from the engine, cleaned with solvent and dried, inspect the bearing journals for uneven wear, pitting and evidence of seizure. If the journals are damaged, the bearing inserts in the block are probably damaged as well. Both the camshaft and bearings will have to be replaced. Replacement of the camshaft bearings requires special tools and techniques which place it beyond the scope of the home mechanic. The block will have to be removed from the vehicle and taken to an automotive machine shop for this procedure.

25   Measure the bearing journals with a micrometer to determine if they are excessively worn or out-of-round **(see illustration)** and compare to this Chapter's Specifications.

26   Check the camshaft lobes for heat discoloration, score marks, chipped areas, pitting and uneven wear. If the lobes are in good condition and if the lobe lift measurements are as specified, the camshaft can be reused.

27   Lubricate the camshaft bearing journals and cam lobes with moly-base grease or

engine assembly lube **(see illustration)**.

28   Slide the camshaft into the engine. Support the cam near the block and be careful not to scrape or nick the bearings.

29   Refer to Section 9 and install the timing chain and sprockets, with the timing marks aligned. **Note:** *If the engine has accumulated enough miles for the camshaft/lifters to have been replaced, this is the time to also replace the timing chain and sprocket set.*

30   Lubricate the lifters with clean engine oil and install them in the block. If the original lifters are being reinstalled, be sure to return them to their original locations. If a new camshaft was installed, be sure to install new lifters as well.

31   Reinstall the oil pump drive assembly, clamp and bolt **(see illustration 11.17)**. Be sure to install a new O-ring on the drive.

32   The remaining installation steps are the reverse of removal.

33   Before starting and running the engine, change the oil and install a new oil filter (see Chapter 1).

12   Oil pan - removal and installation

Refer to illustrations 12.5 and 12.8

Removal

1   Disconnect the cable from the negative terminal of the battery. **Note:** *On models equipped with the Delco Loc II or Theftlock audio system, be sure the lockout feature is turned off before performing any procedure which requires disconnecting the battery.*

2   Raise the front of the vehicle and place it securely on jackstands. Apply the parking brake and block the rear wheels to keep it from rolling off the stands. Drain the engine oil (refer to Chapter 1 if necessary).

3   Refer to Chapter 4 and remove the exhaust crossover pipe and exhaust hanger bolts.

4   Remove the flywheel/driveplate cover and the starter (see Chapter 5). Remove the hood (see Chapter 11).

5   Remove the wiring harness clip from the left side of the pan, disconnect the electrical

14.2 Hold the flywheel/driveplate by jamming a large screwdriver through one of the holes and against the block, while removing the flywheel mounting bolts. There is a retaining ring under the bolts; save it for reassembly

13.2 Oil pump mounting bolt location (arrow)

connector from the oil level sensor, and remove the oil level sensor **(see illustration)**.

6   On automatic transmission equipped models, unclip the cooler lines at the oil pan.

7   Support the engine from above with a hoist, then remove the engine mount through-bolts and raise the engine.

8   Remove the bolts and nuts, then carefully separate the oil pan from the block **(see illustration)**. Don't pry between the block and the pan or damage to the sealing surfaces could occur and oil leaks may develop. Instead, tap the pan with a soft-face hammer to break the gasket seal. **Note:** *If necessary, rotate the crankshaft to reposition the counterweights to allow more room to remove the pan.*

Installation

9   Clean the pan with solvent and remove all old sealant and gasket material from the block and pan mating surfaces. Clean the mating surfaces with lacquer thinner or acetone and make sure the bolt holes in the block are clear. Check the oil pan flange for distortion, particularly around the bolt holes. If necessary, place the pan on a block of wood and use a hammer to flatten and restore the gasket surface.

10   Always use a new gasket whenever the oil pan is installed, and use a small amount of RTV sealant where the pan gasket tabs sit on the rear main cap.

11   Place the oil pan in position on the block and install the nuts/bolts.

12   After the fasteners are installed, tighten them to the torque listed in this Chapter's Specifications. Starting at the center, follow a criss-cross pattern and work up to the final torque in three steps.

13   The remaining steps are the reverse of the removal procedure.

14   Refill the engine with oil, run it until normal operating temperature is reached and check for leaks.

13 Oil pump - removal and installation

Refer to illustration 13.2

1   Remove the oil pan (see Section 12).

2   Unbolt the oil pump and lower it from the engine **(see illustration)**.

3   If the pump is defective, replace it with a new one - don't re-use the original or attempt to rebuild it.

4   Prime the pump by pouring clean engine oil into the pick-up screen while turning the pump driveshaft.

5   To install the pump, turn the hexagonal driveshaft so it mates with the oil pump drive.

6   Install the pump mounting bolt and tighten it to the torque listed in this Chapter's Specifications.

14 Flywheel/driveplate - removal and installation

Refer to illustration 14.2

1   Refer to Chapter 7 and remove the transmission. If your vehicle has a manual transmission, the pressure plate and clutch will also have to be removed (see Chapter 8).

2   Jam a large screwdriver through the driveplate (on automatic transmission models) or in the starter ring gear teeth (on manual transmission models) to keep the crankshaft from turning, then remove the mounting bolts **(see illustration)**. Since it's fairly heavy, support the flywheel as the last bolt is removed. **Warning:** *The flywheel ring-gear teeth are sharp - use gloves or rags to protect your fingers.*

3   If the flywheel doesn't have a locator hole indexed to a dowel-pin on the crank, mark the relationship of the flywheel to the crankshaft flange for later alignment. Remove the flywheel retainer ring and pull straight back on the flywheel/driveplate to detach it

15.2 Pry carefully to remove the old seal - don't nick the crankshaft surface

from the crankshaft.

4   Installation is the reverse of removal. Be sure to align the paint marks made earlier. Use thread-locking compound on the bolt threads and tighten them to the torque listed in this Chapter's Specifications in a criss-cross pattern. **Note:** *If the flywheel is being replaced with a new unit, the bolt-on balance weights from the original flywheel should be installed on the new flywheel in the same locations to preserve engine balance.*

15 Rear main oil seal - replacement

Refer to illustration 15.2

1   Refer to Section 14 to remove the flywheel/driveplate.

2   Insert a screwdriver or seal-removal tool into the lip of the rear seal and pry out all around the end of the crankshaft to remove the old seal. Be careful not to nick the sealing surface of the crankshaft **(see illustration)**.

3   Coat the new seal with engine oil and

16.8a Remove the nut (arrow) on the backside of the through-bolt, then support the engine (right-side mount shown)

16.8b With the pressure off the mounts, pull the through-bolt from the front (lower arrow), then raise the engine and remove the bolts holding the mount to the frame (arrow indicates one of these mount-to-frame bolts)

slip it onto the rear of the crankshaft, keeping the seal square to the crank. Push it on evenly with a large socket or a length of pipe the proper diameter.

16   Engine mounts - check and replacement

1   Engine mounts seldom require attention, but broken or deteriorated mounts should be replaced immediately or the added strain placed on the driveline components may cause damage or wear.

Check

2   During the check, the engine must be raised slightly to remove the weight from the mounts.

3   Raise the vehicle and support it securely on jackstands. Place a floor jack with a wood block on the head of the jack under the oil pan and raise the engine slightly. **Caution:** *Don't apply too much pressure on the jack - the engine oil pan could be damaged.*

4   Check the mounts to see if the rubber is cracked, hardened or separated from the metal plates. Sometimes the rubber will split right down the center.

5   Check for relative movement between the mount plates and the engine or frame (use a large screwdriver or prybar to attempt to move the mounts). If movement is noted, lower the engine and tighten the mount fasteners.

6   Rubber preservative may be applied to the mounts to slow deterioration.

Replacement

Refer to illustrations 16.8a and 16.8b

7   Disconnect the cable from the negative terminal of the battery, then raise the vehicle and support it securely on jackstands (if not already done). **Note:** *On models equipped with the Delco Loc II or Theftlock audio system, be sure the lockout feature is turned off before performing any procedure which requires disconnecting the battery.*

8   Raise the engine slightly with an engine hoist, or by using the method described in Step 3. Remove the through-bolts from the engine bracket and detach the mount from the frame **(see illustrations)**.

9   Installation is the reverse of removal. Use thread-locking compound on the threads and be sure to tighten everything securely.

Chapter 2 Part B 3800 V6 engine

Contents

Specifications

General

Cylinder numbers (drivebelt end-to-transmission end)	
Left bank	1-3-5
Right bank	2-4-6
Firing order	1-6-5-4-3-2

Oil pump

Outer gear-to-housing clearance	0.008 to 0.015 inch
Inner gear-to-outer gear clearance	0.006 inch
Gear end clearance	0.001 to 0.0035 inch
Cover warpage limit	0.002 inch

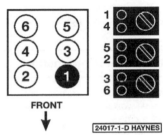

FRONT

24017-1-D HAYNES

Cylinder and coil terminal location

Torque specifications

	Ft-lbs (unless otherwise indicated)
Camshaft position sensor studs	22
Camshaft sprocket bolt	
Step one	74
Step two	Turn an additional 90-degrees
Camshaft thrust plate screws	132 in-lbs
Crankshaft balancer-to-crankshaft bolt	
1998 and earlier	
Step one	111
Step two	Turn an additional 76-degrees
1999 and later	
Step one	111
Step two	Turn an additional 114-degrees
Cylinder head bolts, in sequence **(see illustration 11.19)**	
1998 and earlier	
Step one	37
Step two	Turn an additional 130-degrees
Step three (center 4 bolts only)	Turn an additional 30-degrees
1999 and later	
Step one	37
Step two	Turn an additional 120-degrees
Flywheel/driveplate-to-crankshaft bolts	
Step one	132 in-lbs
Step two	Turn an additional 50-degrees
Exhaust manifold-to-cylinder head bolts/studs	
1998 and earlier	106 in-lbs
1999 and later	
Stud/nuts (front two)	132 in-lbs

Torque specifications (continued)

Ft-lbs (unless otherwise indicated)

Intake manifold fasteners	
1999 and earlier	132 in-lbs
2000 and later	
Lower manifold to head	132 in-lbs
Upper intake vertical bolts	132 in-lbs
Water outlet bolts..	20
Side bolts	22
Oil pan bolts...	125 in-lbs
Oil filter adapter-to-timing chain cover bolts	
1997 and earlier	22
1998 through 2001	
Step 1	132 in-lbs
Step 2	Turn an additional 50-degrees
2002	22
Oil pump	
Cover-to-timing chain cover bolts	98 in-lbs
Pick-up tube and screen assembly bolts.............................	132 in-lbs
Valve cover nuts/bolts	89 in-lbs
Rocker arm pivot bolts	
Step one	132 in-lbs
Step two	Turn an additional 90-degrees
Front cover bolts	
1998 and earlier	
Step one...	132 in-lbs
Step two...	Turn an additional 40-degrees
1999 and later	
Step one...	15
Step two...	Turn an additional 40-degrees
Valve lifter guide bolts	22

1 General information

Warning: *The models covered by this manual are equipped with airbags. Always disable the airbag system before working in the vicinity of the impact sensors, steering column or instrument panel to avoid the possibility of accidental deployment of the airbag(s), which could cause personal injury (see Chapter 12). The yellow wires and connectors routed through the instrument panel and, on 1995 and earlier models, to the front of the vehicle, are for this system. Do not use electrical test equipment on these yellow wires or tamper with them in any way.*

Caution: *On models equipped with a Delco Loc II or Theftlock audio system, be sure the lockout feature is turned off before performing any procedure which requires disconnecting the battery.*

This Part of Chapter 2 is devoted to in-vehicle repair procedures for the 3800 V6 engine (VIN Code K).

Information concerning camshaft bearings, balance shaft and bearings, and engine removal and installation, as well as engine block and cylinder head overhaul, is in Part E of this Chapter.

The following repair procedures are based on the assumption the engine is installed in the vehicle. If the engine has been removed from the vehicle and mounted on a stand, many of the steps included in this Part of Chapter 2 will not apply.

The Specifications included in this Part of Chapter 2 apply only to the procedures in this Part. The Specifications necessary for rebuilding the block and cylinder heads are found in Part E.

2 Repair operations possible with the engine in the vehicle

Many major repair operations can be accomplished without removing the engine from the vehicle.

Clean the engine compartment and the exterior of the engine with some type of pressure washer before any work is done. A clean engine will make the job easier and will help keep dirt out of the internal areas of the engine.

Depending on the components involved, it may be a good idea to remove the hood to improve access to the engine as repairs are performed (refer to Chapter 11 if necessary).

Exterior engine components such as the intake and exhaust manifolds, the oil pan, the oil pump, the water pump, the starter motor, the alternator and the fuel injection system can be removed for repair with the engine in place. The timing chain and sprockets, crankshaft oil seals and cylinder head gaskets are all accessible with the engine in place.

Since the cylinder heads can be removed without pulling the engine, valve component servicing can also be accomplished with the engine in the vehicle, although it is difficult.

In extreme cases caused by a lack of necessary equipment, repair or replacement of piston rings, pistons, connecting rods and rod bearings is possible with the engine in the vehicle. However, this practice is not recommended because of the cleaning and preparation work that must be done to the components involved.

3 Top Dead Center (TDC) for number one piston - locating

Refer to illustration 3.8

1   Top Dead Center (TDC) is the highest point in the cylinder each piston reaches as it travels up-and-down when the crankshaft turns. Each piston reaches TDC on the compression stroke and again on the exhaust stroke, but TDC generally refers to piston position on the compression stroke.

2   Positioning the piston(s) at TDC is an essential part of certain procedures such as timing chain/sprocket removal and camshaft removal.

3   Before beginning this procedure, be sure to place the transmission in Park, apply the parking brake and block the rear wheels.

4   Remove the spark plugs (see Chapter 1). **Warning:** *Make sure the ignition switch is in the Off position.*

5   When looking at the front of the engine, normal crankshaft rotation is clockwise. In order to bring any piston to TDC, the crankshaft must be turned with a socket and ratchet attached to the bolt threaded into the center of the crankshaft balancer on the crankshaft.

6   Have an assistant turn the crankshaft with a socket and ratchet as described above while you hold a finger over the number one spark plug hole. **Note:** *See the cylinder numbering diagram in the Specifications for this Chapter.*

7   When the piston approaches TDC, air

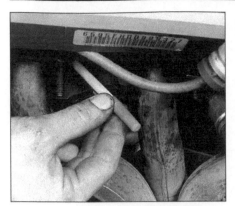

3.8 Insert a soft plastic pen into the hole to detect piston movement

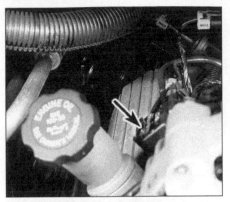

4.4 For right valve cover removal, remove the alternator and the evaporative emissions purge solenoid valve (arrow)

4.9 Remove the ignition coil, module and bracket (arrows)

pressure will be felt at the spark plug hole. Instruct your assistant to turn the crankshaft slowly.

8   Insert a plastic pen into the spark plug hole **(see illustration)**. As the piston rises the pen will be pushed out. Note the point where the pen stops moving out - this is TDC.

9   After the number one piston has been positioned at TDC on the compression stroke, TDC for any of the remaining pistons can be located by repeating the procedure described above and following the firing order. Divide the crankshaft pulley into three equal sections, with chalk marks at three points, indicating each 120-degrees of crankshaft rotation. The TDC position for cylinder 6, for instance, is 120-degrees after TDC for number 1.

4   Valve covers - removal and installation

Refer to illustrations 4.4 and 4.9

Right cover removal

1   Disconnect the negative battery cable from the battery. **Caution:** *On models equipped with a Delco Loc II or Theftlock audio system, be sure the lockout feature is turned off before performing any procedure which requires disconnecting the battery.*

2   Remove the spark plug wires from the spark plugs and clips. Number each wire before removal to ensure correct reinstallation.

3   Remove the serpentine drivebelt (see Chapter 1) and the alternator (see Chapter 5).

4   Remove the evaporative emissions canister purge solenoid valve **(see illustration)**.

5   Unscrew the nut and move the transmission dipstick tube. Also remove the oil fill tube and cap.

6   Loosen the right rear engine lifting bracket. Remove the valve cover mounting bolts.

7   Detach the valve cover. **Note:** *If the cover sticks to the cylinder head, use a soft-face hammer to dislodge it.*

Left cover removal

8   Disconnect the negative battery cable from the battery. **Caution:** *On models equipped with a Delco Loc II or Theftlock*

audio system, be sure the lockout feature is turned off before performing any procedure which requires disconnecting the battery.

9   Remove the ignition coil, module and bracket **(see illustration)**.

10   Unbolt and move the EGR valve, adapter, and pipe to allow removal of the engine lift bracket (see Chapter 6). On 2002 models, remove the nut securing the oil dipstick tube and pull up the tube to remove it.

11   Remove the valve cover mounting bolts/nuts and detach the valve cover. **Note:** *If the cover sticks to the cylinder head, use a soft-face hammer to dislodge it.*

Installation

12   The mating surfaces of the cylinder head and valve cover must be perfectly clean when the covers are installed. Use a gasket scraper to remove all traces of sealant or old gasket, then clean the mating surfaces with lacquer thinner or acetone (if there's sealant or oil on the mating surfaces when the cover is installed, oil leaks may develop). The valve covers are made of aluminum, so be extra careful not to nick or gouge the mating surfaces with the scraper.

13   Apply a non-hardening thread locking compound to the mounting bolt threads. Place the valve cover and new gasket in position, then install the bolts.

14   Tighten the bolts/nuts in several steps to the torque listed in this Chapter's Specifications.

15   Complete the installation by reversing the removal procedure.

16   Start the engine and check for oil leaks at the valve cover-to-cylinder head joints.

5   Rocker arms and pushrods - removal, inspection and installation

Refer to illustration 5.3

Removal

1   Refer to Section 4 and detach the valve covers from the cylinder heads.

2   Loosen the rocker arm pivot bolts one at a time and detach the rocker arms, bolts, piv-

ots and retainer/guide plate. Keep track of the rocker arm positions, since they must be returned to the same locations. Store each set of rocker components separately in a marked plastic bag to ensure that they're reinstalled in their original locations.

3   Remove the pushrods and store them separately to make sure they don't get mixed up during installation **(see illustration)**.

Inspection

4   Check each rocker arm for wear, cracks and other damage, especially where the pushrods and valve stems contact the rocker arm.

5   Check the pivot seat in each rocker arm and the pivot faces. Look for galling, stress cracks and unusual wear patterns. If the rocker arms are worn or damaged, replace them with new ones and install new pivots or shafts as well.

6   Make sure the hole at the pushrod end of each rocker arm is open.

7   Inspect the pushrods for cracks and excessive wear at the ends. Roll each pushrod across a piece of plate glass to see if it's bent (if it wobbles, it's bent).

Installation

8   Lubricate the lower end of each pushrod with clean engine oil or moly-base grease and install them in their original locations.

5.3 If more than one pushrod is being removed, store them in a perforated cardboard box to prevent mix-ups during installation - note the label indicating the front of the engine

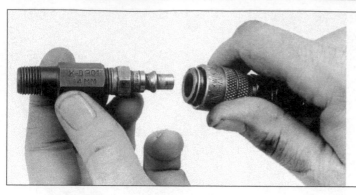

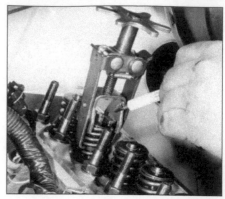

6.4 **This is what the air hose adapter that threads into the spark plug hole looks like - they're commonly available from auto parts stores**

6.7 **After compressing the valve spring, remove the keepers with a magnet or needle-nose pliers**

Make sure each pushrod seats completely in the lifter socket.

9   Apply moly-base grease to the ends of the valve stems, the upper ends of the pushrods and to the pivot faces to prevent damage to the mating surfaces before engine oil pressure builds up.

10   Coat the rocker arm pivot bolts with thread-locking compound. Install the rocker arms, pivots, retainers and bolts. Tighten the bolts to the torque listed in this Chapter's Specifications As the bolts are tightened, make sure the pushrods seat properly in the rocker arms. **Note:** *The rocker arm bolts have a specific "stretch" when tightened. Use only factory bolts - replacement with a weaker or stronger bolt could cause valve train problems.*

11   Install the valve covers (see Section 4).

6   Valve springs, retainers and seals - replacement

Refer to illustrations 6.4, 6.7, 6.8 and 6.16
Note: *Broken valve springs and defective valve stem seals can be replaced without removing the cylinder heads. Two special tools and a compressed air source are normally required to perform this operation, so read through this Section carefully and rent or buy the tools before beginning the job.*

1   Refer to Section 4 and remove the valve cover from the affected cylinder head. If all of the valve stem seals are being replaced, remove both valve covers.

2   Remove the spark plug from the cylinder which has the defective component. If all of

the valve stem seals are being replaced, all of the spark plugs should be removed.

3   Turn the crankshaft until the piston in the affected cylinder is at Top Dead Center on the compression stroke (refer to Section 3 for instructions). If you're replacing all of the valve stem seals, begin with cylinder number one and work on the valves for one cylinder at a time. Move from cylinder-to-cylinder following the firing order sequence (1-6-5-4-3-2).

4   Thread an adapter into the spark plug hole **(see illustration)** and connect an air hose from a compressed air source to it. Most auto parts stores can supply the air hose adapter. **Note:** *Many cylinder compression gauges utilize a screw-in fitting that may work with your air hose quick-disconnect fitting.*

5   Remove the bolt, pivot and rocker arm for the valve with the defective part and pull out the pushrod (see Section 5). If all of the valve stem seals are being replaced, all of the rocker arms and pushrods should be removed.

6   Apply compressed air to the cylinder. The valves should be held in place by the air pressure. If the valve faces or seats are in poor condition, leaks may prevent the air pressure from retaining the valves. **Warning:** *If the cylinder isn't exactly at TDC, air pressure may cause the engine to rotate, do not leave a socket or wrench on the balancer bolt, or you may be injured by the tool.*

7   Stuff shop rags into the cylinder head holes above and below the valves to prevent parts and tools from falling into the engine, then use a valve spring compressor to compress the spring. Remove the keepers with small needle-nose pliers or a magnet **(see**

illustration). **Note:** *because the rear cylinders are under the cowl, a lever-type spring compressor will not work (the clamp-type must be used).*

8   Remove the spring retainer and valve spring, then remove the guide seal **(see illustration)**. **Note:** *If air pressure fails to hold the valve in the closed position during this operation, the valve face or seat is probably damaged. If so, the cylinder head will have to be removed for additional repair operations.*

9   Wrap a rubber band or tape around the top of the valve stem so the valve won't fall into the combustion chamber, then release the air pressure.

10   Inspect the valve stem for damage. Rotate the valve in the guide and check the end for eccentric movement, which would indicate that the valve is bent.

11   Move the valve up-and-down in the guide and make sure it doesn't bind. If the valve stem binds, either the valve is bent or the guide is damaged. In either case, the head will have to be removed for repair.

12   Reapply air pressure to the cylinder to retain the valve in the closed position, then remove the tape or rubber band from the valve stem.

13   Lubricate the valve stem with clean engine oil and install a new valve guide seal. Using a hammer and a deep socket or seal installation tool, gently tap the seal into place until it's completely seated on the guide.

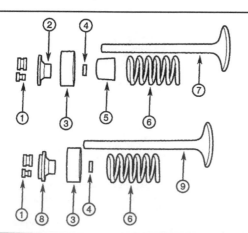

6.8 **Typical valve components**

1   *Keepers*
2   *Retainer*
3   *Oil shield*
4   *O-ring oil seal*
5   *Umbrella seal*
6   *Spring and damper*
7   *Intake valve*
8   *Retainer/rotator*
9   *Exhaust valve*

6.16 **Apply a small dab of grease to each keeper as shown here before installation - it will hold them in place on the valve stem as the spring is released**

7.5 Carefully pry up on a casting protrusion - don't pry between gasket surfaces

7.6 Remove all traces of gasket material, but don't gouge the soft aluminum

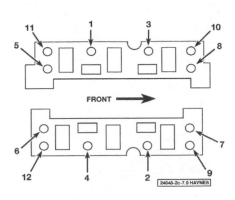

7.10 Intake manifold bolt tightening sequence and gasket arrangement

14   Install the spring in position over the valve.

15   Install the valve spring retainer and compress the valve spring.

16   Position the keepers in the upper groove. Apply a small dab of grease to the inside of each keeper to hold it in place if necessary **(see illustration)**. Remove the pressure from the spring tool and make sure the keepers are seated.

17   Disconnect the air hose and remove the adapter from the spark plug hole.

18   Install the pushrod(s) and rocker arm(s).

19   Install the spark plug(s) and connect the wire(s).

20   Refer to Section 4 and install the valve cover(s).

21   Start and run the engine, then check for oil leaks and unusual sounds coming from the valve cover area.

7   Intake manifold - removal and installation

Refer to illustrations 7.5, 7.6 and 7.10

Removal

1   Relieve the fuel system pressure (see Chapter 4).

2   Disconnect the negative battery cable from the battery. **Caution:** *On models equipped with a Delco Loc II or Theftlock audio system, be sure the lockout feature is turned off before performing any procedure which requires disconnecting the battery.*

3   Remove the air intake duct and the air intake plenum (upper intake manifold) (see Chapter 4). Disconnect the EGR pipe from the intake manifold.

4   Remove the valve covers (see Section 4) and disconnect the engine coolant temperature sensor (see Chapter 3, **illustration 8.1b**).

5   Remove the manifold mounting bolts, and separate the manifold from the engine **(see illustration)**. Do not pry between the manifold and heads, as damage to the gasket sealing surfaces may result. If you're

installing a new manifold, transfer all fittings and sensors to the new manifold.

Installation

Note: *The mating surfaces of the cylinder heads, block and manifold must be perfectly clean when the manifold is installed. Gasket removal solvents are available at most auto parts stores and may be helpful when removing old gasket material that's stuck to the heads and manifold (since the manifold is made of aluminum, aggressive scraping can cause damage). Be sure to follow the directions printed on the container.*

6   Use a gasket scraper to remove all traces of sealant and old gasket material **(see illustration)**, then clean the mating surfaces with lacquer thinner or acetone. If there's old sealant or oil on the mating surfaces when the manifold is installed, oil or vacuum leaks may develop. Use a vacuum cleaner to remove any gasket material that falls into the intake ports or the lifter valley.

7   Use a tap of the correct size to chase the threads in the bolt holes, then use compressed air (if available) to remove the debris from the holes. **Warning:** *Wear safety glasses or a face shield to protect your eyes when using compressed air.*

8   If steel manifold gaskets are used, apply sealant, available at most auto parts stores, to both sides of the gaskets before positioning them on the heads. Apply RTV sealant to the ends of the new manifold-to-block seals, then install them **(see illustration 7.10)**. Make sure the pointed end of the seal fits snugly against both the head and block.

9   Carefully lower the manifold into place. Apply thread lock compound to the mounting bolt threads and install the bolts finger tight.

10   Tighten the mounting bolts in two stages, following the recommended sequence **(see illustration)**, until they're all at the torque listed in this Chapter's Specifications.

11   Install the remaining components in the reverse order of removal.

12   Change the oil and filter and fill the cooling system (see Chapter 1). Start the engine and check for oil and vacuum leaks.

8   Exhaust manifolds - removal and installation

Refer to illustrations 8.3, 8.4 and 8.9
Warning: *Allow the engine to cool completely before beginning this procedure.*
Note: *Exhaust system fasteners are frequently difficult to remove - they get "frozen" in place because of the heating/cooling cycle to which they're constantly exposed. To ease removal, apply penetrating oil to the threads of all exhaust manifold and exhaust pipe fasteners and allow it to soak in.*

Removal

1   Disconnect the negative battery cable.
Caution: *On models equipped with a Delco Loc II or Theftlock audio system, be sure the lockout feature is turned off before performing any procedure which requires disconnecting the battery.*

Left manifold

2   Unbolt the EGR adapter tube from the exhaust manifold.

3   Remove three nuts and the heat shields **(see illustration)**.

4   Raise the vehicle and support it securely on jackstands. Unbolt the exhaust pipe from

8.3 To remove the left manifold, first remove the heat shield (right arrow) and the EGR tube (left arrow)

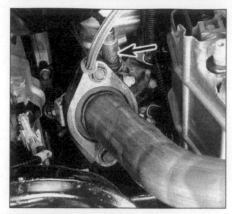

8.4 From underneath, unbolt the pipe flange from the manifold and disconnect the oxygen sensor (arrow)

the manifold **(see illustration)**.
5 Follow the lead from the oxygen sensor and unplug the electrical connector. **Note:** *This connector can be very difficult to reach. If you can't unplug it from below, try unplugging it from above. Feel for the electrical connector behind the left valve cover.*
6 Detach the spark plug wires from the spark plugs (see Chapter 1).
7 Lower the vehicle. Unbolt and remove the exhaust manifold and gasket.

Right manifold

8 Raise the vehicle and support it securely on jackstands. From underneath, disconnect the electrical connector to the oxygen sensor, the exhaust pipe from the manifold, and the spark plug wires from the spark plugs. On 2002 models, disconnect the air injection pipe from the throttle body and valve cover.
9 Lower the car, remove the nuts, heat shields, exhaust manifold studs/bolts, and lift out the exhaust manifold **(see illustration)**.

Installation

10 Clean the mating surfaces of the manifold and cylinder head to remove all traces of old gasket material, then check the manifold for warpage and cracks. If the manifold gasket was blown, take the manifold to an automotive machine shop for resurfacing.
11 Place the manifold in position with a new gasket and install the bolts/studs finger tight.
12 Starting in the middle and working out toward the ends, tighten the mounting bolts/studs a little at a time until all are at the torque listed in this Chapter's Specifications.
13 Install the remaining components in the reverse order of removal.
14 Start the engine and check for exhaust leaks between the manifold and cylinder head and between the manifold and exhaust pipe.

9 Crankshaft front oil seal - replacement

Refer to illustrations 9.4a, 9.4b, 9.5, 9.7 and 9.10
Note: *The crankshaft balancer is serviced as*

8.9 On the right manifold, first remove the nuts retaining the heat shield

9.4b While the flywheel is being held, remove the crankshaft balancer bolt - it is very tight, so watch your fingers

an assembly. Do not attempt to separate the pulley from the balancer hub.

Balancer removal

1 Disconnect the negative cable from the battery. **Caution:** *On models equipped with a Delco Loc II or Theftlock audio system, be sure the lockout feature is turned off before performing any procedure which requires disconnecting the battery.*
2 Remove the serpentine drivebelt (see Chapter 1).
3 Raise the vehicle and support it securely on jackstands.
4 Remove the starter and lower bellhousing cover plate and hold the crankshaft with a special tool, which is available at most auto parts stores, or wedge a large screwdriver in the ring gear teeth **(see illustration)** to keep the crankshaft from turning while an assistant removes the crankshaft balancer bolt **(see illustration)**. The bolt is normally quite tight, so use a large breaker bar and a six-point socket.
5 A three-bolt balancer puller must be used with 1/4-inch bolts to remove the crankshaft balancer. Thread the bolts 1/4-inch through the three slots on the pulley and into the three threaded holes inside. You will need to insert a long Allen bolt into the crankshaft snout for your puller's centerpoint to fit into **(see illustration)**. Leave the Woodruff key in

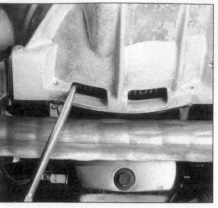

9.4a Use a large screwdriver in the ring gear teeth (this convenient hole is in the manual transmission models - on automatics, you must remove the lower converter shield) to hold the flywheel while removing the crankshaft balancer bolt

9.5 Three 1/4-inch, fine-thread bolts will be needed to adapt your puller to the damper - use a long Allen bolt in the crank snout for the tool's point to fit into

place in the end of the crankshaft. **Note:** *The pulley and balancer are a single assembly - do not attempt to separate them.*
6 If there's a groove worn into the seal contact surface on the crankshaft balancer, sleeves are available that fit over the groove, restoring the contact surface to like-new condition. These sleeves are sometimes included with the seal kit. Check with your parts supplier for details.

Seal replacement

7 Pry the old oil seal out with a seal removal tool or a screwdriver **(see illustration)**. Be very careful not to nick or otherwise damage the crankshaft in the process.
8 Apply a thin coat of RTV-type sealant to the outer edge of the new seal. Lubricate the seal lip with moly-base grease or clean engine oil.
9 Place the seal squarely in position in the bore and press it into place with a seal driver. Make sure the seal enters the bore squarely and seats completely.
10 If a seal driver is unavailable, carefully guide the seal into place with a large socket or piece of pipe and a hammer **(see illustra-**

9.7 Pry out the old seal with a seal removal tool (shown) or a screwdriver, but be careful not to nick the crankshaft

9.10 Gently drive the new seal into place with a hammer and large socket

10.10 Front cover bolt locations

tion). The outer diameter of the socket or pipe should be the same size as the seal outer diameter.

11 Installation is the reverse of removal. Align the keyway with the key and avoid bending the metal tabs. Be sure to apply moly-base grease to the seal contact surface on the back side of the balancer (if it isn't lubricated, the seal lip could be damaged and oil leakage would result).

12 Install the crankshaft balancer. Apply sealant to the threads and tighten the crankshaft balancer bolt to the torque listed in this Chapter's Specifications.

13 Reinstall the remaining parts in the reverse order of removal.

14 Start the engine and check for oil leaks at the seal.

10 Timing chain and sprockets - removal and installation

Front cover removal

Refer to illustration 10.10

1 Disconnect the negative battery cable from the battery. **Caution:** *On models equipped with a Delco Loc II or Theftlock audio system, be sure the lockout feature is turned off before performing any procedure which requires disconnecting the battery.*

2 Set the parking brake and put the transmission in Park. Raise the front of the vehicle and support it securely on jackstands.

3 Drain the engine oil and coolant (see Chapter 1).

4 Remove the crankshaft balancer (see Section 9).

5 Remove the oil pan-to-front cover bolts and lower the vehicle.

6 On 1998 and later vehicles, remove the air cleaner housing and the air intake resonator from the engine compartment (see Chapter 5). Also remove the drive belt tensioner and the power steering pump (see Chapter 10).

7 Remove the coolant hoses from the timing chain cover. Disconnect the electrical

10.12 Turn the crankshaft until the timing marks (arrows) align

connectors from the oil pressure, camshaft and crankshaft sensors.

8 Remove the crankshaft sensor shield and sensor (see Chapter 6).

9 Remove the water pump pulley and the water pump (see Chapter 3).

10 Remove the front cover-to-engine block bolts and three studs **(see illustration)**. Remove the front cover.

Chain and sprocket removal

Refer to illustrations 10.12, 10.13, 10.15 and 10.20

11 The timing chain should be replaced with a new one if the total freeplay midway between the sprockets exceeds one inch. Failure to replace the timing chain may result in erratic engine performance, loss of power and lowered fuel mileage. **Note:** *Replacement timing chains are sold as a set, with the new chain and new camshaft and crankshaft gears; do not attempt to replace parts individually.*

12 Temporarily install the crankshaft balancer bolt and turn the crankshaft clockwise to align the timing marks on the crankshaft and camshaft sprockets directly opposite each other **(see illustration)**.

13 Detach the spring, then remove the bolt and separate the timing chain damper from

10.13 Removing the timing chain damper

the block **(see illustration)**.

14 Remove the camshaft sprocket bolt. **Warning:** *The bolt is very tight. Have an assistant secure the flywheel from turning while loosening this fastener and have your socket and breaker bar securely on the bolt head at all times to avoid injury.*

15 Pull the camshaft sprocket from the cam and remove the sprocket and timing chain together **(see illustration)**.

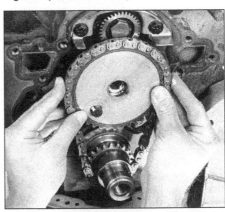

10.15 Remove the camshaft sprocket and timing chain then remove the crankshaft sprocket

10.20 Align the mark on the balance shaft gear (B) with the mark on the drive gear (A)

11.10 Pry carefully - don't force a tool between the gasket surfaces

11.13 Carefully remove all traces of old gasket material

16   Remove the crankshaft sprocket.

17   Use a gasket scraper to remove all traces of old gasket material and sealant from the cover and engine block. The cover is made of aluminum, so be careful not to nick or gouge it. Clean the gasket sealing surfaces with lacquer thinner or acetone.

18   The oil pump cover must be removed and the cavity packed with petroleum jelly as described in Section 14 before the cover is installed.

19   Clean the timing chain components with solvent and dry them with compressed air (if available). **Warning:** *Wear eye protection.*

20   Inspect the components for wear and damage. Look for teeth that are deformed, chipped, pitted, polished or discolored. Behind the camshaft sprocket is the gear that drives the smaller balance shaft gear above it **(see illustration)**. Inspect the two balance shaft gears, and replace them as a set if wear is indicated. **Note:** *Balance shaft removal, inspection and replacement is covered in Chapter 2E.*

Installation

Note: *If the crankshaft has been disturbed, install the new sprocket temporarily and turn the crankshaft until the mark on the crankshaft sprocket is exactly at the top. If the camshaft was disturbed, install the new sprocket temporarily and turn the camshaft until the timing mark is at the bottom, opposite the mark on the crankshaft sprocket* **(see illustration 10.12)**.

21   Install the crankshaft sprocket. Assemble the timing chain to the camshaft sprocket, then slip the sprocket and chain assembly onto the crankshaft sprocket with the timing marks aligned as shown in **illustration 10.12**.

22   Install the camshaft sprocket bolt and tighten it to the torque listed in this Chapter's Specifications.

23   Attach the timing chain damper assembly to the block and install the spring.

24   Rotate the engine through two complete revolutions and check that the timing marks still align.

25   Apply a thin layer of RTV sealant to both sides of the new front cover gasket, then position the gasket on the engine block (the dowel pins should keep it in place). Attach the cover to the engine, making sure the oil pump drive engages with the flats on the crankshaft.

26   Apply thread sealant to the bolt threads, then install them finger tight. Install the crankshaft sensor, but leave the bolts finger tight until Step 27.

27   Follow a criss-cross pattern when tightening the other bolts and work up to the torque listed in this Chapter's Specifications in three steps to avoid warping the cover.

28   The remainder of installation is the reverse of removal.

29   Add oil and coolant, start the engine and check for leaks.

11   Cylinder heads - removal and installation

Refer to illustrations 11.10, 11.13, 11.16 and 11.19

Removal

1   Disconnect the negative battery cable at the battery. **Caution:** *On models equipped with a Delco Loc II or Theftlock audio system, be sure the lockout feature is turned off before performing any procedure which requires disconnecting the battery.*

2   Disconnect the spark plug wires and remove the spark plugs (see Chapter 1). Be sure to label the plug wires to simplify reinstallation.

3   To remove the right cylinder head, remove the rear alternator brace.

4   Remove the valve cover(s) (see Section 4).

5   Remove the intake manifold as described in Section 7.

6   Disconnect all wires and hoses from the cylinder head(s). Be sure to label them to simplify reinstallation.

7   Detach the exhaust manifold(s) from the cylinder head(s) being removed (see Section 8).

8   Remove the rocker arms and pushrods (see Section 5).

9   Loosen the head bolts in 1/4-turn increments until they can be removed by hand. Work from bolt-to-bolt in the reverse of the tightening sequence **(see illustration 11.19)**.

10   Lift the head off the engine. If resistance is felt, don't pry between the head and block as damage to the mating surfaces will result. Recheck for head bolts that may have been overlooked, then use a hammer and block of wood to tap the head and break the gasket seal. Be careful because there are locating dowels in the block which position each head. As a last resort, pry each head up at the rear corner only and be careful not to damage anything **(see illustration)**. After removal, place the head on blocks of wood to prevent damage to the gasket surfaces.

11   Refer to Chapter 2, Part E, for cylinder head disassembly, inspection and valve service procedures.

Installation

12   The mating surfaces of the cylinder heads and block must be perfectly clean when the heads are installed.

13   Use a gasket scraper to remove all traces of carbon and old gasket material **(see illustration)**, then clean the mating surfaces with lacquer thinner or acetone. If there's oil on the mating surfaces when the heads are installed, the gaskets may not seal correctly and leaks may develop. When working on the block, it's a good idea to cover the lifter valley and balancer shaft with shop rags to keep debris out of the engine. Use a shop rag or vacuum cleaner to remove any debris that falls into the cylinders.

14   Check the block and head mating surfaces for nicks, deep scratches and other damage. If damage is slight, it can be removed with a file; if it's excessive, machining may be the only alternative.

15   Use a tap of the correct size to clean up the threads in the head bolt holes. Dirt, corrosion, sealant and damaged threads will affect torque readings.

16   Position the new gaskets over the dowel pins in the block. If steel gaskets are used,

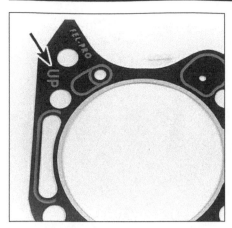

11.16 Look for gasket marks (arrow) to ensure correct installation

11.19 Cylinder head bolt TIGHTENING sequence

apply sealant. Install "non-retorquing" type gaskets dry (no sealant), unless the manufacturer states otherwise. Most gaskets are marked UP or TOP **(see illustration)** because they must be installed a certain way.

17 Carefully position the heads on the block without disturbing the gaskets.

18 Use NEW, factory head bolts - don't reinstall the old ones - and apply sealant to the threads and the undersides of the bolt heads.

19 Tighten the bolts as directed in this Chapter's Specifications in the sequence shown **(see illustration)**. This must be done in three steps, following the sequence each time.

20 The remaining installation steps are the reverse of removal.

21 Change the oil and filter (see Chapter 1).

12 Camshaft and lifters - removal, inspection and installation

Camshaft lobe lift check

Refer to illustration 12.3

1 To determine the extent of cam lobe wear, the lobe lift should be checked prior to camshaft removal. Refer to Section 4 and remove the valve covers.

2 Position the number one piston at TDC on the compression stroke (see Section 3).

3 Beginning with the number one cylinder, mount a dial indicator on the engine and position the plunger against the top surface of the first rocker arm. The plunger should be directly above and in-line with the pushrod **(see illustration)**.

4 Zero the dial indicator, then very slowly turn the crankshaft in the normal direction of rotation until the indicator needle stops and begins to move in the opposite direction. The point at which it stops indicates maximum cam lobe lift.

5 Record this figure for future reference, then reposition the piston at TDC on the compression stroke.

6 Move the dial indicator to the other number one cylinder rocker arm and repeat

the check. Be sure to record the results for each valve.

7 Repeat the check for the remaining valves. Since each piston must be at TDC on the compression stroke for this procedure, work from cylinder-to-cylinder following the firing order sequence (see this Chapter's Specifications).

8 After the check is complete, compare the results to the Chapter 2E Specifications. If camshaft lobe lift is less than specified, cam lobe wear has occurred and a new camshaft should be installed.

9 A noisy valve lifter can be isolated when the engine is idling. Hold a mechanic's stethoscope or a length of hose near the position of each valve while listening at the other end.

10 The most likely causes of noisy valve lifters are dirt trapped between the plunger and the lifter body or lack of oil flow, viscosity or pressure. Before condemning the lifters, we recommend checking the oil for fuel contamination, proper level, cleanliness and correct viscosity.

Lifter removal

Refer to illustration 12.14

11 Remove the valve cover(s) as described in Section 4.

12.3 Position a dial indicator directly above the pushrod to measure camshaft lobe lift

12.14 The roller lifters are kept from turning by guides bolted to the sides of the lifter valley - remove the four bolts to take off the guides

12 Remove the rocker arms and pushrods (see Section 5).

13 Remove the intake manifold as described in Section 7.

14 Roller lifters are kept from turning by guides **(see illustration)**, which must be removed to access the lifters.

15 There are several ways to extract the lifters from the bores. A special tool designed to grip and remove lifters is manufactured by many tool companies and is widely available, but it may not be required in every case. On newer engines without a lot of varnish buildup, the lifters can often be removed with a small magnet or even with your fingers. A machinist's scribe with a bent end can be used to pull the lifters out by positioning the point under the retainer ring in the top of each lifter. **Caution:** *Don't use pliers to remove the lifters unless you intend to replace them with new ones. The pliers will damage the precision machined and hardened lifters, rendering them useless.*

16 Before removing the lifters, arrange to store them in a clearly labeled box to ensure that they're reinstalled in their original locations. Remove the lifters and store them where they won't get dirty.

12.20 With the balance shaft drive gear off the camshaft, remove the three Torx screws and take off the camshaft thrust plate

12.24 The roller must turn freely - check for wear and excessive play as well

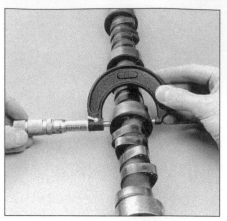

12.28 Measure the camshaft lobes and journals with a micrometer

Camshaft removal

Refer to illustration 12.20

17   To allow room for camshaft removal, the radiator and air conditioning condenser must be removed (see Chapter 3). **Note:** *If the condenser must be removed the system must first be evacuated and the refrigerant recovered by a dealer service department or service station. Do not disconnect any air conditioning lines until the system has been properly depressurized.*

18   Remove the timing chain front cover, the timing chain and camshaft sprocket (see Section 10).

19   Pull the balance shaft drive gear from the front of the camshaft.

20   Remove the camshaft thrust plate bolts and thrust plate **(see illustration)**.

21   Carefully pull the camshaft out. Support the cam near the block so the lobes do not nick or gouge the bearings as the cam is withdrawn. **Caution:** *All the camshaft journals are the same size, so the camshaft must be withdrawn straight out to prevent damage to the bearings or journals.*

Inspection

Refer to illustrations 12.24 and 12.28

22   Parts for valve lifters are not available separately. The work required to remove them from the engine again if repair is unsuccessful outweighs any potential savings from repairing them.

23   Clean the lifters with solvent and dry them thoroughly without mixing them up.

24   Check the rollers carefully for wear and damage and make sure they turn freely without excessive play **(see illustration)**. If the lifter walls are damaged or worn (which is not very likely), inspect the lifter bores in the engine block as well. If the pushrod seats are worn, check the pushrod ends.

25   Used roller lifters can be reinstalled with a new camshaft (provided the lifters are in good condition) and the original camshaft can be used if new lifters are installed (provided the camshaft is in good condition).

26   When reinstalling used lifters, make sure they're replaced in their original bores. Soak new lifters in oil to remove trapped air. Coat all lifters with moly-base grease or engine assembly lube prior to installation.

27   After the camshaft has been removed from the engine, cleaned with solvent and dried, inspect the bearing journals for uneven wear, pitting and evidence of seizure. If the journals are damaged, the bearing inserts in the block are probably damaged as well. Both the camshaft and bearings will have to be replaced. Replacement of the camshaft bearings requires special tools and techniques which place it beyond the scope of the home mechanic. The block will have to be removed from the vehicle and taken to an automotive machine shop for this procedure.

28   Measure the bearing journals with a micrometer to determine if they are excessively worn or out-of-round **(see illustration)** and compare to this Chapter's Specifications.

29   Check the camshaft lobes for heat discoloration, score marks, chipped areas, pitting and uneven wear. If the lobes are in good condition and if the lobe lift measurements are as specified, the camshaft can be reused.

Installation

Refer to illustration 12.30

30   Lubricate the camshaft bearing journals and cam lobes with camshaft installation lube **(see illustration)**.

12.30 Lube the camshaft journals and lobes thoroughly with camshaft installation lube

31   Slide the camshaft into the engine. Support the cam near the block and be careful not to scrape or nick the bearings.

32   Install the camshaft thrust plate, and slip the balance shaft drive gear over the cam, aligning it with the keyway and with drive gear and driven gear timing marks aligned. **Note:** *The camshaft thrust plate is marked TOP - this marking must be at the top and facing the front.*

33   Refer to Section 10 and install the timing chain and sprockets, with the timing marks aligned. **Note:** *If the engine has accumulated enough miles for the camshaft/lifters to have been replaced, this is the time to also replace the timing chain and sprocket set.*

34   Lubricate the lifters with clean engine oil and install them in the block. If the original lifters are being reinstalled, be sure to return them to their original locations.

35   The remaining installation steps are the reverse of removal.

36   Before starting and running the engine, change the oil and install a new oil filter (see Chapter 1).

37   Run the engine and check for leaks.

13   Oil filter adapter and pressure regulator valve - removal and installation

Refer to illustration 13.5

1   Remove the oil filter (see Chapter 1).

2   Remove the air conditioning compressor (if equipped) and position it aside without disconnecting the refrigerant hoses (see Chapter 3). Also remove the air conditioning compressor mounting bracket.

3   Disconnect the electrical connector from the oil pressure sending unit.

4   Remove the four bolts holding the oil filter adapter to the timing chain cover. The cover is spring loaded, so remove the bolts while keeping pressure on the cover, then release the spring pressure carefully.

5   Remove the pressure regulator valve and spring **(see illustration)**. Use a gasket scraper to remove all traces of the old gasket.

6   Clean all parts with solvent and dry them with compressed air (if available).

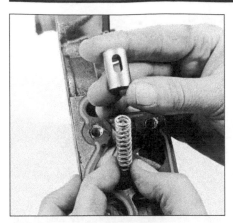

13.5 Remove the pressure regulator valve and spring, then check the valve for wear and damage

Warning: *Wear eye protection. Check for wear, score marks and valve binding.*

7   Installation is the reverse of removal. Be sure to use a new gasket. **Caution:** *If a new front cover is being installed on the engine, make sure the oil pressure relief valve supplied with the new cover is used. If the old style relief valve is installed in a new cover, oil pressure problems will result.*

8   Tighten the bolts to the torque listed in this Chapter's Specifications.

9   Run the engine and check for oil leaks.

---

14   Oil pump - removal, inspection and installation

Refer to illustrations 14.4, 14.10 and 14.11

Removal

1   Remove the oil filter (see Chapter 1).

2   Remove the oil filter adapter, pressure regulator valve and spring (see Section 13).

3   Remove the engine front cover (see Section 10).

4   Remove the oil pump cover bolts **(see illustration)**.

5   Lift out the cover and oil pump gears as an assembly.

Inspection

6   Clean the parts with solvent and dry them with compressed air (if available). **Warning:** *Wear eye protection!*

7   Inspect all components for wear and score marks. Replace any worn out or damaged parts.

8   Refer to Section 13 for pressure regulator valve information.

9   Reinstall the gears in the timing chain cover.

10   Measure the outer gear-to-housing clearance with a feeler gauge **(see illustration)**.

11   Measure the inner gear-to-outer gear clearance at several points **(see illustration)**.

12   Use a dial indicator or straightedge and feeler gauges to measure the gear end clearance (distance from the gear to the gasket surface of the cover). Feeler gauges can be

14.4 The oil pump cover is attached to the inside of the front cover - a T-30 Torx driver is required for removal of the screws

inserted under a straightedge (between the straightedge and the gears) held across the cover to determine the existing clearance.

13   Check for pump cover warpage by laying a precision straightedge across the cover and trying to slip a feeler gauge between the cover and straightedge.

14   Compare the measurements to this Chapter's Specifications. Replace all worn or damaged components with new ones.

Installation

15   Remove the gears and pack the pump cavity with petroleum jelly.

16   Install the gears - make sure petroleum jelly is forced into every cavity. Failure to do so could cause the pump to lose its prime when the engine is started, causing damage from lack of oil pressure.

17   Install the pump cover, using a new gasket only - its thickness is critical for maintaining the correct clearances.

18   Install the pressure regulator spring and valve.

19   Install the front cover.

20   Install the oil filter adapter and filter and check the oil level. Start and run the engine and check for correct oil pressure, then look carefully for oil leaks at the timing chain cover.

---

15   Oil pan - removal and installation

Refer to illustration 15.3

1   Disconnect the cable from the negative battery terminal. **Caution:** *On models equipped with a Delco Loc II or Theftlock audio system, be sure the lockout feature is turned off before performing any procedure which requires disconnecting the battery.*

2   Raise the vehicle and support it securely on jackstands. Drain the engine oil and replace the oil filter (refer to Chapter 1 if necessary).

3   Remove the driveplate inspection cover. Disconnect the electrical connector from the oil level sensor, and remove the oil level sensor and the oil dipstick **(see illustration)**. On 2000 and later models, disconnect the

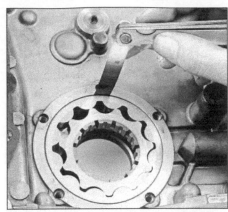

14.10 Measuring the outer gear-to-housing clearance with a feeler gauge

14.11 Measuring the inner gear-to-outer gear clearance with a feeler gauge

mounting bolts on the right shock absorber.

4   Support the engine from above with an engine hoist and raise the engine slightly so the engine mount through-bolts can be removed (see Section 19). Raise the engine. On 2000 and later models, it will be necessary to lower the front crossmember slightly for oil pan removal. Refer to Chapter 10 and disconnect the steering shaft from the rack-and-pinion, then remove the two right crossmember-to-frame bolts and loosen the two left crossmember-to-frame bolts.

5   Remove the oil pan mounting bolts and carefully separate the oil pan from the block.

15.3 Location of the oil level sensor

16.2 Remove the bolts and lower the pick-up tube and screen assembly

Don't pry between the block and the pan or damage to the sealing surfaces may result and oil leaks may develop. Instead, tap the pan with a soft-face hammer to break the gasket seal. Twist the pan while lowering the rear of the pan. **Caution:** *Watch the top-mounted engine components while raising the engine. If it looks like there will be interference, remove components such as the coil pack to prevent them hitting the edge of the cowl.*

6   Clean the pan with solvent and remove all old sealant and gasket material from the block and pan mating surfaces. Clean the mating surfaces with lacquer thinner or acetone and make sure the bolt holes in the block are clear. Check the oil pan flange for distortion, particularly around the bolt holes. If necessary, place the pan on a block of wood and use a hammer to flatten and restore the gasket surface.

7   Always use a new gasket whenever the oil pan is installed.

8   Place the oil pan in position on the block and install the bolts.

9   After the bolts are installed, tighten them to the torque listed in this Chapter's Specifications. Starting at the center, follow a crisscross pattern and work up to the final torque in three steps.

10   The remaining steps are the reverse of the removal procedure.

11   Refill the engine with oil. Run the engine until normal operating temperature is reached and check for leaks.

---

16   Oil pump pick-up tube and screen assembly - removal and installation

Refer to illustration 16.2

1   Remove the oil pan (see Section 15).

2   Unbolt the oil pump pick-up tube and screen assembly and detach it from the engine **(see illustration)**.

3   Clean the screen and housing assembly with solvent and dry it with compressed air, if available. **Warning:** *Wear eye protection.*

4   If the oil screen is damaged or has metal chips in it, replace it. An abundance of metal

18.3 The rear main oil seal is in a large housing on the back of the block which should be removed only for engine overhaul - the seal can be pried out with a screwdriver and installed either with a tool or tapped carefully in place with a large socket

chips indicates a major engine problem which must be corrected.

5   Make sure the mating surfaces of the pipe flange and the engine block are clean and free of nicks and install the screen assembly with a new gasket.

6   Install the oil pan (see Section 15).

7   Be sure to refill the engine with oil before starting it.

---

17   Flywheel/driveplate - removal and installation

This procedure is basically the same as it is for the 3.4L V6. Refer to Chapter 2, Part A and follow the procedure outlined there. However, use the bolt torque figure from this Chapter's Specifications.

---

18   Rear main oil seal - replacement

Refer to illustration 18.3

1   With the engine supported from above with an engine hoist, remove the transmission (see Chapter 7).

2   Remove the driveplate (see Section 17).

3   Using a thin screwdriver or seal removal tool, carefully remove the oil seal from the crankshaft rear oil seal housing **(see illustration)**. Be very careful not to damage the aluminum housing or the crankshaft surface while prying the seal out.

4   Clean the bore in the housing and the seal contact surface on the crankshaft. Check the seal contact surface on the crankshaft for scratches and nicks that could damage the new seal lip and cause oil leaks - if the crankshaft is damaged, the only alternative is a new or different crankshaft. Inspect the seal bore for nicks and scratches. Carefully smooth if with a fine file if necessary, but don't nick the crankshaft in the process.

5   A special tool is recommended to install

19.1 Right side engine mount, through-bolt and nut (arrow)

the new oil seal. Lubricate the lips of the seal with clean engine oil. Slide the seal onto the mandrel until the dust lip bottoms squarely against the collar of the tool. **Note:** *If the special tool isn't available, carefully work the seal lip over the crankshaft and tap it into place with a hammer and a large socket or section of pipe the correct diameter.*

6   Align the dowel pin on the tool with the dowel pin hole in the crankshaft and attach the tool to the crankshaft by hand-tightening the bolts.

7   Turn the tool handle until the collar bottoms against the case, seating the seal.

8   Loosen the tool handle and remove the bolts. Remove the tool.

9   Check the seal and make sure it's seated squarely in the bore.

10   Install the driveplate (see Section 17).

11   Install the transmission (see Chapter 7).

---

19   Engine mounts - check and replacement

Refer to illustrations 19.1 and 19.2

This procedure is basically the same as it is for the 3.4L V6. Refer to Part A of this Chapter and follow the procedure outlined there. However, use the bolt torque figure from this Chapter's Specifications and refer to **illustrations 19.1 and 19.2.**

19.2 Left side engine mount, through-bolt and nut (arrow)

Chapter 2 Part C
5.7L V8 engine - 1997 and earlier models

Contents

Specifications

General

Displacement	
5.7L V8 (1997 and earlier)	350 cubic inches
Bore and stroke	4.000 x 3.480 inches
Cylinder numbers (front-to-rear)	
Left (driver's) side	1-3-5-7
Right side	2-4-6-8
Firing order	1-8-4-3-6-5-7-2
Camshaft	
Bearing journal diameter	1.8677 to 1.8697 inches
Lobe lift	
Intake	0.298 inch
Exhaust	0.306 inch

Torque specifications

Ft-lbs (unless otherwise indicated)

Camshaft sprocket bolts	216 in-lbs
Camshaft retainer plate bolts	105 in-lbs
Crankshaft balancer bolts	63
Crankshaft balancer hub bolt	74

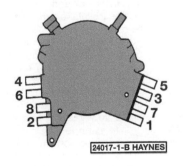

Cylinder and spark plug terminal locations

24017-1-B HAYNES

Torque specifications (continued)

	Ft-lbs (unless otherwise indicated)
Cylinder head bolts - in sequence (see illustration 12.17)	
1995 and earlier	65
1996 through 1997	
Step 1, all	22
Step 2, short bolts	Turn an additional 67-degrees
Step 2, medium and long bolts..	Turn an additional 80-degrees
Exhaust manifold bolts	30
Flywheel/driveplate	74
Intake manifold bolts	
Step 1	71 in-lbs
Step 2	35
Oil deflector nut	30
Oil pan-to-crankcase	
Corner bolt/nut	180 in-lbs
Side-rail bolts	106 in-lbs
Oil pump mounting bolt	65
Oil pump driveshaft bolt	156 in-lbs
Rear oil seal housing-to-block bolts.................................	132 in-lbs
Valve cover bolts	106 in-lbs
Rocker arm studs	50
Front cover bolts...	106 in-lbs

1   General information

Warning: *The models covered by this manual are equipped with airbags. Always disable the airbag system before working in the vicinity of the impact sensors, steering column or instrument panel to avoid the possibility of accidental deployment of the airbag(s), which could cause personal injury (see Chapter 12). The yellow wires and connectors routed through the instrument panel and, on 1995 and earlier models, to the front of the vehicle, are for this system. Do not use electrical test equipment on these yellow wires or tamper with them in any way.*
Caution: *On models equipped with a Delco Loc II or Theftlock audio system, be sure the lockout feature is turned off before performing any procedure which requires disconnecting the battery.*

This Part of Chapter 2 is devoted to in-vehicle repair procedures for the 1997 and earlier 5.7L V8 engine. All information concerning the 1998 and later 5.7L V8 engine is covered in Chapter 2D. All information concerning engine removal and installation and engine block and cylinder head overhaul for either V8 engine can be found in Part E of this Chapter.

Since the repair procedures included in this Part are based on the assumption the engine is still installed in the vehicle, if they are being used during a complete engine overhaul (with the engine already out of the vehicle and on a stand) many of the Steps included here will not apply.

The Specifications included in this Part of Chapter 2 apply only to the procedures found here. The specifications necessary for rebuilding the block and cylinder heads are included in Part E.

The V8 engine used in the 1997 and earlier Camaro and Firebirds displaces 350 cubic inches and is considered a "small-block" V8. The cast-iron block is of 90-degree V design, with aluminum cylinder heads and with the water pump and distributor both at the front of the engine, driven mechanically from the camshaft.

2   Repair operations possible with the engine in the vehicle

Many major repair operations can be accomplished without removing the engine from the vehicle.

Clean the engine compartment and the exterior of the engine with some type of pressure washer before any work is done. A clean engine will make the job easier and will help keep dirt out of the internal areas of the engine.

Depending on the components involved, it may be a good idea to remove the hood to improve access to the engine as repairs are performed (refer to Chapter 11 if necessary).

If oil or coolant leaks develop, indicating a need for gasket or seal replacement, the repairs can generally be made with the engine in the vehicle. The oil pan gasket, the cylinder head gaskets, intake and exhaust manifold gaskets, front cover gaskets and the crankshaft oil seals are accessible with the engine in place, but the rear third of the engine is "tucked" under the cowl, so procedures at that end of the engine can be quite difficult.

Exterior engine components, such as the water pump, the starter motor, the alternator, the distributor and the fuel injection unit, as well as the intake and exhaust manifolds, can be removed for repair with the engine in place.

Since the cylinder heads can be removed without pulling the engine, valve component servicing can also be accomplished with the engine in the vehicle.

Replacement of, repairs to or inspection of the timing chain and sprockets and the oil pump are all possible with the engine in place.

In extreme cases caused by a lack of necessary equipment, repair or replacement of piston rings, pistons, connecting rods and rod bearings is possible with the engine in the vehicle. However, this practice is not recommended because of the cleaning and preparation work that must be done to the components involved.

3   Top Dead Center (TDC) for number one piston - locating

1   Top Dead Center (TDC) is the highest point in the cylinder that each piston reaches as it travels up-and-down when the crankshaft turns. Each piston reaches TDC on the compression stroke and again on the exhaust stroke, but TDC generally refers to piston position on the compression stroke. The cast-in arrow on the crankshaft balancer installed on the front of the crankshaft is referenced to the number one piston at TDC when the arrow is straight up, or at "12 o'clock" (see Section 9).
2   Positioning the pistons at TDC is an essential part of many procedures such as rocker arm removal, valve adjustment, and timing chain and sprocket replacement.
3   In order to bring any piston to TDC, the crankshaft must be turned using one of the methods outlined below. When looking at the front of the engine, normal crankshaft rotation is clockwise. **Warning:** *Before beginning this procedure, be sure to place the transmission in Park or Neutral and disable the ignition system by disconnecting the primary wires from the coil (or by removing the IGNITION fuse from the underhood fuse block).*

a) *The preferred method is to turn the crankshaft with a large socket and breaker bar attached to the balancer hub bolt that is threaded into the front of the crankshaft.*
b) *A remote starter switch, which may save some time, can also be used. Attach the switch leads to the S (switch) and B (battery) terminals on the starter solenoid. Once the piston is close to TDC, use a socket and breaker bar as described in the previous paragraph.*

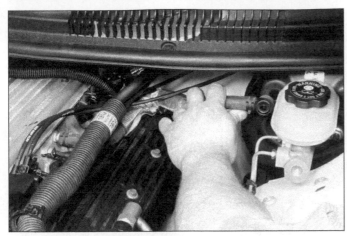

4.5a Remove the vacuum hose from the brake booster and the PCV hose from the left cover

4.5b Remove the air injection fitting from each exhaust manifold (left side shown) and swing it out of the way

c) *If an assistant is available to turn the ignition switch to the Start position in short bursts, you can get the piston close to TDC without a remote starter switch. Use a socket and breaker bar as described in Paragraph 3a to complete the procedure.*

4   Insert a compression gauge (screw-in type with a hose) in the number 1 spark plug hole and zero it. Place the gauge dial where you can see it while turning the balancer hub bolt.

5   Turn the crankshaft until the arrow on the balancer approaches straight up (or 12 o'clock). If you see compression building up on the gauge, you are on the compression stroke for number one. Stop turning when the arrow is straight up. If you did not see compression build up, go one more complete revolution to achieve TDC for number one.

6   After the number one piston has been positioned at TDC on the compression stroke, TDC for any of the remaining cylinders can be located by turning the crankshaft 90-degrees (1/4-turn) at a time and following the firing order (refer to the Specifications). For example, turning 90-degrees past number one TDC would give you TDC for number eight cylinder, the next in the firing order.

4   Valve covers - removal and installation

Refer to illustrations 4.5a, 4.5b and 4.6

Removal

1   Disconnect the negative cable from the battery. **Caution:** *On models equipped with a Delco Loc II or Theftlock audio system, be sure the lockout feature is turned off before performing any procedure which requires disconnecting the battery.*

2   Remove the serpentine drivebelt (see Chapter 1).

3   Disconnect the transmission dipstick tube from the bellhousing.

4   Remove the alternator (see Chapter 5).

5   Pull the PCV hose from the right valve cover, disconnect the brake booster vacuum hose (left side) and disconnect the air injection fitting from the each exhaust manifold **(see illustrations)**. **Note:** *Use penetrating oil on the air injection fittings and allow it to soak in before trying to remove the fittings.*

6   Remove the valve cover mounting bolts **(see illustration)**.

7   Remove the valve cover. **Note:** *If the cover is stuck to the head, bump the cover with a block of wood and a hammer to release it. If it still won't come loose, try to slip a flexible putty knife between the head and cover to break the seal. Do not pry at the cover-to-head joint or damage to the sealing surface and cover flange will result and oil leaks will develop.*

Installation

8   The mating surfaces of each cylinder head and valve cover must be perfectly clean when the covers are installed. Use a gasket scraper to remove all traces of sealant or old gasket, then clean the mating surfaces with lacquer thinner or acetone. If there's sealant or oil on the mating surfaces when the cover is installed, oil leaks may develop.

9   Make sure the threaded holes are clean. Run a tap into them to remove corrosion and restore damaged threads.

10   Mate the new gaskets to the covers before the covers are installed. Apply a thin coat of RTV sealant to the cover flange, then position the gasket inside the cover lip and allow the sealant to set up so the gasket adheres to the cover (if the sealant isn't allowed to set, the gasket may fall out of the cover as it's installed on the engine).

11   Carefully position the cover on the head and install the bolts.

12   Tighten the nuts/bolts in three steps to the torque listed in this Chapter's Specifications.

13   The remaining installation steps are the reverse of removal.

14   Start the engine and check carefully for oil leaks as the engine warms up.

4.6 Unscrew the four mounting bolts (arrows) and detach the cover

5   Rocker arms and pushrods - removal, inspection and installation

Refer to illustrations 5.4, 5.10, 5.11 and 5.13

Removal

Note: *Any valve train components being removed (rocker arms, pivot balls, pushrods or lifters) should be stored in marked containers so they can be returned to their original locations on assembly.*

1   Refer to Section 4 and detach the valve covers from the cylinder heads.

2   Beginning at the front of one cylinder head, loosen and remove the rocker arm stud nuts. **Note:** *If the pushrods are the only items being removed, loosen each nut just enough to allow the rocker arms to be rotated to the side so the pushrods can be lifted out.*

3   Lift off the rocker arms and pivot balls and store them in the marked containers with the nuts (they must be reinstalled in their original locations).

4   Remove the pushrods and store them in

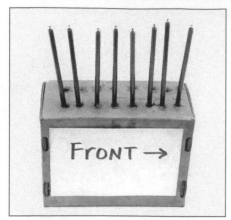

5.4 A perforated cardboard box can be used to store the pushrods to ensure that they're reinstalled in their original locations - note the label indicating the front of the engine

5.10 The ends of the pushrods and the valve stems should be lubricated with moly-base grease prior to installation of the rocker arms

5.11 Moly-base grease applied to the pivot balls will ensure adequate lubrication until oil pressure builds up when the engine is started

order to make sure they don't get mixed up during installation **(see illustration)**.

Inspection

5   Check each rocker arm for wear, cracks and other damage, especially where the pushrods and valve stems contact the rocker arm faces.

6   Make sure the hole at the pushrod end of each rocker arm is open.

7   Check each rocker arm pivot area for wear, cracks and galling. If the rocker arms are worn or damaged, replace them with new ones and use new pivot balls as well.

8   Inspect the pushrods for cracks and excessive wear at the ends. Roll each pushrod across a piece of plate glass to see if it's bent (if it wobbles, it's bent).

Installation

9   Lubricate the lower end of each pushrod with clean engine oil or moly-base grease and install them in their original locations. Make sure each pushrod seats completely in the lifter socket.

10   Apply moly-base grease to the ends of the valve stems and the upper ends of the pushrods before positioning the rocker arms over the studs **(see illustration)**.

11   Set the rocker arms in place, then install the pivot balls and nuts. Apply moly-base grease to the pivot balls to prevent damage to the mating surfaces before engine oil pressure builds up. Be sure to install each nut with the flat side against the pivot ball **(see illustration)**.

Valve adjustment

12   Refer to Section 3 and bring the number one piston to TDC on the compression stroke.

13   Tighten the rocker arm nuts (number one cylinder only) until all play is removed at the pushrods. This can be determined by wiggling the pushrod up-and-down as the nut is tightened **(see illustration)**. When you can no longer feel movement, the play is taken up.

5.13 When all play is removed from the pushrod, tighten the nut an additional 3/4-turn

14   Tighten each nut an additional 3/4-turn to center the lifter plungers in their travel. Valve adjustment for cylinder number one is now complete. A cylinder number illustration and the firing order is included in the Specifications.

15   You can also adjust the number two, five and seven intake valves and the number three, four and eight exhaust valves at this time. After adjusting these valves, turn the crankshaft one complete revolution (360-degrees) and adjust the number three, four, six and eight intake valves and the number two, five, six and seven exhaust valves.

16   Refer to Section 4 and install the valve covers. Start the engine, listen for unusual valve train noises and check for oil leaks at the valve cover joints.

---

6   Valve springs, retainers and seals - replacement

Refer to illustrations 6.4, 6.7 and 6.16
Note: *Broken valve springs and defective valve stem seals can be replaced without removing the cylinder head. Two special tools*

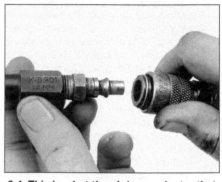

6.4 This is what the air hose adapter that threads into the spark plug hole looks like - they're commonly available from auto parts stores

and a compressed air source are normally required to perform this operation, so read through this Section carefully and rent or buy the tools before beginning the job.

1   Refer to Section 4 and remove the valve cover from the affected cylinder head. If all of the valve stem seals are being replaced, remove both valve covers.

2   Remove the spark plug from the cylinder which has the defective component. If all of the valve stem seals are being replaced, all of the spark plugs should be removed.

3   Turn the crankshaft until the piston in the affected cylinder is at top dead center on the compression stroke (refer to Section 3 for instructions). If you're replacing all of the valve stem seals, begin with cylinder number one and work on the valves for one cylinder at a time. Move from cylinder-to-cylinder following the firing order sequence (see the specifications).

4   Thread an adapter into the spark plug hole and connect an air hose from a compressed air source to it **(see illustration)**. Most auto parts stores can supply an air hose adapter. **Note:** *Many cylinder compression gauges utilize a screw-in fitting that may work with your air hose quick-disconnect fitting.*

5   Remove the nut, pivot ball and rocker arm for the valve with the defective part and

6.7 A screw-type spring compressor must be used in the limited space under the cowl - use a magnet or small pliers to remove the keepers

6.16 Apply a small dab of grease to each keeper as shown here before installation - it'll hold them in place on the valve stem as the spring is released

7.3 Remove the alternator brace from the intake manifold, and remember the location of its stud (arrow)

7.4 Disconnect the accelerator and cruise control cables and the ESC module, then remove the module bracket from the side of the manifold (arrow)

7.7 Remove the bolts and detach the EGR pipe from the back of the intake manifold

pull out the pushrod. If all of the valve stem seals are being replaced, all of the rocker arms and pushrods should be removed (refer to Section 5).

6   Apply compressed air to the cylinder. The valves should be held in place by the air pressure. If the valve faces or seats are in poor condition, leaks may prevent the air pressure from retaining the valves.

7   Stuff shop rags into the cylinder head holes above and below the valves to prevent parts and tools from falling into the engine, then use a valve spring compressor to compress the spring/balancer assembly. Remove the keepers with a pair of small needle-nose pliers or a magnet **(see illustration)**. **Note:** *Because the rear cylinders are under the cowl, a lever-type spring compressor will not work - the clamp-type must be used.*

8   Remove the spring retainer and valve spring assembly with the compressor still attached, then remove the valve stem seal. **Note:** *If air pressure fails to hold the valve in the closed position during this operation, the valve face or seat is probably damaged. If so, the cylinder head will have to be removed for additional repair operations.*

9   Wrap a rubber band or tape around the top of the valve stem so the valve will not fall into the combustion chamber, then release the air pressure. **Note:** *If a rope was used instead of air pressure, turn the crankshaft slightly in the direction opposite normal rotation.*

10   Inspect the valve stem for damage. Rotate the valve in the guide and check the end for eccentric movement, which would indicate that the valve is bent.

11   Move the valve up-and-down in the guide and make sure it doesn't bind. If the valve stem binds, either the valve is bent or the guide is damaged. In either case, the head will have to be removed for repair.

12   Inspect the rocker arm studs for wear. Worn studs should be replaced. Be sure to replace the guide plate if the studs are removed and reinstalled, and use gasket sealer on the studs when threading them into

the head. See this Chapter's Specifications for proper torque.

13   Reapply air pressure to the cylinder to retain the valve in the closed position, then remove the tape or rubber band from the valve stem.

14   Lubricate the valve stem with engine oil and install a new oil seal of the type originally used on the engine.

15   Install the spring and retainer (still in the tool) in position over the valve.

16   Position the keepers in the upper groove. Apply a small dab of grease to the inside of each keeper to hold it in place if necessary **(see illustration)**. Remove the pressure from the spring tool and make sure the keepers are seated.

17   Disconnect the air hose and remove the adapter from the spark plug hole.

18   Refer to Section 5 and install the rocker arms and pushrods.

19   Install the spark plugs and hook up the wires.

20   Refer to Section 4 and install the valve covers.

21   Start and run the engine, then check for oil leaks and unusual sounds coming from the valve cover area.

7   Intake manifold - removal and installation

Removal

Refer to illustrations 7.3, 7.4, 7.7 and 7.8

1   Disconnect the negative cable from the battery, then refer to Chapter 1 and drain the cooling system. **Caution:** *On models equipped with a Delco Loc II or Theftlock audio system, be sure the lockout feature is turned off before performing any procedure which requires disconnecting the battery.*

2   Remove the air intake duct, fuel rails and injectors (see Chapter 4).

3   Remove the alternator rear brace **(see illustration)**.

4   Unbolt and remove the accelerator cable bracket and cable, and the cruise control cable adjuster **(see illustration)**.

5   Label and then disconnect any fuel lines, wires and vacuum hoses from the vehicle to the intake manifold. Lay the wiring harnesses on each side away from the manifold.

6   Disconnect the coolant hoses from the throttle body.

7   Unbolt the EGR pipe from the back of the intake manifold **(see illustration)**.

8   Loosen   the   manifold   mounting

7.8 Pry up at a corner (arrow) of the manifold, being careful not to gouge the soft aluminum in a gasket sealing area

7.9 After covering the lifter valley, use a gasket scraper to remove all traces of sealant and old gasket material from the head and manifold mating surfaces

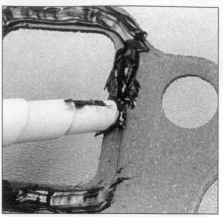

7.11 RTV sealant should be used around the coolant passage holes in the new intake manifold gaskets

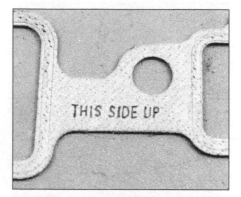

7.12 Be sure to install the gaskets with the marks UP!

bolts/studs in 1/4-turn increments until they can be removed by hand. **Note:** *Mark the manifold to indicate the location of the four studs.* The manifold will probably be stuck to the cylinder heads and force may be required to break the gasket seal. A large prybar can be positioned under the cast-in lug near the thermostat housing to pry up the front of the manifold **(see illustration)**. **Caution:** *Do not pry between the block and manifold or the heads and manifold or damage to the gasket*

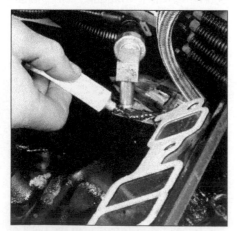

7.13 Apply a bead of sealant to the ends (ridges) of the block

sealing surfaces may result and vacuum leaks could develop.

Installation

Refer to illustrations 7.9, 7.11, 7.12, 7.13 and 7.16

Note: *The mating surfaces of the cylinder heads, block and manifold must be perfectly clean when the manifold is installed. Gasket-removal solvents are available at most auto parts stores and may be helpful when removing old gasket material that is stuck to the heads and manifold. Be sure to follow the directions printed on the container.*

9   Use a gasket scraper to remove all traces of sealant and old gasket material, then wipe the mating surfaces with a cloth saturated with lacquer thinner or acetone. If there is old sealant or oil on the mating surfaces when the manifold is installed, oil or vacuum leaks may develop. Cover the lifter valley with shop rags to keep debris out of the engine **(see illustration)**. Use a vacuum cleaner to remove any gasket material that falls into the intake ports in the heads.

10   Use a tap of the correct size to chase the threads in the bolt holes, then use compressed air (if available) to remove the debris from the holes. **Warning:** *Wear safety glasses*

or a face shield to protect your eyes when using compressed air.

11   Apply a thin coat of RTV sealant **(see illustration)** around the coolant passage holes on the cylinder head side of the new intake manifold gaskets (there is normally one hole at each end).

12   Position the gaskets on the cylinder heads. Make sure all intake port openings, coolant passage holes and bolt holes are aligned correctly and the THIS SIDE UP is visible **(see illustration)**.

13   Apply a bead of RTV sealant at the ends of the block. Apply a 3/16-inch bead of sealant to the front and rear of the block as shown **(see illustration)**. Extend the bead 1/2-inch up each cylinder head to seal and retain the gaskets. Refer to the instructions with the gasket set for further information.

14   Carefully set the manifold in place. Do not disturb the gaskets and do not move the manifold fore-and-aft after it contacts the front and rear seals.

15   Apply a thin coat of a non-hardening sealant to the manifold bolt threads, then install the bolts.

16   While the sealant is still wet, tighten the bolts to the initial torque listed in this Chapter's Specifications following the recom-

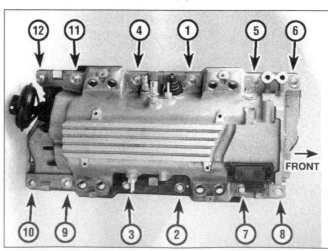

7.16 Intake manifold bolt TIGHTENING sequence

8.4 Unbolt the exhaust pipe-to-manifold flanges (arrow)

8.6 Use care not to round off the fittings when removing air injection pipes (arrow) from the manifolds

9.4 Use a two-pin spanner to hold the balancer while the three balancer bolts are removed (arrows) note the cast-in TDC arrow on the balancer (at top, TDC location)

mended sequence **(see illustration)**, then tighten them to the final torque listed in this Chapter's Specifications, in sequence.
17   The remaining installation steps are the reverse of removal. Start the engine and check carefully for oil, vacuum and coolant leaks at the intake manifold joints.

8   Exhaust manifolds - removal and installation

Refer to illustrations 8.4 and 8.6
Warning: *Allow the engine to cool completely before following this procedure.*

Removal

1   Disconnect the negative cable from the battery. **Caution:** *On models equipped with a Delco Loc II or Theftlock audio system, be sure the lockout feature is turned off before performing any procedure which requires disconnecting the battery.*
2   Set the parking brake and block the rear wheels. Raise the front of the vehicle and support it securely on jackstands. Disconnect the exhaust pipes from the manifolds (see Chapter 4). **Note:** *Penetrating oil is usually required to remove frozen exhaust pipe nuts. Don't apply excessive force to frozen (stuck) nuts - you could shear off the exhaust manifold studs.*
3   Disconnect the electrical connectors from the oxygen sensor in each manifold.
4   Unbolt the exhaust pipe from each exhaust manifold **(see illustration)**, then remove the rearmost three bolts/studs from each manifold and unplug the spark plug boots. Lower the vehicle.

Right manifold

5   Remove the serpentine drivebelt belt (see Chapter 1), and remove the alternator and lower brace (see Chapter 5).
6   Disconnect the air injection pipe from the manifold **(see illustration)**. **Note:** *Use penetrating oil on the air injection fittings and allow it to soak in before trying to remove the fittings.*

7   Remove the remaining manifold bolts/nuts and pull out the manifold.

Left manifold

8   Disconnect the air injection pipe from the manifold (see Section 4). **Note:** *Use penetrating oil on the air injection fittings and allow it to soak in before trying to remove the fittings.*
9   Remove the brake booster vacuum hose (see Section 4).
10   Remove the remaining manifold bolts/nuts and pull out the manifold.

Installation

11   Installation is basically the reverse of the removal procedure. Clean the manifold and head gasket surfaces to remove old gasket material, then install new gaskets. Do not use any gasket cement or sealant on exhaust system gaskets. **Note:** *The exhaust manifold gaskets are sheetmetal and should only go on one way; if put on incorrectly, stamped writing on the gasket will read "installed wrong."* Apply anti-seize compound to the exhaust manifold-to-exhaust pipe studs.
12   Install all the manifold bolts and tighten them to the torque listed in this Chapter's Specifications. Work from the center out and approach the final torque in three steps.

9   Crankshaft front oil seal - replacement

Refer to illustrations 9.4, 9.6a, 9.6b, 9.7 and 9.9
1   Disconnect the negative cable from the battery. **Caution:** *On models equipped with a Delco Loc II or Theftlock audio system, be sure the lockout feature is turned off before performing any procedure which requires disconnecting the battery.* **Note:** *Unless you have a factory puller for the balancer hub, this procedure will be time consuming, as the distributor and water pump may have to be removed (see Step 4).*

2   Remove the serpentine drivebelt belt (see Chapter 1).
3   Raise the vehicle and support it securely on jackstands.
4   Remove the three balancer bolts and the balancer **(see illustration)**. A puller is not required to remove the balancer.
5   Apply matchmarks on the balancer hub and the front cover, and remove the balancer hub bolt. **Caution:** *Once the matchmarks have been applied, do not turn the crankshaft until the balancer has been reinstalled.* When loosening this bolt, prevent the crankshaft from turning by wedging a large screwdriver into the ring gear teeth of the flywheel/driveplate.
6   Use a bolt-type puller to remove the balancer hub **(see illustrations)**. Because the

9.6a The portion of the crank snout exposed inside the hub is too small for a standard puller's beveled tip to fit into, so use a long Allen bolt that fits inside the crank threads (not threaded into them, but smaller) - the puller will then press against the socket-head of the Allen bolt - arrows indicate the TDC mark on the hub and the cover rib it must align with, since the hub has no keyway

9.6b Most standard pullers will have to be adapted with four 5/16-inch bolts and nuts to attach to the balancer hub, with the puller's beveled tip fitting into the Allen bolt

9.7 Carefully pry the old seal out of the front cover - don't damage the crank surface

9.9 Use a special seal-installing tool or drive the seal in carefully and evenly with a large socket and hammer

10.5a On spline-type drives, paint a match mark to line up the splined shaft sticking out of the front cover for faster assembly, although it will only go back into the cover one way

24017-2C-10.5b HAYNES

10.5b The later-type distributor drive uses a long pin on the camshaft which fits into a slot (1) on the back of the distributor - align it with the mark (2) for TDC assembly

bolt holes in the hub are larger than most standard pullers will allow, you will have to use three, four-inch-long (grade 8) 5/16-inch bolts and nuts to use a standard puller, along with a four-inch-long (grade 8) 5/16-inch Allen bolt to fit into the crank snout. Because of the clearance around the distributor, it is safest to remove the water pump (see Chapter 3) and distributor (see Chapter 5) first, which also exposes the portion of the front cover with the "TDC" rib for aligning the keyless hub when reinstalling it.

7   Carefully pry the seal out of the cover with a seal removal tool or a large screwdriver **(see illustration)**. Be careful not to distort the cover or scratch the wall of the seal bore.

8   Clean the bore to remove any old seal material and corrosion. Position the new seal in the bore with the open end of the seal facing IN. A small amount of oil applied to the outer edge of the new seal will make installation easier - don't overdo it!

9   Drive the seal into the bore with a special tool or a large socket and hammer until it's completely seated **(see illustration)**. Select a socket that's the same outside

diameter as the seal.

10   Lubricate the seal lips with engine oil and reinstall the balancer hub, lining up the matchmarks. The remaining installation steps are the reverse of removal. **Note:** *If the balancer is being replaced, be sure to transfer the bolt-on balancer weights from the old balancer to the same positions on the new balancer.*

10 Front cover, timing chain and sprockets - removal, inspection and installation

Removal

Refer to illustrations 10.5a, 10.5b, 10.6 and 10.8

1   Refer to Chapter 3 and remove the water pump.

2   Remove the crankshaft balancer and balancer hub (see Section 9).

3   Refer to Section 3 and position the number one piston at TDC on the compression

stroke. **Caution:** *Once this has been done, do not turn the crankshaft until the timing chain and sprockets have been reinstalled.*

4   The oil pan bolts will have to be loosened and the pan lowered slightly for the front cover to be removed. If the pan has been in place for an extended period of time it's likely the pan gasket will break when the pan is lowered. In this case the pan should be removed and a new gasket installed (see Section 13).

5   Remove the distributor (see Chapter 5). There are two types of distributor drives on the V8 engines: the spline-type (type 1) and the pin-type (type 2) used on 1995 and later models. On the spline-type, the distributor is driven by a splined shaft coming out of the front cover and driven by the camshaft gear. On these models, the short splined shaft may come out of the front cover when the distributor is removed. It will only go back in one way and the distributor will only go back onto it one way, due to key alignment splines. These distributors should not be turned while off the engine. On the later models, the camshaft has a long pin sticking out, which

10.6 Remove the front cover bolts and remove the cover

10.8 Unbolt and remove the camshaft sprocket, lowering it to remove the sprocket and chain as an assembly

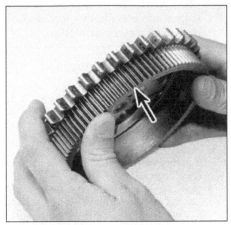

10.11 Examine the camshaft sprocket teeth and the integral water pump drive gear teeth (arrow) on the back for wear

meshes with a slot in the backside of the distributor. Alignment of these types is simpler **(see illustrations)**.

6   Unbolt and remove the front cover **(see illustration)**.

7   On 1996 models, remove the crankshaft position sensor reluctor ring, which is a disc over the crankshaft snout, noting which side faces the engine.

8   Remove the three bolts from the end of the camshaft, then detach the camshaft sprocket and chain as an assembly **(see illustration)**.

9   Using a jaw-type puller, remove the crankshaft sprocket.

Inspection

Refer to illustration 10.11

10   The timing chain should be replaced with a new one if the engine has high mileage, the chain has visible damage, or total freeplay midway between the sprockets exceeds one-inch. Failure to replace a worn timing chain may result in erratic engine performance, loss of power and decreased fuel mileage. Loose chains can "jump" timing. In the worst case, chain "jumping" or breakage will result in severe engine damage.

11   Inspect the timing sprockets and water-pump drive gear (on the back of the cam gear) for worn, non-concentric teeth or wear in the valleys between the teeth **(see illustration)**.

Installation

Refer to illustration 10.15

12   Use a gasket scraper to remove all traces of old gasket material and sealant from the cover and engine block. Stuff a shop rag into the opening at the front of the oil pan to keep debris out of the engine. Clean the cover and block sealing surfaces with lacquer thinner or acetone. **Caution:** *Be careful when scraping the front cover; gouges in the soft aluminum could result in oil leaks.*

13   Replace the O-ring on the outer end of the water-pump driveshaft (see Chapter 3).

14   The timing chain must be replaced as a

set with the camshaft and crankshaft sprockets. Never put a new chain on old sprockets. Align the sprocket with the Woodruff key and press the sprocket onto the crankshaft with a large socket or tap it gently into place until it is completely seated. **Caution:** *If resistance is encountered, do not hammer the sprocket onto the crankshaft. It may eventually move onto the shaft, but it may be cracked in the process and fail later, causing extensive engine damage.*

15   Loop the new chain over the camshaft sprocket, then turn the sprocket until the timing mark is in the 6 o'clock position **(see illustration)**. Mesh the chain with the crankshaft sprocket and position the camshaft sprocket on the end of the cam. If necessary, turn the camshaft so the dowel pin fits into the sprocket hole with the timing mark in the 6 o'clock position. **Note:** *If necessary, turn the water-pump driveshaft slightly to mesh its gear with the water-pump drive gear on the back of the camshaft sprocket.*

16   Apply non-hardening thread locking compound to the camshaft sprocket bolt threads, then install and tighten them to the torque listed in this Chapter's Specifications. Lubricate the chain with clean engine oil.

17   The one-piece oil pan gasket should be checked for cracks and deformation before installing the front cover. If the gasket has deteriorated it must be replaced before reinstalling the cover.

18   Apply a thin layer of RTV sealant to both sides of the new cover gasket, then position it on the engine block. The dowel pins and sealant will hold it in place. **Caution:** *Wrap some plastic tape around the splines on the water-pump driveshaft to protect the cover seal during installation.*

19   Install the front cover on the block, tightening the bolts finger-tight.

20   If the oil pan was removed, reinstall it (see Section 13). If it was only loosened, tighten the oil pan bolts, bringing the oil pan up against the front cover.

21   Tighten the front cover bolts to the torque listed in this Chapter's Specifications.

22   Lubricate the oil seal contact surface of the crankshaft balancer hub with multi-purpose grease or clean engine oil, then install it on the end of the crankshaft. The keyway in the balancer hub must be aligned with the Woodruff key in the crankshaft nose. If the hub cannot be seated by hand, slip a large washer over the bolt, install the bolt and tighten it to push the hub into place. Remove the large washer and tighten the bolt to the torque listed in this Chapter's Specifications.

23   If plastic tape was used to install the water-pump-driveshaft O-ring (see Chapter 3) remove the plastic tape from the water-pump driveshaft before installing the water pump. The remaining installation steps are the reverse of removal. **Caution:** *When reinstalling spline-type distributors, the distributor must go on smoothly and without much effort. It is **possible** to install the spline-type with improper alignment and force it on and tighten the bolts, but this will cause expensive damage to the parts. Wiggle the distributor slightly until you feel the spline actually engage fully before bolting down the distributor.*

10.15 Align the timing marks (arrows) before bolting the camshaft sprocket in place - larger arrow indicates the splined distributor driveshaft in place on the cam gear

11 Camshaft and lifters - removal, inspection and installation

Camshaft lobe lift check

Refer to illustration 11.3

1   To determine the extent of cam lobe wear, the lobe lift should be checked prior to camshaft removal. Refer to Section 4 and remove the valve covers.

2   Position the number one piston at TDC on the compression stroke (see Section 3).

3   Beginning with the number one cylinder, mount a dial indicator on the engine and position the plunger against the top surface of the first rocker arm. The plunger should be directly above and in-line with the pushrod **(see illustration)**.

4   Zero the dial indicator, then very slowly turn the crankshaft in the normal direction of rotation until the indicator needle stops and begins to move in the opposite direction. The point at which it stops indicates maximum cam lobe lift.

5   Record this figure for future reference, then reposition the piston at TDC on the compression stroke.

6   Move the dial indicator to the other number one cylinder rocker arm and repeat the check. Be sure to record the results for each valve.

7   Repeat the check for the remaining valves. Since each piston must be at TDC on the compression stroke for this procedure, work from cylinder-to-cylinder following the firing order sequence. Turn the crankshaft 90-degrees to reach TDC for each of the cylinders after Number 1.

8   After the check is complete, compare the results to the Chapter 2E Specifications. If camshaft lobe lift is less than specified, cam lobe wear has occurred and a new camshaft should be installed.

Removal

Refer to illustrations 11.10, 11.11 and 11.12

9   Refer to the appropriate Sections and remove the intake manifold, the rocker arms,

11.3 When checking the camshaft lobe lift, the dial indicator plunger must be positioned directly above and in-line with the pushrod

the pushrods and the timing chain and camshaft sprocket. The radiator must be removed as well (see Chapter 3). **Note:** *If the vehicle is equipped with air conditioning it will be necessary to remove the air conditioning condenser. If the condenser must be removed, the system must first be evacuated by a dealer service department or service station, and the refrigerant recovered. Do not disconnect any air conditioning lines until the system has been properly evacuated.*

10   In the engine's lifter valley, remove the bolts securing the lifter guide retainer and take out the retainer **(see illustration)**. Remove the lifter guides (one guide for every pair of lifters) and keep them in a clearly labeled box to ensure that they are reinstalled in their original locations, along with their corresponding lifters.

11   There are several ways to extract the lifters from the bores. A special tool designed to grip and remove lifters is manufactured by many tool companies and is widely available, but it may not be required in every case **(see illustration)**. On newer engines without a lot of varnish buildup, the lifters can often be removed with a small magnet or even with

11.10 Remove the three bolts holding the lifter guide retainer to the block

your fingers. A machinist's scribe with a bent end can be used to pull the lifters out by positioning the point under the retainer ring inside the top of each lifter. **Caution:** *Do not use pliers to remove the lifters unless you intend to replace them with new ones (along with the camshaft). The pliers will damage the precision machined and hardened lifters, rendering them useless.*

12   Remove the camshaft retainer plate **(see illustration)**. Thread six-inch long 5/16 - 18 bolts into the camshaft sprocket bolt holes to use as a handle when removing the camshaft from the block.

13   Remove the oil pump drive from inside the lifter valley at the right rear and carefully pull the camshaft out. Support the cam near the block so the lobes do not nick or gouge the bearings as the cam is withdrawn.

Inspection

Refer to illustrations 11.15 and 11.17

Camshaft and bearings

14   After the camshaft has been removed from the engine, cleaned with solvent and dried, inspect the bearing journals for uneven wear, pitting and evidence of seizure. If the

11.11 A lifter removal tool may be necessary with high-mileage components - store the lifters in an organized container

11.12 Using a Torx bit, unbolt the camshaft retainer plate

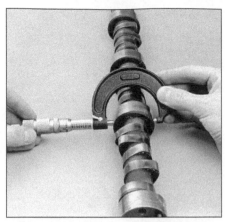

11.15 The camshaft bearing journal diameter is checked to pinpoint excessive wear and out-of-round conditions

11.17 The rollers on the lifters must turn freely - check for wear and excessive play as well

11.19 Coat the lobes and journals with special camshaft installation lube

journals are damaged, the bearing inserts in the block are probably damaged as well. Both the camshaft and bearings will have to be replaced. Replacement of the camshaft bearings requires special tools and techniques which place it beyond the scope of the home mechanic. The block will have to be removed from the vehicle and taken to an automotive machine shop for this procedure.

15 Measure the bearing journals with a micrometer to determine if they are excessively worn or out-of-round **(see illustration)**.

16 Check the camshaft lobes for heat discoloration, score marks, chipped areas, pitting and uneven wear. If the lobes are in good condition and if the lobe lift measurements are as specified, the camshaft can be re-used.

17 Check the lifter rollers carefully for wear and damage and make sure they turn freely without excessive play **(see illustration)**.

18 Used roller lifters can be reinstalled with a new camshaft and the original camshaft can be used if new lifters are installed, but the factory does recommend replacing both lifters and cam at the same time, especially if the engine has high mileage.

Installation

Refer to illustration 11.19

19 Lubricate the camshaft bearing journals and cam lobes with camshaft installation

lubricant **(see illustration)**.

20 Slide the camshaft into the engine. Support the cam near the block and be careful not to scrape or nick the bearings.

21 Position the camshaft retainer plate and tighten the bolts to the torque listed in this Chapter's Specifications.

22 Refer to Section 10 and install the timing chain and sprockets.

23 Lubricate the lifters with clean engine oil and install them in the block. If the original lifters are being reinstalled, be sure to return them to their original locations. If a new camshaft was installed, be sure to install new lifters as well.

24 Reinstall the oil pump drive at the back of the lifter valley and tighten it's bolt. The remaining installation steps are the reverse of removal.

25 Before starting and running the engine, change the oil and install a new oil filter (see Chapter 1).

12 Cylinder heads - removal and installation

Refer to illustrations 12.6, 12.7 and 12.17

Removal

1 Refer to Section 4 and remove the valve covers.

2 Refer to Section 7 and remove the intake manifold. Note that the cooling system must be drained (see Chapter 1) to prevent coolant from getting into internal areas of the engine when the manifold and heads are removed.

3 Refer to Section 8 and detach both exhaust manifolds.

4 Refer to Section 5 and remove the rocker arms and pushrods.

5 Remove the ignition coil (see Chapter 5), spark plugs and disconnect the spark plug wire harnesses from their clips.

6 At the rear of the cylinder heads there is a tubular steel water pipe connecting the back of both heads **(see illustration)**. It is retained to each head with a "banjo" bolt, which can be difficult to reach, as they are very close to the firewall. However, it is important to get a firm purchase on these bolts when removing them, to avoid rounding them off.

7 Loosen the head bolts in 1/4-turn increments until they can be removed by hand. Work from bolt-to-bolt in a pattern that's the reverse of the tightening sequence **(see illustration 12.17)**. Note: *Don't overlook the row of bolts on the lower edge of each head, near the spark plug holes. Store the bolts in the cardboard holder as they're removed. This*

12.6 Remove these two banjo bolts (arrows) holding this water transfer tube to the back of the heads (shown removed from the vehicle for clarity)

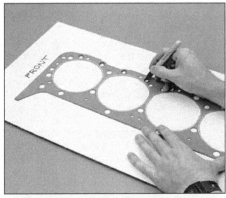

12.7 To avoid mixing up the head bolts, use a new gasket to transfer the bolt hole pattern to a piece of cardboard, then punch holes to accept the bolts

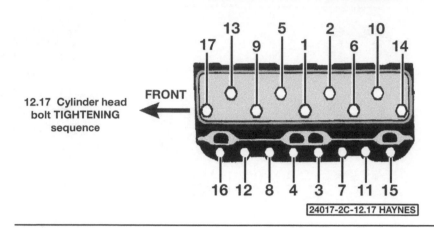

12.17 Cylinder head bolt TIGHTENING sequence

FRONT

24017-2C-12.17 HAYNES

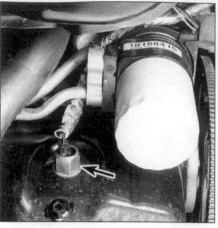

13.8 Disconnect the electrical connection from the oil level sensor (arrow), then remove the sensor - the lower bellhousing cover has been removed here also

will ensure the bolts are reinstalled in their original holes **(see illustration)**.

8   Lift the heads off the engine. If resistance is felt, do not pry between the head and block as damage to the mating surfaces will result. To dislodge the head, place a block of wood against the end of it and strike the wood block with a hammer. Store the heads on blocks of wood to prevent damage to the gasket sealing surfaces.

9   Cylinder head disassembly and inspection procedures are covered in detail in Chapter 2, Part E.

Installation

10   The mating surfaces of the cylinder heads and block must be perfectly clean when the heads are installed.

11   Use a gasket scraper to remove all traces of carbon and old gasket material, then clean the mating surfaces with lacquer thinner or acetone. If there's oil on the mating surfaces when the heads are installed, the gaskets may not seal correctly and leaks could develop. When working on the block, cover the lifter valley with shop rags to keep debris out of the engine. Use a vacuum cleaner to remove any debris that falls into the cylinders.

12   Check the block and head mating surfaces for nicks, deep scratches and other damage. If damage is slight, it can be removed with a file. If it's excessive, machining may be the only alternative.

13   Use a tap of the correct size to chase the threads in the head bolt holes in the block. Mount each bolt in a vise and run a die down the threads to remove corrosion and restore the threads. Dirt, corrosion, sealant and damaged threads will affect torque readings.

14   Position the new gaskets over the dowel pins in the block. The yellow tabs on each should be UP.

15   Carefully position the heads on the block without disturbing the gaskets.

16   Before installing the head bolts, coat the threads with a non-hardening sealant such as Permatex no. 2, and paint the heads of the shorter bolts to identify them for proper torquing.

17   Install the bolts in their original locations and tighten them finger-tight. Following the recommended sequence, tighten the bolts in several steps to the torque listed in this Chapter's Specifications **(see illustration)**.

18   The remaining installation steps are the reverse of removal.

13   Oil pan - removal and installation

Refer to illustrations 13.8 and 13.13

Removal

1   Disconnect the negative cable from the battery. **Caution:** *On models equipped with a Delco Loc II or Theftlock audio system, be sure the lockout feature is turned off before performing any procedure which requires disconnecting the battery.*

2   Raise the vehicle and support it securely on jackstands.

3   Drain the engine oil and remove the oil filter (see Chapter 1) and the oil dipstick.

4   Unbolt the catalytic converter from the left exhaust manifold and the flanged joint under the transmission bellhousing (see Chapter 4).

5   Remove the lower bellhousing cover.

6   Remove the starter, if necessary for clearance (see Chapter 5).

7   Remove the transmission fluid lines from the clip at the oil pan.

8   Disconnect the electrical connector from the oil level sensor in the pan, and remove the sensor **(see illustration)**. **Note:** *The sensor can be damaged if left in the pan during pan removal.*

9   Attach an engine hoist to the engine from above and raise the engine slightly to allow removal of the engine mount throughbolts (see Section 17).

10   Raise the engine approximately three inches. **Caution:** *When lifting the engine, check that no upper engine components or wire harnesses are hitting the edge of the cowl.*

11   Place blocks of wood between the crossmember and the engine block in the area of the motor mounts to hold the engine in the raised position.

12   Turn the crankshaft until the timing arrow on the balancer is at the bottom.

13   Remove the oil pan bolts **(see illustration)** and reinforcements. Note that studs and nuts are used in some positions, and mark their locations.

14   Remove the pan by tilting the rear down and working it away from the crankshaft throws, oil pump pick-up and front crossmember.

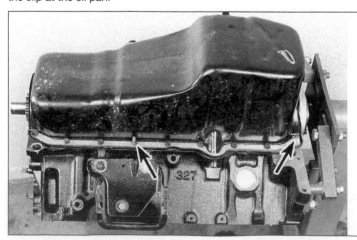

13.13 When removing the oil pan bolts/nuts, keep track of where the studs go (two are shown here with arrows) - do not forget the metal reinforcement strips (between the bolts/nuts and the pan) during reassembly

14.2 Remove the single bolt holding the oil pump to the rear main cap and remove the oil pump

14.3 Make sure the nylon sleeve is in place between the oil pump and driveshaft

Installation

15 Clean the sealing surfaces with lacquer thinner or acetone. Make sure the bolt holes in the block are clean.

16 Check the oil pan flange for distortion, particularly around the bolt holes. If necessary, place the pan on a block of wood and use a hammer to flatten and restore the gasket surface.

17 Apply a one-inch long bead of RTV sealant to the corners of the timing-cover-to-block and rear-oil-seal-housing-to-block junctions.

18 The one-piece rubber gasket should be checked carefully and replaced with a new one if damage is noted. Apply a small amount of RTV sealant to the corners of the semi-circular cutouts at both ends of the pan, then attach the rubber gasket to the pan.

19 Carefully position the pan against the block and install the bolts/nuts finger-tight (don't forget the reinforcement strips). Make sure the seals and gaskets haven't shifted, then tighten the bolts/nuts in three steps to the torque listed in this Chapter's Specifications. Start at the center of the pan and work

out toward the ends in a spiral pattern.

20 The remaining steps are the reverse of removal. **Caution:** *Don't forget to refill the engine with oil before starting it* (see Chapter 1).

14 Oil pump - removal and installation

Refer to illustrations 14.2 and 14.3

1 Remove the oil pan as described in Section 13.

2 While supporting the oil pump, remove the pump-to-rear main bearing cap bolt **(see illustration)**.

3 Lower the pump and remove it along with the pump driveshaft. Note that on most models a hard nylon sleeve is used to align the oil pump driveshaft and the oil pump shaft. Make sure this sleeve is in place on the oil pump driveshaft **(see illustration)**. If it is not there, check the oil pan for the pieces of the sleeve, clean them out of the pan, then get a new sleeve for the oil pump driveshaft.

4 If a new oil pump is installed, make sure

the pump driveshaft is mated with the shaft inside the pump.

5 Position the pump on the engine and make sure the slot in the upper end of the driveshaft is aligned with the tang on the lower end of the oil pump drive. It is absolutely essential that the components mate properly.

6 Install the mounting bolt and tighten it to the torque listed in this Chapter's Specifications.

7 Install the oil pan.

15 Flywheel/driveplate - removal and installation

Refer to illustration 15.1

This procedure is basically the same as it is for the V6 engines. Refer to Part A of this Chapter and follow the procedure outlined there. However, use the bolt torque value from this Chapter's Specifications **(see illustration)**.

If the flywheel/driveplate is being replaced, be sure to transfer the bolt-on flywheel weights from the old flywheel to the same positions on the new flywheel.

16 Rear main oil seal - replacement

Refer to illustrations 16.2

1 The one-piece rear main oil seal is installed in a bolt-on housing. Replacing this seal requires removal of the transmission, clutch assembly and flywheel (manual transmission) or torque converter and driveplate (automatic transmission). Refer to Chapter 7 for the transmission removal procedures and Section 15 of this Chapter for flywheel removal.

2 Insert a screwdriver blade into the notches in the seal housing and pry out the old seal **(see illustration)**. Be sure to note how far it's recessed into the housing bore before removal so the new seal can be

15.1 On the V8 engine the flywheel/driveplate only goes on one way because the crankshaft has a long locator pin (arrow) corresponding to a locator hole in the flywheel

16.2 The rear main oil seal retainer has notches where a screwdriver can be used to pry out the old seal

17.1a Location of the right engine mount bolt

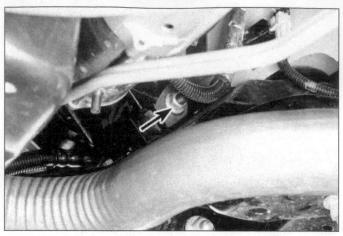

17.1b Location of the left engine mount bolt

installed to the same depth. Although the seal can be removed this way, installation with the housing still mounted on the block requires the use of a special tool, available at most auto parts stores, which attaches to the threaded holes in the crankshaft flange and then presses the new seal into place.

3   If the special installation tool is not available, remove the oil pan (see Section 13) and the bolts securing the housing to the block, then detach the housing and gasket. Whenever the housing is removed from the block a new seal and gasket must be installed. **Note:** *The oil pan must be removed because the*

two rearmost oil pan fasteners are studs that are part of the rear seal retainer, and the pan must come away enough to clear these studs to allow retainer removal and installation.

4   Clean the housing thoroughly, then apply a thin coat of engine oil to the new seal. Set the seal squarely into the recess in the housing, then, using two pieces of wood, one on each side of the housing, use a hammer to press the seal into place.

5   Carefully slide the seal over the crankshaft and bolt the seal housing to the block. Be sure to use a new gasket, but don't use any gasket sealant.

6   The remainder of installation is the reverse of the removal procedure.

---

17   Engine mounts - check and replacement

Refer to illustrations 17.1a and 17.1b

This procedure is basically the same as it is for the 3.4L V6. Refer to Part A of this Chapter and follow the procedure outlined there. However, use the accompanying illustrations and the bolt torque value from this Chapter's Specifications **(see illustrations)**.

Chapter 2 Part D
5.7L V8 engine - 1998 and later models

Contents

Specifications

General

Displacement	
5.7L V8 (1998 and later)	346 cubic inches
Bore and stroke	3.898 x 3.622 inches
Cylinder numbers (front-to-rear)	
Left (driver's) side	1-3-5-7
Right side	2-4-6-8
Firing order	1-8-7-2-6-5-4-3
Cylinder compression pressure	
Minimum	100 psi
Maximum variation between cylinders	25 percent from the highest reading

Camshaft

Journal diameters	2.164 to 2.166 inches
Camshaft endplay	0.001 to 0.012 inch

Lobe lift

To 2000	
Intake	0.292 inch
Exhaust	0.292 inch
2001 and later	
Intake	0.274 inch
Exhaust	0.281 inch

24017-1-B HAYNES

Cylinder numbering

Torque specifications*

Ft-lbs (unless otherwise indicated)

Camshaft sprocket bolts	26
Camshaft retainer bolts	18
Crankshaft balancer bolt	
Step one (use old bolt)	240
Step two (use new bolt)	37
Step three (use new bolt)	Turn an additional 140 degrees
Cylinder head bolts (in sequence - **see illustration 9.17**)	
Step one	
All 11mm bolts	22
Step two	
All 11mm bolts	Turn an additional 90 degrees
Step three	
11mm bolts (1 through 8)	Turn an additional 90 degrees
11mm bolts (9 and 10)	Turn an additional 50 degrees
Step four	
All 8 mm bolts	22
Engine mount retaining bolts	37
Exhaust manifold bolts	
Step one	132 in-lbs
Step two	18
Exhaust manifold heat shield bolt	80 in-lbs
Exhaust pipe flange nuts	20 to 25
Flywheel/driveplate bolts	
Step one	15
Step two	37
Step three	74
Intake manifold bolts	
Step one	44 in-lbs
Step two	89 in-lbs
Lifter retainer bolts	106 in-lbs
Oil pan baffle bolts	106 in-lbs
Oil pan drain plug	18
Oil pan rear access plugs	80 in-lbs
Oil pan bolts	18
Oil pan-to-rear-cover bolts	106 in-lbs
Oil pump cover bolts	106 in-lbs
Oil pump mounting bolts	18
Rocker arm bolts	22
Front timing chain cover bolts	18
Valve cover bolts	106 in-lbs
Vapor vent pipe bolts	106 in-lbs

*Note: *Refer to Part E for additional specifications.*

1   General information

Warning: *The models covered by this manual are equipped with airbags. Always disable the airbag system before working in the vicinity of the impact sensors, steering column or instrument panel to avoid the possibility of accidental deployment of the airbag(s), which could cause personal injury (see Chapter 12).* **Caution:** *On models equipped with a Theftlock audio system, be sure the lockout feature is turned off before performing any procedure which requires disconnecting the battery.*

This Part of Chapter 2 is devoted to in-vehicle repair procedures for the 1998 and later 5.7L V8 engine. All information concerning the 1997 and earlier 5.7L V8 engine is covered in Chapter 2C. All information concerning engine removal and installation and engine block and cylinder head overhaul on either V8 engine can be found in Part E of this Chapter.

Since the repair procedures included in this Part are based on the assumption the engine is still installed in the vehicle, if they are being used during a complete engine overhaul (with the engine already out of the vehicle and on a stand) many of the Steps included here will not apply.

The Specifications included in this Part of Chapter 2 apply only to the procedures found here. The specifications necessary for rebuilding the block and cylinder heads are included in Part E.

The V8 engine used in the 1998 and later Camaro and Firebirds displaces 346 cubic inches and is considered the new generation modular V8. These engines utilize aluminum blocks with eight cylinders arranged in a "V" shape at a 90-degree angle between the two banks. The aluminum cylinder cylinder heads utilize an overhead valve arrangement with pressed-in valve guides and hardened valve seats, Hydraulic roller lifters actuate the valves through tubular pushrods and rocker arms. The oil pump is mounted at the front of the engine behind the timing chain cover and is driven by the crankshaft.

2   Repair operations possible with the engine in the vehicle

Many major repair operations can be accomplished without removing the engine from the vehicle.

Clean the engine compartment and the exterior of the engine with some type of pressure washer before any work is done. A clean engine will make the job easier and will help keep dirt out of the internal areas of the engine.

Depending on the components involved, it may be a good idea to remove the hood to improve access to the engine as repairs are performed (refer to Chapter 11 if necessary).

If oil or coolant leaks develop, indicating a need for gasket or seal replacement, the repairs can generally be made with the engine in the vehicle. The oil pan gasket, the cylinder head gaskets, intake and exhaust manifold gaskets, timing chain cover gaskets and the crankshaft oil seals are all accessible with the engine in place.

Exterior engine components, such as

the water pump, the starter motor, the alternator, the power steering pump and the fuel injection components, as well as the intake and exhaust manifolds, can be removed for repair with the engine in place.

Since the cylinder heads can be removed without removing the engine, valve component servicing can also be accomplished with the engine in the vehicle.

Replacement of, repairs to or inspection of the timing chain and sprockets and the oil pump are all possible with the engine in place.

In extreme cases caused by a lack of necessary equipment, repair or replacement of piston rings, pistons, connecting rods and rod bearings is possible with the engine in the vehicle. However, this practice is not recommended because of the cleaning and preparation work that must be done to the components involved.

3   Top Dead Center (TDC) for number one piston - locating

Refer to illustration 3.6

1   Top Dead Center (TDC) is the highest point in the cylinder that each piston reaches as it travels up the cylinder bore. Each piston reaches TDC on the compression stroke and again on the exhaust stroke, but TDC generally refers to piston position on the compression stroke.

2   Positioning the piston(s) at TDC is an essential part of many procedures such as distributor and timing chain/sprocket removal.

3   Before beginning this procedure, be sure to place the transmission in Neutral and apply the parking brake or block the rear wheels. Also, disable the ignition system by disconnecting the primary electrical connectors at the ignition coil packs, then remove the spark plugs (see Chapter 1).

4   In order to bring any piston to TDC, the crankshaft must be turned using one of the methods outlined below. When looking at the front of the engine, normal crankshaft rotation is clockwise.

a)   *The preferred method is to turn the crankshaft with a socket and ratchet attached to the bolt threaded into the front of the crankshaft. Apply pressure on the bolt in a clockwise direction only. Never turn the bolt counterclockwise.*

b)   *A remote starter switch, which may save some time, can also be used. Follow the instructions included with the switch. Once the piston is close to TDC, use a socket and ratchet as described in the previous paragraph.*

c)   *If an assistant is available to turn the ignition switch to the Start position in short bursts, you can get the piston close to TDC without a remote starter switch. Make sure your assistant is out of the vehicle, away from the ignition switch, then use a socket and ratchet as described in Paragraph (a) to complete the procedure.*

3.6 A long screwdriver inserted in the number one spark plug hole can be used to determine the highest point reached by that piston – make sure to wrap the tip of the screwdriver with tape to avoid scratching the top of the piston or the cylinder walls

5   Place your finger partially over the number one spark plug hole and rotate the crankshaft using one of the methods described above until air pressure is felt at the spark plug hole. Air pressure at the spark plug hole indicates that the cylinder has started the compression stroke. Once the compression stroke has begun, TDC for the number one cylinder is obtained when the piston reaches the top of the cylinder on the compression stroke.

6   To bring the piston to the top of the cylinder, insert a long screwdriver into the number one spark plug hole until it touches the top of the piston. **Note:** *Make sure to wrap the tip of the screwdriver with tape to avoid scratching the top of the piston and the cylinder walls.* Use the screwdriver (as a feeler gauge) to tell where the top of the piston is located in the cylinder while slowly rotating the crankshaft **(see illustration)**. As the piston rises the screwdriver will be pushed out. The point at which the screwdriver stops moving outward is TDC. **Note:** *Always hold the screwdriver upright while the engine is being rotated so that the screwdriver will not get wedged as the piston travels upward.*

7   If you go past TDC, rotate the crankshaft counterclockwise until the piston is approximately one inch below TDC, then slowly rotate the crankshaft clockwise again until TDC is reached.

8   After the number one piston has been positioned at TDC on the compression stroke, TDC for any of the remaining pistons can be located by repeating the procedure described above and following the firing order.

4   Valve covers - removal and installation

Removal

Refer to illustration 4.12

1   Disconnect the cable from the negative terminal of the battery. **Caution:** *On models equipped with the Theftlock audio system, be sure the lockout feature is turned off before performing any procedure which requires disconnecting the battery (see the front of this manual).*

2   Remove the secondary air injection hoses from the air injection shut-off valve on the left valve cover and any remaining support brackets or clamps securing the hoses to the engine, then remove the air injection check valve and pipe assembly from the exhaust manifold on the side from which you wish to remove the valve cover (see Chapter 6). If both valve covers are being removed, both air injection check valves and pipe assemblies must be removed.

3   Remove the air cleaner and resonator assembly to allow access to the front of the engine (see Chapter 4).

Right side

4   If necessary, detach the air conditioning evaporator tube from the right side of the engine to allow access to the right valve cover. Be sure to properly evacuate the air conditioning system before doing so and follow the **Cautions** outlined in Chapter 3.

5   Remove the heater hoses and the bracket bolt and move the heater hoses aside. Disconnect the electrical connectors from the ignition coils and the EGR valve. Unclip the wiring harness from the ignition coil bracket and lay it aside. On 2000 and later models, disconnect the AIR pipe from the exhaust manifold.

6   Remove the ignition coils from the valve cover (see Chapter 5). Be sure each plug wire is labeled before removal to ensure correct reinstallation. Also detach the PCV valve pipe from the intake manifold.

7   Remove the valve cover bolts, then detach the cover from the cylinder head **(see illustration 4.12)**. **Note:** *If the cover is stuck to the cylinder head, bump one end with a block of wood and a hammer to jar it loose. If that doesn't work, try to slip a flexible putty knife between the cylinder head and cover to break the gasket seal. Don't pry at the cover-to-head joint or damage to the sealing surfaces may occur (leading to oil leaks in the future).*

Left side

8   Relieve the fuel system pressure and detach the fuel feed and return lines from the fuel rail and the vapor purge valve (see Chapter 4).

4.12 Valve cover mounting bolts (arrows) - arrow to the far right indicates location of the PCV valve (left valve cover shown)

4.15 Position the new gasket in the valve cover lip

9   Remove the power brake booster vacuum hose from the power brake booster, then remove the secondary air injection solenoid tube from the air injection shut off valve. Detach the shut-off valve mounting bracket from the valve cover and the rear of the cylinder head and detach the valve from the engine. **Note:** *The bolt on the rear of the cylinder head only needs to be loosened several turns to allow removal of the air injection shut-off valve bracket.*

10   Remove the ignition coils from the valve cover (see Chapter 5). Be sure each plug wire is labeled before removal to ensure correct reinstallation.

11   Disconnect the PCV valve from the valve cover.

12   Remove the valve cover bolts **(see illustration)**, then detach the cover from the cylinder head. **Note:** *If the cover is stuck to the cylinder head, bump one end with a block of wood and a hammer to jar it loose. If that doesn't work, try to slip a flexible putty knife between the cylinder head and cover to break the gasket seal. Don't pry at the cover-to-head joint or damage to the sealing surfaces may occur (leading to oil leaks in the future).*

Installation

Refer to illustration 4.15

13   The mating surfaces of each cylinder head and valve cover must be perfectly clean when the covers are installed. Use a gasket scraper to remove all traces of sealant and old gasket material, then clean the mating surfaces with lacquer thinner or acetone. If there's sealant or oil on the mating surfaces when the cover is installed, oil leaks may develop.

14   Clean the mounting bolt threads with a die to remove any corrosion and restore damaged threads. Make sure the threaded holes in the cylinder head are clean - run a tap into them to remove corrosion and restore damaged threads.

15   The gaskets should be mated to the covers before the covers are installed. Position the gasket inside the cover lip **(see illustration)**. If the gasket will not stay in place in the cover lip, apply a thin coat of RTV sealant to the cover flange, then and allow the sealant to set up so the gasket adheres to the cover.

16   Inspect the valve cover bolt grommets for damage. If the grommets aren't damaged they can be reused. Carefully position the valve cover(s) on the cylinder head and install the bolts and grommets.

17   Tighten the bolts in three or four steps to the torque listed in this Chapter's Specifications.

18   The remaining installation steps are the reverse of removal.

19   Start the engine and check carefully for oil leaks as the engine warms up.

5   Rocker arms and pushrods - removal, inspection and installation

Removal

Refer to illustrations 5.2 and 5.3

1   Refer to Section 4 and detach the valve covers from the cylinder heads.

2   Loosen the rocker arm pivot bolts one at a time and detach the rocker arms and bolts, then detach the pivot support pedestal **(see illustration)**. Keep track of the rocker arm positions, since they must be returned to the same locations. Store each set of rocker components separately in a marked plastic bag to ensure that they're reinstalled in their original locations.

3   Remove the pushrods and store them separately to make sure they don't get mixed up during installation **(see illustration)**.

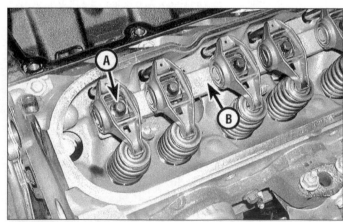

5.2 Remove the mounting bolts (A) and rocker arms, then remove the pivot support pedestal (B)

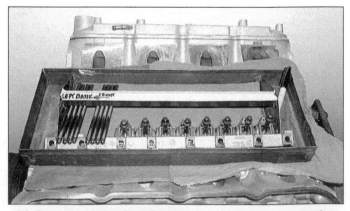

5.3 Store the pushrods and rocker arms in order to ensure they are reinstalled in their original locations - note the arrow indicating the front of the engine

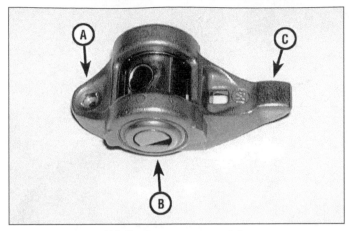

5.4 Rocker arm wear points

A   *Pushrod socket*   C   *Valve stem contact point*
B   *Pivot bearings*

5.9 Lubricate the pushrod ends and the valve stems with engine assembly lube before installing the rocker arms

Inspection

Refer to illustration 5.4

4   Check each rocker arm for wear, cracks and other damage, especially where the pushrods and valve stems contact the rocker arm **(see illustration)**.
5   Check the pivot bearings for binding and roughness. If the bearings are worn or damaged, replacement of the entire rocker arm will be necessary. **Note:** *Keep in mind that there is no valve adjustment on these engines, so excessive wear or damage in the valve train can easily result in excessive valve clearance, which in turn will cause valve noise when the engine is running.* Also check the rocker arm pivot support pedestal for cracks and other obvious damage.
6   Make sure the hole at the pushrod end of each rocker arm is open.
7   Inspect the pushrods for cracks and excessive wear at the ends, also check that the oil hole running through each pushrod is not clogged. Roll each pushrod across a piece of plate glass to see if it's bent (if it wobbles, it's bent).

Installation

Refer to illustration 5.9

8   Lubricate the lower end of each pushrod with clean engine oil or engine assembly lube and install them in their original locations. Make sure each pushrod seats completely in the lifter socket.
9   Apply engine assembly lube to the ends of the valve stems and to the upper ends of the pushrods to prevent damage to the mating surfaces on initial start-up **(see illustration)**. Also apply clean engine oil to the pivot shaft and bearing of each rocker arm and install the rocker arms loosely in their original locations. DO NOT tighten the bolts at this time!
10   Rotate the crankshaft until the number one piston is at TDC (see Section 3). With the number one piston is at TDC, tighten the intake valve rocker arms for the Number 1, 3, 4 and 5 cylinders and the exhaust rocker

arms for the Number 1, 2, 7 and 8 cylinders. Tighten each of the specified rocker arm bolts to the torque listed in this Chapter's Specifications.
11   Rotate the crankshaft 360 degrees. Tighten the intake valve rocker arms for the Number 2, 6, 7 and 8 cylinders and the exhaust rocker arms for the Number 3, 4, 5 and 6 cylinders. Tighten each of the rocker arm bolts to the torque listed in this Chapter's Specifications.
12   Refer to Section 4 and install the valve covers. Start the engine, listen for unusual valve train noses and check for oil leaks at the valve cover gaskets.

6   Valve springs, retainers and seals - replacement

Refer to illustrations 6.5, 6.8, 6.10, 6.15a, 6.15b and 6.19
Note: *Broken valve springs and defective valve stem seals can be replaced without removing the cylinder head. Two special tools and a compressed air source are normally required to perform this operation, so read*

through this Section carefully and rent or buy the tools before beginning the job.
1   Remove the spark plugs (see Chapter 1).
2   Remove the valve covers (see Section 4).
3   Rotate the crankshaft until the number one piston is at top dead center on the compression stroke (see Section 3).
4   Remove the rocker arms for the number 1 piston.
5   Thread an adapter into the spark plug hole and connect an air hose from a compressed air source to it **(see illustration)**. Most auto parts stores can supply the air hose adapter. **Note:** *Many cylinder compression gauges utilize a screw-in fitting that may work with your air hose quick-disconnect fitting. If a cylinder compression gauge fitting is used it will be necessary to remove the schrader valve from the end of the fitting before using it in this procedure.*
6   Apply compressed air to the cylinder. The valves should be held in place by the air pressure. **Warning:** *If the cylinder isn't exactly at TDC, air pressure may force the piston down, causing the engine to quickly rotate. DO NOT leave a wrench on the crankshaft balancer bolt or you may be injured by the tool.*

6.5 This is what the air hose adapter that fits into the spark plug hole looks like - they're commonly available from auto parts stores

6.8 Once the spring is depressed, the keepers can be removed with a small magnet or needle-nose pliers (a magnet is preferred to prevent dropping the keepers)

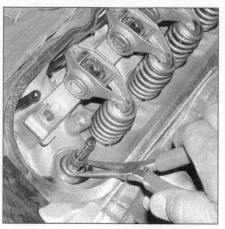

6.10 Use a pair of needle-nose pliers to remove the valve seals

6.15a Be sure to install the seals on the correct valve stems

1   *Intake valve seal*
2   *Exhaust valve seal*

6.15b Install the intake and exhaust valve seals to the specified depth - measure from the spring seat to the top edge of the valve seal

6.19 Apply small dab of grease to each keeper as shown here before installation - it'll hold them in place on the valve stem as the spring is released

7   Stuff shop rags into the cylinder head holes around the valves to prevent parts and tools from falling into the engine.

8   Using a socket and a hammer gently tap on the top of each valve spring retainer several times (this will break the seal between the valve keeper and the spring retainer and allow the keeper to separate from the valve spring retainer as the valve spring is compressed), then use a valve-spring compressor to compress the spring. Remove the keepers with small needle-nose pliers or a magnet **(see illustration)**. **Note:** *Several different types of tools are available for compressing the valve springs with the head in place. One type grips the lower spring coils and presses on the retainer as the knob is turned, while the lever-type shown here utilizes the rocker arm bolt for leverage. Both types work very well, although the lever type is usually less expensive.*

9   Remove the valve spring and retainer. **Note:** *If air pressure fails to retain the valve in the closed position during this operation, the valve face or seat may be damaged. If so, the cylinder head will have to be removed for repair.*

10   Remove the old valve stem seals, noting differences between the intake and exhaust seals **(see illustration)**.

11   Wrap a rubber band or tape around the top of the valve stem so the valve won't fall into the combustion chamber, then release the air pressure.

12   Inspect the valve stem for damage. Rotate the valve in the guide and check the end for eccentric movement, which would indicate that the valve is bent.

13   Move the valve up-and-down in the guide and make sure it does not bind. If the valve stem binds, either the valve is bent or the guide is damaged. In either case, the head will have to be removed for repair.

14   Reapply air pressure to the cylinder to retain the valve in the closed position, then remove the tape or rubber band from the valve stem.

15   If you're working on an exhaust valve, install the new exhaust valve seal on the valve stem and press it down over the valve guide to the specified depth. Don't force the seal against the top of the guide **(see illustrations)**. **Note 1:** *On aluminum heads, be sure to take this measurement from the steel spring seat to the top edge of the intake and exhaust valve seals, not from the aluminum seat on the head!* **Note 2:** *On 2001 and later models, the valve stem seals and spring seat shims are one-piece.*

16   If you're working on an intake valve, install a new intake valve stem seal over the valve stem and press it down over the valve guide to the specified depth. Don't force the intake valve seal against the top of the guide. **Caution:** *Do not install an exhaust valve seal on an intake valve, as high oil consumption will result.*

17   Install the spring and retainer in position over the valve.

18   Compress the valve spring assembly only enough to install the keepers in the valve stem.

19   Position the keepers in the valve stem groove. Apply a small dab of grease to the inside of each keeper to hold it in place if necessary **(see illustration)**. Remove the pressure from the spring tool and make sure the keepers are seated.

20   Disconnect the air hose and remove the adapter from the spark plug hole.

21   Repeat the above procedure on the remaining cylinders, following the firing order sequence (see this Chapter's Specifications). Bring each piston to top dead center on the compression stroke before applying air pressure (see Section 3).

22   Reinstall the rocker arm assemblies and the valve covers (see Sections 4 and 5).

23   Start the engine, then check for oil leaks and unusual sounds coming from the valve cover area. Allow the engine to idle for at least five minutes before revving the engine.

7   Intake manifold - removal and installation

Warning: *Wait until the engine is completely cool before starting this procedure.*

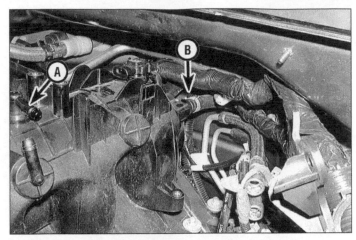

7.7a Disconnect the coolant hoses (A) and the crankcase breather hose (B) from the throttle body, then remove the EVAP canister purge tube (C) and valve from the left side of the intake manifold

7.7b Detach the PCV hose (A) and the power brake booster vacuum hose (B) from the rear of the intake - PCV hose already removed in this photo

Removal

Refer to illustrations 7.7a, 7.7b, 7.8a, 7.8b and 7.8c

1   Disconnect the cable from the negative terminal of the battery. **Caution:** *On models equipped with the Theftlock audio system, be sure the lockout feature is turned off before performing any procedure which requires disconnecting the battery (see the front of this manual).*

2   Clamp off the coolant hoses leading to the throttle body.

3   Remove the air cleaner and the air intake resonator assembly, then relieve the fuel system pressure (see Chapter 4).

4   Disconnect the accelerator linkage (see Chapter 4) and, if equipped, the cruise control linkage.

5   Disconnect the electrical connectors from the fuel injectors, EGR valve, EVAP solenoid, the MAP sensor and from the sensors on the throttle body. Label each connector clearly to aid in the reassembly process. Also disconnect the knock sensor electrical connector and detach the ground straps at the rear of the left cylinder head. Detach any remaining wiring harness brackets from the top of the intake manifold and lay the harness aside.

6   Remove the fuel rails and injectors as an assembly (see Chapter 4). The two fuel rails can be pulled straight up with the injectors still attached, but it will take some force to dislodge the injectors from the intake manifold. **Note:** *This Step is not absolutely necessary but it will help prevent subsequent damage to the fuel injectors as the intake manifold is removed.*

7   Disconnect the coolant hoses from the throttle body. Also disconnect any vacuum hoses attached to the intake manifold or throttle body such as the power brake booster, the PCV and the EVAP purge control valve. **(see illustrations).**

8   Remove the EGR valve and pipe assembly from the engine **(see Illustrations).**

9   Disconnect any remaining electrical

7.8a EGR pipe to intake manifold mounting bolt (arrow) pull straight up and out to remove it from the intake manifold

7.8b EGR valve mounting bracket to cylinder head mounting bolts (arrows)

connectors or vacuum hoses connected to the intake manifold or throttle body.

10   Loosen the intake manifold mounting bolts in 1/4-turn increments in the reverse order of the tightening sequence until they can be removed by hand **(see illustration 7.16)**. The manifold will probably be stuck to the cylinder heads and force may be required

to break the gasket seal. A pry bar can be positioned between the front of the manifold and the valley tray to break the bond made by the gasket. **Caution:** *Do not pry between the manifold and the heads or damage to the gasket sealing surfaces may result and vacuum leaks could develop. Also, don't use too much force - the manifold is made of a plastic composite and could crack.*

7.8c EGR pipe to exhaust manifold mounting bolts (arrows)

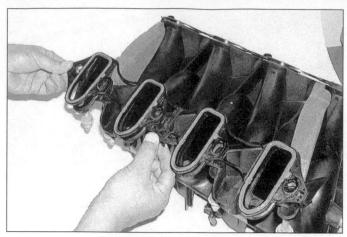

7.14 Align the tabs on the intake gaskets with the tabs on the manifold and snap the gasket into place

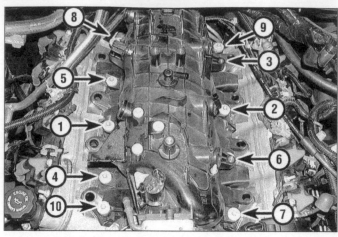

7.16 Intake manifold bolt tightening sequence

11   Remove the intake manifold. As the manifold is lifted from the engine, be sure to check for and disconnect anything still attached to the manifold.

Installation

Refer to illustrations 7.14 and 7.16

Note: *The mating surfaces of the cylinder heads, block and manifold must be perfectly clean when the manifold is installed.*

12   Carefully remove all traces of old gasket material. Note that the intake manifold is made of a composite material and the cylinder heads are made of aluminum, therefore aggressive scraping is not suggested and will damage the sealing surfaces. After the gasket surfaces are cleaned and free of any gasket material wipe the mating surfaces with a cloth saturated with safety solvent. If there is old sealant or oil on the mating surfaces when the manifold is installed, oil or vacuum leaks may develop. Use a vacuum cleaner to remove any gasket material that falls into the intake ports in the heads.

13   Use a tap of the correct size to chase the threads in the bolt holes, then use compressed air (if available) to remove the debris from the holes. **Warning:** *Wear safety glasses or a face shield to protect your eyes when using compressed air.*

14   Position the new gaskets on the intake manifold **(see illustration)**. Note that the gaskets are equipped with installation tabs that must snap into place on the intake manifold. The words "Manifold Side" may appear on the gasket, If so, this will ensure proper installation. Make sure the gaskets snap into place and all intake port openings align.

15   Carefully set the manifold in place.

16   Apply medium-strength threadlocking compound to the threads of the bolts. Install the bolts and tighten them following the recommended sequence **(see illustration)** to the torque listed in this Chapter's Specifications. Do not overtighten the bolts or gasket leaks may develop.

17   The remaining installation steps are the reverse of removal. Check the coolant level, adding as necessary (see Chapter 1). Start the engine and check carefully for vacuum leaks at the intake manifold joints.

---

8   Exhaust manifolds - removal and installation

Removal

Refer to illustrations 8.4, 8.8 and 8.9

Warning: *Use caution when working around the exhaust manifolds - the sheetmetal heat shields can be sharp on the edges. Also, the engine should be cold when this procedure is followed.*

1   Disconnect the cable from the negative terminal of the battery. **Caution:** *On models equipped with the Theftlock audio system, be sure the lockout feature is turned off before performing any procedure which requires disconnecting the battery (see the front of this manual).*

2   Raise the vehicle and support it securely on jackstands.

3   Working under the vehicle, apply penetrating oil to the exhaust pipe-to-manifold studs and nuts (they're usually rusty). Disconnect the electrical connector for oxygen sensor.

4   Remove the nuts retaining the exhaust pipe(s) to the manifold(s) **(see illustration)**.

5   Detach the spark plug wires and remove the spark plugs from the side being worked on (see Chapter 1), If both manifolds are being removed, detach all the spark plug wires and remove all the spark plugs.

6   Remove the secondary air injection pipe (if equipped) from the exhaust manifold being removed (see Chapter 6).

7   On models through 2000, remove the valve cover from the side being worked on (see section 4). If both manifolds are being removed, detach the both valve covers.

Right side manifold

8   Remove the oil dipstick, unbolt the dipstick tube bracket and move the dipstick tube **(see illustration)**.

9   Remove the EGR valve and pipe assembly **(see illustrations 7.8a, 7.8b and 7.8c)**. Remove the mounting bolts and separate the exhaust manifold from the cylinder head **(see illustration)**. Remove the heat shields from the manifold after the manifold has been removed.

Left side manifold

10   Disconnect the electrical connector from the Engine Coolant Temperature (ECT) sensor (see Chapter 6).

11   Remove the mounting bolts and separate the exhaust manifold from the cylinder head. Remove the heat shields from the manifold after the manifold has been removed.

Installation

12   Check the manifold for cracks and make sure the bolt threads are clean and undamaged. The manifold and cylinder head mating surfaces must be clean before the manifolds are reinstalled - use a gasket scraper to remove all carbon deposits and gasket material. **Note:** *The cylinder heads are made of aluminum, therefore aggressive scraping is not suggested and will damage the sealing surfaces.*

8.4 Remove the exhaust pipe-to-manifold nuts

8.8 Remove the oil dipstick tube mounting bolt (A) and tube - (B) indicates the secondary air injection check valve and pipe assembly on the right manifold

8.9 Exhaust manifold fastener locations (right side shown, left side similar)

13   Install the heat shields, then install the bolts and gaskets onto the manifold. Retaining tabs surrounding the gasket bolt holes should hold the assembly together as the manifold is installed.

14   Starting at the fourth thread, apply a 1/4-inch wide band of medium-strength threadlocking compound to the threads of the bolts. **Note:** *The manufacturer recommends not applying threadlocking compound on the first three threads.*

15   Place the manifold on the cylinder head and install the mounting bolts finger tight.

16   When tightening the mounting bolts, work from the center to the ends and be sure to use a torque wrench. Tighten the bolts in two steps to the torque listed in this Chapter's Specifications. If required, bend the exposed end of the exhaust manifold gasket back against the cylinder head.

17   The remaining installation steps are the reverse of removal. Always use new O-rings and gaskets on the EGR valve and pipe assembly.

18   Start the engine and check for exhaust leaks.

9   Cylinder heads - removal and installation

Note: *It will be necessary to purchase a new set of 11 mm head bolts before or during this procedure.*

Removal

Refer to illustrations 9.2 and 9.8

1   Disconnect the cable from the negative terminal of the battery and drain the cooling system (see Chapter 1). **Caution:** *On models equipped with the Theftlock audio system, be sure the lockout feature is turned off before performing any procedure which requires disconnecting the battery (see the front of this manual).*

2   Remove the intake manifold (see Section 7) and the vapor vent pipe **(see illustration)**.

3   Detach both exhaust manifolds from the cylinder heads (see Section 8).

4   Remove the valve covers (see Section 4).

5   Remove the rocker arms and pushrods (see Section 5). **Caution:** *Again, as mentioned in Section 5, keep all the parts in order so they are reinstalled in the same location.*

6   Disconnect the wiring from the back of the alternator, then remove the power steering pump and the power steering pump rear mounting bracket from the left cylinder head. Lay the pump and brackets aside without disconnecting the lines from the steering pump (see Chapter 10). Then remove the wiring harness retaining bolt and the transmission dipstick tube (automatic transmission only) from the rear of the right cylinder head. On 2001 and later models, remove the heater hoses.

7   Loosen the head bolts in 1/4-turn increments in the reverse order of the tightening sequence **(see illustration 9.17)** until they can be removed by hand. **Note:** *There will be different length and size head bolts for different locations. Make a note of the different sizes and lengths and where they go when removing the bolts to ensure correct installation of the new bolts.*

8   Lift the heads off the engine. If resistance is felt, do not pry between the head and block as damage to the mating surfaces will result. To dislodge the head, place a pry bar or long screwdriver into the intake port and carefully pry the head off the engine **(see illustration)**. Store the heads on blocks of wood to prevent damage to the gasket seal-

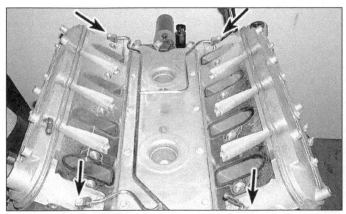

9.2 The vapor vent pipe is retained by two bolts at the front and two bolts at the rear of the cylinder heads (arrows)

9.8 Using a prybar inserted into an intake port to break the head loose - do not use excessive force or damage to the head may result

9.14 Position the head gasket over the dowels at each end of the cylinder head with the mark (arrow) facing the front of the vehicle

9.17 Cylinder head bolt tightening sequence - 1998 and later V8 engines

ing surfaces.

9   Cylinder head disassembly and inspection procedures are covered in detail in Chapter 2, Part E.

Installation

Refer to illustrations 9.14, 9.17 and 9.18

10   The mating surfaces of the cylinder heads and block must be perfectly clean when the heads are installed. Gasket removal solvents are available at auto parts stores and may prove helpful.

11   Use a gasket scraper to remove all traces of carbon and old gasket material, then wipe the mating surfaces with a cloth saturated with lacquer thinner or acetone. **Note:** *The cylinder heads are made of aluminum, therefore aggressive scraping is not suggested and will damage the sealing surfaces.* If there is oil on the mating surfaces when the heads are installed, the gaskets may not seal correctly and leaks may develop. When working on the block, use a vacuum cleaner to remove any debris that falls into the cylinders.

12   Check the block and head mating surfaces for nicks, deep scratches and other damage. If damage is slight, it can be removed with emery cloth. If it is excessive, machining may be the only alternative.

13   Use a tap of the correct size to chase the threads in the head bolt holes in the block. If a tap is not available, spray a liberal amount of brake cleaner into each hole. Use compressed air (if available) to remove the debris from the holes. **Warning:** *Wear safety glasses or a face shield to protect your eyes when using compressed air.* Use a wire brush to remove the threadlocking compound from the 8 mm head bolts. Corrosion, sealant and damaged threads will affect torque readings, so be sure the threads are clean.

14   Position the new gaskets over the dowels in the block **(see illustration)**.

15   Carefully position the heads on the block without disturbing the gaskets.

16   Before installing the 8mm head bolts, coat the threads with a medium-strength

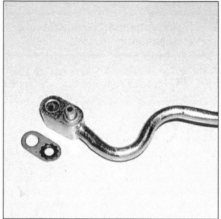

9.18 Be sure to use new gaskets at each cylinder head-to-vent pipe joint - position the O-ring seal over the vent pipe nipple

threadlocking compound. Then install the 8mm head bolts (bolts 11 through 15) finger tight.

17   Install **New** 11 mm head bolts (bolts 1 through 10) and tighten them finger tight. Following the recommended sequence **(see illustration)**, tighten the bolts in four steps to the torque and angle of rotation listed in this Chapter's Specifications. **Warning:** *DO NOT reuse 11mm head bolts - always replace them with new ones.*

18   Install the vapor vent pipe, using new gaskets, onto the cylinder heads **(see illustration)**. Tighten the bolts to the torque listed in this Chapter's Specifications.

19   The remaining installation steps are the reverse of removal.

20   Add coolant and change the oil and filter (see Chapter 1). Start the engine and check for proper operation and coolant or oil leaks.

10   Crankshaft balancer - removal and installation

Refer to illustrations 10.5, 10.6 and 10.9

Note: *This procedure requires a special bal-*

10.5 Use a strap wrench to hold the crankshaft balancer while removing the center bolt (a chain-type wrench may be used if you wrap a section of old drivebelt or rag around the balancer first)

ancer installation tool that is available through specialized tool manufacturers only and a new crankshaft balancer bolt. Read through the entire procedure and obtain the tool and materials before proceeding.

1   Disconnect the cable from the negative terminal of the battery. **Caution:** *On models equipped with the Theftlock audio system, be sure the lockout feature is turned off before performing any procedure which requires disconnecting the battery (see the front of this manual).*

2   Raise the front of the vehicle and support it securely on jackstands. Then apply the parking brake.

3   Remove the drivebelt (see Chapter 1).

4   Working under the vehicle, remove the transmission oil cooler lines and the power steering cooler lines from the radiator (if equipped).

5   Use a strap wrench around the crankshaft pulley to hold it while using a breaker bar and socket to remove the crankshaft pulley center bolt **(see illustration)**.

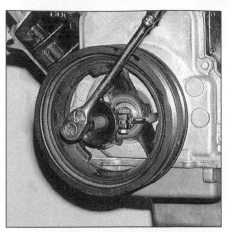

10.6 The use of a three jaw puller will be necessary to remove the crankshaft balancer - always place the puller jaws around the hub, not the outer ring

10.9 Before the new crankshaft bolt is installed and tightened, the balancer must be measured for proper installation - when properly installed, the balancer hub should extend 3/32 to 11/64-inch past the crankshaft snout

11.2 Carefully pry the old seal out of the timing chain cover - don't damage the crankshaft in the process

6   Pull the balancer off the crankshaft with a puller **(see illustration)**. **Caution:** *The jaws of the puller must only contact the hub of the balancer - not the outer ring.* **Note:** *A long Allen-head bolt should be inserted into the crankshaft nose for the puller's tapered tip to push against to prevent damage to the crankshaft threads.*

7   Position the crankshaft pulley/balancer on the crankshaft and slide it on as far as it will go. Note that the slot (keyway) in the hub must be aligned with the Woodruff key in the end of the crankshaft.

8   Using the specialized crankshaft balancer installation tool, press the crankshaft pulley/balancer onto the crankshaft.

9   Install the old crankshaft balancer bolt and tighten the crankshaft bolt to 240 ft-lbs. Remove the old bolt and measure the distance from the snout of the crankshaft to the balancer hub **(see illustration)**. When properly installed, the balancer hub should extend 3/32 to 11/64-inch past the crankshaft snout. If the measurement is incorrect, reinstall the balancer installation tool and press the balancer on the crankshaft until the measurement is correct.

10   Install a **New** crankshaft balancer bolt and tighten it in two steps to the torque and angle of rotation listed in this Chapter's Specifications.

11   The remaining installation steps are the reverse of removal.

11 Crankshaft front oil seal - removal and installation

Refer to illustrations 11.2, 11.4 and 11.5

1   Remove the crankshaft balancer (see Section 10).

2   Note how the seal is installed - the new one must be installed to the same depth and facing the same way. Carefully pry the oil seal out of the cover with a seal puller or a large screwdriver **(see illustration)**. Be very careful

11.4 Drive the new seal into place with a large socket and hammer

not to distort the cover or scratch the crankshaft! Wrap electrician's tape around the tip of the screwdriver to avoid damage to the crankshaft.

3   If the seal is being replaced with the timing chain cover removed, support the cover on top of two blocks of wood and drive the seal out from the backside with a hammer and punch. **Caution:** *Be careful not to scratch, gouge or distort the area that the seal fits into or a leak will develop.*

4   Apply clean engine oil or multi-purpose grease to the outer edge of the new seal, then install it in the cover with the lip (spring side) facing IN. Drive the seal into place **(see illustration)** with a large socket and a hammer (if a large socket isn't available, a piece of pipe will also work). Make sure the seal enters the bore squarely and stop when the front face is at the proper depth.

5   Check the surface on the balancer hub that the oil seal rides on. If the surface has been grooved from long-time contact with the seal, a press-on sleeve may be available

11.5 If the sealing surface of the pulley hub has a wear groove from contact with the seal, repair sleeves are available at most auto parts stores

to renew the sealing surface **(see illustration)**. This sleeve is pressed into place with a hammer and a block of wood and is commonly available at auto parts stores for various applications.

6   Lubricate the balancer hub with clean engine oil and reinstall the crankshaft balancer as described in Section 10.

7   The remainder of installation is the reverse of the removal.

12 Timing chain - removal, inspection and installation

Removal and inspection

Refer to illustrations 12.6, 12.9 and 12.12

1   Disconnect the cable from the negative terminal of the battery. **Caution:** *On models equipped with the Theftlock audio system, be sure the lockout feature is turned off before performing any procedure which requires disconnecting the battery (see the front of this manual).*

12.6 Timing chain cover mounting bolts (arrows)

12.9 Timing chain alignment marks (arrows) - when properly aligned, the crankshaft gear should be in the 12 o'clock position, the camshaft gear should be in the 6 o'clock position and the number one piston should be at TDC

12.12 The sprocket on the crankshaft can be removed with a two or three-jaw puller

2   Refer to Chapter 1 and drain the cooling system and engine oil.
3   Refer to Chapter 4 and remove the air cleaner and the air intake resonator from the engine compartment, then refer to Chapter 3 and remove the upper and lower radiator hoses, the cooling fans, drivebelt and the water pump.
4   Remove the crankshaft balancer (see Section 10).
5   Remove the oil pan (see Section 14).
6   Remove the timing chain cover mounting bolts and separate the timing chain cover from the block **(see illustration)**. The cover may be stuck; if so, use a putty knife to break the gasket seal. Since the cover is made of aluminum it can easily be damaged, so DO NOT attempt to pry it off.
7   Remove the oil pick-up tube and the oil pump (see Section 15).
8   Measure the timing chain freeplay. If it is more than 5/8 inch, the chain and both sprockets should be replaced.
9   Loosen the camshaft sprocket bolts one turn, then screw the crankshaft balancer bolt into the end of the crankshaft and rotate the crankshaft in the normal direction of rotation (clockwise) until the timing marks align **(see illustration)**. Verify that the number one piston is at TDC.
10   Remove the three bolts from the end of the camshaft, then detach the camshaft sprocket and chain as an assembly.
11   Inspect the camshaft and crankshaft sprockets for damage or wear.
12   If replacement of the timing chain is necessary, remove the sprocket on the crankshaft with a two-or three-jaw puller, but be careful not to damage the threads in the end of the crankshaft **(see illustration)**.

Installation

Refer to illustrations 12.15, 12.19 and 12.20
Note: *Timing chains must be replaced as a set with the camshaft and crankshaft sprockets. Never put a new chain on old sprockets.*
13   Use a gasket scraper to remove all traces of old gasket material and sealant from

the cover and engine block.
14   Align the crankshaft sprocket with the Woodruff key and press the sprocket onto the crankshaft (if removed) with the vibration damper bolt, a large socket and some washers or tap it gently into place until it is completely seated. **Caution:** *If resistance is encountered, do not hammer the sprocket onto the crankshaft. It may eventually move onto the shaft, but it may be cracked in the process and fail later, causing extensive engine damage.*
15   Loop the new chain over the camshaft sprocket, then turn the sprocket until the timing mark is at the bottom **(see illustration)**. Mesh the chain with the crankshaft sprocket and position the camshaft sprocket on the end of the camshaft. If necessary, turn the camshaft so the dowel in the camshaft fits into the hole in the sprocket with the timing mark in the 6 o'clock position **(see illustra-**

tion 12.9). When the chain is installed, the timing marks MUST align as shown.
16   Apply a thread locking compound to the camshaft sprocket bolt threads and tighten the bolts to the torque listed in this Chapter's Specifications.
17   Lubricate the chain with clean engine oil.
18   Install the oil pump and the oil pick up tube onto the engine (see Section 15). Now would be a good time to replace the crankshaft front oil seal (see Section 11).
19   Install the timing chain cover on the engine loosely using a new gasket **(see illustration)**.
20   The manufacturer recommends that it is mandatory to use a front cover alignment tool when installing the front cover, however at the time this manual was printed the tools were unobtainable. If the tool is not available, follow the alternate procedure outlined below:
  a) *Install the crankshaft balancer on the engine as described in Section 10. This Step will align the front oil seal with the balancer hub.*

12.15 Slip the chain and camshaft sprocket in place over the crankshaft sprocket with the camshaft sprocket timing mark (arrow) at the bottom

12.19 Install the front cover with a new gasket onto the engine block LOOSELY - the cover must be aligned properly before final installation

12.20 With the crankshaft balancer in place and the front cover bolts installed LOOSELY, measure the distance between the oil pan rail and the front cover sealing surface on each side (arrows) - then adjust the cover so the measurements are even on both sides before tightening the cover bolts

13.2a The roller lifters are held in place by retainers - remove the retainer bolts, then remove the retainers and the lifters as an assembly - note that each retainer houses four individual lifters and they must be installed back in their original locations if they're going to be reused

13.2b Once the lifters and retainers are removed from the block they can be marked (for location and installation purposes) and inspected

b) *Place a straightedge on the engine block oil pan rail. Measure the distance on each side of the block from the oil pan rail to the timing chain cover with a feeler gauge (see illustration). This Step measures the difference between the sealing surface of the oil pan and the sealing surface of the timing chain cover in relationship to each other.*

c) *Tilt the front timing cover as necessary to achieve an even measurement on each side. This Step properly aligns the front timing cover to oil pan sealing surfaces. Typically 0.000 to 0.020 inch is an acceptable tolerance. Note: Ideally the timing chain cover should be flush with the oil pan rail, but because of the differences in seal thickness, this may not always be obtainable. That is why there is a tolerance of 0.000 to 0.020 inch. Always let the front seal center itself around the crankshaft balancer hub and tilt the cover from side to side to even up the measurement at both oil pan rails. Never push downward on the front timing cover in an attempt to make the oil pan sealing surface flush, as this will distort the front oil seal and eventually lead to an oil leak!*

d) *With the timing chain cover properly aligned, tighten the cover bolts to the torque listed in this Chapter's Specifications.*

21   Apply a thin layer of RTV sealant to the areas where the timing chain cover and cylinder block meet, then install the oil pan as described in Section 14.

22   The remaining installation steps are the reverse of removal.

23   Add coolant and oil to the engine (see Chapter 1). Run the engine and check for oil and coolant leaks.

13 Camshaft and lifters - removal and installation

Note 1: *The camshaft should always be thoroughly inspected before installation and camshaft endplay should always be checked prior to camshaft removal. Camshaft and lifter inspection procedures for the 2nd generation V8 engine are identical to inspection procedures on the 1st generation V8 engine, therefore refer to Chapter 2C for the Camshaft and lifter inspection procedures.*

Note 2: *If the camshaft is being replaced, always install new lifters as well. Do not use old lifters on a new camshaft.*

Removal

Refer to illustrations 13.2a, 13.2b and 13.4

1   Refer to the appropriate Sections and remove the intake manifold, valve covers, rocker arms, pushrods, timing chain and the cylinder heads. Also remove the radiator and air conditioning condenser (see Chapter 3) and the camshaft position sensor (see Chapter 6).

2   Before removing the lifters, arrange to store them in a clearly labeled box to ensure that they're reinstalled in their original locations. Remove the lifter retainers and lifters and store them where they won't get dirty (see illustrations). DO NOT attempt to withdraw the camshaft with the lifters in place.

3   If the lifters are built up with gum and varnish they may not come out with the retainer. If so, there are several ways to extract the lifters from the bores. A special tool designed to grip and remove lifters is

manufactured by many tool companies and is widely available, but it may not be required in every case. On newer engines without a lot of varnish buildup, the lifters can often be removed with a small magnet or even with your fingers. A machinist's scribe with a bent end can be used to pull the lifters out by positioning the point under the retainer ring in the top of each lifter. **Caution:** *Don't use pliers to remove the lifters unless you intend to replace them with new ones. The pliers will damage the precision machined and hardened lifters, rendering them useless.*

4   Remove the bolts and the camshaft retainer plate, noting which direction faces the block (see illustration).

13.4 Remove the bolts (arrows) and take off the camshaft retainer plate, noting which side faces the block

5   Thread three 6-inch long bolts into the camshaft sprocket bolt holes to use as a "handle" when removing the camshaft from the block.

6   Carefully pull the camshaft out. Support the cam near the block so the lobes don't nick or gouge the bearings as it's withdrawn.

Installation

Refer to illustration 13.7

7   Lubricate the camshaft bearing journals and cam lobes with camshaft and lifter assembly lube **(see illustration)**.

8   Slide the camshaft into the engine. Support the cam near the block and be careful not to scrape or nick the bearings.

9   Turn the camshaft until the dowel pin is in the 3 o'clock position, and install the camshaft thrust plate, tighten the bolts to the torque listed in this Chapter's Specifications. Make sure the gasket surface on the camshaft thrust plate and the engine block are free from oil and dirt.

10   Install the timing chain and sprockets (see Section 12). Also install the camshaft position sensor (see Chapter 6).

11   Lubricate the lifters with clean engine oil and install them in the lifter retainers. Be sure to align the flats on the lifters with the flats in the lifter retainers. Install the retainer and lifters into the engine block as an assembly. If the original lifters are being reinstalled, be sure to return them to their original locations. If a new camshaft is being installed, install new lifters as well. Tighten the lifter retainer bolts to the torque listed in this Chapter's Specifications.

12   The remaining installation steps are the reverse of removal.

13   Before starting and running the engine, change the oil and install a new oil filter (see Chapter 1).

---

14   Oil pan - removal and installation

Removal

Refer to illustration 14.5

1   Disconnect the cable from the negative terminal of the battery. **Caution:** *On models equipped with a Delco Loc II or Theftlock audio system, be sure the lockout feature is turned off before performing any procedure which requires disconnecting the battery.*

2   Raise the vehicle and support it securely on jackstands, then refer to Chapter 1 and drain the engine oil and remove the oil filter. Also remove the air cleaner housing and the air intake resonator (see Chapter 4).

3   Disconnect the front exhaust Y pipe from the engine and the exhaust system and remove it from the vehicle. This step is not absolutely necessary, but it will help facilitate removal of the oil pan.

4   Remove the starter motor (see Chapter 5). Also remove the plastic bellhousing covers on each side of the bellhousing.

5   Remove any wiring harness brackets

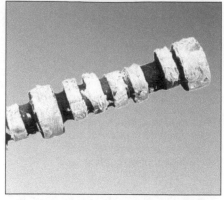

13.7 Be sure to apply camshaft assembly lube to the cam lobes and bearing journals before installing the camshaft

connected to the oil pan and the brackets on the passenger side of the oil pan securing the transmission oil cooler lines (if equipped). Also disconnect the electrical connector from the oil level sensor **(see illustration)**.

6   Attach an engine hoist to the engine from above and raise the engine slightly to allow removal of the engine mount through bolts or the engine mount to cradle bolts (see Section 18).

7   Raise the engine approximately three inches. **Caution:** *When lifting the engine check that no upper engine components or wiring harnesses are hitting the edge of the cowl.*

8   Unbolt the shock absorbers from the lower control arms, then loosen the steering shaft coupler from the steering gear (see Chapter 10). Loosen the (six) front suspension cradle bolts four to five turns. This will lower the front suspension cradle away from the body and will allow clearance for removal of the oil pan. **Caution:** *Do not loosen the front suspension cradle bolts more than five turns without the use of floor jack or jackstands to support the cradle or injury may occur.* **Note:** *On 2002 models, it will be necessary to lower the crossmember further, using a jack underneath it. Refer to Chapter 10 and disconnect the stabilizer bar links, remove the rack-and-pinion gear, and disconnect any fuel or brake line mounting brackets from the crossmember. Refer to Chapter 9 and remove the front brake calipers and support them to the body with mechanic's wire. An alternative method is to disconnect and cap the brake lines at the brake pressure modulator, and leave the calipers and brake lines attached to the crossmember.*

9   Remove the transmission to oil pan bolts (see Chapter 7).

10   If the vehicle is equipped with an engine oil cooler, remove the engine oil cooler lines and adapter from the driver's side of the oil pan.

11   Remove the access plugs covering the nuts at the rear of the oil pan (if equipped).

12   Remove all the oil pan bolts, then lower the pan from the engine. The pan will probably stick to the engine, so strike the pan with a rub-

14.5 The oil level sensor is located on the passenger side of the oil pan

ber mallet until it breaks the gasket seal. **Caution:** *Before using force on the oil pan, be sure all the bolts have been removed. Carefully slide the oil pan down and out, to the rear.*

Installation

Refer to illustrations 14.13 and 14.16

13   Drill out the rivets securing the oil pan gasket to the oil pan and remove the old gasket **(see illustration)**. Wash out the oil pan with solvent.

14   Thoroughly clean the mounting surfaces of the oil pan and engine block of old gasket material and sealer. Wipe the gasket surfaces clean with a rag soaked in lacquer thinner, acetone or brake system cleaner.

15   Apply a 3/16-inch wide, one inch long bead of RTV sealant to the corners of the block where the front cover and the rear cover meet the engine block. Then attach the new gasket to the pan, install the pan and tighten the bolts finger-tight. Be sure the oil gallery passages in the pan and the gasket are aligned properly.

16   The alignment of the rear face of the aluminum pan to the rear of the block is important. Measure between the rear face of the pan and the front face of the transmission bellhousing with feeler gauges. Clearance should ideally be flush, but a gap of up to 0.010-inch is allowable. If the clearance is

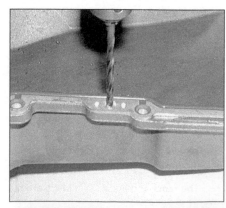

14.13 The manufacturer uses rivets to hold the gasket to the oil pan during assembly - carefully drill them out (it isn't necessary to rivet the new gasket to the oil pan)

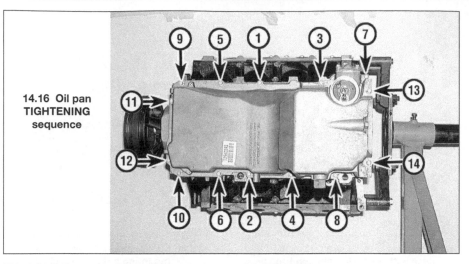

14.16 Oil pan TIGHTENING sequence

15.2a Oil pick-up tube-to-main stud retaining nuts (arrows)

OK, tighten the pan bolts/studs in sequence to the torque listed in this Chapter's Specifications **(see illustration)**. If the clearance is not acceptable, install the two lower oil pan-to-bellhousing bolts and tighten them finger tight. This should draw the oil pan flush with the bellhousing.

17 The remainder of installation is the reverse of removal.

18 Add the proper type and quantity of oil (see Chapter 1), start the engine and check for leaks before placing the vehicle back in service.

15 Oil pump - removal, inspection and installation

Removal

Refer to illustrations 15.2a, 15.2b and 15.3

1 Refer to the Section 12, Steps 1 through 6 and remove the timing chain cover.

2 Remove the oil pump pick-up tube mounting nuts and bolts and lower the pick-up tube and screen assembly from the vehicle **(see illustrations)**.

3 Remove the oil pump retaining bolts and slide the pump off the end of the crankshaft **(see illustration)**.

Inspection

Refer to illustration 15.4

4 Remove the oil pump cover and withdraw the rotors from the pump body **(see illustration)**. Clean the components with solvent, dry them thoroughly and inspect for any obvious damage. Also check the bolt holes for damaged threads and the splined surfaces on the crankshaft sprocket for any apparent damage. If any of the components are scored, scratched or worn, replace the entire oil pump assembly. There are no serviceable parts currently available.

Installation

Refer to illustration 15.8

5 Prime the pump by pouring clean motor oil into the pick-up tube hole, while turning the pump by hand.

6 Position the oil pump over the end of the crankshaft and align the teeth on the crankshaft sprocket with the teeth on the oil pump drive gear. Making sure the pump is fully seated against the block.

7 Install the oil pump mounting bolts and tighten them to the torque listed in this Chapter's Specifications.

8 Install a new O-ring on the oil pump pick-up tube, then fasten it to the oil pump

15.2b Remove the bolt (arrow) securing the oil pick-up tube to the oil pump and remove it from the engine

and the engine block main studs **(see illustration)**. **Caution:** *Be absolutely certain that the pick-up tube-to-oil pump bolts are properly tightened so that no air can be sucked into the oiling system at this connection.*

9 Install and align the timing chain cover, then install the oil pan. Refer to Sections 12 and 14 for the installation procedures.

10 The remainder of installation is the reverse of removal.

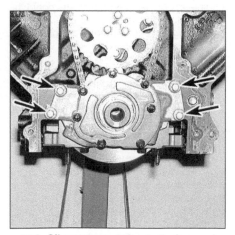

15.3 Oil pump mounting bolts (arrows)

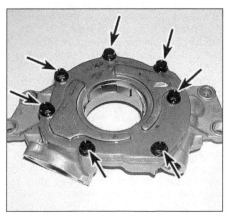

15.4 Oil pump cover-to-oil pump housing mounting bolts (arrows)

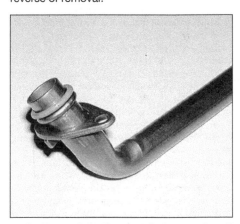

15.8 Always install a new O-ring on the oil pump pick up tube

17.3 Carefully pry the old seal out with a screwdriver at the notches provided in the rear cover

17.4 The rear oil seal can be pressed into place with a seal installation tool, a section of pipe or a blunt object (shown here) - in any case be sure the seal is installed squarely into the seal bore and flush with the rear cover

18.11 Engine mount-to-engine block mounting bolts (arrows)

11   Add oil and coolant as necessary. Run the engine and check for oil and coolant leaks. Also check the oil pressure as described in Chapter 2E.

---

16 Flywheel/driveplate - removal and installation

The flywheel/driveplate replacement for 2nd generation V8 engine is identical to the flywheel/driveplate replacement procedure for the V6 engines. Refer to Chapter 2 Part A for the procedure and use the torque figures in this Chapter's Specifications.

---

17 Rear main oil seal - replacement

Refer to illustrations 17.3 and 17.4
Note: *If you're installing a new rear seal during a complete engine overhaul, refer to the procedure in Chapter 2E.*
1   Remove the transmission (see Chapter 7).
2   Remove the flywheel/driveplate (see Section 16).
3   Pry the oil seal from the rear cover with a screwdriver **(see illustration)**. Be careful not to nick or scratch the crankshaft or the seal bore. Be sure to note how far it's recessed into the housing bore before removal so the new seal can be installed to the same depth. Thoroughly clean the seal bore in the block with a shop towel. Remove all traces of oil and dirt.
4   Lubricate the outside diameter of the seal and install the seal over the end of the crankshaft. Make sure the lip of the seal points toward the engine. Preferably, a seal installation tool (available at most auto parts store) should be used to press the new seal back into place. If the proper seal installation tool is unavailable, use a large socket, section of pipe or a blunt tool and carefully drive the new seal squarely into the seal bore and

flush with the rear cover **(see illustration)**.
5   Install the flywheel/driveplate (see Section 16).
6   Install the transmission (see Chapter 7).

---

18 Engine mounts - check and replacement

1   Engine mounts seldom require attention, but broken or deteriorated mounts should be replaced immediately or the added strain placed on the driveline components may cause damage.

Check

2   During the check, the engine must be raised slightly to remove the weight from the mounts.
3   Raise the vehicle and support it securely on jackstands, then position the jack under the engine oil pan. Place a large block of wood between the jack head and the oil pan, then carefully raise the engine just enough to take the weight off the mounts. Do not use the jack to support the entire weight of the engine.
4   Check the mounts to see if the rubber is cracked, hardened or separated from the metal plates. Sometimes the rubber will split right down the center. Rubber preservative or WD-40 can be applied to the mounts to slow deterioration.
5   Check for relative movement between the mount plates and the engine or frame (use a large screwdriver or prybar to attempt to move the mounts). If movement is noted, check the tightness of the mount fasteners first before condemning the mounts. Usually when engine mounts are broken, they are very obvious as the engine will easily move away from the mount when pried or under load.

Replacement

Refer to illustration 18.11
6   Disconnect the cable from the negative terminal of the battery, then raise the vehicle and support it securely on jackstands. **Caution:** *On models equipped with the Theftlock audio system, be sure the lockout feature is turned off before performing any procedure which requires disconnecting the battery (see the front of this manual).*
7   Remove the air cleaner housing and the air intake resonator from the engine compartment (see Chapter 4).
8   If you're removing the drivers side mount remove the engine mount-heat shield from the engine compartment. If you're removing the passenger side mount, disconnect the oil level sensor electrical connector from the oil pan, then remove the starter and detach the negative battery cable from the engine block (see Chapter 5).
9   Working below the vehicle, remove the engine mount to front suspension cradle bolts. There are four bolts securing each mount to the cradle.
10   Attach an engine hoist to the top of the engine for lifting; do not use a jack under the oil pan to support the entire weight of the engine or the oil pump pick-up could be damaged. **Note:** *If a hoist is not available, casting lugs on each side of the engine block can be used to support the entire weight of the engine while the engine mounts are being replaced.*
11   Raise the engine slightly until the engine mount can be unbolted from the block. Unbolt the mount from the engine block and remove it from the vehicle **(see illustration)**. Separate the upper and lower mount on the bench.
12   Installation is the reverse of removal. Use non-hardening thread-locking compound on the mount bolts and be sure to tighten them to the torque listed in this Chapter's Specifications.

Chapter 2 Part E
General engine overhaul procedures

Contents

Specifications

3.4L V6 engine

General

RPO Code	L32
VIN code	S
Displacement	204 cubic inches
Cylinder compression pressure	100 psi minimum
Maximum variation between cylinders	30-percent from highest reading
Firing order	1-2-3-4-5-6
Oil pressure, minimum	15 psi at 1100 rpm

Cylinder head

Warpage limit	0.005 inch
Maximum allowable cut	0.010 inch

Valves and related components

Valve face angle	45-degrees
Valve seat angle	46-degrees
Valve stem-to-guide clearance	
Intake	0.0014 to 0.0027 inch
Exhaust	0.0015 to 0.0029 inch
Valve spring free length (intake and exhaust)	1.91 inches
Valve spring pressure	
Closed	80 lbs @ 1.61 inches
Open	190 lbs @ 1.20 inches
Installed height	1.61 inches

Crankshaft and connecting rods

Connecting rod journal	
Diameter	1.9987 to 1.9994 inches
Bearing oil clearance	0.0011 to 0.0032 inch
Connecting rod side clearance (endplay)	0.007 to 0.017 inch
Main bearing journal	
Diameter	2.6473 to 2.6483 inches
Bearing oil clearance	0.0012 to 0.0030 inch
Taper/out of round limit	0.0002 inch
Crankshaft endplay (at thrust bearing)	0.0024 to 0.0083 inch

3.4L V6 engine (continued)

Engine block

Cylinder bore
 Diameter .... 3.6228 to 3.6235 inches
 Out-of-round limit .... 0.0003 inch

Pistons and rings

Piston-to-bore clearance .... 0.0011 to 0.0024 inch
Piston ring end gap
 Top compression ring .... 0.007 to 0.016 inch
 Second compression ring .... 0.019 to 0.029 inch
 Oil control ring .... 0.010 to 0.030 inch
Piston ring side clearance
 Compression rings .... 0.0020 to 0.0035 inch
 Oil control ring .... 0.008 inch maximum

Camshaft

Bearing journal diameter .... 1.868 to 1.871 inches
Bearing oil clearance .... 0.001 to 0.004 inch
Lobe lift
 Intake .... 0.2626 inch
 Exhaust .... 0.2732 inch

Torque specifications*

Ft-lbs (unless otherwise indicated)

Main bearing cap bolts
 Step 1 37
 Step 2 .... Rotate an additional 77-degrees
Connecting rod cap nuts 37

*Note: *Refer to Part A for additional torque specifications.*

3800 V6 engine

General

RPO Code L36
VIN code .... K
Displacement .... 231 cubic inches
Cylinder compression pressure .... 100 psi minimum
Maximum variation between cylinders .... 30-percent from highest reading
Firing order 1-6-5-4-3-2
Oil pressure .... 60 psi at 1850 rpm

Cylinder head

Warpage limit .... 0.003 inch per 6 inch span/0.006 inch overall
Maximum allowable cut .... 0.010 inch

Valves and related components

Margin width (minimum) 0.025-inch
Seat angle 45-degrees
Face angle 45-degrees
Stem-to-guide clearance, all .... 0.0015 to 0.0032 inch
Spring free length
 To 1999 .... 1.981 inches
 2000 and later .... 1.960 inches
Spring pressure
 To 1999
  Closed .... 80 lbs @ 1.75 inches
  Open .... 210 lbs @ 1.315 inches
 2000 and later
  Closed .... 75 lbs @ 1.72 inches
  Open .... 228 lbs @ 1.277 inches
Spring installed height
 To 2001 .... 1.690 to 1.720 inches
 2002 .... 1.690 to 1.750 inches

Crankshaft and connecting rods

Connecting rod journal
Diameter .... 2.2487 to 2.2499 inches
Bearing oil clearance
To 1999 .... 0.0008 to 0.0022 inch
2000 and later .... 0.0005 to 0.0026 inch
Maximum taper .... 0.0003 inch
Connecting rod side clearance (end play)
To 1999 .... 0.003 to 0.015 inch
2000 and later .... 0.004 to 0.020 inch
Main bearing journal
Diameter .... 2.4988 to 2.4998 inches
Bearing oil clearance
To 1999 .... 0.0008 to 0.0022 inch
2000 and later
Number 1 .... 0.0007 to 0.0016 inch
Numbers 2, 3 and 4 .... 0.0009 to 0.0018 inch

Engine block

Cylinder bore
Diameter .... 3.800 inches
Out-of-round limit .... 0.0004 inch
Taper limit .... 0.0005 inch

Pistons and rings

Piston-to-bore clearance limits (1.5 inches from top of piston) .... 0.0004 to 0.0020 inch
To 1999
Top ring .... 0.012 to 0.022 inch
Second ring .... 0.030 to 0.040 inch
2000 and later
Top ring .... 0.010 to 0.018 inch
Second ring .... 0.023 to 0.033 inch
Oil control ring .... 0.010 to 0.030 inch
Piston ring side clearance
Compression rings .... 0.0013 to 0.0031 inch
Oil control ring .... 0.0009 to 0.0079 inch

Camshaft

Bearing journal diameter
To 1999 .... 1.7865 to 1.7885 inches
2000 and later .... 1.8462 to 1.8448 inches
Bearing oil clearance
To 1999 .... 0.0005 to 0.0035 inch
2000 and later .... 0.0016 to 0.0047 inch
Lobe lift
Intake
To 1999 .... 0.250 inch
2000 and later .... 0.258 inch
Exhaust
To 2001 .... 0.255 inch
2002 .... 0.258 inch

Balance shaft

Endplay .... 0.008 inch maximum
Radial play
Front .... 0.0011 inch maximum
Rear .... 0.0005 to 0.0047 inch
Drive gear lash .... 0.002 to 0.005 inch
Bearing bore diameter
Front .... 2.0462 to 2.0472 inches
Rear
To 1999 .... 1.950 to 1.952 inches
2000 and later .... 1.8735 to 8745 inches

3800 V6 engines (continued)
Torque specifications*

	Ft-lbs (unless otherwise indicated)
Main bearing caps	See text for procedure (Section 24)
Main cap side bolts	
Step 1	132 in-lbs
Step 2	Rotate an additional 45-degrees
Connecting rod bolts	
Step one	20
Step two	Rotate an additional 50-degrees
Balance shaft gear bolt	
Step one	16
Step two	Rotate an additional 70-degrees
Balance shaft retainer bolts	22

* Note: *Refer to Part B for additional torque specifications.*

5.7L V8 engine - 1997 and earlier

General

RPO Code	LT1
VIN code	P
Displacement	350 cubic inches
Cylinder compression pressure	100 psi minimum
Maximum variation between cylinders	30-percent from highest reading
Firing order	1-8-4-3-6-5-7-2
Oil pressure	
@1000 RPM	6 psi minimum
@2000 RPM	18 psi minimum
@4000 RPM	24 psi minimum

Cylinder head

Warpage limit	0.003 inch per 6 inch span/0.006 inch overall
Maximum allowable cut	0.010 inch

Valves and related components

Seat angle	46-degrees
Face angle	45-degrees
Stem-to-guide clearance, maximum	
Intake	0.0037 inch
Exhaust	0.0047 inch
Spring free length	2.02 inches
Spring pressure	
Closed	81 to 89 lbs @ 1.75 inches
Open	245 to 265 lbs @ 1.33 inches
Spring installed height, all	1.78 inches

Crankshaft and connecting rods

Connecting rod journal	
Diameter	2.0978 to 2.0998 inches
Bearing oil clearance	0.0008 to 0.0022 inch
Maximum taper	0.0003 inch
Connecting rod side clearance (endplay)	0.003 to 0.015 inch
Main bearing journal	
Diameter	2.4988 to 2.4998 inches
Bearing oil clearance	0.001 to 0.003 inch
Maximum taper/out-of-round	0.003 inch
Crankshaft endplay (at thrust bearing)	0.002 to 0.008 inch

Engine block

Cylinder bore	
Diameter	4.0007 to 4.0017 inches
Out-of-round limit	0.0002 inch
Taper limit	0.001 inch

Pistons and rings

Piston-to-bore clearance limits	0.0010 to 0.0027 inch
Piston ring end gap	
Top ring	0.010 to 0.016 inch
Second ring	0.018 to 0.026 inch
Oil control ring	0.010 to 0.030 inch
Piston ring side clearance	
Compression rings	0.019 to 0.035 inch
Oil control ring	0.002 to 0.008 inch

Camshaft

Bearing journal diameter	1.8677 to 1.78697 inches
End clearance	0.004 to 0.012 inch
Lobe lift	
Intake	0.298 inch
Exhaust	0.306 inch

Torque specifications*   **Ft-lbs** (unless otherwise indicated)

Main bearing caps	78
Connecting rod nuts/bolts	
1993 through 1995	45
1996 and 1997	
Step 1	20
Step 2	Tighten an additional 55-degrees

* Note: *Refer to Part C for additional torque specifications.*

5.7L V8 engine - 1998 and later

General

RPO Code	LS1
VIN code	G
Displacement	346 cubic inches
Cylinder compression pressure	100 psi minimum
Maximum variation between cylinders	25-percent from highest reading
Firing order	1-8-7-2-6-5-4-3
Oil pressure	
@1000 RPM	6 psi minimum
@2000 RPM	18 psi minimum
@4000 RPM	24 psi minimum

Cylinder head

Warpage limit	0.003 inch per 6 inches
Cylinder head height	4.732 inches (measured from deck surface to valve cover rail)

Valves and related components

Valve face angle (intake and exhaust)	45-degrees
Valve seat angle (intake and exhaust)	46-degrees
Valve margin (minimum)	0.050 inch
Valve seat width	
Intake	0.040 to 0.060 inch
Exhaust	0.070 to 0.080 inch
Valve stem-to-guide clearance (maximum)	
Intake	0.0037 inch
Exhaust	0.0041 inch
Valve spring free length	2.08 inches
Valve spring installed height	1.80 inches

Crankshaft and connecting rods

Rod journal diameter	2.0991 to 2.100 inches
Rod journal taper (maximum)	0.0007 inch
Rod journal out-of-round (maximum)	0.0004 inch
Rod bearing oil clearance	
Desired	0.0006 to 0.0025 inch
Allowable	0.0006 to 0.003 inch
Connecting rod side clearance (endplay)	0.0043 to 0.020 inch
Main bearing journal diameter	2.558 to 2.5593 inches
Main bearing journal taper (maximum)	0.0007 inch
Main bearing journal out-of-round (maximum)	0.0003 inch
Main bearing oil clearance	0.0007 to 0.0021 inch
Crankshaft endplay	0.0015 to 0.0078 inch

5.7L V8 engine - 1998 and later (continued)
Engine block

Bore diameter	3.897 to 3.898 inch
Taper (maximum)	0.001 inch
Out-of-round (maximum)	0.002 inch
Deck (head gasket surface) warpage limit	0.003 inch per 6 inches
Lifter bore diameter	0.843 to 0.844 inch

Pistons and rings

Piston-to-bore clearance
Through 2001	0.0007 to 0.0021 inch

2002

Production
Coated pistons	0.001 to 0.0011 inch
Uncoated pistons	0.0005 to 0.0019 inch
Service limit	0.0196 inch

Piston ring end gap

Top compression ring

To 2001
Production	0.009 to 0.0149 inch
Service limit	0.018 inch

2002
Production	0.009 to 0.017 inch
Service limit	0.0196 inch

Second compression ring
Production	0.018 to 0.025 inch
Service limit	0.028 inch

Oil ring
Production	0.007 to 0.027 inch
Service limit	0.030 inch

Piston ring groove clearance

Top compression ring
Production	0.0016 to 0.0031 inch
Maximum	0.0033 inch

Second compression ring
Production	0.0016 to 0.0031 inch
Maximum	0.0033 inch

Oil ring (steel rails)
Production	0.0004 to 0.008 inch
Maximum	0.008 inch

Camshaft

Bearing journal diameters	2.164 to 2.166 inches
Camshaft endplay	0.001 to 0.012 inch

Lobe lift

To 2000
Intake	0.292 inch
Exhaust	0.292 inch

2001 and later
Intake	0.274 inch
Exhaust	0.281 inch

Torque specifications*   **Ft-lbs** (unless otherwise indicated**)**

Main bearing cap bolts (in sequence - **see illustration 24.13**)

Inner bolts
Step one	15
Step two	Turn an additional 80 degrees

Outer stud nuts
Step one	15
Step two	Turn an additional 50 degrees
Side bolts	18

Connecting rod cap bolts or nuts

Models through 1999, and 2000 and later with 1st design bolts
Step 1	15
Step 2	turn an additional 60 degrees

2000 and later, 2nd design bolts
Step 1	15
Step 2	turn an additional 75 degrees

Crankshaft rear oil seal housing bolts	18
Engine valley cover bolts	18
Windage tray nuts	18

* **Note:** *Refer to Part D for additional torque specifications.*

2.4a The oil pressure sending unit on the 3.4L V6 engine is located on the left front of the block between the oil filter and the starter

2.4b The oil pressure sending unit on the 3800 V6 engine is located on top of the oil filter housing (arrow) - remove the sending unit

2.4c The oil pressure sending unit on V8 engines is located behind the intake manifold - 1st generation V8 engine shown

1 General information

Included in this portion of Chapter 2 are the general overhaul procedures for the cylinder heads and internal engine components.

The information ranges from advice concerning preparation for an overhaul and the purchase of replacement parts to detailed, step-by-step procedures covering removal and installation of internal engine components and the inspection of parts.

The following Sections have been written based on the assumption the engine has been removed from the vehicle. For information concerning in-vehicle engine repair, as well as removal and installation of the external components necessary for the overhaul, see Parts A, B, C and D of this Chapter and Section 7 of this Part.

The Specifications included in this Part are only those necessary for the inspection and overhaul procedures which follow. Refer to Parts A, B,C and D for additional Specifications.

2 Engine overhaul - general information

Refer to illustrations 2.4a, 2.4b and 2.4c

It's not always easy to determine when, or if, an engine should be completely overhauled, as a number of factors must be considered.

High mileage isn't necessarily an indication an overhaul is needed, while low mileage doesn't preclude the need for an overhaul. Frequency of servicing is probably the most important consideration. An engine that's had regular and frequent oil and filter changes, as well as other required maintenance, will most likely give many thousands of miles of reliable service. Conversely, a neglected engine may require an overhaul very early in its life.

Excessive oil consumption is an indication that piston rings, valve seals and/or valve guides are in need of attention. Make sure oil leaks aren't responsible before deciding the rings and/or guides are bad. Perform a cylinder compression check to determine the extent of the work required (see Section 3).

Remove the oil pressure sending unit and check the oil pressure with a gauge installed in its place. On V6 models, the sending unit is located on the oil filter housing, and behind the intake manifold on V8 models **(see illustrations)**. Compare the results to this Chapter's Specifications. As a general rule, engines should have ten psi oil pressure for every 1,000 rpm. If the pressure is extremely low, the bearings and/or oil pump are probably worn out.

Loss of power, rough running, knocking or metallic engine noises, excessive valve-train noise and high fuel consumption rates may also point to the need for an overhaul, especially if they're all present at the same time. If a complete tune-up doesn't remedy the situation, major mechanical work is the only solution.

An engine overhaul involves restoring the internal parts to the specifications of a new engine. During an overhaul, the piston rings are replaced and the cylinder walls are reconditioned (rebored and/or honed). If a rebore is done by an automotive machine shop, new oversize pistons will also be installed. The main bearings, connecting rod bearings and camshaft bearings are generally replaced with new ones and, if necessary, the crankshaft may be reground to restore the journals. Generally, the valves are serviced as well, since they're usually in less-than-perfect condition at this point. While the engine is being overhauled, other components, such as the starter and alternator, can be rebuilt as well. The end result should be a like-new engine that will give many trouble free miles. **Note:** *Critical cooling system components such as the hoses, drivebelts, thermostat and*

water pump MUST be replaced with new parts when an engine is overhauled. The radiator should be checked carefully to ensure it isn't clogged or leaking (see Chapter 3). Also, we don't recommend overhauling the oil pump on the 3.4L V6 or the V8 engine - always install a new one when an engine is rebuilt.

Before beginning the engine overhaul, read through the entire procedure to familiarize yourself with the scope and requirements of the job. Overhauling an engine isn't particularly difficult, if you follow all of the instructions carefully, have the necessary tools and equipment and pay close attention to all specifications; however, it can be time consuming. Plan on the vehicle being tied up for a minimum of two weeks, especially if parts must be taken to an automotive machine shop for repair or reconditioning. Check on availability of parts and make sure any necessary special tools and equipment are obtained in advance. Most work can be done with typical hand tools, although a number of precision measuring tools are required for inspecting parts to determine if they must be replaced. Often an automotive machine shop will handle the inspection of parts and offer advice concerning reconditioning and replacement. **Note:** *Always wait until the engine has been completely disassembled and all components, especially the engine block, have been inspected before deciding what service and repair operations must be performed by an automotive machine shop. Since the block's condition will be the major factor to consider when determining whether to overhaul the original engine or buy a rebuilt one, never purchase parts or have machine work done on other components until the block has been thoroughly inspected. As a general rule, time is the primary cost of an overhaul, so it doesn't pay to install worn or substandard parts.*

As a final note, to ensure maximum life and minimum trouble from a rebuilt engine, everything must be assembled with care in a spotlessly clean environment.

3 Cylinder compression check

Refer to illustration 3.6

1 A compression check will tell you what mechanical condition the upper end (pistons, rings, valves, head gaskets) of the engine is in. Specifically, it can tell you if the compression is down due to leakage caused by worn piston rings, defective valves and seats or a blown head gasket. **Note:** *The engine must be at normal operating temperature and the battery must be fully charged for this check.*

2 Begin by cleaning the area around the spark plugs before you remove them. Compressed air should be used, if available, otherwise a small brush or even a bicycle tire pump will work. The idea is to prevent dirt from getting into the cylinders as the compression check is being done.

3 Remove all of the spark plugs from the engine (see Chapter 1).

4 Block the throttle wide open.

5 Disable the fuel and ignition systems by removing the ECM IGN fuse from the instrument panel fuse block and the IGNITION fuse from the underhood fuse block on 1993 models, the PCM IGN fuse from the instrument panel fuse block and the IGNITION fuse from the underhood fuse block on 1994 and 1995 models, the PCM BATT fuse from the instrument panel fuse block and the IGNITION fuse from the underhood fuse block on 1996 and 1997 models. On 1998 and later models, disable the ignition system by disconnecting the primary electrical connectors from the ignition coil packs, then disable the fuel system by removing the fuel pump relay from the engine compartment fuse block.

6 Install the compression gauge in the number one spark plug hole **(see illustration)**.

7 Crank the engine over at least seven compression strokes and watch the gauge. The compression should build up quickly in a healthy engine. Low compression on the first stroke, followed by gradually increasing pressure on successive strokes, indicates worn piston rings. A low compression reading on the first stroke, which doesn't build up during successive strokes, indicates leaking valves or a blown head gasket (a cracked head could also be the cause). Deposits on the undersides of the valve heads can also cause low compression. Record the highest gauge reading obtained.

8 Repeat the procedure for the remaining cylinders, turning the engine over for the same length of time for each cylinder, and compare the results to this Chapter's Specifications.

9 If the readings are below normal, add some engine oil (about three squirts from a plunger-type oil can) to each cylinder, through the spark plug hole, and repeat the test.

10 If the compression increases after the oil is added, the piston rings are definitely worn. If the compression doesn't increase significantly, the leakage is occurring at the valves or head gasket. Leakage past the valves may be caused by burned valve seats and/or faces or warped, cracked or bent valves.

11 If two adjacent cylinders have equally low compression, there's a strong possibility the head gasket between them is blown. The appearance of coolant in the combustion chambers or the crankcase would verify this condition.

12 If one cylinder is about 20-percent lower than the others, and the engine has a slightly rough idle, a worn exhaust lobe on the camshaft could be the cause.

13 If the compression is unusually high, the combustion chambers are probably coated with carbon deposits. If that's the case, the cylinder heads should be removed and decarbonized.

14 If compression is way down or varies greatly between cylinders, it would be a good idea to have a leak-down test performed by an automotive repair shop. This test will pinpoint exactly where the leakage is occurring and how severe it is.

15 Install the fuses and or relays and drive the vehicle to restore the "block learn" memory.

4 Vacuum gauge diagnostic checks

A vacuum gauge provides valuable information about what is going on in the engine at a low cost. You can check for worn rings or cylinder walls, leaking head or intake manifold gaskets, incorrect carburetor adjustments, restricted exhaust, stuck or burned valves, weak valve springs, improper ignition or valve timing and ignition problems.

Unfortunately, vacuum gauge readings are easy to misinterpret, so they should be used in conjunction with other tests to confirm the diagnosis.

Both the gauge readings and the rate of needle movement are important for accurate interpretation. Most gauges measure vacuum in inches of mercury (in-Hg). As vacuum increases (or atmospheric pressure decreases), the reading will increase. Also, for every 1,000-foot increase in elevation above sea level, the gauge readings will decrease about one inch of mercury.

Connect the vacuum gauge directly to intake manifold vacuum, not to ported (carburetor) vacuum. Be sure no hoses are left disconnected during the test or false readings will result.

Before you begin the test, allow the engine to warm up completely. Block the wheels and set the parking brake. With the transmission in Park, start the engine and allow it to run at normal idle speed.

Read the vacuum gauge; an average, healthy engine should normally produce about 17 to 22 inches of vacuum with a fairly steady needle. Refer to the following vacuum gauge readings and what they indicate about the engine's condition:

1 A low, steady reading usually indicates a leaking gasket between the intake manifold and carburetor or throttle body, a leaky vacuum hose, late ignition timing or incorrect camshaft timing. Eliminate all other possible causes, utilizing the tests provided in this Chapter before you remove the timing chain cover to check the timing marks.

2 If the reading is three to eight inches below normal and it fluctuates at that low reading, suspect an intake manifold gasket leak at an intake port.

3 If the needle has regular drops of about two to four inches at a steady rate, the valves are probably leaking. Perform a compression or leak-down test to confirm this.

4 An irregular drop or down-flick of the needle can be caused by a sticking valve or an ignition misfire. Perform a compression or leak-down test and read the spark plugs.

5 A rapid vibration of about four inches-Hg vibration at idle combined with exhaust smoke indicates worn valve guides. Perform a leak-down test to confirm this. If the rapid vibration occurs with an increase in engine speed, check for a leaking intake manifold gasket or head gasket, weak valve springs, burned valves or ignition misfire.

6 A slight fluctuation, say one inch up and down, may mean ignition problems. Check all the usual tune-up items and, if necessary, run the engine on an ignition analyzer.

7 If there is a large fluctuation, perform a compression or leak-down test to look for a weak or dead cylinder or a blown head gasket.

8 If the needle moves slowly through a wide range, check for a clogged PCV system, incorrect idle fuel mixture, throttle body or intake manifold gasket leaks.

9 Check for a slow return after revving the engine by quickly snapping the throttle open until the engine reaches about 2,500 rpm and let it shut. Normally the reading should drop to near zero, rise above normal idle reading (about 5 in-Hg over) and then return to the previous idle reading. If the vacuum returns slowly and doesn't peak when the throttle is snapped shut, the rings may be worn. If there

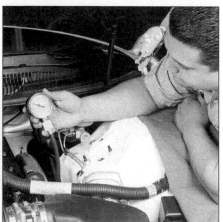

3.6 A compression gauge with a threaded fitting for the spark plug hole is preferred over the type that requires hand pressure to maintain the seal - be sure to open the throttle valve as far as possible during the compression check

is a long delay, look for a restricted exhaust system (often the muffler or catalytic converter). An easy way to check this is to temporarily disconnect the exhaust ahead of the suspected part and re-test.

5 Engine removal - methods and precautions

If you've decided the engine must be removed for overhaul or major repair work, several preliminary steps should be taken. Locating a suitable place to work is extremely important. Adequate work space, along with storage space for the vehicle, will be needed.

Cleaning the engine compartment and engine before beginning the removal procedure will help keep tools clean and organized. An engine hoist will also be necessary. Safety is of primary importance, considering the potential hazards involved in removing the engine from this vehicle.

If the engine is being removed by a novice, a helper should be available. Advice and aid from someone more experienced would also be helpful. There are many instances when one person cannot simultaneously perform all of the operations required when lifting the engine out of the vehicle.

Plan the operation ahead of time. Arrange for or obtain all of the tools and equipment you'll need prior to beginning the job. Some of the equipment necessary to perform engine removal and installation safely and with relative ease in addition to a hydraulic jack, jack stands and an engine hoist) are a complete sets of wrenches and sockets as described in the front of this manual, wooden blocks and plenty of rags and cleaning solvent for mopping up spilled oil, coolant and gasoline.

Plan for the vehicle to be out of use for quite a while. A machine shop will be required to perform some of the work which the do-it-yourselfer can't accomplish without special equipment. These shops often have a busy schedule, so it would be a good idea to consult them before removing the engine in order to accurately estimate the amount of time required to rebuild or repair components that may need work.

Always be extremely careful when removing and installing the engine. Serious injury can result from careless actions. Plan ahead, take your time and a job of this nature, although major, can be accomplished successfully. **Note:** *Because it may be some time before you reinstall the engine, it is very helpful to make sketches or take photos of various accessory mountings and wiring hookups before removing the engine.*

6 Engine - removal and installation

Warning 1: *The models covered by this manual are equipped with airbags. Always disable the airbag system before working in the vicin-*

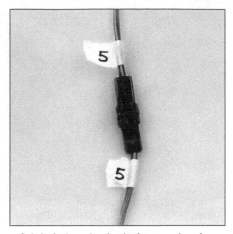

6.4 Label each wire before unplugging the connector

ity of the impact sensors, steering column or instrument panel to avoid the possibility of accidental deployment of the airbag(s), which could cause personal injury (see Chapter 12). The yellow wires and connectors routed through the instrument panel and, on 1995 and earlier models, to the front of the vehicle, are for this system. Do not use electrical test equipment on these yellow wires or tamper with them in any way.*
Warning 2: *Gasoline is extremely flammable, so take extra precautions when you work on any part of the fuel system. Don't smoke or allow open flames or bare light bulbs near the work area, and don't work in a garage where a gas-type appliance (such as a water heater or a clothes dryer) is present. Since gasoline is carcinogenic, wear latex gloves when there's a possibility of being exposed to fuel, and, if you spill any fuel on your skin, rinse it off immediately with soap and water. Mop up any spills immediately and do not store fuel-soaked rags where they could ignite. The fuel system is under constant pressure, so, if any fuel lines are to be disconnected, the fuel pressure in the system must be relieved first. When you perform any kind of work on the fuel system, wear safety glasses and have a Class B type fire extinguisher on hand.*
Warning 3: *The air conditioning system is under high pressure - have a dealer service department or service station evacuate the system and recapture the refrigerant before disconnecting any of the hoses or fittings.*
Caution: *On models equipped with a Delco Loc II or Theftlock audio system, be sure the lockout feature is turned off before performing any procedure which requires disconnecting the battery.*

Removal

Refer to illustrations 6.4, 6.8, 6.11 and 6.20
Note: *The factory method for engine removal in the Camaro and Firebird is out the bottom of the vehicle, by means of raising the vehicle and taking down the entire front suspension/engine cradle. Without a garage-type lift, the home mechanic has no **safe** way to raise a vehicle high enough to utilize this method of*

6.8 While the vehicle is raised, disconnect any wiring harnesses attached to the block

engine removal. The procedure outlined below illustrates the necessary steps to remove the engine traditionally, i.e. from above with an engine hoist.*
1   Relieve the fuel system pressure (see Chapter 4), then disconnect the negative cable from the battery.
2   Cover the fenders and cowl. Special pads are available to protect the fenders, but an old bedspread or blanket will also work. Remove the hood (see Chapter 11).
3   Remove the air intake duct assembly (see Chapter 4).
4   To ensure correct reassembly, label each vacuum line, emission system hose, electrical connector, ground strap and fuel line. Pieces of masking tape with numbers or letters written on them prevent confusion at assembly time **(see illustration)**. Or sketch the engine compartment routing of lines, hoses and wires.
5   Drain the cooling system (see Chapter 1) and label and detach all coolant hoses from the engine.
6   Disconnect the throttle and cruise control cables (see Chapter 4) and TV cable (see Chapter 7B).
7   Disconnect the air conditioning hoses from the compressor, hang them out of the way, detach the air conditioning compressor from its mounting bracket and remove it from the engine compartment (see Chapter 3).
8   Raise the vehicle and suitably support it on jackstands. While raised, perform the disassembly procedures that can be done only from underneath, such as disconnecting the exhaust, and disconnecting the transmission cooler lines where they are held by a clip to one of the oil pan bolts (automatic transmission models) and disconnecting wiring harnesses attached to the block **(see illustration)**.
9   Drain the engine oil and remove the filter (see Chapter 1).
10   Remove the starter (see Chapter 5). Remove the driveplate access cover.
11   On models with an automatic transmission, make an alignment mark between the driveplate and the torque converter **(see**

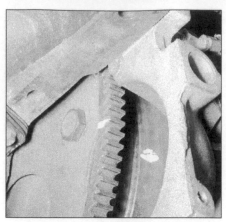

6.11 Paint or scribe alignment marks on the driveplate and the torque converter to ensure that the two components are still in balance when they're reassembled

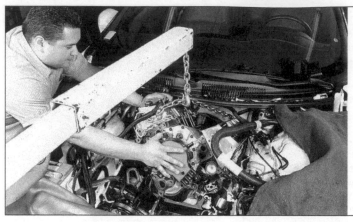

6.20 With the accessories tied out of the way and the engine mount through-bolts removed, tilt, twist and raise the engine with the hoist until it clears everything

illustration), then rotate the crankshaft and remove the driveplate bolts.

12   Remove the lower bellhousing to-engine bolts.

13   Lower the vehicle and place a floor jack under the transmission, then support the engine with an engine hoist from above. If the vehicle is equipped with an automatic transmission, be sure to place a block of wood on the head of the jack to protect the fluid pan.

14   Remove the radiator and air conditioning condenser (see Chapter 3).

15   Unbolt the power steering pump and bracket and tie them out of the way.

16   Remove the intake manifold plenum on V6 models, or the whole intake manifold on the V8 engines.

17   Raise the engine enough to take the weight off the engine mounts and remove the engine mount through-bolts (see Part A, B, C or D of this Chapter).

18   Working at the back of the engine, remove the remaining bellhousing bolts. **Note:** *On the V8 engines, the upper left bellhousing bolt will not come out of the bellhousing all the way, as it is too close to the firewall. Just unscrew it completely and leave the bolt in the bellhousing.*

19   Raise the engine and pull it forward to free it from the torque converter snout (automatic transmissions) or transmission input shaft (manual transmissions).

20   Tie the wiring harnesses out of the way, raise the engine with the hoist, and with a combination of tilting, twisting and raising, move the engine around any obstructions and pull it out of the vehicle, raising it high enough to clear the front of the body **(see illustration)**.

21   Remove the driveplate or flywheel (see Section 28) and mount the engine on an engine stand.

Installation

22   While the engine is out, check the engine mounts and the transmission mount (see Chapter 7). If they're worn or damaged, replace them.

23   Carefully lower the engine, twisting it to clear any harnesses or obstructions, until the transmission input shaft or converter snout lines up and the bellhousing-to-engine bolts can be inserted. **Caution:** *DO NOT use the transmission-to-engine bolts to force the transmission and engine together. Take great care when mating the torque converter to the driveplate, following the procedure outlined in Chapter 7 Part B. Make sure the alignment marks you made on the driveplate and the torque converter during removal are lined up.*

24   On models with an automatic transmission, install the driveplate-to-torque converter bolts and tighten them to the torque listed in this Chapter's Specifications.

25   Reinstall the remaining components in the reverse order of removal. Double-check to make sure everything is hooked up right, using the sketches or photos taken earlier to go by. On V8 engines, be sure the metal tubing that connects the air pump ports from one side of the engine to the other is installed right after the engine is installed, and before the other accessories are bolted on.

26   Add coolant, oil, power steering and transmission fluid as needed.

27   Run the engine and check for leaks and proper operation of all accessories, then install the hood and test drive the vehicle.

---

7   Engine rebuilding alternatives

The home mechanic is faced with a number of options when performing an engine overhaul. The decision to replace the engine block, piston/connecting rod assemblies and crankshaft depends on a number of factors, with the number one consideration being the condition of the block. Other considerations are cost, access to machine shop facilities, parts availability, time required to complete the project and the extent of prior mechanical experience.

Some of the rebuilding alternatives include:

Individual parts - If the inspection procedures reveal the engine block and most engine components are in reusable condition, purchasing individual parts may be the most economical alternative. The block, crankshaft and piston/connecting rod assemblies should all be inspected carefully. Even if the block shows little wear, the cylinder bores should be surface-honed.

Short-block - A short-block consists of an engine block with a crankshaft and piston/connecting rod assemblies already installed. All new bearings are incorporated and all clearances will be correct. The existing camshaft, valve train components, cylinder heads and external parts can be bolted to the short block with little or no machine shop work necessary. Some rebuilding companies include a new timing chain, camshaft and lifters with their short-block assemblies.

Long-block - A long-block consists of a short block plus an oil pump, oil pan, cylinder heads, rocker arm covers, camshaft and valve train components, timing sprockets and chain and timing chain cover. All components are installed with new bearings, seals and gaskets incorporated throughout. The installation of manifolds and external parts is all that's necessary. Give careful thought to which alternative is best for you and discuss the situation with local automotive machine shops, auto parts dealers and experienced rebuilders before ordering or purchasing replacement parts.

Used Engine – Money can often be saved by purchasing a complete used engine from an auto wrecking yard. Make sure you get the same year and model engine, that the donor vehicle's mileage is low, and that the wrecking yard offers a warranty on the used engine.

---

8   Engine overhaul - disassembly sequence

1   It's much easier to disassemble and work on the engine if it's mounted on a portable engine stand. A stand can often be rented quite cheaply from an equipment rental yard. Before it's mounted on a stand, the flywheel/driveplate should be removed from the engine.

2   If a stand isn't available, it's possible to disassemble the engine with it blocked up on the floor. Be extra careful not to tip or drop

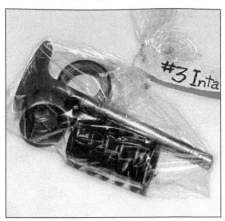

9.2 A small plastic bag, with an appropriate label, can be used to store the valve train components so they can be kept together and reinstalled in the original positions

9.3 Use a valve spring compressor to compress the spring, then remove the keepers from the valve stem

9.4 If the valve won't pull through the guide, deburr the edge of the stem end and the area around the top of the keeper groove with a file or whetstone

the engine when working without a stand.

3   If you're going to obtain a rebuilt engine, all external components must come off first, to be transferred to the replacement engine, just as they will if you're doing a complete engine overhaul yourself. These include:

Alternator and brackets
Emissions control components
Ignition coil/module assembly, spark plug wires and spark plugs
Thermostat and housing cover
Water pump
Engine front cover
Fuel injection components
Intake/exhaust manifolds
Oil filter
Engine mounts
Flywheel/driveplate
Rear main oil seal housing

Note: *When removing the external components from the engine, pay close attention to details that may be helpful or important during installation. Note the installed position of gaskets, seals, spacers, pins, brackets, washers, bolts and other small items.*

4   If you're obtaining a short-block, then the cylinder heads, oil pan and oil pump will have to be removed as well. See *Engine rebuilding alternatives* for additional information regarding the different possibilities to be considered.

5   If you're planning a complete overhaul, the engine must be disassembled and the internal components removed in the following general order:

Valve covers
Intake and exhaust manifolds
Rocker arms and pushrods
Cylinder heads
Valve lifters
Oil pan
Timing chain cover and oil pump
Timing chain and sprockets
Camshaft
Balance shaft (3800 engine only)
Water pump driveshaft (1st generation V8 engine only)

Windage tray (2nd generation V8 engine only)
Piston/connecting rod assemblies
Rear main oil seal retainer
Crankshaft and main bearings

6   Before beginning the disassembly and overhaul procedures, make sure the following items are available. Also, refer to *Engine overhaul - reassembly sequence* for a list of tools and materials needed for engine reassembly.

Common hand tools
Small cardboard boxes or plastic bags for storing parts
Gasket scraper
Ridge reamer
Engine balancer puller
Micrometers
Telescoping gauges
Dial indicator set
Valve spring compressor
Cylinder surfacing hone
Piston ring groove-cleaning tool
Electric drill motor
Tap and die set
Wire brushes
Oil gallery brushes
Cleaning solvent

9   Cylinder head - disassembly

Refer to illustrations 9.2, 9.3 and 9.4
Note: *New and rebuilt cylinder heads are commonly available for most engines at dealerships and auto parts stores. Due to the fact that some specialized tools are necessary for the disassembly and inspection procedures, and replacement parts aren't always readily available, it may be more practical and economical for the home mechanic to purchase replacement heads rather than taking the time to disassemble, inspect and recondition the originals.*

1   Cylinder head disassembly involves removal of the intake and exhaust valves and related components. Remove the rocker arm bolts, pivots and rocker arms from the cylinder heads. Label the parts or store them separately so they can be reinstalled in their orig-

inal locations.

2   Before the valves are removed, arrange to label and store them, along with their related components, so they can be kept separate and reinstalled in their original locations **(see illustration)**.

3   Compress the springs on the first valve with a spring compressor and remove the keepers **(see illustration)**. Carefully release the valve spring compressor and remove the retainer, the spring and the spring seat (if used).

4   Pull the valve out of the head, then remove the oil seal from the guide. If the valve binds in the guide (won't pull through), push it back into the head and deburr the area around the keeper groove with a fine file or whetstone **(see illustration)**.

5   Repeat the procedure for the remaining valves. Remember to keep all the parts for each valve together so they can be reinstalled in the same locations.

6   Once the valves and related components have been removed and stored in an organized manner, the heads should be thoroughly cleaned and inspected. If a complete engine overhaul is being done, finish the engine disassembly procedures before beginning the cylinder head cleaning and inspection process.

10   Cylinder head - cleaning and inspection

1   Thorough cleaning of the cylinder heads and related valve train components, followed by a detailed inspection, will enable you to decide how much valve service work must be done during the engine overhaul. **Note:** *If the engine was severely overheated, the cylinder head is probably warped* (see Step 12).

Cleaning

2   Scrape all traces of old gasket material and sealant off the head gasket, intake manifold and exhaust manifold mating surfaces.

10.12 Check the cylinder head gasket surface for warpage by trying to slip a feeler gauge under the straightedge (see this Chapter's Specifications for the maximum warpage allowed and use a feeler gauge of that thickness)

10.14 A dial indicator can be used to determine the valve stem-to-guide clearance (move the valve stem as indicated by the arrows)

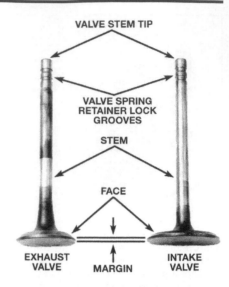

10.15 Check for valve wear at the points shown here

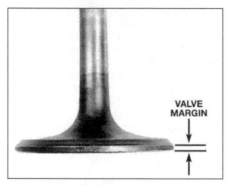

10.16 The margin width on the valve must be as specified (if no margin exists, the valve cannot be re-used)

10.17 Measure the free length of each valve spring with a dial or vernier caliper

Be very careful not to gouge the cylinder head, especially on the V8 engine, which has aluminum heads. Special gasket removal solvents that soften gaskets and make removal much easier are available at auto parts stores.

3   Remove all built-up scale from the coolant passages.

4   Run a stiff wire brush through the various holes to remove deposits that may have formed in them.

5   Run an appropriate-size tap into each of the threaded holes to remove corrosion and thread sealant that may be present. If compressed air is available, use it to clear the holes of debris produced by this operation. **Warning:** *Wear eye protection when using compressed air!*

6   Clean the rocker arm pivot stud or bolt threads with a wire brush.

7   Clean the cylinder head with solvent and dry it thoroughly. Compressed air will speed the drying process and ensure that all holes and recessed areas are clean. **Note:** *Decarbonizing chemicals are available and may prove very useful when cleaning cylinder heads and valve train components. They're very caustic and should be used with caution. Be sure to follow the instructions on the container.*

8   Clean the rocker arms, pivots, bolts and pushrods with solvent and dry them thoroughly (don't mix them up during the cleaning process). Compressed air will speed the drying process and can be used to clean out the oil passages.

9   Clean all the valve springs, keepers and retainers with solvent and dry them thoroughly. Do the components from one valve at a time to avoid mixing up the parts.

10   Scrape off any heavy deposits that may have formed on the valves, then use a motorized wire brush to remove deposits from the valve heads and stems. Again, make sure the valves don't get mixed up.

Inspection

Note: *Be sure to perform all of the following inspection procedures before concluding machine shop work is required. Make a list of the items that need attention.*

Cylinder head

Refer to illustrations 10.12 and 10.14

11   Inspect the head very carefully for cracks, evidence of coolant leakage and other damage. If cracks are found, check with an automotive machine shop concerning repair. If repair isn't possible, a new cylinder head must be obtained.

12   Using a straightedge and feeler gauge, check the head gasket mating surface for warpage **(see illustration)**. If the warpage exceeds the limit in this Chapter's Specifications, it can be resurfaced at an automotive machine shop. **Note:** *If the heads are resurfaced, the intake manifold flanges will also require machining.*

13   Examine the valve seats in each of the combustion chambers. If they're pitted, cracked or burned, the head will require valve service that's beyond the scope of the home mechanic.

14   Check the valve stem-to-guide clearance by measuring the lateral movement of

the valve stem with a dial indicator attached securely to the head **(see illustration)**. The valve must be in the guide and approximately 1/16-inch off the seat. The total valve stem movement indicated by the gauge needle must be divided by two to obtain the actual clearance. After this is done, if there's still some doubt regarding the condition of the valve guides, they should be checked by an automotive machine shop (the cost should be minimal).

Valves

Refer to illustrations 10.15 and 10.16

15   Carefully inspect each valve face for uneven wear, deformation, cracks, pits and burned areas. Check the valve stem for scuffing and galling and the neck for cracks. Rotate the valve and check for any obvious indication that it's bent. Look for pits and excessive wear on the end of the stem. The presence of any of these conditions **(see illustration)** indicates the need for valve service by an automotive machine shop.

16   Measure the margin width on each valve **(see illustration)**. Any valve with a margin narrower than specified in this Chapter will have to be replaced with a new one.

10.18 Check each valve spring for squareness

Valve components

Refer to illustrations 10.17 and 10.18

17 Check each valve spring for wear (on the ends) and pits. Measure the free length and compare it to this Chapter's Specifications **(see illustration)**. Any springs that are shorter than specified have sagged and shouldn't be re-used. The tension of all springs should be checked with a special fixture before deciding they're suitable for use in a rebuilt engine (take the springs to an automotive machine shop for this check).

18 Stand each spring on a flat surface and check it for squareness **(see illustration)**. If any of the springs are distorted or sagged, replace all of them with new parts.

19 Check the spring retainers and keepers for obvious wear and cracks. Any questionable parts should be replaced with new ones, as extensive damage will occur if they fail during engine operation.

Rocker arm components

20 Check the rocker arm faces (the areas that contact the pushrod ends and valve stems) for pits, wear, galling, score marks and rough spots. Check the rocker arm pivot contact areas and pivots as well. Look for cracks in each rocker arm and bolt.

21 Inspect the pushrod ends for scuffing and excessive wear. Roll each pushrod on a flat surface, like a piece of plate glass, to determine if it's bent. Check the pushrod guide plates for signs of excessive wear.

22 Check the rocker arm bolt holes or studs in the cylinder heads for damaged threads.

23 Any damaged or excessively worn parts must be replaced with new ones.

All components

24 If the inspection process indicates the valve components are in generally poor condition and worn beyond the limits specified, which is usually the case in an engine that's being overhauled, reassemble the valves in the cylinder head (see Section 11 for valve servicing recommendations).

11 Valves - servicing

1 Because of the complex nature of the job and the special tools and equipment needed, servicing of the valves, the valve seats and the valve guides, commonly known as a valve job, should be done by a professional.

2 The home mechanic can remove and disassemble the head, do the initial cleaning and inspection, then reassemble and deliver it to a dealer service department or an automotive machine shop for the actual service work. Doing the inspection will enable you to see what condition the head and valvetrain components are in and will ensure that you know what work and new parts are required when dealing with an automotive machine shop.

3 The dealer service department, or automotive machine shop, will remove the valves and springs, recondition or replace the valves and valve seats, recondition the valve guides, check and replace the valve springs, rotators, spring retainers and keepers (as necessary), replace the valve seals with new ones, reassemble the valve components and make sure the installed spring height is correct. The cylinder head gasket surface will also be resurfaced if it's warped.

4 After the valve job has been performed by a professional, the head will be in like new condition. When the head is returned, be sure to clean it again before installation on the engine to remove any metal particles and abrasive grit that may still be present from the valve service or head resurfacing operations. Use compressed air, if available, to blow out all the oil holes and passages.

12 Cylinder head - reassembly

1 Regardless of whether or not the head was sent to an automotive repair shop for valve servicing, make sure it's clean before beginning reassembly.

2 If the head was sent out for valve servicing, the valves and related components will already be in place. Begin the reassembly procedure with Step 8.

3 Beginning at one end of the head, lubricate and install the first valve. Apply moly-base grease or clean engine oil to the valve stem.

4 Install the valve spring seat (aluminum heads only) and/or the spring shims, if originally installed, before the valve seals.

5 Install new seals on each of the valve guides. Gently tap each seal into place until it's completely seated on the guide. **Note:** *On 2nd generation V8 engines (1998 and later) be sure to position the valve seals at the proper depth on the guide or damage to the seals will occur* **(see illustration 6.15b in Chapter 2D).** Many seal sets come with a plastic installer, but use hand pressure. Do not hammer on the seals or they could be driven down too far and subsequently leak. Don't twist or cock the seals during installation or they won't seal properly on the valve stems.

V6 engines

Refer to illustration 12.7

6 **Note:** *On V6 engines, the intake valves have an oil seal O-ring and a stem seal, while the exhaust valves have only the O-ring seal, but the exhaust valves have an oil shield or "shedder" between the retainer and the valve spring.*

7 The V6 components may be installed in the following order **(see illustration):**

Shims
Seals (intake only)
Valves, followed by the stem O-rings
Spring dampers
Springs
Oil shedder shields (exhaust only)
Retainers
Keepers

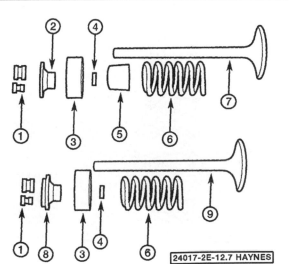

12.7 Typical valve components

1 *Keepers*
2 *Retainer*
3 *Oil shield*
4 *O-ring oil seal*
5 *Umbrella seal*
6 *Spring and damper*
7 *Intake valve*
8 *Retainer/rotator*
9 *Exhaust valve*

24017-2E-12.7 HAYNES

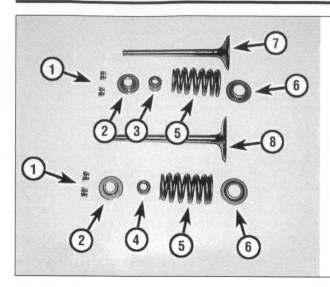

12.8 Valves and related components (2nd generation V8 engine shown, 1st generation V8 similar)

1   Valve keeper
2   Retainer
3   Exhaust valve seal
4   Intake valve seal
5   Valve spring
6   Valve spring seat (aluminum heads)
7   Exhaust valve
8   Intake valve

12.9 Apply a small dab of grease to each keeper as shown here before installation - it'll hold them in place on the valve stem as the spring is released

V8 engines

8   Install the components in the following order **(see illustration)**:

Spring seat
Shims (if necessary)
Valve, followed by the stem seals
Springs
Retainers
Keepers

All engines

Refer to illustration 12.9

9   Compress the springs with a valve spring compressor and carefully install the keepers in the groove, then slowly release the compressor and make sure the keepers seat properly. Apply a small dab of grease to each keeper to hold it in place if necessary **(see illustration)**. Tap the valve stem tips with a plastic hammer to seat the keepers, if necessary.

10   Repeat the procedure for the remaining valves. Be sure to return the components to

their original locations - don't mix them up!

11   Check the installed valve spring height with a ruler graduated in 1/32-inch increments or a dial caliper. If the head was sent out for service work, the installed height should be correct (but don't automatically assume it is). The measurement is taken from the top of each spring seat or top shim to the bottom of the retainer. If the height is greater than specified in this Chapter, shims can be added under the springs to correct it. **Caution:** Do not, under any circumstances, shim the springs to the point where the installed height is less than specified.

12   If the rocker arm studs had been previously removed from the head (3.4L V6 and 1st generation V8 engines), install the guide plates, coat the bottom threads of the rocker studs with non-hardening sealant and torque them to Specifications.

13   Apply moly-base grease to the rocker arm faces and the pivots/balls, then install the rocker arms and pivots/balls on the cylinder heads. Tighten the bolts/nuts finger-tight.

13 Camshaft, balance shaft and bearings - removal and inspection

Camshaft removal

Refer to illustrations 13.1a and 13.1b

1   Refer to Part A, B, C or D for the camshaft and lifter removal and inspection procedures. Disregard the Steps that do not apply, since the engine is already removed from the vehicle. If the cylinder heads and camshaft have already been removed, an alternate method of checking camshaft lobe lift is outlined below. When performing an overhaul on 2nd generation V8 engines, it will be necessary to remove the engine valley cover to aid in the removal of the camshaft. Remove the bolts and detach the cover from the engine block. Be sure to install a new gasket and knock sensor oil seals upon installation **(see illustrations)**. It is assumed that the cylinder heads and the coolant vapor vent tube is already removed at this point.

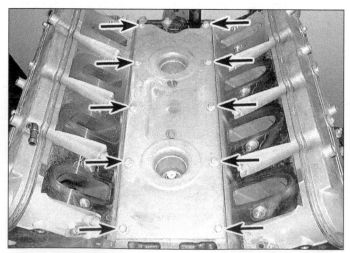

13.1a Valley cover mounting bolts - 2nd generation V8 engine

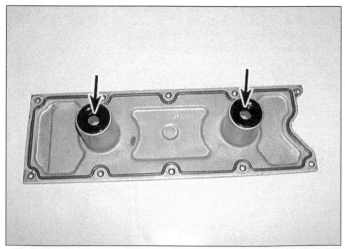

13.1b Be sure to replace the valley cover gasket and the knock sensor oil seals (arrows) upon installation

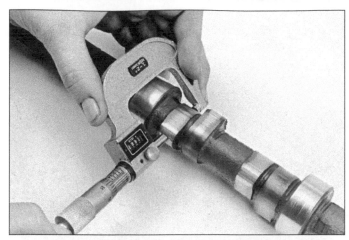

13.2 To verify camshaft lobe lift, measure the major and the minor diameters - subtract each minor diameter from the major diameter to arrive at the lobe lift

13.3 Measure the radial runout of the balance shaft at the center of the shaft

13.5a Remove the balance shaft retainer

13.5b A slide hammer must be threaded into the front of the balance shaft to pull the balance shaft front bearing out of the block

Alternate method for camshaft lobe-lift check

Refer to illustration 13.2

2   Using a micrometer, measure the lobe at its highest point **(see illustration)**. Then measure the base circle perpendicular (90-degrees) to the lobe. Do this for each lobe and record the results. Subtract the base circle measurement from the lobe height. The difference is the lobe lift.

Balance shaft inspection and removal (3800 V6 only)

Refer to illustrations 13.3, 13.5a and 13.5b
Note: *If the balance shaft is to be removed, the balance shaft gear bolt should be removed before the timing chain is removed, so the shaft is held while the bolt is loosened.*
3   Using a dial indicator against the center of the balance shaft, within the lifter valley of the block, measure the radial runout of the balance shaft and compare the figures to this Chapter's Specifications **(see illustration)**. The balance shaft rides in a bearing at the

front of the block and a bushing at the rear. If the runout exceeds Specifications, replace the balance shaft, bearing and bushing as a set.
4   The rear bushing must be installed by a machine shop with the proper tools, to the proper depth and with the oil hole lined up.
5   Remove the two bolts and the balance shaft retainer at the front of the block **(see illustration)**. The balance shaft's front bearing fits tightly into the block, and a slide hammer must be threaded into the front of the balance shaft to pull it out **(see illustration)**.

Camshaft bearing inspection

6   Check the camshaft bearings in the block for wear and damage. Look for galling, pitting and discolored areas.
7   The inside diameter of each bearing can be determined with a small hole gauge and outside micrometer or an inside micrometer. Subtract the camshaft bearing journal diameter(s) from the corresponding bearing inside diameter(s) to obtain the bearing oil clearance. If it's excessive, new bearings will be

required regardless of the condition of the originals.
8   Balance shaft and camshaft bearing replacement requires special tools and expertise that place it outside the scope of the home mechanic. Take the block to an automotive machine shop to ensure the job is done correctly.

14   Pistons and connecting rods - removal

Refer to illustrations 14.1, 14.3, 14.4 and 14.6
Note: *Prior to removing the piston/connecting rod assemblies, remove the cylinder heads, the oil pan and the oil pump by referring to the appropriate Sections in Part A, B, C or D of Chapter 2. On 2nd generation (1998 and later) V8 engines, it will also be necessary to remove the crankshaft windage tray.*
1   Use your fingernail to feel if a ridge has formed at the upper limit of ring travel (about 1/4-inch down from the top of each cylinder).

14.1 A ridge reamer is required to remove the ridge from the top of each cylinder - do this before removing the pistons!

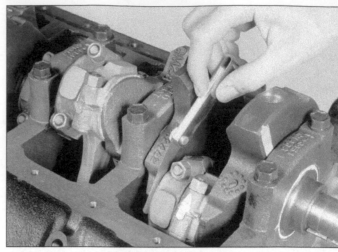

14.3 Check the connecting rod side clearance with a feeler gauge as shown

If carbon deposits or cylinder wear have produced ridges, they must be completely removed with a special tool **(see illustration)**. Follow the manufacturer's instructions provided with the tool. Failure to remove the ridges before attempting to remove the piston/connecting rod assemblies may result in piston breakage.

2   After the cylinder ridges have been removed, turn the engine upside-down so the crankshaft is facing up.

3   Before the connecting rods are removed, check the endplay with feeler gauges. Slide them between the first connecting rod and the crankshaft throw until the play is removed **(see illustration)**. The endplay is equal to the thickness of the feeler gauge(s). If the endplay exceeds the service limit, new connecting rods will be required. If new rods (or a new crankshaft) are installed, the endplay may fall under the minimum specified in this Chapter (if it does, the rods will have to be machined to restore it - consult an automotive machine shop for advice if

necessary). Repeat the procedure for the remaining connecting rods.

4   Check the connecting rods and caps for identification marks. If they aren't plainly marked, use a small center-punch **(see illustration)** to make the appropriate number of indentations on each rod and cap (1, 2, 3, etc., depending on the cylinder they're associated with).

5   Loosen each of the connecting rod cap nuts 1/2-turn at a time until they can be removed by hand. Remove the number one connecting rod cap and bearing insert. Don't drop the bearing insert out of the cap.

6   Slip a short length of plastic or rubber hose over each connecting rod cap bolt to protect the crankshaft journal and cylinder wall as the piston is removed **(see illustration)**. **Note:** *On 2nd generation V8 engines, the rod bolts are removed with the caps, so it may be helpful to make a set of connecting rod guides. Find several bolts that fit the rods at your local hardware store, or purchase a couple of new rod bolts. Cut the heads of the*

bolts off with a hacksaw or similar tool and slip a short length of hose over the end of each bolt to make a pair connecting rod guides. Screw the guides into the connecting rods during removal and installation of the piston/connecting rod from the engine block.

7   Remove the bearing insert and push the connecting rod/piston assembly out through the top of the engine. Use a wooden or plastic hammer handle to push on the upper bearing surface in the connecting rod. If resistance is felt, double-check to make sure all of the ridge was removed from the cylinder.

8   Repeat the procedure for the remaining cylinders.

9   After removal, reassemble the connecting rod caps and bearing inserts in their respective connecting rods and install the cap nuts finger tight. Leaving the old bearing inserts in place until reassembly will help prevent the connecting rod bearing surfaces from being accidentally nicked or gouged.

10   Don't separate the pistons from the connecting rods.

14.4 Mark the rod bearing caps in order from the front of the engine to the rear (one mark for the front cap, two for the second one and so on)

14.6 To prevent damage to the crankshaft journals and cylinder walls, slip sections of rubber or plastic hose over the rod bolts before removing the pistons/rods

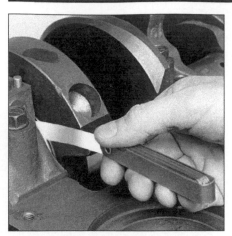

15.3 Checking crankshaft endplay with a feeler gauge

15.4 The arrow on the main bearing cap indicates the front of the engine

16.4a A hammer and large punch can be used to knock the core plugs sideways in their bores

15 Crankshaft - removal

Refer to illustrations 15.3 and 15.4

Note: *The crankshaft can be removed only after the engine has been removed from the vehicle. It's assumed the flywheel/driveplate, crankshaft balancer, timing chain, oil pan, oil pump, windage tray (if equipped) and the piston/connecting rod assemblies have already been removed. The rear main oil seal retainer must also be removed first (3800 V6 and V8 engines).*

1   Before the crankshaft is removed, check the endplay. Mount a dial indicator with the stem in line with the crankshaft and touching one of the crank throws.

2   Push the crankshaft all the way to the rear and zero the dial indicator. Next, pry the crankshaft to the front as far as possible and check the reading on the dial indicator. The distance it moves is the endplay. If it's greater than listed in this Chapter's Specifications, check the crankshaft thrust surfaces for wear. If no wear is evident, new main bearings should correct the endplay.

3   If a dial indicator isn't available, feeler gauges can be used. Gently pry or push the crankshaft all the way to the front of the engine. Slip feeler gauges between the crank-shaft and the front face of the thrust main bearing to determine the clearance **(see illustration)**. **Note:** *The thrust bearing is located at the number two (3800), number three (3.4L and 2nd generation V8) or number five (1st generation V8) main bearing cap.*

4   Check the main bearing caps to see if they're marked to indicate their locations. They should be numbered consecutively from the front of the engine to the rear. If they aren't, mark them with number stamping dies or a center-punch. Main bearing caps generally have a cast-in arrow, which points to the front of the engine **(see illustration)**. Loosen the main bearing cap bolts 1/4-turn at a time each, until they can be removed by hand. Note if any stud bolts are used and make sure they're returned to their original locations when the crankshaft is reinstalled.

Note: *On the 3800 V6 and 2nd generation V8 engines, there are side bolts on the main caps that must be removed before removing the caps.*

5   Gently tap the caps with a soft-face hammer, then separate them from the engine block. If necessary, use the bolts as levers to remove the caps. Try not to drop the bearing inserts if they come out with the caps. **Note:** *Main caps on the 3800 V6 are "press-fit" into the block, and it is recommended using a special tool, available at most auto parts stores, which is a steel plate that attaches with two bolts to the main cap (after the main cap bolts are removed). To this plate you attach a slide hammer and gently yank the cap straight out of the block.*

6   On 3800 V6 and V8 engines, remove the rear main oil seal retainer plate before removing the crankshaft.

7   Carefully lift the crankshaft straight out of the engine. It may be a good idea to have an assistant available, since the crankshaft is quite heavy. With the bearing inserts in place in the engine block and main bearing caps, return the caps to their respective locations on the engine block and tighten the bolts finger tight.

16 Engine block - cleaning

Refer to illustrations 16.4a, 16.4b, 16.8 and 16.10

1   Remove the main bearing caps and separate the bearing inserts from the caps and the engine block. Tag the bearings, indicating which cylinder they were removed from and whether they were in the cap or the block, then set them aside.

2   Using a gasket scraper, remove all traces of gasket material from the engine block. Be very careful not to nick or gouge the gasket sealing surfaces.

3   Remove all of the covers and threaded oil gallery plugs from the block. The plugs are usually very tight - they may have to be drilled out and the holes retapped. Use new plugs when the engine is reassembled.

16.4b Pull the core plugs from the block with pliers

4   Remove the core plugs from the engine block. To do this, knock one side of the plugs into the block with a hammer and punch, then grasp them with large pliers and pull them out **(see illustrations)**.

5   If the engine is extremely dirty, it should be taken to an automotive machine shop to be cleaned. **Note:** *If the block is cleaned in a caustic-solution hot tank, this will ruin any camshaft or balance-shaft bearings left in the block. If the engine is being rebuilt, these bearings should be replaced anyway.*

6   After the block is returned, clean all oil holes and oil galleries one more time. Brushes specifically designed for this purpose are available at most auto parts stores. Flush the passages with warm water until the water runs clear, dry the block thoroughly and wipe all machined surfaces with a light, rust preventive oil. If you have access to compressed air, use it to speed the drying process and blow out all the oil holes and galleries. **Warning:** *Wear eye protection when using compressed air!*

7   If the block isn't extremely dirty or sludged up, you can do an adequate cleaning job with hot soapy water and a stiff brush. Take plenty of time and do a thorough job.

16.8 All bolt holes in the block - particularly the main bearing cap and head bolt holes - should be cleaned and restored with a tap (be sure to remove debris from the holes after this is done)

16.10 A large socket on an extension can be used to drive the new core plugs into the bores

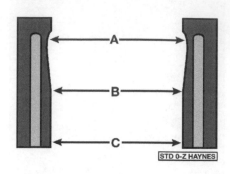

17.4a Measure the diameter of each cylinder just under the wear ridge (A), at the center (B) and at the bottom (C)

Regardless of the cleaning method used, be sure to clean all oil holes and galleries very thoroughly, dry the block completely and coat all machined surfaces with light oil.

8   The threaded holes in the block must be clean to ensure accurate torque readings during reassembly. Run the proper size tap into each of the holes to remove rust, corrosion, thread sealant or sludge and restore damaged threads **(see illustration)**. If possible, use compressed air to clear the holes of debris produced by this operation. Now is a good time to clean the threads on the head bolts and the main bearing cap bolts as well (not necessary on the 3800 V6, since these head bolts must be replaced during an overhaul).

9   Reinstall the main bearing caps and tighten the bolts finger tight.

10   After coating the sealing surfaces of the new core plugs with a non-hardening sealant (such as Permatex no. 2), install them in the engine block **(see illustration)**. Make sure they're driven in straight and seated properly or leakage could result. Special tools are available for this purpose, but a large socket, with an outside diameter that will just slip into the core plug, a 1/2-inch drive extension and a hammer will work just as well.

11   Apply non-hardening sealant (such as Permatex no. 2 or Teflon pipe sealant) to the new oil gallery plugs and thread them into the holes in the block. Make sure they're tightened securely.

12   If the engine isn't going to be reassembled right away, cover it with a large plastic trash bag to keep it clean.

17   Engine block - inspection

Refer to illustrations 17.4a, 17.4b and 17.4c
Note: *The manufacturer recommends checking the block deck for warpage and the main bearing bore concentricity and alignment. Since special measuring tools are needed,*

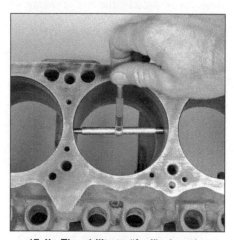

17.4b The ability to "feel" when the telescoping gauge is at the correct point will be developed over time, so work slowly and repeat the check until you're satisfied the bore measurement is accurate

the checks should be done by an automotive machine shop.

1   Before the block is inspected, it should be cleaned as described in Section 16.

2   Visually check the block for cracks, rust and corrosion. Look for stripped threads in the threaded holes. It's also a good idea to have the block checked for hidden cracks by an automotive machine shop that has the special equipment to do this type of work. If defects are found, have the block repaired, if possible, or replaced.

3   Check the cylinder bores for scuffing and scoring.

4   Check the cylinders for taper and out-of-round conditions as follows **(see illustrations)**:

5   Measure the diameter of each cylinder at the top (just under the ridge area), center and bottom of the cylinder bore, parallel to the crankshaft axis.

6   Next, measure each cylinder's diameter at the same three locations perpendicular to the crankshaft axis.

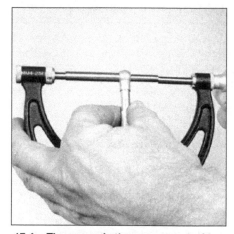

17.4c The gauge is then measured with a micrometer to determine the bore size

7   The taper of each cylinder is the difference between the bore diameter at the top of the cylinder and the diameter at the bottom. The out-of-round specification of the cylinder bore is the difference between the parallel and perpendicular readings. Compare your results to this Chapter's Specifications.

8   If the cylinder walls are badly scuffed or scored, or if they're out-of-round or tapered beyond the limits given in this Chapter's Specifications, have the engine block rebored and honed at an automotive machine shop.

9   If a rebore is done, oversize pistons and rings will be required.

10   Using a precision straightedge and feeler gauge, check the block deck (the surface the cylinder heads mate with) for distortion as you did with the cylinder heads (see Section 10). If it's distorted beyond the specified limit, the block decks can be resurfaced by an automotive machine shop.

11   If the cylinders are in reasonably good condition and not worn to the outside of the limits, and if the piston-to-cylinder clearances can be maintained properly, they don't have to be rebored. Honing is all that's necessary (see Section 18).

18.3a A "bottle brush" hone will produce better results if you've never honed cylinders before

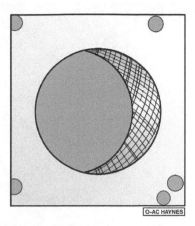

18.3b The cylinder hone should leave a smooth, crosshatch pattern with the lines intersecting at approximately a 60-degree angle

19.4a The piston ring grooves can be cleaned with a special tool, as shown here . . .

18 Cylinder honing

Refer to illustrations 18.3a and 18.3b

1   Prior to engine reassembly, the cylinder bores must be honed so the new piston rings will seat correctly and provide the best possible combustion chamber seal. **Note:** *If you don't have the tools or don't want to tackle the honing operation, most automotive machine shops will do it for a reasonable fee.*

2   Before honing the cylinders, install the main bearing caps and tighten the bolts to the torque listed in this Chapter's Specifications.

3   Two types of cylinder hones are commonly available - the flex hone or "bottle brush" type and the more traditional surfacing hone with spring-loaded stones. Both will do the job, but for the less experienced mechanic the "bottle brush" hone will probably be easier to use. You'll also need some honing oil (kerosene will work if honing oil isn't available), rags and an electric drill motor. Proceed as follows:

a) *Mount the hone in the drill motor, compress the stones and slip it into the first cylinder* **(see illustration)**. *Be sure to wear safety goggles or a face shield!*

b) *Lubricate the cylinder with plenty of honing oil, turn on the drill and move the hone up-and-down in the cylinder at a pace that will produce a fine crosshatch pattern on the cylinder walls, and with the drill square and centered with the bore. Ideally, the crosshatch lines should intersect at approximately a 45-60-degree angle* **(see illustration)**. *Be sure to use plenty of lubricant and don't take off any more material than is absolutely necessary to produce the desired finish.* **Note:** *Piston ring manufacturers may specify a different crosshatch angle - read and follow any instructions included with the new rings.*

c) *Don't withdraw the hone from the cylinder while it's running. Instead, shut off the drill and continue moving the hone up-and-down in the cylinder until it*

comes to a complete stop, then compress the stones and withdraw the hone. If you're using a "bottle brush" type hone, stop the drill motor, then turn the chuck in the normal direction of rotation while withdrawing the hone from the cylinder.

d) *Wipe the oil out of the cylinder and repeat the procedure for the remaining cylinders.*

4   After the honing job is complete, chamfer the top edges of the cylinder bores with a small file so the rings won't catch when the pistons are installed. Be very careful not to nick the cylinder walls with the end of the file.

5   The entire engine block must be washed again very thoroughly with warm, soapy water to remove all traces of the abrasive grit produced during the honing operation. **Note:** *The bores can be considered clean when a lint-free white cloth - dampened with clean engine oil - used to wipe them out doesn't pick up any more honing residue, which will show up as gray areas on the cloth. Be sure to run a brush through all oil holes and galleries and flush them with running water.*

6   After rinsing, dry the block and apply a coat of light rust preventive oil to all machined surfaces. Wrap the block in a plastic trash bag to keep it clean and set it aside until reassembly.

19 Pistons and connecting rods - inspection

Refer to illustrations 19.4a, 19.4b, 19.10 and 19.11

1   Before the inspection process can be carried out, the piston/connecting rod assemblies must be cleaned and the original piston rings removed from the pistons. **Note:** *Always use new piston rings when the engine is reassembled.*

2   Using a piston ring installation tool, carefully remove the rings from the pistons.

19.4b . . . or a section of broken ring

Be careful not to nick or gouge the pistons in the process.

3   Scrape all traces of carbon from the top of the piston. A hand-held wire brush or a piece of fine emery cloth can be used once the majority of the deposits have been scraped away. Do not, under any circumstances, use a wire brush mounted in a drill motor to remove deposits from the pistons. The piston material is soft and may be eroded away by the wire brush.

4   Use a piston ring groove-cleaning tool to remove carbon deposits from the ring grooves. If a tool isn't available, a piece broken off the old ring will do the job. Be very careful to remove only the carbon deposits - don't remove any metal and do not nick or scratch the sides of the ring grooves **(see illustrations)**.

5   Once the deposits have been removed, clean the piston/rod assemblies with solvent and dry them with compressed air (if available). **Warning:** *Wear eye protection. Make sure the oil return holes in the back sides of the ring grooves are clear.*

6   If the pistons and cylinder walls aren't damaged or worn excessively, and if the

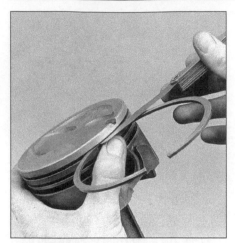

19.10 Check the ring side clearance with a feeler gauge at several points around the groove

19.11 Measure the piston diameter at a 90-degree angle to the piston pin and at the specified height

20.1 The oil holes should be chamfered so sharp edges don't gouge or scratch the new bearings

engine block isn't rebored, new pistons won't be necessary. Normal piston wear appears as even vertical wear on the piston thrust surfaces and slight looseness of the top ring in its groove. New piston rings, however, should always be used when an engine is rebuilt.

7   Carefully inspect each piston for cracks around the skirt, at the pin bosses and at the ring lands.

8   Look for scoring and scuffing on the thrust faces of the skirt, holes in the piston crown and burned areas at the edge of the crown. If the skirt is scored or scuffed, the engine may have been suffering from overheating and/or abnormal combustion, which caused excessively high operating temperatures. The cooling and lubrication systems should be checked thoroughly. A hole in the piston crown is an indication that abnormal combustion (preignition) was occurring. Burned areas at the edge of the piston crown are usually evidence of spark knock (detonation). If any of the above problems exist, the causes must be corrected or the damage will occur again. The causes may include intake air leaks, incorrect fuel/air mixture, low octane fuel, ignition timing and EGR system malfunctions.

9   Corrosion of the piston, in the form of small pits, indicates coolant is leaking into the combustion chamber and/or the crankcase. Again, the cause must be corrected or the problem may persist in the rebuilt engine.

10   Measure the piston ring side clearance by laying a new piston ring in each ring groove and slipping a feeler gauge in beside it **(see illustration)**. Check the clearance at three or four locations around each groove. Be sure to use the correct ring for each groove - they are different. If the side clearance is greater than specified in this Chapter, new pistons will have to be used.

11   Check the piston-to-bore clearance by measuring the bore (see Section 17) and the piston diameter. Make sure the pistons and

bores are correctly matched. Measure the piston across the skirt, at a 90-degree angle to the piston pin **(see illustration)**. The measurement must be taken at a specific point, depending on the engine type, to be accurate:

a) *The piston diameter on both V6 engines is measured directly in line with the piston pin centerline.*

b) *On the V8 engines, pistons are measured 0.450-inch from the bottom of the skirt, at right angles to the piston pin. Measure the V8 cylinder bore 2.5-inches from the top for comparison with the piston measurement.*

12   Subtract the piston diameter from the bore diameter to obtain the clearance. If it's greater than specified, the block will have to be rebored and new pistons and rings installed.

13   Check the piston-to-rod clearance by twisting the piston and rod in opposite directions. Any noticeable play indicates excessive wear, which must be corrected. The piston/connecting rod assemblies should be taken to an automotive machine shop to have the pistons and rods re-sized and new pins installed.

14   If the pistons must be removed from the connecting rods for any reason, they should be taken to an automotive machine shop. While they are there, have the connecting rods checked for bend and twist, since automotive machine shops have special equipment for this purpose. **Note:** *Unless new pistons and/or connecting rods must be installed, do not disassemble the pistons and connecting rods.*

15   Check the connecting rods for cracks and other damage. Temporarily remove the rod caps, lift out the old bearing inserts, wipe the rod and cap bearing surfaces clean and inspect them for nicks, gouges and scratches. After checking the rods, replace the old bearings, slip the caps into place and tighten the nuts finger tight. **Note:** *If the*

engine is being rebuilt because of a connecting rod knock, be sure to install new rods.

20 Crankshaft - inspection

Refer to illustrations 20.1, 20.2, 20.5 and 20.7

1   Remove all burrs from the crankshaft oil holes with a stone, file or scraper **(see illustration)**.

2   Clean the crankshaft with solvent and dry it with compressed air (if available). **Warning:** *Wear eye protection when using compressed air.* Be sure to clean the oil holes with a stiff brush **(see illustration)** and flush them with solvent.

3   Check the main and connecting rod bearing journals for uneven wear, scoring, pits and cracks.

4   Check the rest of the crankshaft for cracks and other damage. It should be Magnafluxed to reveal hidden cracks - an automotive machine shop will handle the procedure.

20.2 Use a wire or stiff plastic bristle brush to clean the oil passages in the crankshaft

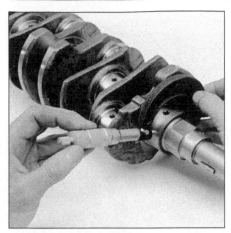

20.5 Measure the diameter of each crankshaft journal at several points to detect taper and out-of-round conditions

5   Using a micrometer, measure the diameter of the main and connecting rod journals and compare the results to this Chapter's Specifications **(see illustration)**. By measuring the diameter at a number of points around each journal's circumference, you'll be able to determine whether or not the journal is out-of-round. Take the measurement at each end of the journal, near the crank throws, to determine if the journal is tapered.

6   If the crankshaft journals are damaged, tapered, out-of-round or worn beyond the limits given in the Specifications, have the crankshaft reground by an automotive machine shop. Be sure to use the correct-size bearing inserts if the crankshaft is reconditioned.

7   Check the oil seal journals at each end of the crankshaft for wear and damage. If the seal has worn a groove in the journal, or if it's nicked or scratched **(see illustration)**, the new seal may leak when the engine is reassembled. In some cases, an automotive machine shop may be able to repair the journal by pressing on a thin sleeve. If repair isn't feasible, a new or different crankshaft should be installed.

8   Examine the main and rod bearing inserts (see Section 21).

---

21   Main and connecting rod bearings - inspection

Refer to illustration 21.1

1   Even though the main and connecting rod bearings should be replaced with new ones during the engine overhaul, the old bearings should be retained for close examination, as they may reveal valuable information about the condition of the engine **(see illustration)**.

2   Bearing failure occurs because of lack of lubrication, the presence of dirt or other foreign particles, overloading the engine and corrosion. Regardless of the cause of bearing failure, it must be corrected before the engine is reassembled to prevent it from happening again.

20.7 If the seals have worn grooves in the crankshaft journals, or if the seal contact surfaces are nicked or scratched, the new seals will leak

3   When examining the bearings, remove them from the engine block, the main bearing caps, the connecting rods and the rod caps and lay them out on a clean surface in the same general position as their location in the engine. This will enable you to match any bearing problems with the corresponding crankshaft journal.

4   Dirt and other foreign particles get into the engine in a variety of ways. It may be left in the engine during assembly, or it may pass through filters or the PCV system. It may get into the oil, and from there into the bearings. Metal chips from machining operations and normal engine wear are often present. Abrasives are sometimes left in engine components after reconditioning, especially when parts aren't thoroughly cleaned using the proper cleaning methods. Whatever the source, these foreign objects often end up embedded in the soft bearing material and are easily recognized. Large particles won't embed in the bearing and will score or gouge the bearing and journal. The best prevention for this cause of bearing failure is to clean all parts thoroughly and keep everything spotlessly clean during engine assembly. Frequent and regular engine oil and filter changes are also recommended.

5   Lack of lubrication (or lubrication breakdown) has a number of interrelated causes. Excessive heat (which thins the oil), overloading (which squeezes the oil from the bearing face) and oil leakage or throw off (from excessive bearing clearances, worn oil pump or high engine speeds) all contribute to lubrication breakdown. Blocked oil passages, which usually are the result of misaligned oil holes in a bearing shell, will also oil starve a bearing and destroy it. When lack of lubrication is the cause of bearing failure, the bearing material is wiped or extruded from the steel backing of the bearing. Temperatures may increase to the point where the steel

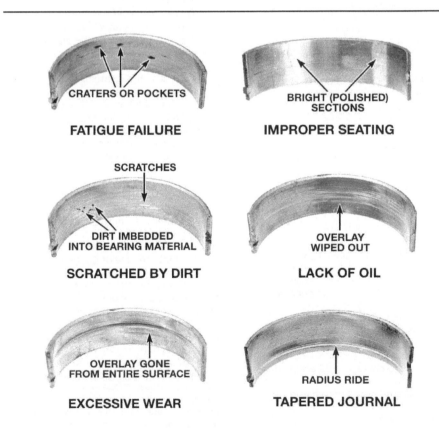

21.1 Typical bearing failures

backing turns blue from overheating.

6   Driving habits can have a definite effect on bearing life. Low speed operation in too high a gear (lugging the engine) puts very high loads on bearings, which tends to squeeze out the oil film. These loads cause the bearings to flex, which produces fine cracks in the bearing face (fatigue failure). Eventually the bearing material will loosen in pieces and tear away from the steel backing. Short trip driving leads to corrosion of bearings because insufficient engine heat is produced to drive off the condensed water and corrosive gases. These products collect in the engine oil, forming acid and sludge. As the oil is carried to the engine bearings, the acid attacks and corrodes the bearing material.

7   Incorrect bearing installation during engine assembly will lead to bearing failure as well. Tight-fitting bearings leave insufficient oil clearance and will result in oil starvation. Dirt or foreign particles trapped behind a bearing insert result in high spots on the bearing which lead to failure.

22   Engine overhaul - reassembly sequence

1   Before beginning engine reassembly, make sure you have all the necessary new parts, gaskets and seals as well as the following items on hand:

> Common hand tools
> Torque wrench (1/2-inch drive) with
>   angle-torque gauge
> Piston ring installation tool
> Piston ring compressor
> Crankshaft balancer installation tool
> Short lengths of rubber or plastic hose to
>   fit over connecting rod bolts
> Plastigage
> Feeler gauges
> Fine-tooth file
> New engine oil
> Engine assembly lube or moly-base
>   grease
> Gasket sealant
> Thread locking compound

2   In order to save time and avoid problems, engine reassembly must be done in the following general order:

> Crankshaft and main bearings
> Piston/connecting rod assemblies
> Balance shaft (3800 engine only)
> Water pump drive gear (1st generation
>   V8 engine only)
> Windage tray (2nd generation
>   V8 engine only)
> Camshaft
> Rear main oil seal and retainer
> Water pump driveshaft (V8 engine only)
> Timing chain and sprockets
> Timing chain cover and oil pump
> Oil pan
> Cylinder heads
> Valve lifters
> Rocker arms and pushrods
> Driveplate

23.3 When checking piston ring end gap, the ring must be square in the cylinder bore (this is done by pushing the ring down with the top of a piston as shown)

Assembled after engine installation

> Intake and exhaust manifolds
> Valve covers

23   Piston rings - installation

Refer to illustrations 23.3, 23.4, 23.5, 23.9a, 23.9b and 23.12

1   Before installing the new piston rings, the ring end gaps must be checked. It's assumed the piston ring side clearance has been checked and verified correct (see Section 19).

2   Lay out the piston/connecting rod assemblies and the new ring sets so the ring sets will be matched with the same piston and cylinder during the end gap measurement and engine assembly.

3   Insert the top (number one) ring into the first cylinder and square it up with the cylinder walls by pushing it in with the top of the piston **(see illustration)**. The ring should be near the bottom of the cylinder, at the lower limit of ring travel.

4   To measure the end gap, slip feeler gauges between the ends of the ring until a gauge equal to the gap width is found **(see illustration)**. The feeler gauge should slide between the ring ends with a slight amount of drag. Compare the measurement to this Chapter's Specifications. If the gap is larger or smaller than specified, double-check to make sure you have the correct rings before proceeding.

5   If the gap is too small, it must be enlarged or the ring ends may come in contact with each other during engine operation, which can cause serious engine damage. The end gap can be increased by filing the ring ends very carefully with a fine file. Mount the file in a vise equipped with soft jaws, slip the ring over the file with the ends contacting the file teeth and slowly move the ring to remove material from the ends. When performing this

23.4 With the ring square in the cylinder, measure the end gap with a feeler gauge

operation, file only from the outside in **(see illustration)**. **Note:** *When you have the end gap correct, remove any burrs from the filed ends of the rings with a whetstone.*

6   Excess end gap isn't critical unless it's greater than 0.040-inch. Again, double-check to make sure you have the correct rings for the engine. If the engine block has been bored oversize, necessitating oversize pistons, matching oversize rings are required.

7   Repeat the procedure for each ring that will be installed in the first cylinder and for each ring in the remaining cylinders. Remember to keep rings, pistons and cylinders matched up.

8   Once the ring end gaps have been checked/corrected, the rings can be installed on the pistons.

9   The oil control ring (lowest one on the piston) is usually installed first. It's composed of three separate components. Slip the spacer/expander into the groove **(see illustration)**. If an anti-rotation tang is used, make sure it's inserted into the drilled hole in the ring groove. Next, install the lower side rail. Don't use a piston ring installation tool on the

23.5 If the end gap is too small, clamp a file in a vise and file the ring ends (from the outside in only) to enlarge the gap slightly

23.9a Installing the spacer/expander in the oil control ring groove

23.9b DO NOT use a piston ring installation tool when installing the oil ring side rails

23.12 Installing the compression rings with a ring expander - the mark (arrow) must face up

oil ring side rails, as they may be damaged. Instead, place one end of the side rail into the groove between the spacer/expander and the ring land, hold it firmly in place and slide a finger around the piston while pushing the rail into the groove **(see illustration)**. Next, install the upper side rail in the same manner.

10   After the three oil ring components have been installed, check to make sure both the upper and lower side rails can be turned smoothly in the ring groove.

11   The number two (middle) ring is installed next. It's usually stamped with a mark, which must face up, toward the top of the piston. **Note:** *Always follow the instructions printed on the ring package or box - different manufacturers may require different approaches. Don't mix up the top and middle rings, as they have different cross-sections.*

12   Use a piston ring installation tool and make sure the identification mark is facing the top of the piston, then slip the ring into the middle groove on the piston **(see illustration)**. Don't expand the ring any more than necessary to slide it over the piston.

13   Install the number one (top) ring in the same manner. Make sure the mark is facing up. Be careful not to confuse the number one and number two rings.

14   Repeat the procedure for the remaining pistons and rings.

24   Crankshaft - installation and main bearing oil clearance check

1   Crankshaft installation is the first step in engine reassembly. It's assumed at this point that the engine block and crankshaft have been cleaned, inspected and repaired or reconditioned.

2   Position the engine with the bottom facing up.

3   Remove the main bearing cap bolts and lift out the caps (see Section 15 for the procedure on the 3800 V6). Lay them out in the proper order to ensure correct installation.

4   If they're still in place, remove the origi-

nal bearing inserts from the block and the main bearing caps. Wipe the bearing surfaces of the block and caps with a clean, lint-free cloth. They must be kept spotlessly clean.

Main bearing oil clearance check

Refer to illustrations 24.11, 24.13 and 24.15
Note: *Don't touch the faces of the new bearing inserts with your fingers. Oil and acids from your skin can etch the bearings.*

5   Clean the back sides of the new main bearing inserts and lay one in each main bearing saddle in the block. If one of the bearing inserts from each set has a large groove in it, make sure the grooved insert is installed in the block. Lay the other bearing from each set in the corresponding main bearing cap. Make sure the tab on the bearing insert fits into the recess in the block or cap, neither higher than the cap's edge nor lower. **Caution:** *The oil holes in the block must line up with the oil holes in the bearing inserts. Do not hammer the bearing into place and don't nick or gouge the bearing faces. No lubrication should be used at this time.*

6   The flanged thrust bearing must be installed in the number three cap and saddle on 3.4L and 2nd generation (1998 and later) V8 engines, the number two cap and saddle of a 3800 engine (counting from the front of the engine), and the last main cap on 1st generation (1997 and earlier) V8 engines. **Caution:** *Some engines have a 0.008-inch oversize rear main bearing. Check your crankshaft for a marking on the last counterweight, and check the backside of the old bearing insert for a similar marking. If they are marked oversize, an oversize rear bearing will be required.*

7   Clean the faces of the bearings in the block and the crankshaft main bearing journals with a clean, lint-free cloth.

8   Check or clean the oil holes in the crankshaft, as any dirt here can go only one way - straight through the new bearings.

9   Once you're certain the crankshaft is

clean, carefully lay it in position in the main bearings.

10   Before the crankshaft can be permanently installed, the main bearing oil clearance must be checked.

11   Cut several pieces of the appropriate size Plastigage (they should be slightly shorter than the width of the main bearings) and place one piece on each crankshaft main bearing journal, parallel with the journal axis **(see illustration)**.

12   Clean the faces of the bearings in the caps and install the caps in their original locations (don't mix them up) with the arrows pointing toward the front of the engine. Don't disturb the Plastigage.

13   Starting with the center main and working out toward the ends, tighten the main bearing cap bolts to the proper torque and angle of rotation: On the 3.4L V6, tighten all to 37 ft-lbs, then an additional 77-degrees rotation. On the 1st generation V8 engines, tighten all to 78 ft-lbs in three steps. On the 3800 V6, use the following sequence:

a) *Tighten all caps to 52 ft-lbs to seat the caps*

24.11 Lay the Plastigage strips on the main bearing journals, parallel to the crankshaft centerline

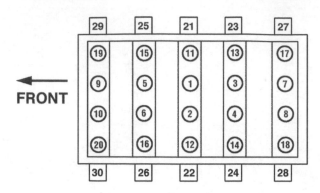

24.13 Main bearing cap TIGHTENING sequence - 2nd generation (1998 and later) V8 engine

24.15 Measuring the width of the crushed Plastigage to determine the main bearing oil clearance (be sure to use the correct scale - standard and metric ones are included)

b) *Loosen all caps one full turn*
c) *Tighten all to 177 in-lbs*
d) *Tighten all to 29.5 ft-lbs*
e) *Tighten all an additional 35-degrees*
f) *Tighten all an additional 35-degrees*
g) *Tighten all an additional 40-degrees*
h) *Insert and tighten the side bolts to 132 in-lbs*
i) *Tighten the side bolts an additional 45-degrees*

On 2nd generation V8 engines, use the following sequence **(see illustration)**:
a) *Tighten the inner bolts (1 through 10) to 15 ft-lbs*
b) *Tighten the inner bolts (1 through 10) an additional 80 degrees*
c) *Tighten the outer bolts (11 through 20) to 15 ft-lbs*
d) *Tighten the outer bolts (11 through 20) an additional 50 degrees*
e) *Tighten the **NEW** side bolts (21 through 30) to 18 ft-lbs*
f) *DO NOT use the new side bolts during the oil clearance check as this will damage the O-ring on the new side bolts.*

Don't rotate the crankshaft at any time during this operation, and do not tighten one cap completely - tighten all caps equally. **Caution:** *On 3.4L V6 and V8 engines, the main caps can be seated using light taps with a brass or plastic mallet, but on the 3800 V6 they must be placed squarely on the block and brought into place with the main cap bolts, tightening each side a little at a time. Do not use a hammer on 3800 main caps.*

14   Remove the bolts/studs and carefully lift off the main bearing caps. Keep them in order. Don't disturb the Plastigage or rotate the crankshaft. If any of the main bearing caps are difficult to remove, tap them gently from side-to-side with a soft-face hammer to loosen them. **Note:** *Refer to Section 15 for cap removal on 3800 V6 engines.*

15   Compare the width of the crushed Plastigage on each journal to the scale printed on the Plastigage envelope to obtain the main bearing oil clearance **(see illustration)**. Check the Specifications to make sure it's correct.

16   If the clearance is not as specified, the bearing inserts may be the wrong size (which means different ones will be required). Before deciding different inserts are needed, make sure no dirt or oil was between the bearing inserts and the caps or block when the clearance was measured. If the Plastigage was wider at one end than the other, the journal may be tapered (see Section 20).

17   Carefully scrape all traces of the Plastigage material off the main bearing journals and/or the bearing faces. Use your fingernail or the edge of a credit card - don't nick or scratch the bearing faces.

Final crankshaft installation

18   Carefully lift the crankshaft out of the engine.

19   Clean the bearing faces in the block, then apply a thin, uniform layer of moly-base grease or engine assembly lube to each of the bearing surfaces. Be sure to coat the thrust faces as well as the journal face of the thrust bearing.

20   Make sure the crankshaft journals are clean, then lay the crankshaft back in place in the block.

21   Clean the faces of the bearings in the caps, then apply lubricant to them.

22   Install the caps in their original locations with the arrows pointing toward the front of the engine. **Note 1:** *On V6 engines, apply a small dab of sealant on either side of the rear cap-to-block mating surface.* **Note 2:** *On the 3800 V6 only, the rear cap must be installed 0.010-inch ahead of the rear face of the block to avoid interference with the flywheel. Use a precision straightedge and feeler gauge to check this on assembly, and check again after torquing down the caps.*

23   With all caps in place and bolts just started, tap the ends of the crankshaft forward and backward with a lead or brass hammer to line up the main bearing and crankshaft thrust surfaces.

24   Following the procedures outlined in Step 13, retighten all main bearing cap bolts to the torque in this Chapter's Specifications, starting with the center main and working out toward the ends.

25   Rotate the crankshaft a number of times by hand to check for any obvious binding.

26   The final step is to check the crankshaft endplay with feeler gauges or a dial indicator as described in Section 15. The endplay should be correct if the crankshaft thrust faces aren't worn or damaged and new bearings have been installed.

25   Camshaft and balance shaft - installation

Camshaft

Refer to illustrations 25.1 and 25.4

1   Lubricate the camshaft bearing journals and cam lobes with a special camshaft installation lubricant **(see illustration)**.

2   Slide the camshaft into the engine, using three six-inch-long, 5/16 x 18 bolts screwed into the front of the camshaft as a

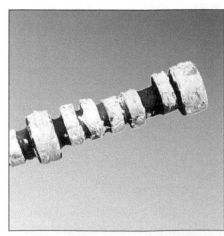

25.1 Be sure to prelube the camshaft bearing journals and lobes before installation

25.4 On 1st generation (1997 and earlier) V8 engines, install the water pump drive gear and retainer using a Torx driver

25.6 Drive the front balance shaft bearing into the block just until the bearing retainer can be bolted in place

"handle". **Note:** *This technique won't work on the 3800 V6 engine.* Support the cam near the block and be careful not to scrape or nick the bearings. Install the camshaft retainer plate and tighten the bolts to the torque listed in the Part A , B, C or D Specifications.

3   On 3.4L V6 and 1st generation V8 engines, dip the gear portion of the oil pump driveshaft in engine oil and insert it into the block. It should be flush with its mounting boss before inserting the retaining bolt (V8) or bolt and clamp (V6). **Note:** *Position a new O-ring on the oil pump driveshaft before installation.*

4   On either generation V8 engines, install the camshaft retainer plate at the front of the block and a new camshaft rear plug, if it had been removed during the overhaul. Coat the plug with non-hardening sealant before installation. On 1st generation V8 engines, reinstall the water pump drive gear **(see illustration)**.

5   Complete the installation of the timing chain and sprockets (3.4L engine: see Part A; 3800 engine: see Part B; 1st generation V8 engine: see Part C; 2nd generation V8 engine: see Part D). On 3800 engines, perform the fol-

lowing procedure on balance shaft installation before installing the timing chain and sprockets.

Balance shaft (3800 V6 only)

Refer to illustrations 25.6, 25.9 and 25.10

6   Lubricate the front bearing and rear journal of the balance shaft with engine oil and insert the balance shaft carefully into the block. When the front bearing approaches the insert in the front of the block, use a hammer and an appropriate-size socket to drive the front bearing into its insert. Drive it in just enough to allow installation of the balance shaft bearing retainer **(see illustration)**.

7   Install the balance shaft gear and its bolt.

8   Turn the camshaft so that with the camshaft sprocket temporarily installed, the timing mark is straight down.

9   With the camshaft sprocket and the camshaft gear removed, turn the balance shaft so the timing mark on the gear points straight down **(see illustration)**.

10   Install the camshaft gear (that drives the balance shaft) onto the cam, aligning it with

the keyway and align the marks on the balance shaft gear and the camshaft gear **(see illustration)** by turning the balance shaft.

11   Tighten the balance shaft retainer bolts, and after the camshaft sprocket, crankshaft sprocket and timing chain have been installed, tighten the balance shaft sprocket bolt.

26   Rear main oil seal - replacement

3.4L V6 engine

1   A special tool is recommended for installing the new seal. Lubricate the lip of the new seal with grease or clean engine oil, then slide the seal onto the mandril until the dust lip bottoms squarely against the collar of the tool. **Note:** *If the special tool isn't available, carefully work the seal over the crankshaft and tap it evenly into place with a hammer and punch.*

2   Align the dowel pin on the tool with the dowel pin hole in the crankshaft and attach the tool to the crankshaft by hand-tightening the bolts.

3   Turn the tool handle until the collar bottoms against the case, seating the seal.

25.9 The balance shaft gear mark (arrow) should point straight down

25.10 Align the marks (arrows) on both balance shaft gears as shown here

26.5 On these engines it will be necessary to install the new seal in the housing first, then install a new gasket and the seal housing on the block - be sure to lubricate the seal lip and carefully work the seal over the crankshaft with a blunt tool

26.6 With the rear seal housing in place and the bolts installed LOOSELY, measure the distance between the oil pan rail and the seal housing on each side (arrows) - then adjust the housing so the measurements are even on both sides before tightening the cover bolts

3800 V6 and 1st generation V8 engines

Refer to illustration 26.5

4   Lubricate the new seal with clean engine oil and install it in the rear seal retainer plate, either with a seal-installer mandrel or tap it in evenly with a hammer and punch.

5   Install the seal retainer plate to the back of the engine block, using a new gasket and tighten the bolts to the torque listed in this Chapter's Specifications **(see illustration)**.

2nd generation V8 engines

Refer to illustration 26.6

6   Lubricate the new seal with clean engine oil and install it into the rear seal housing on a work bench. Make sure the seal is installed squarely into the housing bore, then install the rear seal and housing over the end of the crankshaft. Install the rear seal housing retaining bolts on the engine loosely and align the rear housing as follows:

a) *Place a straightedge on the engine block oil pan rail. Measure the distance on each side of the block from the oil pan rail to the rear housing with a feeler gauge* **(see illustration)**. *This Step measures the difference between the sealing surface of the oil pan and the sealing surface of the rear seal housing in relationship to each other.*

b) *Tilt the rear seal housing as necessary to achieve an even measurement on each side. This Step properly aligns the rear seal housing to oil pan sealing surfaces. Typically 0.000 to 0.020 inch is an acceptable tolerance.* **Note:** *Ideally the rear seal housing should be flush with the oil pan rail, but because of the differences in seal thickness, this may not always be obtainable. That is why there is a tolerance of 0.000 to 0.020 inch. Always let the rear seal center itself around the crankshaft and tilt the cover from side-to-side to even up the mea-*

27.1a On 3.4L V6 and 1st generation V8 engines, the flanges on the rod (arrow) must be positioned correctly with the notch on the piston to have proper oiling of the cylinder walls - left bank piston shown, right bank pistons are opposite

surement at both oil pan rails. Never push downward on the rear seal housing in an attempt to make the oil pan sealing surface flush, as this will distort the rear oil seal and eventually lead to an oil leak!

c) *With the rear seal housing properly aligned, tighten the housing bolts to the torque listed in this Chapter's Specifications.*

27 Pistons/connecting rods - installation and rod bearing oil clearance check

Refer to illustrations 27.1a and 27.1b,

1   Before installing the piston/connecting rod assemblies, the cylinder walls must be perfectly clean, the top edge of each cylinder must be chamfered, and the crankshaft must be in place. If new pistons or connecting rods

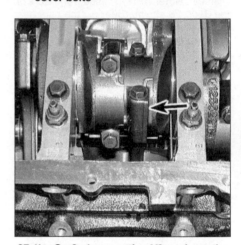

27.1b On 2nd generation V8 engines, the flat side of the rods (arrow) and the notch in the top of the piston must face the front of the engine

were installed, check the orientation of the connecting rod to the piston for correct installation by the machine shop. On 3.4L V6 and 1st generation V8 engines, the left bank connecting rod flanges must face the front of the engine and the right bank connecting rod flanges must face the rear of the engine, with the mark on the top of the pistons facing the front of the engine **(see illustration)**. You should have three pistons with the connecting rod flanges and the mark on the pistons aligned on the same side (these are the left bank pistons) and three pistons with the connecting rod flanges opposite the mark on the pistons (these are the right bank pistons). On 3800 V6 engine the connecting rod can be installed on the piston in either direction. On 2nd generation V8 engines, the flat side of the connecting rods and the marks on the top of each piston must all face the front of the engine **(see illustration)**. So, the pistons and connecting rod assemblies can be installed in either the right or left bank if new rods were installed.

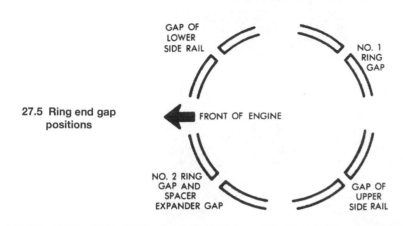

27.5 Ring end gap positions

GAP OF LOWER SIDE RAIL

NO. 1 RING GAP

FRONT OF ENGINE

NO. 2 RING GAP AND SPACER EXPANDER GAP

GAP OF UPPER SIDE RAIL

27.9 The notch or arrow on each piston must face the front end of the engine as the pistons are installed

2   Remove the cap from the end of the number one connecting rod (check the marks made during removal). Remove the original bearing inserts and wipe the bearing surfaces of the connecting rod and cap with a clean, lint-free cloth. They must be kept spotlessly clean.

Piston installation and rod bearing oil clearance check

Refer to illustrations 27.5, 27.9, 27.11, 27.13 and 27.17

Note: *Don't touch the faces of the new bearing inserts with your fingers. Oil and acids from your skin can etch the bearings.*

3   Clean the back side of the new upper bearing insert, then lay it in place in the connecting rod. Make sure the tab on the bearing fits into the recess in the rod. Don't hammer the bearing insert into place and be very careful not to nick or gouge the bearing face. Don't lubricate the bearing at this time.

4   Clean the back side of the other bearing insert and install it in the rod cap. Again, make sure the tab on the bearing fits into the recess in the cap, and don't apply any lubricant. It's critically important that the mating surfaces of the bearing and connecting rod are perfectly clean and oil free when they're assembled.

5   Stagger the piston ring gaps around the piston **(see illustration)**.

6   Slip a section of plastic or rubber hose over each connecting rod cap bolt.

7   Lubricate the piston and rings with clean engine oil and attach a piston ring compressor to the piston. Leave the skirt protruding about 1/4-inch to guide the piston into the cylinder. The rings must be compressed until they're flush with the piston.

8   Rotate the crankshaft until the number one connecting rod journal is at BDC (bottom dead center) and apply a coat of engine oil to the cylinder walls.

9   With the mark or notch on top of the piston **(see illustration)** facing the front of the engine, gently insert the piston/connecting rod assembly into the number one cylinder bore and rest the bottom edge of the ring compressor on the engine block.

10   Tap the top edge of the ring compressor to make sure it's contacting the block around its entire circumference.

11   Gently tap on the top of the piston with the end of a wooden or plastic hammer handle **(see illustration)** while guiding the end of the connecting rod into place on the crankshaft journal. The piston rings may try to pop out of the ring compressor just before entering the cylinder bore, so keep some pressure down on the ring compressor. Work

slowly, and if any resistance is felt as the piston enters the cylinder, stop immediately. Find out what's hanging up and fix it before proceeding. Do not, for any reason, force the piston into the cylinder - you might break a ring and/or the piston.

12   Once the piston/connecting rod assembly is installed, the connecting rod bearing oil clearance must be checked before the rod cap is permanently bolted in place.

13   Cut a piece of the appropriate size Plastigage slightly shorter than the width of the connecting rod bearing and lay it in place on the number one connecting rod journal, parallel with the journal axis **(see illustration)**.

14   Clean the connecting rod cap bearing face, remove the protective hoses from the connecting rod bolts and install the rod cap. Make sure the mating mark on the cap is on the same side as the mark on the connecting rod. **Note:** *2000 and later model 5.7L engines have two types of connecting rod bolts, with different torque specifications. The 1st design has a smooth shank, while the 2nd design has a sleeve on an undersize shank.*

15   Install the nuts and tighten them to the torque listed in this Chapter's Specifications.

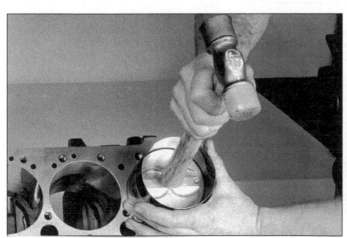

27.11 Drive the piston into the cylinder bore with the end of a wooden or plastic hammer handle

27.13 Lay the Plastigage strips on each rod bearing journal, parallel to the crankshaft centerline

Work up to it in three steps. **Note:** *Use a thin-wall socket to avoid erroneous torque readings that can result if the socket is wedged between the rod cap and nut. If the socket tends to wedge itself between the nut and the cap, lift up on it slightly until it no longer contacts the cap. Do not rotate the crankshaft at any time during this operation.*

16   Remove the nuts and detach the rod cap, being very careful not to disturb the Plastigage.

17   Compare the width of the crushed Plastigage to the scale printed on the Plastigage envelope to obtain the oil clear-ance **(see illustration)**. Compare it to this Chapter's Specifications to make sure the clearance is correct.

18   If the clearance is not as specified, the bearing inserts may be the wrong size (which means different ones will be required). Before deciding different inserts are needed, make sure no dirt or oil was between the bearing inserts and the connecting rod or cap when the clearance was measured. Also, recheck the journal diameter. If the Plastigage was wider at one end than the other, the journal may be tapered (see Section 20).

Final connecting rod installation

19   Carefully scrape all traces of the Plastigage material off the rod journal and/or bearing face. Be very careful not to scratch the bearing - use your fingernail or the edge of a credit card.

20   Make sure the bearing faces are perfectly clean, then apply a uniform layer of clean moly-base grease or engine assembly lube to both of them. You'll have to push the piston into the cylinder to expose the face of the bearing insert in the connecting rod - be sure to slip the protective hoses over the rod bolts first.

21   Slide the connecting rod back into place on the journal, remove the protective hoses from the rod cap bolts, install the rod cap and tighten the nuts to the torque listed in this Chapter's Specifications. Again, work up to the torque in three steps.

22   Repeat the entire procedure for the remaining pistons/connecting rods.

23   The important points to remember are . . .

  a) *Keep the back sides of the bearing inserts and the insides of the connecting rods and caps perfectly clean when assembling them.*

  b) *Make sure you have the correct piston/rod assembly for each cylinder.*

  c) *The arrow or mark on the piston must face the front of the engine.*

  d) *Lubricate the cylinder walls with clean oil.*

  e) *Lubricate the bearing faces when installing the rod caps after the oil clearance has been checked.*

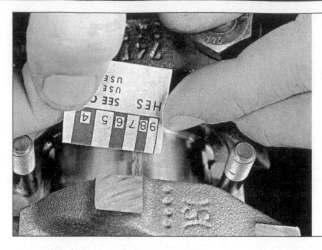

27.17 Measuring the width of the crushed Plastigage to determine the rod bearing oil clearance (be sure to use the correct scale - standard and metric ones are included)

24   After all the piston/connecting rod assemblies have been properly installed, rotate the crankshaft a number of times by hand to check for any obvious binding.

25   As a final step, the connecting rod end-play must be checked (see Section 14).

26   Compare the measured endplay to this Chapter's Specifications to make sure it's correct. If it was correct before disassembly and the original crankshaft and rods were reinstalled, it should still be right. If new rods or a new crankshaft were installed, the endplay may be inadequate. If so, the rods will have to be removed and taken to an automotive machine shop for re-sizing.

28   Flywheel/driveplate - removal and installation

For flywheel/driveplate removal and installation procedures, refer to Part A for the 3.4L V6, Part B for the 3800 V6, and Part C for the 1st generation V8 engine or Part D for the 2nd generation V8 engine.

29   Initial start-up and break-in after overhaul

Warning: *Have a fire extinguisher handy when starting the engine for the first time.*

1   Once the engine has been installed in the vehicle, double-check the oil and coolant levels.

2   With the spark plugs out of the engine, disable the fuel and ignition systems by removing the ECM IGN fuse from the instrument panel fuse block and the IGNITION fuse from the underhood fuse block on 1993 models, the PCM IGN fuse from the instrument panel fuse block and the IGNITION fuse from the underhood fuse block on 1994 and 1995 models, the PCM BATT fuse from the instrument panel fuse block and the IGNITION fuse from the underhood fuse block on 1996 and

1997 models. On 1998 and later models, disable the ignition system by disconnecting the primary electrical connectors from the ignition coil packs, then disable the fuel system by removing the fuel pump relay from the engine compartment fuse block. Crank the engine until oil pressure registers on the gauge or the light goes out.

3   Install the spark plugs, hook up the plug wires and install the fuses.

4   Start the engine. It may take a few moments for the fuel system to build up pressure, but the engine should start without a great deal of effort. **Note:** *If the engine keeps backfiring, recheck the valve timing and spark plug wire routing.*

5   After the engine starts, it should be allowed to warm up to normal operating temperature. While the engine is warming up, make a thorough check for fuel, oil and coolant leaks.

6   Shut the engine off and recheck the engine oil and coolant levels.

7   Drive the vehicle to an area with no traffic, accelerate from 30 to 50 mph, then allow the vehicle to slow to 30 mph with the throttle closed. Repeat the procedure 10 or 12 times. This will load the piston rings and cause them to seat properly against the cylinder walls. Check again for oil and coolant leaks.

8   Drive the vehicle gently for the first 500 miles (no sustained high speeds) and keep a constant check on the oil level. It isn't unusual for an engine to use oil during the break-in period.

9   At approximately 500 to 600 miles, change the oil and filter.

10   For the next few hundred miles, drive the vehicle normally. Don't pamper it or abuse it.

11   After 2000 miles, change the oil and filter again and consider the engine broken in.

Chapter 3
Cooling, heating and air conditioning systems

Contents

Specifications

General

Coolant capacity	See Chapter 1
Drivebelt tension	See Chapter 1
Radiator pressure cap rating	18 psi
Thermostat opening temperature	185 to 205-degrees F

Torque specifications
Ft-lbs (unless otherwise indicated)

Thermostat housing bolts	
All engines (except 1998 and later V8)	21
1998 and later V8 engine	132 in-lbs
Water pump attaching bolts	
3.4L V6 **(see illustration 7.17)**	
Long bolts	33
Short bolts	192 in-lbs
Small bolts	89 in-lbs
3800 V6	
Step 1	132 in-lbs
Step 2	Rotate an additional 80-degrees
5.7L V8	
1997 and earlier	30
1998 and 1999	22
2000 and later	
Step 1	132 in-lbs
Step 2	22
Water pump pulley bolts	
3.4L V6	216 in-lbs
3800 V6	115 in-lbs

1 General information

The cooling system consists of a radiator and coolant reserve system, a radiator pressure cap, a thermostat, an electric fan (two on V8 models equipped with air conditioning), and a water pump. On V6 and 1998 and later V8 engines, the water pump has a pulley that is driven by the serpentine belt, while the 1997 and earlier V8 engine has a water pump driven mechanically from the camshaft by a splined coupling.

The cooling fan is mounted on the engine-side of the radiator and is controlled by the PCM and a fan relay.

The system is pressurized by a spring-loaded radiator cap, which, by maintaining pressure, increases the boiling point of the coolant. If the coolant temperature goes above this increased boiling point, the extra pressure in the system forces the radiator cap valve off its seat and exposes the overflow pipe or hose. The overflow pipe/hose leads to a coolant recovery system. This consists of a plastic reservoir into which the coolant that normally escapes due to expansion is retained. When the engine cools, the excess coolant is drawn back into the radiator by the vacuum created as the system cools, maintaining the system at full capacity. This is a continuous process and provided the level in the reservoir is correctly maintained, it is not necessary to add coolant to the radiator.

Coolant in the right tank of the radiator is drawn by the water pump, which forces it through the water passages in the cylinder block. The coolant then travels up into the cylinder head, circulates around the combustion chambers and valve seats, travels out of the cylinder head past the open thermostat into the upper radiator hose and back into the radiator. On V8 engines, the water pump circulates coolant through the cylinder heads first, before circulating it through the engine block.

When the engine is cold, the thermostat restricts the circulation of coolant to the engine. When the minimum operating temperature is reached, the thermostat begins to open, allowing coolant to return to the radiator.

2.6 An inexpensive hydrometer can be used to test the condition of your coolant

3.9a Thermostat cover location (arrow) - 3.4L V6 engine

3.9b Thermostat cover location - 3800 V6 engine

Automatic transmission-equipped models have a cooler element incorporated into the right tank of the radiator to cool the transmission fluid, and on some models with the V8, an optional engine oil cooler is connected to the left side of the radiator.

The heating system works by directing air through the heater core mounted in the dash and then to the interior of the vehicle by a system of ducts. Temperature is controlled by mixing heated air with fresh air, using a system of doors in the ducts, and a blower motor.

Air conditioning is an optional accessory, consisting of an evaporator core located under the dash, a condenser in front of the radiator, a receiver-drier (3800 V6) or accumulator (3.4L V6 and V8 engines) in the engine compartment and a belt-driven compressor mounted at the front of the engine.

2   Antifreeze - general information

Refer to illustration 2.6

Warning: *Do not allow antifreeze to come in contact with your skin or painted surfaces of the vehicle. Rinse off spills immediately with plenty of water. Antifreeze is highly toxic if ingested. Never leave antifreeze lying around in an open container or in puddles on the floor; children and pets are attracted by it's sweet smell and may drink it. Check with local authorities about disposing of used antifreeze. Many communities have collection centers which will see that antifreeze is disposed of safely. Never dump used anti-freeze on the ground or pour it into drains.*

Note: *Non-toxic antifreeze is now available at most auto parts stores, but even these types should be disposed of properly.*

The cooling system should be filled with a water/ethylene glycol based antifreeze solution which will prevent freezing down to at least -20-degrees F (even lower in cold climates). It also provides protection against corrosion and increases the coolant boiling point.

The cooling system should be drained, flushed and refilled at least every other year (see Chapter 1). The use of antifreeze solutions for periods of longer than two years is likely to

3.9c Thermostat cover location (arrow) - 1997 and earlier V8 engine

cause damage and encourage the formation of rust and scale in the system. However, 1996 and later models are filled with a new, long-life coolant called "Dex-Cool", which the factory claims is good for five years.

Before adding antifreeze to the system, check all hose connections. Antifreeze can leak through very minute openings.

The exact mixture of antifreeze to water which you should use depends on the relative weather conditions. The mixture should contain at least 50-percent antifreeze, but should never contain more than 70-percent antifreeze. Consult the mixture ratio chart on the antifreeze container before adding coolant. Hydrometers are available at most auto parts stores to test the coolant **(see illustration)**. Use antifreeze which meets the vehicle manufacturer's specifications.

3   Thermostat - check and replacement

Refer to illustrations 3.9a, 3.9b, 3.9c, 3.9d, 3.10 and 3.13

Warning: *The engine must be completely cool when this procedure is performed.*

Note: *Don't drive the vehicle without a thermostat! The computer may stay in open loop mode and emissions and fuel economy will suffer.*

3.9d Thermostat cover location (arrow) - 1998 and later V8 engine

Check

1   Before condemning the thermostat, check the coolant level, drivebelt tension and temperature gauge (or light) operation.

2   If the engine takes a long time to warm up, the thermostat is probably stuck open. Replace the thermostat.

3   If the engine runs hot, check the temperature of the upper radiator hose. If the hose isn't hot, the thermostat is probably stuck shut. Replace the thermostat.

4   If the upper radiator hose is hot, it means the coolant is circulating and the thermostat is open. Refer to the *Troubleshooting* section for the cause of overheating.

5   If an engine has been overheated, you may find damage such as leaking head gaskets, scuffed pistons and warped or cracked cylinder heads.

Replacement

6   Drain coolant (about 1 gallon) from the radiator, until the coolant level is below the thermostat housing.

7   Remove the air intake duct (see Chapter 4).

8   Disconnect the radiator hose from the thermostat cover. **Note 1:** *Most models do not have a thermostat housing gasket, and it isn't necessary to remove the hose unless it is being replaced.* **Note 2:** *On 2001 and later models with 5.7L engine, the thermostat and*

3.10 Note how it's installed, then remove the thermostat (the spring end points toward the engine)

3.13 Install a new rubber seal around the thermostat

4.1a Locations of the fan connectors (arrows) on twin cooling fan applications (3800 V6 and V8 models)

4.1b Disconnect the electrical connector from the fan and apply fused battery power and ground to test the fan

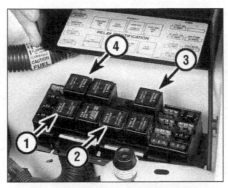

4.2 Location of the cooling fan and air conditioning relays in the underhood fuse/relay box on 1997 and earlier vehicles:

1   Air conditioning compressor relay
2   Cooling fan relay #1
3   Cooling fan relay #2
4   Cooling fan relay #3

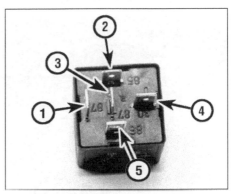

4.3 Test for continuity between relay terminals 1 and 4 - there should be no continuity until power is applied to terminal 5 and ground to terminal 2

thermostat housing must be replaced as a unit.
9   Remove the bolts and lift the cover off **(see illustrations)**. It may be necessary to tap the cover with a soft-face hammer to break the gasket seal.
10   Note how it's installed, then remove the thermostat **(see illustration)**. Be sure to use a replacement thermostat with the correct opening temperature (see this Chapter's Specifications).
11   If a gasket was used, use a scraper or putty knife to remove all traces of old gasket material and sealant from the mating surfaces. **Caution:** Be careful not to gouge or damage the gasket surfaces, because a leak could develop after assembly. Make sure no gasket material falls into the coolant passage; it's a good idea to stuff a rag in the passage. Wipe the mating surfaces with a rag saturated with lacquer thinner or acetone.
12   Install the thermostat and make sure the correct end faces out - the spring is directed toward the engine **(see illustration 3.10).**
13   Most models will not have a traditional gasket, but rather a rubber ring around the thermostat. If so, replace this ring and install the thermostat cover without gasket sealant **(see illustration)**. **Note:** If a gasket was used, apply a thin coat of RTV sealant to both sides of the new gasket and position it on the

engine side, over the thermostat, and make sure the gasket holes line up with the bolt holes in the housing.
14   Carefully position the cover and install the bolts. Tighten them to the torque listed in this Chapter's Specifications - do not over-tighten the bolts or the cover may crack or become distorted.
15   Reattach the radiator hose to the cover and tighten the clamp - now may be a good time to check and replace the hoses and clamps (see Chapter 1).
16   Refer to Chapter 1 and refill the system, then run the engine and check carefully for leaks.
17   Repeat steps 1 through 5 to be sure the repairs corrected the previous problem(s).

4   Engine cooling fan(s) and circuit - check, removal and installation

Warning: The models covered by this manual are equipped with airbags. Always disable the airbag system before working in the vicinity of the impact sensors, steering column or instrument panel to avoid the possibility of accidental deployment of the airbag(s), which could cause personal injury (see Chapter 12).

The yellow wires and connectors routed through the instrument panel and, on 1995 and earlier models, to the front of the vehicle, are for this system. Do not use electrical test equipment on these yellow wires or tamper with them in any way.
Warning: Keep hands, tools and clothing away from the fan. To avoid injury or damage DO NOT operate the engine with a damaged fan. Do not attempt to repair fan blades - replace a damaged fan with a new one.

Check

Refer to illustrations 4.1b, 4.1b, 4.2, 4.3 and 4.6
Note: Some air-conditioned models have two electric fans. The following procedures apply to both fans.
1   To test a fan motor, unplug the electrical connector at the motor **(see illustrations)** and use fused jumper wires to connect the fan directly to the battery. If the fan still does-n't work, replace the motor.
2   If the motor tests OK, check the cooling fan relays, located in the underhood fuse/relay panel **(see illustration)**. There are three of them.
3   Remove cooling fan relay no. 1 and test for continuity between terminals 1 and 4 **(see illustration)**. There should be no continuity.

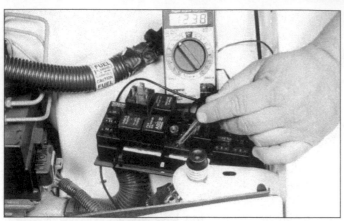

4.6 Probing with a grounded test light or voltmeter, check socket no.1 on the relay panel - there should be power at all times, and only with the key On at socket no. 5

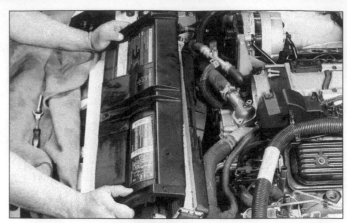

4.11 Remove the bolts to the upper radiator panel and remove the panel

4.13 Tilt and lower the fan assembly from under the vehicle

4.14 Remove the nut (arrow) on the motor shaft to separate the fan blades from the motor

4   Apply power to terminal 5, and ground to terminal 2, and test for continuity again between 1 and 4. There should now be continuity; if not, replace the relay. This test applies to all three relays.

5   An additional test for relay no. 3 involves testing for continuity between terminals 3 and 4 when no power is applied to the relay. There should be continuity.

6   Test the number 1 socket (on the fuse/relay panel) for each relay **(see illustration)**. There should be power at all times, probing with a grounded test light. The number 5 socket should have power only with the key ON.

7   If the relays check out and the fan motor works, check the circuit back to the PCM (computer).

8   If the circuit checks OK but the fan(s) still don't come on, refer to Chapter 6 and check the engine coolant temperature sensor.

Replacement

Refer to illustrations 4.11, 4.13 and 4.14
Note: *This procedure applies to either fan.*
9   Disconnect the cable from the negative terminal of the battery. **Caution:** *On models equipped with a Delco Loc II or Theftlock audio system, be sure the lockout feature is*

turned off before performing any procedure which requires disconnecting the battery.
10   Remove the air intake duct and air cleaner (see Chapter 4). On 2000 and later models, disconnect the air intake temperature sensor (and MAF sensor on 5.7L engines) before removing the air intake duct.
11   Remove the upper radiator panel **(see illustration)** and disconnect the electrical connectors from the fan(s) **(see illustration 4.1a)**.
12   Pull the fan assembly up slightly to dislodge its tabs from the radiator, then raise and support the vehicle.
13   Guide the fan assembly out from the bottom, making sure that all wiring clips are disconnected **(see illustration)**. Be careful not to contact the radiator cooling fins
14   To detach the fan from the motor, remove the motor shaft nut **(see illustration)**.
15   To detach the fan motor from its mount, remove the screws.
16   Installation is the reverse of removal.

---

5   Radiator and coolant reservoir - removal and installation

---

Warning 1: *The models covered by this manual are equipped with airbags. Always disable*

the airbag system before working in the vicinity of the impact sensors, steering column or instrument panel to avoid the possibility of accidental deployment of the airbag(s), which could cause personal injury (see Chapter 12). The yellow wires and connectors routed through the instrument panel and, on 1995 and earlier models, to the front of the vehicle, are for this system. Do not use electrical test equipment on these yellow wires or tamper with them in any way.
Warning 2: *The engine must be completely cool when this procedure is performed.*

Radiator

Refer to illustrations 5.4a, 5.4b, 5.6a, 5.6b and 5.7
1   Disconnect the cable from the negative terminal of the battery. **Caution:** *On models equipped with a Delco Loc II or Theftlock audio system, be sure the lockout feature is turned off before performing any procedure which requires disconnecting the battery.*
2   Drain the cooling system as described in Chapter 1. Refer to the coolant warning in Section 2 of this Chapter.
3   Remove the air intake duct and air cleaner. Raise the vehicle and support it securely on jackstands.

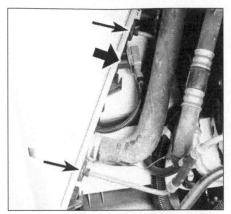

5.4a On the right side, disconnect the transmission cooler lines (small arrows), coolant level sensor (large arrow) and radiator hose (above lower cooler line)

5.4b On the left side, disconnect the engine oil cooler line (arrow) if equipped

5.6a Disconnect the two small hoses at the radiator neck

5.6b On the upper left side of the radiator, remove the upper hose (arrow)

5.7 Working from the bottom, carefully angle the radiator out to avoid damaging any of the cooling fins

5.13 The coolant reservoir is mounted underneath the battery tray - remove the tray and tank together

4   If equipped, disconnect the coolant level indicator module. If equipped with an automatic transmission cooler or engine oil cooler, remove the cooler lines from the radiator **(see illustrations)** - be careful not to damage the lines or fittings. Plug the ends of the disconnected lines to prevent leakage and stop dirt from entering the system. Have a drip pan ready to catch any spills.

5   Remove the upper radiator panel and the cooling fan assembly (see Section 4).

6   Disconnect the upper and lower radiator hoses from the radiator, and disconnect the two small hoses at the radiator neck **(see illustrations)**. **Note:** On 1998 and later it will be necessary to evacuate the air conditioning system and disconnect the fittings from the condenser (see Section 16) then remove the radiator and the condenser together.

7   Raise and suitably support the vehicle. The radiator can be removed from the bottom with some wiggling to get the lower right corner to clear the lower refrigerant hose to the condenser **(see illustration)**.

8   Prior to installation of the radiator, replace any damaged hose clamps and radiator hoses.

9   If leaks have been noticed or there have been cooling problems, have the radiator cleaned and tested at a radiator shop.

10   Installation is the reverse of removal.

11   After installation, fill the system with the proper mixture of antifreeze, bleed the air from the cooling system as described in Chapter 1, and also check the automatic transmission fluid level, where applicable (see Chapter 1).

Coolant reservoir

Refer to illustration 5.13

12   Remove the battery and disconnect the coolant reservoir hose from the radiator.

13   Unbolt the battery tray and remove the battery tray and coolant reservoir together **(see illustration)**.

14   Prior to installation make sure the reservoir is clean and free of debris which could be drawn into the radiator (wash it with soapy water and a brush if necessary, then rinse thoroughly).

15   Installation is the reverse of removal.

6   Water pump - check

Refer to illustration 6.2

Warning: *The models covered by this manual are equipped with airbags. Always disable the airbag system before working in the vicinity of the impact sensors, steering column or instrument panel to avoid the possibility of accidental deployment of the airbag(s), which could cause personal injury (see Chapter 12). The yellow wires and connectors routed through the instrument panel and, on 1995 and earlier models, to the front of the vehicle, are for this system. Do not use electrical test equipment on these yellow wires or tamper with them in any way.*

1   Water pump failure can cause overheating and serious damage to the engine. There are three ways to check the operation of the water pump while it is installed on the engine. If any one of the following quick-checks indicates water pump problems, it should be replaced immediately.

2   A seal protects the water pump impeller shaft bearing from contamination by engine coolant. If this seal fails, a weep hole in the

6.2 Check the weep hole (arrow) for leakage (pump removed for clarity)

7.6 Locations of the water pump mounting bolts (arrows) - 3800 V6 engine

water pump snout will leak coolant **(see illustration)** (an inspection mirror can be used to look at the underside of the pump if the hole isn't on top). If the weep hole is leaking, shaft bearing failure will follow. Replace the water pump immediately.

3   Besides contamination by coolant after a seal failure, the water pump impeller shaft bearing can also be prematurely worn out by an improperly tensioned drivebelt (on 1997 and earlier V8 engines the water pump is mechanically driven). When the bearing wears out, it emits a high-pitched squealing sound. If such a noise is coming from the water pump during engine operation, the shaft bearing has failed - replace the water pump immediately. **Note:** *Do not confuse belt noise with bearing noise.*

4   To identify excessive water pump bearing wear on V6 and 1998 and later V8 engines, grasp the water pump pulley and try to force it up-and-down or from side-to-side. If the pulley can be moved either horizontally or vertically, the bearing is nearing the end of its service life. Replace the water pump. Don't mistake drivebelt slippage, which causes a squealing sound, for water pump bearing failure.

5   It is possible for a water pump to be bad, even if it doesn't howl or leak water. Sometimes the fins on the back of the impeller can corrode away until the pump is no longer effective. The only way to check for this is to remove the pump for examination.

7   Water pump - removal and installation

Warning 1: *The models covered by this manual are equipped with airbags. Always disable the airbag system before working in the vicinity of the impact sensors, steering column or instrument panel to avoid the possibility of accidental deployment of the airbag(s), which could cause personal injury (see Chapter 12). The yellow wires and connectors routed through the instrument panel and, on 1995 and earlier models, to the front of the vehicle, are for this system. Do not use electrical test equipment on these yellow wires or tamper with them in any way.*

7.8 Unbolt and move the ignition coils on 1997 and earlier V8 models

Warning 2: *Wait until the engine is completely cool before starting this procedure.*

Removal

1   Disconnect the cable from the negative terminal of the battery. **Caution:** *On models equipped with a Delco Loc II or Theftlock audio system, be sure the lockout feature is turned off before performing any procedure which requires disconnecting the battery.*

2   Drain the coolant (see Chapter 1). Also remove the air cleaner housing and the air intake resonator on 1998 and later vehicles (see Chapter 4).

3   Loosen the water pump pulley bolts on V6 models, then remove the serpentine drivebelt (see Chapter 1).

V6 models

Refer to illustration 7.6

4   Unbolt and move the power steering pump bracket (see Chapter 10) and one coil of the coil pack (3.4L engine only).

5   Detach the coolant hoses from the water pump.

6   Remove the water pump pulley and unbolt the water pump **(see illustration)**. It may be necessary to tap the pump with a soft-face hammer to break the gasket seal. Inspect the impeller blades on the backside of the pump for corrosion. If any blades are missing or badly corroded, replace the pump with a new one.

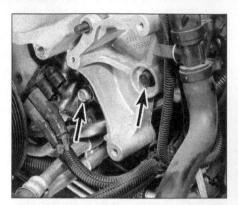

7.9 After removing the air injection pump on 1997 and earlier V8 models, remove these two bolts and the pump bracket - it covers the water pump stud

V8 models

Refer to illustrations 7.8, 7.9, 7.12a and 7.12b

6   Refer to Section 4 and remove the electric cooling fan assembly.

7   Remove the upper and lower radiator hoses and heater hoses from the water pump.

8   On 1997 and earlier models, disconnect the electrical connector from the coolant temperature sensor on top of the water pump (see Chapter 6), then unclip its wire harness from the two clips attached to the face of the water pump. Also unbolt the ignition coil bracket and move it out of the way **(see illustration)**. **Caution:** *Be careful when removing the coil mounting nuts - the ground wire ring-terminals under the nuts are fragile, and can break if they are twisted too far.*

9   On 1997 and earlier models, unbolt and move the air injection pump (see Chapter 6). Remove the two bolts holding the air injection pump bracket **(see illustration)**.

10   On 1998 and later models, remove the drivebelt tensioner from the water pump housing.

11   On 1998 and later models, detach the throttle body heater hose from the throttle body and position it aside.

12   Remove the water pump mounting bolts and detach it. If necessary, strike the pump with a soft-face hammer or a block of wood or wooden hammer handle to break the gas-

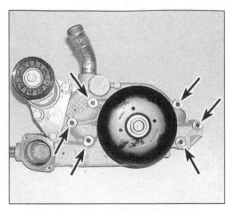

7.12a Locations of the water pump mounting bolts (arrows) - 1998 and later V8 engine

7.12b Use a soft face hammer to break the gasket seal on the pump - arrows indicate hoses to be disconnected on 1997 and earlier V8 engines

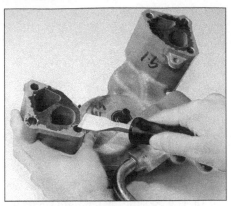

7.13 Remove all traces of old gasket material - use care to avoid gouging the soft aluminum

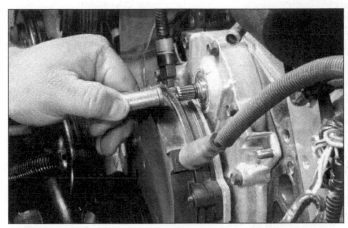

7.15a On 1997 and earlier V8 models, the water pump is mechanically driven by this splined shaft from the timing cover. Remove the splined coupler from the drive gear on the engine . . .

7.15b . . . and replace the O-ring seals - one at the pump splines, one at the drive splines - use something tapered to slip the seal evenly around the shafts

ket seal (see illustrations). Do not pry between the pump and the block.

Installation

Refer to illustrations 7.13, 7.15a, 7.15b and 7.16

13   Clean the sealing surfaces of all gasket material on both the water pump and block (see illustration). Wipe the mating surfaces with a rag saturated with lacquer thinner or acetone.

14   Apply a thin layer of RTV sealant to both sides of the new gasket and install the gasket on the water pump.

15   On 1997 and earlier V8 engines, pull the water pump driveshaft coupling (see illustration) from the driveshaft (the short, splined shaft coming out of the block above the distributor). Use something tapered like a marking-pen top and slip lightly-oiled, new O-ring seals over the splines at the driven-gear end and the splines at the back of the water pump (see illustration).

16   Place the water pump in position and install the bolts finger tight. Use caution to ensure that the gasket doesn't slip out of position. Remember to replace any mounting

brackets secured by the water pump mounting bolts/studs. Tighten the bolts to the torque listed in this Chapter's Specifications. **Note 1:** *On 1997 and earlier, V8 engines, install the water pump driveshaft coupling over the driveshaft and align the pump's splines with the coupling when putting the pump into place on the engine.* **Note 2:** *There are three different bolt torque specifications*

for the 3.4L engine water pump (see illustration).

17   The remainder of the installation procedure is the reverse of removal.

18   Add coolant to the specified level (see Chapter 1) and start the engine and check for the proper coolant level and the water pump and hoses for leaks. Bleed the cooling system of air as described in Chapter 1.

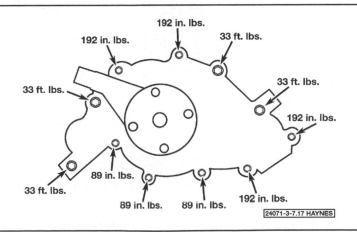

7.16 There are three different torque specifications for the bolts on 3.4L V6 water pumps

8.1a The coolant-temperature sending unit on 3.4L V6 engines (arrow) is located near the front of the left cylinder head, behind the power steering pump

8.1b The dual-purpose (PCM and gauge) coolant temperature sensor (arrow) on the 3800 V6 engine is located under the throttle body

8.1c Location of the coolant-temperature sending unit on V8 engines through 1999 (arrow, exhaust manifold removed for clarity) - on 2000 and later models the sensor is at the front of the left cylinder head

8   Coolant temperature sending unit - check and replacement

Refer to illustrations 8.1a, 8.1b and 8.1c
Warning: *Wait until the engine is completely cool before beginning this procedure.*

Check

1   The coolant temperature indicator system is composed of a temperature gauge mounted in the dash and a coolant temperature sending unit mounted on the engine **(see illustrations)**. Some vehicles have more than one sending unit, but only one is used for the indicator system and the other is used to send temperature information to the PCM. The 3800 V6 uses a combination sensor that sends signals to both the PCM and the temperature gauge.
2   If an overheating indication occurs, check the coolant level in the system and then make sure the wiring between the gauge and the sending unit is secure and all fuses are intact. **Note:** *The temperature gauge does not have equal divisions of temperature. The middle position on the gauge equals 210 to 220-degrees F.*
3   Check the circuit and gauge operation by grounding the wire to the sending unit while the ignition is On (engine NOT running). If the gauge deflects full scale, the circuit and gauge are OK. The problem lies in the sending unit.
4   To confirm the sending unit is defective, check the resistance of the unit when the engine is cold (100-degrees F or lower). Resistance should be high - approximately 1365 ohms at 100-degrees F. Next, run the engine until it is fully warmed up and check the resistance of the sending unit again. The resistance should now be low - approximately 55 ohms at 260-degrees F. If it doesn't respond at close to these numbers, replace the sending unit.

Replacement

5   Make sure the engine is cool before removing the defective sending unit. There will be some coolant loss as the unit is removed, so be prepared to catch it. Refer to the coolant Warning in Section 2.

6   Prepare the new sending unit by wrapping the threads with Teflon tape or by coating them with sealant. Disconnect the electrical connector and unscrew the sensor. Install the new sensor as quickly as possible to minimize coolant loss.
7   Check the coolant level after the replacement unit has been installed and top up the system, if necessary (see Chapter 1). Check now for proper operation of the gauge and sending unit.

9   Blower motor and circuit - check

Refer to illustrations 9.4 and 9.7
Warning: *The models covered by this manual are equipped with airbags. Always disable the airbag system before working in the vicinity of the impact sensors, steering column or instrument panel to avoid the possibility of accidental deployment of the airbag(s), which could cause personal injury (see Chapter 12). The yellow wires and connectors routed through the instrument panel and, on 1995 and earlier models, to the front of the vehicle, are for this system. Do not use electrical test equipment on these yellow wires or tamper with them in any way.*
1   Check the fuse (marked HVAC) and all

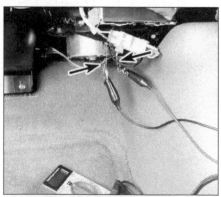

9.4 Connect a voltmeter to the heater blower motor connector (arrows) by backprobing, and check the running voltage at each blower switch position

connections in the circuit for looseness and corrosion. Make sure the battery is fully charged.
2   With the transmission in Park, the parking brake securely set, turn the ignition switch to the Run position. It isn't necessary to start the vehicle.
3   Remove the lower right dash insulator panel (below the glove box) for access to the blower motor.
4   Backprobe the blower motor electrical connector with two small paper clips (straightened out) and connect a voltmeter to the blower motor connector and ground **(see illustration)**.
5   Move the blower switch through each of its positions and note the voltage readings. Changes in voltage indicate that the motor speeds will also vary as the switch is moved to the different positions.
6   If there is voltage present, but the blower motor does not operate, the blower motor is probably faulty. Disconnect the blower motor connector and hook one side to a chassis ground and the other to a fused source of battery voltage. If the blower doesn't operate, it is faulty.

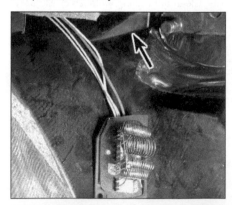

9.7 The blower motor resistor is very hard to access, situated underneath the heater/air conditioning box (arrow) just to the right of the transmission tunnel - it is retained by one screw (it's been removed for clarity in this photo)

10.1 Remove the three screws retaining the blower motor to the housing

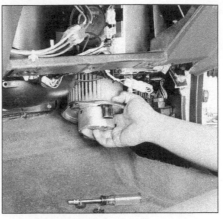

10.2 Lower the blower motor and fan assembly straight down

7   If there was no voltage present at the blower motor at one or more speeds, and the motor itself tested OK, check the blower motor resistor. Pull the carpeting back from around the heater module to access the blower resistor **(see illustration)**.

8   Backprobe the connector to the resistor while it is connected and look for varying voltages at each position.

9   Remove the single resistor mounting screw, pull down the resistor and disconnect the electrical connector from the blower motor resistor. With the ignition on, check for voltage at each of the terminals in the connector as the blower speed switch is moved to the different positions. If the voltmeter responds correctly to the switch then the resistor is probably faulty. If there is no voltage present from the switch, then the switch, control panel or related wiring is probably faulty.

10 Blower motor - removal and installation

Refer to illustrations 10.1 and 10.2
Warning: *The models covered by this manual are equipped with airbags. Always disable the airbag system before working in the vicinity of the impact sensors, steering column or instrument panel to avoid the possibility of accidental deployment of the airbag(s), which could cause personal injury (see Chapter 12). The yellow wires and connectors routed through the instrument panel and, on 1995 and earlier models, to the front of the vehicle, are for this system. Do not use electrical test equipment on these yellow wires or tamper with them in any way.*

1   Remove the lower right instrument panel insulator panel below the glove box. Disconnect the electrical connector from the blower motor and remove the three screws from the blower housing **(see illustration)**.

2   Pull the blower motor and fan straight down **(see illustration)**.

3   To remove the fan from the blower motor, squeeze the spring clip together and slip the fan off the shaft. **Note:** *On some models, there is no clip or nut retaining the*

fan to the shaft; in this case the fan can be pried off with two screwdrivers.

4   Install the fan onto the motor and install the blower motor into the heater housing.

11 Heater core - removal and installation

Warning 1: *The models covered by this manual are equipped with airbags. Always disable the airbag system before working in the vicinity of the impact sensors, steering column or instrument panel to avoid the possibility of accidental deployment of the airbag(s), which could cause personal injury (see Chapter 12). The yellow wires and connectors routed through the instrument panel and, on 1995 and earlier models, to the front of the vehicle, are for this system. Do not use electrical test equipment on these yellow wires or tamper with them in any way.*

Warning 2: *The air conditioning system is under high pressure. DO NOT loosen any fittings or remove any components until after the system has been discharged. Air conditioning refrigerant should be properly discharged into an EPA-approved container at a dealership service department or an automotive air conditioning facility. Always wear eye protection when disconnecting air conditioning system fittings.*

Note: *On 2002 models, the entire instrument panel and cowl bracing must be removed to*

11.4 A small screw (arrow) retains a clamp over the two heater core tubes

access the HVAC module, which must be removed from the car to access the heater core. Due to the number of electrical connections and sometimes difficult to locate fasteners, we don't recommend this procedure for the home mechanic.

Removal

Refer to illustrations 11.3, 11.4, 11.5, 11.7, 11.8a, 11.8b and 11.9

1   Disconnect the battery cable at the negative battery terminal. **Caution:** *On models equipped with a Delco Loc II or Theftlock audio system, be sure the lockout feature is turned off before performing any procedure which requires disconnecting the battery.*

2   Drain the cooling system (see Chapter 1).

3   Disconnect the heater hoses at the heater core inlet and outlet on the engine side of the firewall (passenger side) and plug the open fittings **(see illustration)**. If the hoses are stuck to the pipes, cut them off.

4   Remove the heater core pipe clamp-bolt and clamp **(see illustration)**.

5   From the inside of the car, remove the glove box (open it and squeeze the sides in, then pull out, remove the hinge screws) for access to the upper heater core cover mounting screw **(see illustration)**. Below the glove box, remove the dash insulating panel

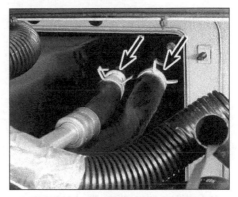

11.3 Loosen the hose clamps (arrows) and disconnect the heater hoses from the heater core tubes at the firewall

11.5 Disconnect the electrical connector (right arrow) from the heater core cover, and remove the two retaining screws (left arrows)

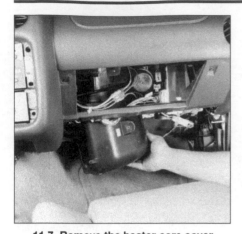

11.7 Remove the heater core cover

11.8a Remove this one screw and the heater core retaining plate . . .

11.8b . . . then pull the heater core down and out

for access to the lower screw.

6   If equipped with air conditioning, remove the evaporator temperature sensor wiring clip **(see illustration 11.5)** from the heater core cover.

7   Remove the heater core cover **(see illustration)**.

8   Remove the heater core clamp bolt and clamp **(see illustration)**, then slide the heater core out carefully **(see illustration)**.

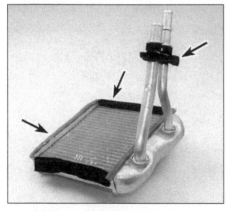

11.9 When reinstalling the heater core, make sure all of the sealing material (arrows) is in place

Installation

9   Installation is the reverse of removal. **Note:** *When reinstalling the heater core, make sure any original insulating/sealing materials are in place around the heater core pipes and around the core* **(see illustration)**.

10   Refill the cooling system (see Chapter 1).

11   Start the engine and check for proper operation.

---

12   Heater and air conditioning control assembly - removal and installation

Warning: *The models covered by this manual are equipped with airbags. Always disable the airbag system before working in the vicinity of the impact sensors, steering column or instrument panel to avoid the possibility of accidental deployment of the airbag(s), which could cause personal injury (see Chapter 12). The yellow wires and connectors routed through the instrument panel and, on 1995 and earlier models, to the front of the vehicle, are for this system. Do not use electrical test equipment on these yellow wires or tamper with them in any way.*

Removal

Refer to illustrations 12.3, 12.4a, 12.4b and 12.4c

1   Disconnect the battery cable from the negative battery terminal. **Caution:** *On models equipped with a Delco Loc II or Theftlock audio system, be sure the lockout feature is turned off before performing any procedure which requires disconnecting the battery.*

2   Remove the center trim panel from the instrument panel (see Chapter 11).

3   Removal of the bezel allows access to the heater/air conditioning control mounting screws **(see illustration)**.

4   Remove the control assembly retaining screws and pull the unit from the dash **(see illustration)**. It can be pulled out just far enough to allow disconnecting the control cable end, electrical connections and vacuum lines (on air conditioning-equipped models) from the control head. **Note:** *Use a small screwdriver to release the clips holding the control cable* **(see illustrations)**.

Installation

5   To install the control assembly, reverse the removal procedure. **Caution:** *When reconnecting vacuum lines to the control assembly, do not use any lubricant to make*

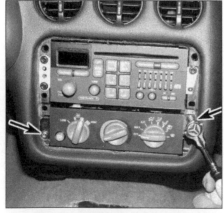

12.3 Remove the screws (arrows) to allow removal of the heater/air conditioning control panel

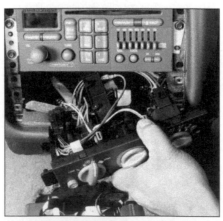

12.4a Pull the bezel forward - for complete removal, detach the wiring and vacuum lines

12.4b Unsnap the vacuum block (arrow) and lines from behind the mode 3 control switch

12.4c Use a small screwdriver to pry apart the snaps (arrow) that retain the control cable (upper arrow)

13.8 With the system on, feel the receiver/drier and the smaller diameter line to the evaporator - the surface of the receiver/drier should feel cool and the line should feel warm

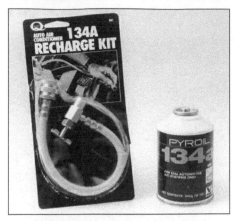

13.10 A basic charging kit for 134A systems is available at most auto parts stores - it must say 134A (not R-12) and so should the 12-ounce can of refrigerant

them slip on easier, it can affect vacuum operation. If necessary, use a drop of plain water to make reconnection easier.

13 Air conditioning and heating system - check and maintenance

Warning: The air conditioning system is under high pressure. DO NOT loosen any fittings or remove any components until after the system has been discharged. Air conditioning refrigerant should be properly discharged into an EPA-approved recovery container at a dealership service department or an automotive air conditioning repair facility. Always wear eye protection when disconnecting air conditioning system fittings.

1   The following maintenance steps should be performed on a regular basis to ensure that the air conditioner continues to operate at peak efficiency:

a) Check the drivebelt (see Chapter 1).
b) Check the condition of the hoses. Look for cracks, hardening and deterioration. Look at potential leak areas (hoses and fittings) for signs of refrigerant oil leaking out. **Warning:** Do not replace air conditioning hoses until the system has been discharged by a dealership or air conditioning repair facility.
c) Check the fins of the condenser for leaves, bugs and other foreign material. A soft brush and compressed air can be used to remove them.
d) Check the wire harness for correct routing, broken wires, damaged insulation, etc. Make sure the electrical connectors are clean and tight.
e) Maintain the correct refrigerant charge.

2   The system should be run for about 10 minutes at least once a month. This is particularly important during the winter months because long-term non-use can cause hardening of the internal seals.
3   Because of the complexity of the air conditioning system and the special equip-

ment required to effectively work on it, accurate troubleshooting of the system should be left to a certified air conditioning technician.
4   If the air conditioning system doesn't operate at all, check the fuse panel. Check the HVAC fuse and the air conditioning compressor relay.
5   The most common cause of poor cooling is simply a low system refrigerant charge. If a noticeable drop in cool air output occurs, the following quick check will help you determine if the refrigerant level is low. For more complete information on the air conditioning system, refer to the *Haynes Automotive Heating and Air Conditioning Manual*.

Checking the refrigerant charge

Refer to illustration 13.8

6   Warm the engine up to normal operating temperature.
7   Place the air conditioning temperature selector at the coldest setting and the blower at the highest setting. Open the doors (to make sure the air conditioning system doesn't cycle off as soon as it cools the passenger compartment).
8   With the compressor engaged - the clutch will make an audible click and the center of the clutch will rotate - feel the surface of the receiver/drier and the small diameter line to the evaporator **(see illustration)**. **Note:** The 3800 V6 uses a type of compressor that does not cycle; it changes its stroke to match system demand. If the smaller diameter line feels warm and the receiver/drier feels cool, the system is properly charged **(see illustration)**. On 3800 V6 models, feel the pipe on either side of the expansion tube - there should be a noticeable difference in temperature.
9   Place a thermometer in the dashboard vent nearest the evaporator and add refrigerant to the system until the indicated temperature is around 40 to 45-degrees F. If the ambient (outside) air temperature is very high,

say 110-degrees F, the duct air temperature may be as high as 60-degrees F, but generally the air conditioning is 30 to 50-degrees F cooler than the ambient air. **Note:** Humidity of the ambient air also affects the cooling capacity of the system. Higher ambient humidity lowers the effectiveness of the air conditioning system.

Adding refrigerant

Refer to illustrations 13.10 and 13.13

10   Buy an automotive charging kit at an auto parts store. A charging kit includes a 14-ounce can of refrigerant, a tap valve and a short section of hose that can be attached between the tap valve and the system low side service valve **(see illustration)**. Because one can of refrigerant may not be sufficient to bring the system charge up to the proper level, it's a good idea to buy an extra can. Make sure that one of the cans contains red refrigerant dye. If the system is leaking, the red dye will leak out with the refrigerant and help you pinpoint the location of the leak. **Caution:** There are two types of refrigerant, R-12, used on vehicles up to 1992, and the more environmentally friendly R-134a used in all the models covered by this book. These two refrigerants (and their appropriate refrigerant oils) are not compatible and must never be mixed or components will be damaged. Use only R-134a refrigerant in the models covered by this book. **Warning:** Never add more than two cans of refrigerant to the system.
11   Hook up the charging kit by following the manufacturer's instructions. **Warning:** DO NOT hook the charging kit hose to the system high side! The fittings on the charging kit are designed to fit **only** on the low side of the system.
12   Back off the valve handle on the charging kit and screw the kit onto the refrigerant can, making sure first that the O-ring or rubber seal inside the threaded portion of the kit is in place. **Warning:** Wear protective eyewear when dealing with pressurized refrigerant cans.

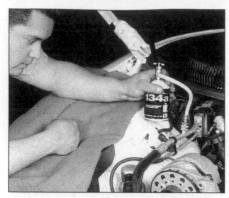

13.13 Add R-134A refrigerant to the low-side port only - the procedure is easier if you wrap the can with a warm, wet towel to prevent icing

13 Remove the dust cap from the low-side charging connection and attach the quick-connect fitting on the kit hose **(see illustration)**.

14 Warm up the engine and turn on the air conditioner. Keep the charging kit hose away from the fan and other moving parts. **Note:** *The charging process requires the compressor to be running. Your compressor may cycle off if the pressure is low due to a low charge.* If the clutch cycles off, you can pull the low-pressure cycling switch plug from the evaporator inlet line and attach a jumper wire across the terminals. This will keep the compressor ON.

15 Turn the valve handle on the kit until the stem pierces the can, then back the handle out to release the refrigerant. You should be able to hear the rush of gas. Add refrigerant to the low side of the system until both the accumulator surface (on 3800 engines; others use a receiver-drier) and the evaporator inlet pipe feel about the same temperature . Allow stabilization time between each addition.

16 If you have an accurate thermometer, you can place it in the center air conditioning duct inside the vehicle and keep track of the "conditioned" air temperature. A charged system that is working properly should put out air that is 40-degrees F. If the ambient (outside) air temperature is very high, say 110-degrees F, the duct air temperature may be as high as 60-degrees F, but generally the air conditioning is 30 to 50-degrees-F cooler than the ambient air.

17 When the can is empty, turn the valve handle to the closed position and release the connection from the low-side port. Replace the dust cap.

18 Remove the charging kit from the can and store the kit for future use with the piercing valve in the UP position, to prevent inadvertently piercing the can on the next use.

Heating systems

19 If the carpet under the heater core is damp, or if antifreeze vapor or steam is coming through the vents, the heater core is leaking. Remove it (see Section 11) and install a new unit (most radiator shops will not repair a leaking heater core).

20 If the air coming out of the heater vents isn't hot, the problem could stem from any of the following causes:

a) *The thermostat is stuck open, preventing the engine coolant from warming up enough to carry heat to the heater core. Replace the thermostat (see Section 3).*

b) *A heater hose is blocked, preventing the flow of coolant through the heater core. Feel both heater hoses at the firewall. They should be hot. If one of them is cold, there is an obstruction in one of the hoses or in the heater core, or the heater control valve is shut (located in the hot water hose to the heater core, along the passenger side of the engine on 1980 and earlier models). Detach the hoses and back flush the heater core with a water hose. If the heater core is clear but circulation is impeded, remove the two hoses and flush them out with a water hose.*

c) *If flushing fails to remove the blockage from the heater core, the core must be replaced (see Section 11).*

14 Air conditioning compressor - removal and installation

Refer to illustration 14.4

Removal

Warning: *The air conditioning system is under high pressure. DO NOT loosen any fittings or remove any components until after the system has been discharged. Air conditioning refrigerant should be properly discharged into an EPA-approved container at a dealership service department or an automotive air conditioning repair facility. Always wear eye protection when disconnecting air conditioning system fittings.*

Note: *The receiver/drier or accumulator (see Section 15) should be replaced whenever the compressor is replaced.*

1 Have the air conditioning system discharged (see **Warning** above). Disconnect the cable from the negative terminal of the

14.4 Disconnect the electrical connector (arrow) from the air conditioning compressor

battery. **Caution:** *On models equipped with a Delco Loc II or Theftlock audio system, be sure the lockout feature is turned off before performing any procedure which requires disconnecting the battery.*

2 Clean the compressor thoroughly around the refrigerant line fittings.

3 Remove the serpentine drivebelt (see Chapter 1).

4 Disconnect the electrical connector from the air conditioning compressor **(see illustration)**. Raise the vehicle and support it securely on jackstands. On 2000 and later models with 3.8L engine, disconnect the negative battery cable from its clip at the engine and move it out of the way.

5 Disconnect the suction and discharge lines from the compressor. Both lines are mounted to the back of the compressor with a plate secured by one bolt. Plug the open fittings to prevent the entry of dirt and moisture, and discard the seals between the plate and compressor.

6 Unbolt and remove the rear compressor mount.

7 Remove the compressor-to-front-bracket bolts and nuts and lower the compressor from the engine compartment.

Installation

8 If a new compressor is being installed, pour the oil from the old compressor into a graduated container and add that exact amount of new refrigerant oil to the new compressor. Also follow any directions included with the new compressor. **Note:** *Some replacement compressors come with refrigerant oil in them.* Follow the directions with the compressor regarding the draining of excess oil prior to installation. **Caution:** *The oil used must be labeled as compatible with R-134a refrigerant systems.*

9 Installation is basically the reverse of the disassembly. When installing the line fitting bolt to the compressor, use new seals lubricated with clean refrigerant oil (seals and oil must be compatible with R-134a refrigerant), and tighten the bolt securely.

10 Reconnect the battery cable to the negative battery terminal.

11 Have the system evacuated, recharged and leak tested by a dealership service department or an automotive air conditioning repair facility.

15 Air conditioning accumulator or receiver/drier - removal and installation

Removal

Refer to illustration 15.2

Warning: *The air conditioning system is under high pressure. DO NOT loosen any fittings or remove any components until after the system has been discharged. Air conditioning refrigerant should be properly discharged into an EPA-approved container at a dealership service department or an automo-*

15.2 Use back-up wrenches when disconnecting the fittings (arrows) at the accumulator (3800 V6 shown)

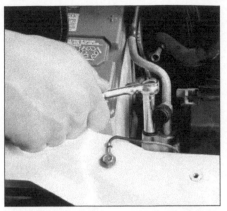

16.2a Near the right side of the radiator top, remove the one bolt securing the line to the top of the condenser . . .

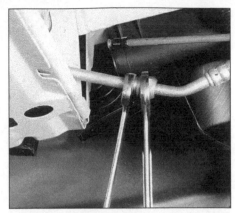

16.2b . . . and from underneath, use two wrenches to disconnect the lower line

tive air conditioning repair facility. Always wear eye protection when disconnecting air conditioning system fittings.

1   Have the air conditioning system discharged (see **Warning** above). Disconnect the cable from the negative terminal of the battery. **Caution:** On models equipped with a Delco Loc II or Theftlock audio system, be sure the lockout feature is turned off before performing any procedure which requires disconnecting the battery.

2   Disconnect the refrigerant inlet and outlet lines **(see illustration)**, using back-up wrenches. Cap or plug the open lines immediately to prevent the entry of dirt or moisture.

3   Loosen the clamp bolt on the mounting bracket and slide the accumulator (3800 models) or receiver/drier assembly (3.4L V6 and V8 models) up and out of the compartment.

Installation

4   If you are replacing the accumulator or receiver-drier with a new one, add one ounce of fresh refrigerant oil to the new unit (oil must be R-134a compatible).

5   Place the new accumulator or receiver/drier into position in the bracket.

6   Install the inlet and outlet lines, using clean refrigerant oil on the new O-rings. Tighten the mounting bolt securely.

7   Connect the cable to the negative terminal of the battery.

8   Have the system evacuated, recharged and leak tested by a dealership service department or an automotive air conditioning repair facility.

---

16   Air conditioning condenser - removal and installation

Warning 1: The models covered by this manual are equipped with airbags. Always disable the airbag system before working in the vicinity of the impact sensors, steering column or instrument panel to avoid the possibility of accidental deployment of the airbag(s), which could cause personal injury (see Chapter 12).

The yellow wires and connectors routed through the instrument panel and, on 1995 and earlier models, to the front of the vehicle, are for this system. Do not use electrical test equipment on these yellow wires or tamper with them in any way.

Warning 2: The air conditioning system is under high pressure. DO NOT loosen any fittings or remove any components until after the system has been discharged. Air conditioning refrigerant should be properly discharged into an EPA-approved container at a dealership service department or an automotive air conditioning repair facility. Always wear eye protection when disconnecting air conditioning system fittings.

Removal

Refer to illustrations 16.2a, 16.2b, 16.4a and 16.4b

1   Have the air conditioning system discharged (see **Warning** above). Disconnect the cable from the negative terminal of the battery. **Caution:** On models equipped with a Delco Loc II or Theftlock audio system, be sure the lockout feature is turned off before performing any procedure which requires disconnecting the battery.

16.4a Pull the condenser up and out, with the plastic front shrouding still attached

2   Disconnect the refrigerant line fittings from the right side of the condenser and cap the open fittings to prevent the entry of dirt and moisture. One bolt secures the condenser-to-accumulator line to the top of the condenser, while the compressor-to-condenser line is attached with fittings to the bottom of the condenser **(see illustrations)**.

3   Remove the upper radiator shroud (see Section 4).

4   Pull the condenser up from the brackets on the front of the radiator **(see illustration)**. **Note:** The condenser is secured to the radiator by tangs that fit into clips on the radiator **(see illustration)**. **Caution:** The condenser is made of aluminum - be careful not to damage it during removal.

Installation

5   Installation is the reverse of removal. Be sure to use new, compatible O-rings on the refrigerant line fittings (lubricate the O-rings with clean refrigerant oil, compatible with R-134a refrigerant). If a new condenser is installed, add 1 ounce of new refrigerant oil to the system.

6   Have the system evacuated, recharged and leak tested by a dealership service department or an automotive air conditioning repair facility.

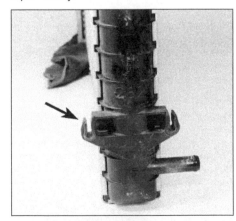

16.4b The tangs (arrow) on the front of the radiator hold the bottom of the condenser

17.3 One bolt (arrow) retains the inlet and outlet lines at the expansion valve (3.4L V6 and V8 engines)

17.5 Arrow indicates the single screw retaining the blend air door cable to the blend air door housing (housing removed for clarity) - with the housing installed use a mirror underneath to see the screw location

17.6 After disconnecting the cable, remove the screws and pull the blend air door housing down and out

17 Air conditioning evaporator - removal and installation

Warning 1: *The models covered by this manual are equipped with airbags. Always disable the airbag system before working in the vicinity of the impact sensors, steering column or instrument panel to avoid the possibility of accidental deployment of the airbag(s), which could cause personal injury (see Chapter 12). The yellow wires and connectors routed through the instrument panel and, on 1995 and earlier models, to the front of the vehicle, are for this system. Do not use electrical test equipment on these yellow wires or tamper with them in any way.*

Warning 2: *The air conditioning system is under high pressure. DO NOT loosen any fittings or remove any components until after the system has been discharged. Air conditioning refrigerant should be properly discharged into an EPA-approved container at a dealership service department or an automotive air conditioning repair facility. Always wear eye protection when disconnecting air conditioning system fittings.*

Note: *On 2002 models, the entire instrument panel and cowl bracing must be removed to access the HVAC module, which must be removed from the car to access the evaporator core. Due to the number of electrical connections and sometimes difficult to locate fasteners, we don't recommend this procedure for the home mechanic.*

Removal

Refer to illustrations 17.3, 17.5, 17.6, 17.7 and 17.9

1   Have the air conditioning system discharged (see **Warning** above). Disconnect the cable from the negative terminal of the battery. **Caution:** *On models equipped with a Delco Loc II or Theftlock audio system, be sure the lockout feature is turned off before performing any procedure which requires disconnecting the battery.*

2   Drain the cooling system (see Chapter 1).

3   Disconnect the air conditioning lines at

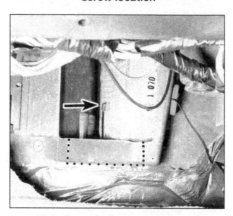

17.7 Arrow indicates location of temperature sensor which must be pulled straight out of the evaporator core - dotted lines indicate portion of plastic housing that must be cut out for evaporator removal, but save the cutout portion - it will be epoxied back in later

the passenger-side of the firewall. On models with the 3800 V6 engine the lines have individual fittings, while on the 3.4L V6 and V8 engines both lines connect to an expansion valve that is retained by one bolt (see illustration).

4   Follow the procedures in Section 11 for removing the heater core cover, heater core, and disconnecting the evaporator temperature sensor's electrical connector.

5   Remove the one screw retaining the blend air door cable (see illustration).

6   Remove the blend-air door housing (see illustration). There are two screws inside the car and one screw from the engine side of the firewall.

7   Using needle-nose pliers, pull the evaporator temperature sensor straight out of the evaporator core. There are lines in the plastic housing that indicate where the plastic must be cut out with a small hand-saw (see illustration), to allow evaporator removal.

17.9 A plastic "fin comb" can be used to straighten bent fins - the tool has a head for various fins-per-inch spacings

8   Slide the evaporator core out of the case.

9   Check the core over carefully for signs of leaks. The efficiency can be improved slightly if any bent fins are straightened with a "fin comb" (see illustration).

10   If a new evaporator core is to be installed, save all of the sealing gaskets from the original unit and transfer them.

Installation

11   Installation is the reverse of the removal procedure. Be sure to re-fill the cooling system (see Chapter 1). Use new R-134a-compatible O-rings on all connections, lubricated by R-134a-compatible refrigerant oil.

12   When reinstalling the expansion valve, use new gaskets, being careful not to scratch the sealing surfaces or dent the gaskets during assembly. Bolt the expansion valve to the evaporator coil first, then the lines to the expansion valve.

13   If a new evaporator has been installed, add 1 ounce of refrigerant oil. Have the system evacuated, recharged and leak tested by a dealership service department or an automotive air conditioning repair facility.

Chapter 4
Fuel and exhaust systems

Contents

Specifications

Sequential Electronic Fuel Injection (SEFI) system

Fuel pressure
  Key On, engine not running
    1997and earlier .... 41 to 47 psi
    1998 and later
      V6 engine .... 49 to 55 psi
      V8 engine .... 55 to 60 psi
    Engine running (at idle) .... Pressure should decrease by 3 to 10 psi
Injector resistance .... 11.0 to 14.0 ohms

Torque specifications   **Ft-lbs** (unless otherwise indicated)

Exhaust pipe-to-manifold nuts .... 15 to 22
Air intake plenum mounting bolts/nuts
  3.4L engines 18
  3800 engines **(see illustration 13.41)**
    Vertical bolts (1 through10) .... 132 in-lbs
    Coolant outlet bolts (11) 20
    Side bolt (12 and 13) 22
Fuel rail bolts/nuts
  3.4L engine 18
  3800 engine .... 75 to 84 in-lbs
  5.7L engines .... 84 to 89 in-lbs
Throttle body bolts/nuts
  3.4L engine 20
  3800 engine .... 84 to 89 in-lbs
  5.7L engines
    1997 and earlier 18
    1998 and later .... 106 in-lbs
IAC valve screws
  3.4L engine .... 30 in-lbs
  3800 engine .... 27 in-lbs
  5.7L engines .... 28 in-lbs

1   General information

The fuel system consists of a fuel tank, an electric fuel pump and fuel pump relay, an air cleaner assembly and a Sequential Electronic Fuel Injection (SEFI) system.

Sequential Electronic Fuel Injection (SEFI) system

These models are equipped with Sequential Electronic Fuel Injection (SEFI) systems. Sequential fuel injection pulses the injectors once per crankshaft revolution. Fuel injection provides optimum mixture ratios at all stages of combustion. Combined with its immediate response characteristics, fuel injection permits the engine to run on the leanest possible air/fuel mixture, which greatly reduces exhaust gas emissions.

The fuel injection system is controlled directly by the vehicle's Electronic Control Module (ECM/PCM) or Powertrain Control Module (PCM), which automatically adjusts the air/fuel mixture in accordance with engine load and performance. Because each cylinder is equipped with an injector mounted immediately adjacent to the intake valve, much better control of the fuel/air mixture ratio is possible.

Fuel pump and lines

Fuel is circulated from the fuel tank to the fuel injection system, and back to the fuel tank, through a pair of lines running along the underside of the vehicle. An electric fuel pump is attached to the fuel sending unit

2.3 Attach a fuel pressure gauge to the test port and open the valve to drain the excess fuel out of the drain tube (arrow) and into an approved fuel container

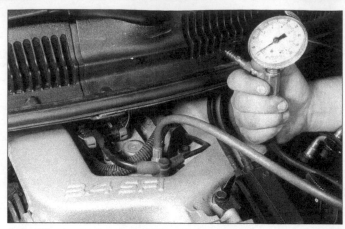

3.2a Attach a fuel pressure gauge to the test port on the fuel rail, turn the ignition key ON (engine not running) and check the fuel pressure (3.4L engine shown)

inside the fuel tank. A return system routes all vapors and excess fuel back to the fuel tank through separate return lines.

Exhaust system

The exhaust system includes an exhaust manifold fitted with an exhaust oxygen sensor, a catalytic converter, an exhaust pipe, and a muffler. The catalytic converter is an emission control device added to the exhaust system to reduce pollutants. A single-bed converter is used in combination with a three-way (reduction) catalyst. Refer to Chapter 6 for more information regarding the catalytic converter.

2   Fuel pressure relief procedure

Refer to illustration 2.3
Warning: *Gasoline is extremely flammable, so take extra precautions when you work on any part of the fuel system. Don't smoke or allow open flames or bare light bulbs near the work area, and don't work in a garage where a gas-type appliance (such as a water heater or a clothes dryer) is present. Since gasoline is carcinogenic, wear latex gloves when there's a possibility of being exposed to fuel, and, if you spill any fuel on your skin, rinse it off immediately with soap and water. Mop up any spills immediately and do not store fuel-soaked rags where they could ignite. The fuel system is under constant pressure, so, if any fuel lines are to be disconnected, the fuel pressure in the system must be relieved first. When you perform any kind of work on the fuel system, wear safety glasses and have a Class B type fire extinguisher on hand.*
Note: *After the fuel pressure has been relieved. it's a good idea to lay a shop towel over any fuel connection to be disassembled, to absorb the residual fuel that may leak out when servicing the fuel system.*
1   Before servicing any fuel system component, you must relieve the fuel pressure to minimize the risk of fire or personal injury.
2   Remove the fuel filler cap - this will

relieve any pressure built up in the tank.
3   Use one of the two following methods:
  a) *Attach a fuel pressure gauge, available at most auto parts stores, to the Schrader valve on the fuel rail (see illustration). Place the gauge bleeder hose in an approved fuel container. Open the valve on the gauge to relieve pressure, then disconnect the cable from the negative terminal of the battery.*
  b) *Locate the fuel pressure test port and carefully place several shop towels around the test port and the fuel rail. Remove the cap and, using the tip of a screwdriver, depress the Schrader valve and let the fuel drain into the shop towels. Be careful to catch any fuel that might spray upward by using another shop towel.*
4   Unless this procedure is followed before servicing fuel lines or connections, fuel spray (and possible injury) may occur.

3   Fuel pump/fuel system pressure - check

Warning: *Gasoline is extremely flammable, so take extra precautions when you work on any part of the fuel system. Don't smoke or allow open flames or bare light bulbs near the work area, and don't work in a garage where a gas-type appliance (such as a water heater or a clothes dryer) is present. Since gasoline is carcinogenic, wear latex gloves when there's a possibility of being exposed to fuel, and, if you spill any fuel on your skin, rinse it off immediately with soap and water. Mop up any spills immediately and do not store fuel-soaked rags where they could ignite. The fuel system is under constant pressure, so, if any fuel lines are to be disconnected, the fuel pressure in the system must be relieved first (see Section 2). When you perform any kind of work on the fuel system, wear safety glasses and have a Class B type fire extinguisher on hand.*

3.2b Checking the fuel pressure on the 5.7L engine

Note 1: *On Board Diagnostic (OBD-I) systems may set a trouble code 44 or 64. Refer to the code extracting procedure in Chapter 6 for additional fuel system diagnostic information.*
Note 2: *The following checks assume the fuel filter is in good condition. If you doubt its condition, install a new one (see Chapter 1).*
1   Check that there is adequate fuel in the fuel tank. Relieve the fuel pressure (see Section 2).

Fuel pump output and pressure check

Refer to illustrations 3.2a, 3.2b, 3.5a, 3.5b and 3.6
2   Install a fuel pressure gauge at the Schrader valve on the fuel rail **(see illustrations)**.
3   Turn the ignition switch ON (engine not running) with the air conditioning off. The fuel pump should run for about two seconds - note the reading on the gauge. After the pump stops running the pressure should hold steady. It should be within the range listed in this Chapter's Specifications.
4   Start the engine and let it idle at normal operating temperature. The pressure should be lower by 3 to 10 psi. If all the pressure readings are within the limits listed in this Chapter's Specifications, the system is operating properly.

3.5a Install a vacuum pump to the fuel pressure regulator and check for fuel pressure without vacuum applied

3.5b Next, apply vacuum to the fuel pressure regulator and check the fuel pressure - the fuel pressure should decrease as the vacuum increases

3.6 Checking vacuum to the fuel pressure regulator on a 5.7L engine

3.12 Checking for battery voltage on the fuel pressure relay connector

5   If the pressure did not drop by 3 to 10 psi after starting the engine, apply 12 to 14 inches of vacuum to the pressure regulator (see illustrations). If the pressure drops, repair the vacuum source to the regulator. If the pressure does not drop, replace the regulator.

6   If the fuel pressure is not within specifications, check the following:

a) If the pressure is higher than specified, check for vacuum to the fuel pressure regulator (see illustration). Vacuum must fluctuate with the increase or decrease in the engine rpm. If vacuum is present, check for a pinched or clogged fuel return hose or pipe. If the return line is OK, replace the regulator.

b) If the pressure is lower than specified, inspect the fuel filter - make sure it's not clogged. If the filter is good and there are no fuel leaks but the pressure is still low, install a fuel line shut-off adapter between the pressure regulator and the return line. With the valve open, start the engine (if possible) and slowly close the valve. If the pressure rises above 47 psi, replace the regulator (see Section 13). **Warning:** *Don't allow the fuel pressure to exceed 60 psi. Also, don't attempt to restrict the return line by pinching it, as*

the nylon fuel line will be damaged.

c) If the pressure is still low with the fuel return line restricted, an injector (or injectors) may be leaking (see Section 13) or the fuel pump may be faulty.

7   After the testing is done, relieve the fuel pressure (see Section 2) and remove the fuel pressure gauge.

8   If there are no problems with any of the above-listed components, check the fuel pump electrical circuits (see below).

Fuel pump electrical circuit check

Refer to illustration 3.12

9   If you suspect a problem with the fuel pump, verify the pump actually runs. Have an assistant turn the ignition switch to ON - you should hear a brief whirring noise as the pump comes on and pressurizes the system. Have the assistant start the engine. This time you should hear a constant whirring sound from the pump (but it's more difficult to hear with the engine running).

10   If the pump does not come on (makes no sound), proceed to the next Step.

11   Check the fuel pump fuse. **Note:** *On 1993 through 1995 models the fuel pump*

fuse is located in the passenger compartment fuse block. On 1996 and later models the fuel pump fuse is located in the engine compartment Power Distribution Center (see Chapter 12). If the fuse is blown, replace the fuse and see if the pump works. If the pump still does not work, go to the next step.

12   Check the fuel pump relay circuit. If the pump does not run, check for an open circuit between the relay and the fuel pump. With the ignition key ON (engine not running), check for battery voltage at the relay connector (see illustration). The fuel pump relay is located behind the driver's side kick panel.

13   If battery voltage exists, replace the relay with a known good relay and retest. If necessary, have the relay checked by a qualified automotive parts store.

14   If the fuel pump does not activate, check for power to the fuel pump at the fuel tank. Access to the fuel pump is difficult but it is possible to check for battery voltage at the electrical connector near the tank.

4   Fuel lines and fittings - repair and replacement

Refer to illustrations 4.10, 4.12 and 4.13
Warning: *Gasoline is extremely flammable, so take extra precautions when you work on any part of the fuel system. Don't smoke or allow open flames or bare light bulbs near the work area, and don't work in a garage where a gas-type appliance (such as a water heater or a clothes dryer) is present. Since gasoline is carcinogenic, wear latex gloves when there's a possibility of being exposed to fuel, and, if you spill any fuel on your skin, rinse it off immediately with soap and water. Mop up any spills immediately and do not store fuel-soaked rags where they could ignite. The fuel system is under constant pressure, so, if any fuel lines are to be disconnected, the fuel pressure in the system must be relieved first (see Section 2). When you perform any kind of work on the fuel system, wear safety glasses and have a Class B type fire extinguisher on hand.*

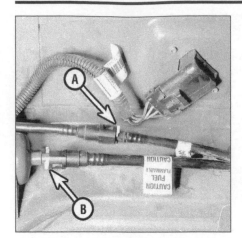

4.10 Some models are equipped with fuel lines that can be disconnected by pinching the tabs and separating each connector

A   Fuel return line
B   Fuel feed line

4.12 On quick-connect fuel lines, use a special tool and push them into the connector (arrows) to separate the fuel lines

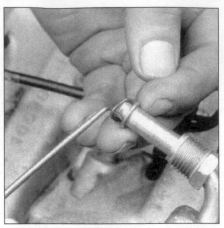

4.13 Always replace the fuel line O-rings (if equipped)

1   Always relieve the fuel pressure before servicing fuel lines or fittings on fuel-injected vehicles (see Section 2).

2   The fuel feed, return and vapor lines extend from the fuel tank to the engine compartment. The lines are secured to the underbody with clip and screw assemblies. These lines must be occasionally inspected for leaks, kinks and dents.

3   If evidence of dirt is found in the system or fuel filter during disassembly, the line should be disconnected and blown out. Check the fuel strainer on the fuel gauge sending unit (see Section 7) for damage and deterioration.

Steel and nylon tubing

4   Because fuel lines used on fuel-injected vehicles are under high pressure, they require special consideration.

5   If replacement of a metal fuel line or emission line is called for, use welded steel tubing . Don't use copper or aluminum tubing to replace steel tubing. These materials cannot withstand normal vehicle vibration.

6   If is becomes necessary to replace a section of nylon fuel line, replace it only with the correct part number - don't use any substitutes.

7   Most fuel lines have threaded fittings with O-rings. Any time the fittings are loosened to service or replace components:

a) *Use a backup wrench while loosening and tightening the fittings.*

b) *Check all O-rings for cuts, cracks and deterioration. Replace any that appear worn or damaged.*

c) *If the lines are replaced, always use original equipment parts, or parts that meet the manufacturer's standards specified in this Section.*

Rubber hose

Warning: *These models are equipped with sequential electronic fuel injection, use only original equipment replacement hoses or*

their equivalent. Others may fail from the high pressures of this system.

8   When rubber hose is used to replace a metal line, use reinforced, fuel resistant hose with the word "*Fluoroelastomer*" imprinted on it. Hose(s) not clearly marked like this could fail prematurely and could fail to meet Federal emission standards. Hose inside diameter must match line outside diameter.

9   Don't use rubber hose within four inches of any part of the exhaust system or within ten inches of the catalytic converter. Metal lines and rubber hoses must never be allowed to chafe against the frame. A minimum of 1/4-inch clearance must be maintained around a line or hose to prevent contact with the frame.

Removal and installation

Note: *The following procedure and accompanying illustrations are typical for vehicles covered by this manual. On quick-disconnect (non-threaded) fittings, clean off the fittings before disconnection to prevent dirt from getting in the fittings. After disconnection, clean the fittings with compressed air and apply a few drops of oil. On later models there are two types of quick-connect fittings in the fuel system. Both the metal-collar and plastic-collar connectors can be disconnected with a simple plastic tool available at auto parts stores.*

10   Relieve the fuel pressure (see Section 2) and disconnect the fuel feed, return or vapor line at the fuel tank **(see illustration)**. **Note:** *Some fuel line connections may be threaded. Be sure to use a back-up wrench when separating the connections.*

11   Remove all fasteners attaching the lines to the vehicle body.

12   Detach the fitting(s) that attach the fuel hoses to the engine compartment metal lines **(see illustration)**. Twisting them back and forth will allow them to separate more easily.

13   Installation is the reverse of removal. Be sure to use new O-rings at the threaded fittings, if equipped **(see illustration)**.

5   Fuel tank - removal and installation

Refer to illustrations 5.5, 5.6a, 5.6b, 5.6c, 5.7a, 5.7b, 5.8a, 5.8b, 5.9, 5.11, 5.13 and 5.14

Warning: *Gasoline is extremely flammable, so take extra precautions when you work on any part of the fuel system. Don't smoke or allow open flames or bare light bulbs near the work area, and don't work in a garage where a gas-type appliance (such as a water heater or a clothes dryer) is present. Since gasoline is carcinogenic, wear latex gloves when there's a possibility of being exposed to fuel, and, if you spill any fuel on your skin, rinse it off immediately with soap and water. Mop up any spills immediately and do not store fuel-soaked rags where they could ignite. The fuel system is under constant pressure, so, if any fuel lines are to be disconnected, the fuel pressure in the system must be relieved first (see Section 2). When you perform any kind of work on the fuel system, wear safety glasses and have a Class B type fire extinguisher on hand.*

Note: *Don't begin this procedure until the fuel gauge indicates the tank is empty or nearly empty. If the tank must be removed when it's full (for example, if the fuel pump malfunctions), siphon any remaining fuel from the tank prior to removal.*

1   Unless the vehicle has been driven far enough to completely empty the tank, it's a good idea to siphon the residual fuel out before removing the tank from the vehicle. **Warning:** *DO NOT start the siphoning action by mouth! Use a siphoning kit, available at most auto parts stores.*

2   Relieve the fuel system pressure (see Section 2).

3   Detach the cable from the negative terminal of the battery. **Caution:** *On models equipped with a Delco Loc II or Theftlock audio system, be sure the lockout feature is turned off before performing any procedure which requires disconnecting the battery.*

4   Raise the vehicle and support it securely

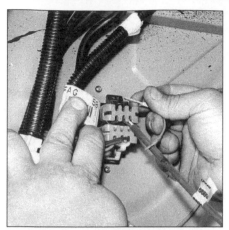

5.5 Pull the plastic retainer from the locking tab mechanism to release the connector from the harness

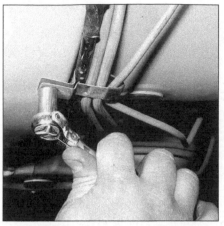

5.6a Remove the clamp from the fuel return, feed and vapor lines

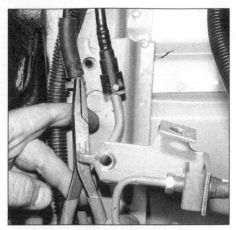

5.6b Use pliers to release the clamp from the fuel vapor line

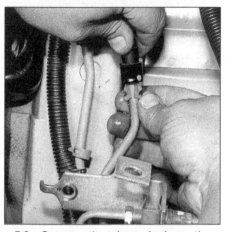

5.6c Squeeze the tabs and release the fuel return line from the collar

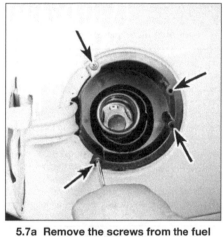

5.7a Remove the screws from the fuel filler protector and . . .

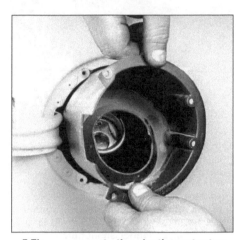

5.7b . . . separate the plastic protector from the body

on jackstands placed underneath the jacking points.

5   Locate the electrical connector for the electric fuel pump and fuel gauge sending unit below the fuel tank and unplug it **(see illustration)**.

6   Disconnect the fuel feed and return lines and the vapor return line **(see illustrations)**.

Note: *Remove the fuel feed line at the fuel filter* (see Chapter 1).

7   Remove the plastic guard from the fuel filler neck **(see illustrations)** and vent tubes. On 2000 and later models, remove the plastic access panel in the trunk and unbolt the fuel filler pipe.

8   On models through 1999, remove the

Panhard rod (see Chapter 10). Remove the Panhard rod brace **(see illustrations)**.

9   Support the rear axle with a floor jack and unbolt the lower ends of the shock absorbers from the rear axle (see Chapter 10). Partially lower the rear axle **(see illustration)**. **Caution:** *Don't lower the axle far enough to stretch the brake hose.*

5.8a Remove the Panhard rod brace from the left side of the vehicle

5.8b Use back-up wrenches to remove the bolt and nut from the right side of the Panhard rod brace

5.9 Partially lower the rear differential to allow the exhaust system and fuel tank the necessary clearance

5.11 Remove the fuel tank protector plate screws (arrows) and lower the protector plate

5.13 Remove the fuel tank strap bolts from the tank

5.14 Have an assistant angle the fuel tank into the right side fenderwell while making sure the filler neck is not hung-up on the upper fender sheet metal

10   Disconnect the rear section of the exhaust system and remove it from the vehicle (see Section 14).
11   Remove the protector plate from the bottom side of the fuel tank **(see illustration)**.
12   Support the fuel tank with a floor jack.
13   Disconnect both fuel tank retaining straps **(see illustration)**.
14   Lower the tank enough to disconnect the wires and ground strap from the fuel pump/fuel gauge sending unit **(see illustration)**.

15   Remove the tank from the vehicle.
16   Installation is the reverse of removal.

6   Fuel tank cleaning and repair - general information

1   All repairs to the fuel tank or filler neck should be carried out by a professional who has experience in this critical and potentially dangerous work. Even after cleaning and flushing of the fuel system, explosive fumes can remain and ignite during repair of the tank.
2   If the fuel tank is removed from the vehicle, it should not be placed in an area where sparks or open flames could ignite the fumes coming out of the tank. Be especially careful inside garages where a gas-type appliance is located, because it could cause an explosion.

7   Fuel pump - removal and installation

Refer to illustrations 7.4, 7.5, 7.6, 7.8, 7.9a and 7.9b
Warning: *Gasoline is extremely flammable, so take extra precautions when you work on any part of the fuel system. Don't smoke or allow open flames or bare light bulbs near the work area, and don't work in a garage where a gas-type appliance (such as a water heater*

or a clothes dryer) is present. Since gasoline is carcinogenic, wear latex gloves when there's a possibility of being exposed to fuel, and, if you spill any fuel on your skin, rinse it off immediately with soap and water. Mop up any spills immediately and do not store fuel-soaked rags where they could ignite. The fuel system is under constant pressure, so, if any fuel lines are to be disconnected, the fuel pressure in the system must be relieved first (see Section 2). When you perform any kind of work on the fuel system, wear safety glasses and have a Class B type fire extinguisher on hand.
1   Relieve the fuel system pressure (see Section 2).
2   Disconnect the cable from the negative battery terminal. **Caution:** *On models equipped with a Delco Loc II or Theftlock audio system, be sure the lockout feature is turned off before performing any procedure which requires disconnecting the battery.*
3   Remove the fuel tank (see Section 5).
4   Disconnect the fuel feed, return and vapor line as an assembly from the top of the fuel tank **(see illustration)**.
5   The fuel pump/sending unit assembly is located inside the fuel tank. On most models, it is held in place by five nuts **(see illustration)** attached to the upper ring washer.

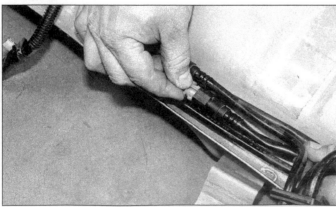

7.4 Squeeze the return line fuel connector and separate the fuel return, feed and vapor line as an assembly

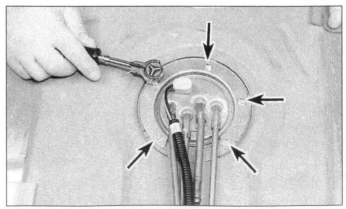

7.5 Remove the five nuts (arrows) from the fuel pump/sending unit retaining ring

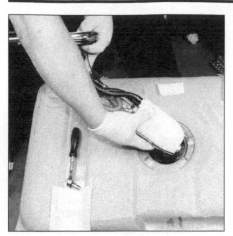

7.6 Lift the fuel pump assembly from the fuel tank

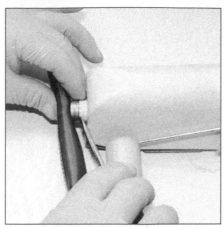

7.8 Pry on the collar to detach the strainer from the fuel pump assembly

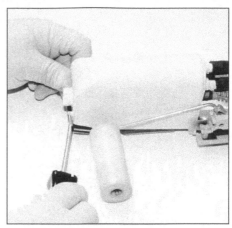

7.9a On 1997 and earlier models, press the locking tab to release the fuel pump cover from the frame . . .

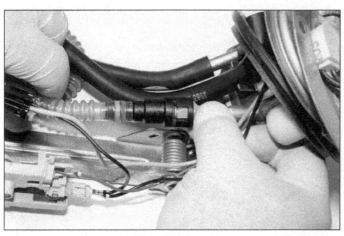

7.9b . . . then squeeze the tabs to release the fuel line from the collar

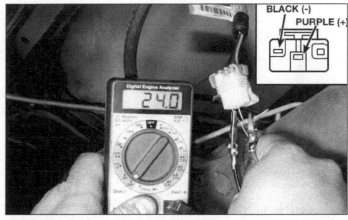

8.2 Connect the probes of the ohmmeter to the black (-) wire and the purple (+) wire and check the resistance of the fuel level sending unit - 1997 and earlier model connector shown

Remove the fuel pump/sending unit assembly. On later models, the assembly is attached to the tank with a retaining ring that twists off. Strike the corners of the ring with a hammer and piece of wood to rotate the ring off.

6   Lift the fuel pump/sending unit assembly from the fuel tank **(see illustration)**. **Caution:** *The fuel level float and sending unit are delicate. Do not bump them against the tank during removal or the accuracy of the sending unit may be affected.* **Warning:** *On 1998 and later models, some fuel may remain in the module reservoir and spill as the module is removed. Have several shop towels ready and a drain pan nearby to place the module in.* **Note:** *The electric fuel pump on 1998 and later models, is not serviced separately. In the event of failure, the complete module assembly must be replaced. Transfer the fuel pressure sensor and fuel level sending unit to the new fuel pump module assembly, if necessary (see Section 8).*

7   Inspect the condition of the rubber gasket around the opening of the tank on 1997 and earlier models or the fuel pump module on 1998 and later models. If it is dried, cracked or deteriorated, replace it.

8   Remove the strainer from the lower end

of the fuel pump **(see illustration)**. If it is dirty, clean it with a suitable solvent and blow it out with compressed air. If it is too dirty to be cleaned, replace it.

9   On 1997 and earlier models, lift the locking tabs to release the fuel pump cover **(see illustration)**. If it is necessary to separate the fuel pump and sending unit, disconnect the electrical connectors at the pump, noting their position, then disconnect the fuel line from the pump **(see illustration)**. Separate the fuel pump from the assembly.

10   Installation is the reverse of removal. Reassemble the fuel pump to the sending unit bracket assembly and insert the assembly into the fuel tank.

11   Install the fuel tank (see Section 5).

8   Fuel level sending unit - check and replacement

Warning: *Gasoline is extremely flammable, so take extra precautions when you work on any part of the fuel system. Don't smoke or allow open flames or bare light bulbs near the work area, and don't work in a garage where*

a gas-type appliance (such as a water heater or a clothes dryer) is present. Since gasoline is carcinogenic, wear latex gloves when there's a possibility of being exposed to fuel, and, if you spill any fuel on your skin, rinse it off immediately with soap and water. Mop up any spills immediately and do not store fuel-soaked rags where they could ignite. The fuel system is under constant pressure, so, if any fuel lines are to be disconnected, the fuel pressure in the system must be relieved first (see Section 2). When you perform any kind of work on the fuel system, wear safety glasses and have a Class B type fire extinguisher on hand.

Check

Refer to illustrations 8.2 and 8.5

1   Raise the vehicle and secure it on jackstands.

2   Position the probes of an ohmmeter onto the fuel level sending unit electrical connector terminals **(see illustration)** and check for resistance.

3   First, check the resistance of the sending unit with the fuel tank completely full. The resistance of the sending unit should be about 95 ohms.

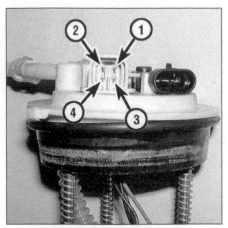

8.5 Fuel pump module terminal identification - 1998 and later models

1   *Fuel level sending unit signal*
2   *Fuel pump 12-volt supply (from fuel pump relay)*
3   *Ground*
4   *Fuel level sending unit ground*

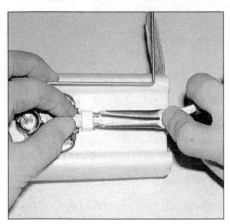

8.9a On 1998 and later models, remove the sending unit retaining clip . . .

4   Wait until the tank is nearly empty and check the resistance of the unit again. The resistance should be 2 ohms.
5   If the readings are incorrect or there is very little change in resistance as the float travels from full to empty, replace the fuel level sending unit assembly. **Note:** *Another check for the fuel level sending unit is to remove the fuel pump from the tank (see Section 7) and check the resistance while moving the float from full (arm at highest point of travel) to empty (arm at lowest point of travel* **(see illustration)**.

Replacement

Refer to illustrations 8.7, 8.8, 8.9a and 8.9b

6   Remove the fuel tank (see Section 5) and the fuel pump/module (see Section 7) from the vehicle.
7   Disconnect the fuel level sending unit electrical connector from the module cover **(see illustration)**. Be sure to note the routing of the wiring for installation.

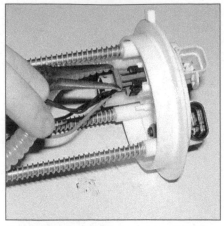

8.7 Disconnect the fuel pump level sending unit electrical connector - 1998 and later model shown, 1997 and earlier similar

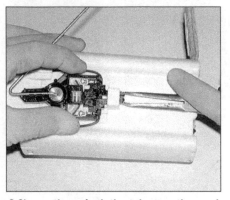

8.9b . . . then pinch the tabs together and remove the sending unit from the module

8   On 1997 and earlier models, remove the Torx bolts from the fuel level sending unit **(see illustration)** and separate it from the assembly.
9   On 1998 and later models, remove the sending unit retaining clip, then pinch the tabs together and slide the fuel level sending unit off the module **(see illustrations)**.
10   Installation is the reverse of removal.

9   Air cleaner housing - removal and installation

1   Detach the cable from the negative terminal of the battery. **Caution:** *On models equipped with a Delco Loc II or Theftlock audio system, be sure the lockout feature is turned off before performing any procedure which requires disconnecting the battery.*
2   Disconnect the electrical connector to the air intake sensor and the connector for the mass airflow sensor (see Chapter 6).
3   Remove the clamps that retain the air intake duct/resonator assembly to the throttle body and the air cleaner housing. Detach the air intake duct/resonator assembly and remove from the engine compartment.
4   Follow the air filter removal procedure in Chapter 1 and separate the air filter and the upper cover from the air cleaner housing.

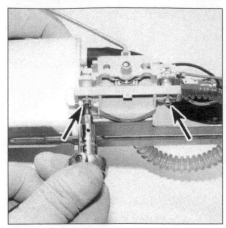

8.8 On 1997 and earlier models, remove the Torx bolts from the fuel level sending unit

5   Detach the air cleaner housing retaining bolts and remove it from the engine compartment.
6   Installation is the reverse of removal.

10   Accelerator cable - removal and installation

Refer to illustrations 10.2, 10.4, 10.5 and 10.6
Note: *On 2000 and later models the throttle control system is electronic. There is no mechanical cable, but rather a throttle position sensor at the pedal and an electric motor at the throttle body.*

Removal

1   Detach the screws and the clip retaining the lower instrument panel trim and lower the trim (if necessary) (see Chapter 11).
2   Detach the accelerator cable from the accelerator pedal **(see illustration)**.
3   Squeeze the accelerator cable cover tangs and push the cable through the firewall into the engine compartment.
4   Remove the throttle valve plastic cover next to the throttle body **(see illustration)**.
5   Detach the accelerator cable-to-throttle

10.2 Slide the accelerator cable through the slot in the pedal arm

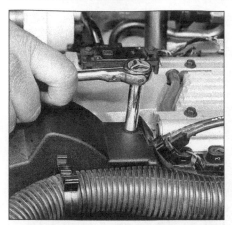

10.4 Remove the throttle valve protector to gain access to the accelerator cable

10.5 Remove the accelerator cable end through the slot in the throttle lever

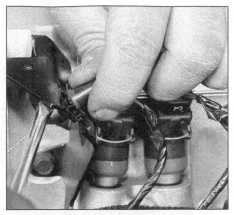

10.6 Squeeze the accelerator cable retaining tabs and pull the assembly through the bracket housing

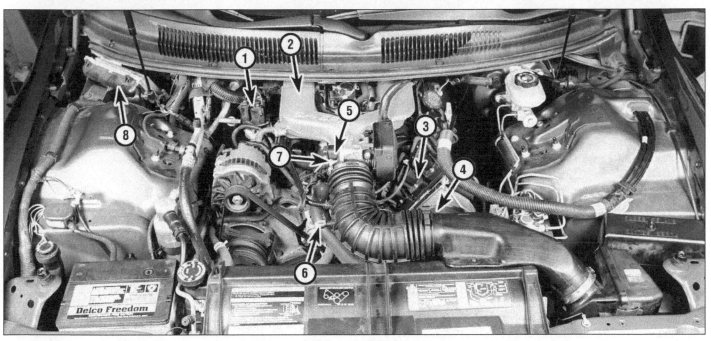

11.1a Fuel injection component locations on the 3.4L V6 engine

1	Digital EGR valve
2	Air Intake Plenum
3	Ignition coil packs
4	AIR pump (below air intake duct)
5	IAC valve
6	IAT sensor
7	TPS sensor
8	Electronic Control Module (ECM/PCM)

lever retainer and detach the accelerator cable from the throttle lever **(see illustration)**.

6   Squeeze the accelerator cable retaining tangs and push the cable through the accelerator cable bracket **(see illustration)**.

Installation

7   Installation is the reverse of removal. **Note:** *To prevent possible interference, flexible components (hoses, wires, etc.) must not be routed within two inches of moving parts, unless routing is controlled.*

8   Operate the accelerator pedal and check for any binding condition by completely opening and closing the throttle.

9   At the engine compartment side of the firewall, apply sealant around the accelerator cable.

11 Fuel injection system - general information

Refer to illustrations 11.1a, 11.1b and 11.1c

General Information

These models are equipped with Sequential Electronic Fuel Injection (SEFI) systems **(see illustrations)**. Sequential fuel injection pulses the injectors once per crankshaft revolution. Fuel injection provides optimum mixture ratios at all stages of combustion. Combined with its immediate response characteristics, fuel injection permits the engine to run on the leanest possible air/fuel mixture, which greatly reduces exhaust gas emissions.

The fuel injection system is controlled

directly by the vehicle's Electronic Control Module (ECM/PCM) or Powertrain Control Module (PCM), which automatically adjusts the air/fuel mixture in accordance with engine load and performance.

Sequential Electronic Fuel Injection (SEFI)

Sequential Electronic Fuel Injection (SEFI) consists of an air intake manifold, the throttle body, the injectors, the fuel rail assembly, an electric fuel pump and associated plumbing.

Air is drawn through the air cleaner and throttle body. A Mass Air Flow (MAF) sensor or a Manifold Absolute Pressure (MAP) sensor compensates for temperature and pressure variations. 5.7L engines are equipped with both the MAF and MAP sensors.

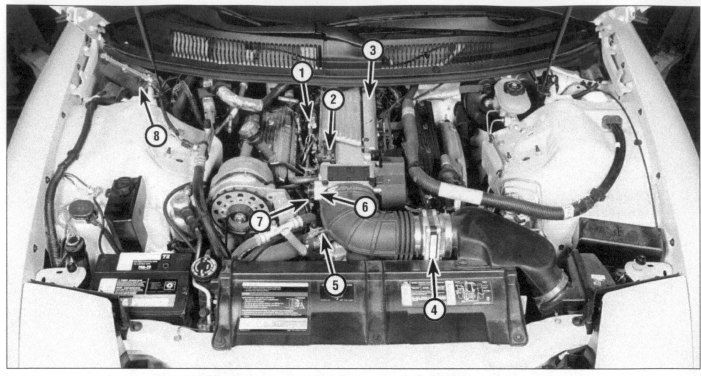

11.1b Fuel injection component locations on the 1997 and earlier 5.7L V8 engine

1	Fuel injector	5	IAT sensor
2	MAP sensor	6	IAC valve
3	Fuel rail and test port (located under cowl)	7	TPS sensor
4	MAF sensor	8	Electronic Control Module (ECM/PCM)

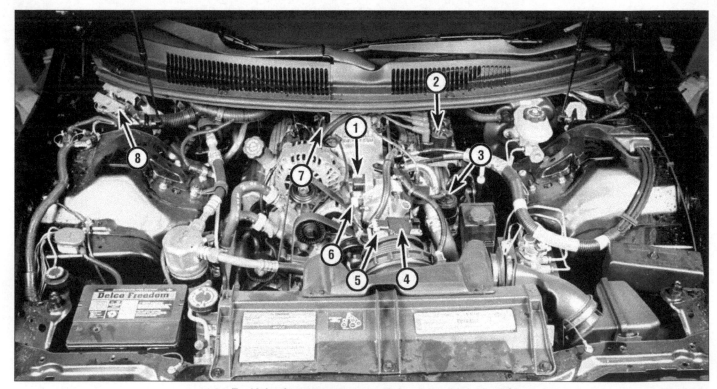

11.1c Fuel injection component locations on the 3800 V6 engine

1	IAC valve	4	MAF sensor	6	TPS sensor
2	Ignition coil packs	5	Coolant temperature sensor (located under throttle body)	7	MAP sensor
3	Linear EGR valve			8	Powertrain Control Module (PCM)

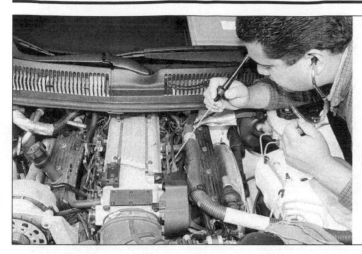

12.11 Use a stethoscope or screwdriver to determine if the injectors are working properly - they should make a steady clicking sound that rises and falls with engine speed changes

While the engine is running, the fuel constantly circulates through the fuel rail, which removes vapors and keeps the fuel cool while maintaining sufficient pressure to the injectors under all running conditions.

SEFI injection system is controlled by the ECM/PCM so that it works in conjunction with the rest of the vehicle functions to provide optimum driveability and emissions control.

Because the SEFI system meters fuel and air precisely, it is important to the proper operation of the vehicle that the fuel and air filters be changed at the specified intervals.

The ECM/PCM has a learning capability for certain performance conditions. If the battery is disconnected, part of the ECM/PCM memory is erased, which makes it necessary to "re-learn" the computer. This is done by thoroughly warming up the engine and operating the vehicle at part throttle, stop and go and idle conditions.

A fuel pump relay is used to control the electric fuel pump operation. When the ignition is turned on, the fuel pump relay immediately supplies current to the fuel pump to pressurize the fuel system. If the engine doesn't start after two seconds, the fuel pump will automatically shut off. If the fuel pump relay fails, the fuel pump will still operate after the

ECM/PCM receives pulses from the distributor or about four pounds of oil pressure has built up, depending on the model.

The throttle stop screw, used to regulate the minimum idle speed, is adjusted at the factory and sealed with a plug to discourage unnecessary adjustment.

12 Fuel injection system - check

Warning: *Gasoline is extremely flammable, so take extra precautions when you work on any part of the fuel system. Don't smoke or allow open flames or bare light bulbs near the work area, and don't work in a garage where a gas-type appliance (such as a water heater or a clothes dryer) is present. Since gasoline is carcinogenic, wear latex gloves when there's a possibility of being exposed to fuel, and, if you spill any fuel on your skin, rinse it off immediately with soap and water. Mop up any spills immediately and do not store fuel-soaked rags where they could ignite. The fuel system is under constant pressure, so, if any fuel lines are to be disconnected, the fuel pressure in the system must be relieved first (see Section 2). When you perform any kind*

of work on the fuel system, wear safety glasses and have a Class B type fire extinguisher on hand.
Note: *The following procedure is based on the assumption that the fuel pump is working and the fuel pressure is adequate (see Section 3).*

Preliminary checks

1   Check all electrical connectors that are related to the system. Loose electrical connectors and poor grounds can cause many problems that resemble more serious malfunctions.
2   Check to see that the battery is fully charged, as the control unit and sensors depend on an accurate supply voltage in order to properly meter the fuel.
3   Check the air filter element - a dirty or partially blocked filter will severely impede performance and economy (see Chapter 1).
4   If a blown fuse is found, replace it and see if it blows again. If it does, search for a grounded wire in the harness to the fuel pump.

System checks

Refer to illustrations 12.11, 12.12 and 12.13
5   Check the ground wire connections on the intake manifold for tightness. Check all electrical connectors that are related to the system. Loose connectors and poor grounds can cause many problems that resemble more serious malfunctions.
6   Check to see that the battery is fully charged, as the control unit and sensors depend on an accurate supply voltage in order to properly meter the fuel.
7   Check the air filter element - a dirty or partially blocked filter will severely impede performance and economy (see Chapter 1).
8   If a blown fuse is found, replace it and see if it blows again. If it does, search for a grounded wire in the harness to the fuel pump.
9   Check the air intake duct to the intake manifold for leaks, which will result in an excessively lean mixture. Also check the condition of all vacuum hoses connected to the intake manifold.
10   Remove the air intake duct from the throttle body and check for dirt, carbon or other residue build-up. If it's dirty, clean it with carburetor cleaner and a toothbrush.
11   With the engine running, place an automotive stethoscope against each injector, one at a time, and listen for a clicking sound, indicating operation **(see illustration)**. If you don't have a stethoscope, place the tip of a screwdriver against the injector and listen through the handle.
12   Unplug the injector electrical connectors and test the resistance of each injector **(see illustration)**. Compare the values to the Specifications listed in this Chapter.
13   Install an injector test light ("noid" light) into each injector electrical connector, one at a time **(see illustration)**. Crank the engine over. Confirm that the light flashes evenly on each connector. This will test for voltage to the injectors.
14   The remainder of the system checks can be found in the following Sections.

12.12 Measure the resistance of each injector - it should be approximately 11 to 14 ohms

12.13 Install the "noid" light into each injector electrical connector and confirm that it blinks when the engine is cranking or running

13.2 Clean the throttle body with carburetor cleaner to remove sludge deposits

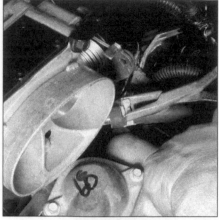

13.9a Disconnect the coolant lines from the throttle body

13.9b The lower coolant line is very difficult to access so remove the throttle body and angle the assembly to remove the clamp

13 Multi-Port Fuel Injection (MPFI) - component check and replacement

Warning: *Gasoline is extremely flammable, so take extra precautions when you work on any part of the fuel system. Don't smoke or allow open flames or bare light bulbs near the work area, and don't work in a garage where a gas-type appliance (such as a water heater or a clothes dryer) is present. Since gasoline is carcinogenic, wear latex gloves when there's a possibility of being exposed to fuel, and, if you spill any fuel on your skin, rinse it off immediately with soap and water. Mop up any spills immediately and do not store fuel-soaked rags where they could ignite. The fuel system is under constant pressure, so, if any fuel lines are to be disconnected, the fuel pressure in the system must be relieved first (see Section 2). When you perform any kind of work on the fuel system, wear safety glasses and have a Class B type fire extinguisher on hand.*

Throttle body

Check

Refer to illustration 13.2
Note: *On 2000 and later models the throttle control system is electronic. There is no mechanical cable, but rather a throttle position sensor at the pedal and an electric motor at the throttle body. Diagnosis of this system is best performed with a scan tool.*
1   Detach the air intake duct from the throttle body and move the duct out of the way.
2   On models through 1999, have an assistant depress the throttle pedal while you watch the throttle valve. Check that the throttle valve moves smoothly when the throttle is moved from closed (idle position) to fully open (wide open throttle). **Note:** *Spray carburetor cleaner into the throttle body, especially around the shaft area* **(see illustration)** *to free-up any binding caused by the accumulation of carbon deposits or sludge buildup.*

3   Wiggle the throttle lever while watching the throttle shaft inside the bore. If it appears worn (loose), replace the throttle body unit.

Removal

Refer to illustrations 13.9a, 13.9b and 13.10
Warning: *Wait until the engine is completely cool before beginning this procedure.*
4   Disconnect the cable from the negative terminal of the battery. **Caution:** *On models equipped with a Delco Loc II or Theftlock audio system, be sure the lockout feature is turned off before performing any procedure which requires disconnecting the battery.*
5   Detach the air intake duct. Drain the cooling system.
6   Unplug the Idle Air Control (IAC) valve and the Throttle Position Sensor (TPS) electrical connectors (see Chapter 6). On 2000 and later models, disconnect the throttle body actuator connector.
7   Mark and disconnect any vacuum hoses connected to the throttle body. Also detach the breather hose, if equipped.
8   On 1999 and earlier models, disconnect the accelerator cable from the throttle lever, then detach the cable housing from its bracket (see Section 10).
9   Loosen the clamps and disconnect the coolant hoses from the underside of the throttle body **(see illustration)**. Be prepared for some coolant spillage and plug the ends

of the hoses. **Note:** *On some models you can't get to these hoses until after the throttle body has been unbolted from the plenum* **(see illustration)**.
10   Remove the throttle body bolts and detach the throttle body **(see illustration)**.

Installation

11   Clean off all traces of old gasket material from the throttle body and the plenum.
12   Install the throttle body and a new gasket and tighten the bolts to the torque listed in this Chapter's Specifications.
13   The rest of the procedure is the reverse of removal. Be sure to check the coolant level (see Chapter 1) and add, if necessary.

Idle Air Control (IAC) valve

Check

Refer to illustrations 13.15, 13.16, 13.17a and 13.17b
14   The idle air control valve (IAC) controls the engine idle speed. This output actuator is mounted on the throttle body and is controlled by voltage pulses sent from the ECM/PCM (computer). The IAC valve pintle moves in or out allowing more or less intake air into the system according to the engine conditions. To increase idle speed, the ECM/PCM retracts the IAC valve pintle away

13.10 Throttle body mounting bolts on a 1997 and earlier V8 engine

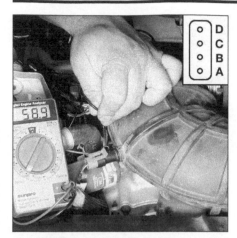

13.15 First measure the resistance across terminals D and C, then across terminals A and B

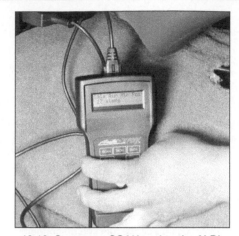

13.16 Connect a SCAN tool to the ALDL and observe the position of the IAC motor by observing the steps (counts) as the throttle is opened then closed

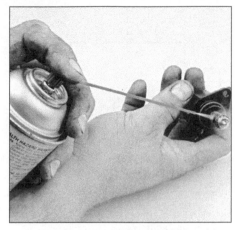

13.17a Clean the IAC valve pintle with carburetor cleaner to remove carbon deposits

from the seat and allows more air to bypass the throttle bore. To decrease idle speed, the ECM/PCM extends the IAC valve pintle towards the seat, reducing the air flow.

15 To check the IAC valve, unplug the electrical connector and, using an ohmmeter, measure the resistance across terminals A and B, then terminals C and D. Each resistance check should indicate 40 to 80 ohms **(see illustration)**. If not, replace the IAC valve.

16 There is an alternate method for testing the IAC valve. Various SCAN tools are available from auto parts stores and specialty tool companies that can be plugged into the ALDL (diagnostic connector) for the purpose of monitoring the sensors. Connect the SCAN tool and switch to the Idle Air Motor Position mode and monitor the steps (motor winding position) **(see illustration)**. The SCAN tool should indicate between 10 to 200 steps depending upon the rpm range. Allow the engine to idle for several minutes and while observing the count reading, snap the throttle to achieve high rpm (under 3,500). Repeat the procedure several times and observe the SCAN tool steps (counts) when the engine goes back to idle. The readings should be within 5 to 10 steps each time. If the readings

fluctuate greatly, replace the IAC valve. **Note:** *When the IAC valve electrical connector is disconnected for testing, the ECM/PCM will have to "relearn" its idle mode. In other words, it will take a certain amount of time before the idle motor resets for the correct idle speed. Make sure the idle is smooth and not misfiring before plugging in the SCAN tool.*

17 Next, remove the valve (see Step 18) and inspect it:

a) *Check the pintle for excessive carbon deposits. If necessary, clean it with carburetor cleaner spray* **(see illustration)**. *Also clean the IAC valve housing to remove any deposits* **(see illustration)**.

b) *Check the IAC valve electrical connections. Make sure the pins are not bent and make good contact with the connector terminals.*

Removal

Refer to illustration 13.19

18 Unplug the electrical connector from the Idle Air Control (IAC) valve.

19 Unscrew the valve or remove the two IAC valve attaching screws and withdraw the valve **(see illustration)**.

20 Check the condition of the rubber O-ring. If it's hardened or deteriorated, replace it. On models equipped with a gasket, remove the gasket.

21 Clean the sealing surface and the bore of the idle air/vacuum signal housing assembly to ensure a good seal. **Caution:** *The IAC valve itself is an electrical component and must not be soaked in any liquid cleaner, as damage may result.*

22 Before installing the IAC valve, the position of the pintle must be checked. If the pintle is extended too far, damage to the assembly may occur.

Installation

23 Measure the distance from the flange or gasket mounting surface of the IAC valve to the tip of the pintle. If the distance is greater than 1-1/8 inch, reduce the distance by applying firm pressure onto the pintle to retract it. Try some side-to-side motion in the event the pintle binds.

24 Position the new O-ring or gasket on the IAC valve. Lubricate the O-ring with a light film of engine oil. If the IAC valve is the screw-in type, apply a light film of RTV sealant to the threads of the valve. Install the

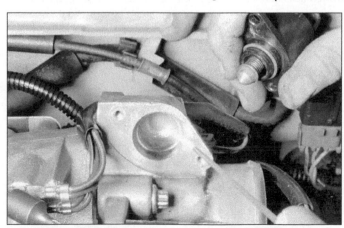

13.17b Spray carburetor cleaner into the IAC valve housing and check for clogged air passages in the air intake plenum

13.19 Remove the IAC motor screws (arrows) and separate the IAC from the throttle body

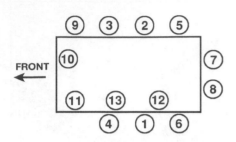

13.41 Air intake plenum TIGHTENING sequence - 3800 V6 engine

IAC valve and tighten the valve or the mounting screws securely.
25   Plug in the electrical connector at the IAC valve assembly. **Note:** *No adjustment is made to the IAC assembly after reinstallation. The IAC resetting is controlled by the ECM/PCM when the engine is started.*

Throttle Position Sensor (TPS)
Check
26   Check for stored trouble codes in the ECM/PCM using the On Board Diagnosis system (see Chapter 6).
27   To check the operation and replacement of the TPS, refer to the *Information Sensors* in Chapter 6.

Air intake plenum (3.4L and 3800 V6 engines only)
Removal
Note: *This component is sometimes referred to as the upper intake manifold.*
28   Disconnect the cable from the negative terminal of the battery. **Caution:** *On models equipped with a Delco Loc II or Theftlock audio system, be sure the lockout feature is turned off before performing any procedure which requires disconnecting the battery.*
29   Detach the air intake duct from the throttle body.
30   Drain the coolant (see Chapter 1).
31   Disconnect the accelerator cable, transmission control cable and cruise control cable (if equipped) from the throttle lever (see Section 10). Unbolt the accelerator cable bracket and position the bracket and cables aside.
32   Detach the fuel injector electrical connectors from each injector and the fuel rail assembly.
33   Detach any hoses and electrical connectors from the throttle body and plenum such as the IAC valve, IAT, TPS, MAP and EVAP canister control electrical connectors. If necessary, mark them with pieces of numbered tape to avoid confusion during reassembly.
34   Remove the serpentine drivebelt on the 3800 engine (see Chapter 1). On 2000 and later models, remove the ignition coil pack (see Chapter 5).

35   Relieve the fuel pressure (see Section 2) and remove the fuel lines from the fuel rail

13.44 Use a special fuel line disconnect tool to release the fuel line coupler seal from the fuel rail

(see Steps 43 through 45).
36   Remove the fuel rail from the air intake plenum.
37   Remove the EGR flexible pipe and bolts from the manifold (see Chapter 6).
38   Remove the plenum bolts and lift the plenum from the intake manifold. **Note:** *On 3800 engines, remove the thermostat housing (see Chapter 3) to gain access to the "hidden" bolt that secures the upper and lower intake manifolds together. If the plenum sticks, use a block of wood and a hammer to dislodge it. Don't pry between the sealing flanges, as this will damage the machined surfaces and could cause vacuum leaks to develop.*
39   Remove all traces of old gasket material from the plenum and intake manifold mating surfaces. It's a good idea to stuff rags into the intake manifold openings to prevent debris from falling in.

Installation
Refer to illustration 13.41
40   Install the new gasket(s) and set the plenum into position.
41   Install the plenum bolts and tighten them to the torque listed in this Chapter's Specifications in a criss-cross pattern. **Note:** *On 3800 engines, follow the torque sequence* **(see illustration)**.
42   The remainder of installation is the

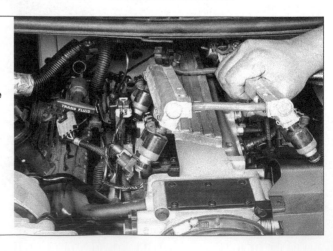

13.48 ... and separate the fuel rail from the intake manifold (1997 and earlier 5.7L engine shown)

13.47 Remove the bolts (arrows) . . .

reverse or removal.

Fuel rail and injectors
Refer to illustrations 13.44, 13.47, 13.48, 13.49a, 13.49b, 13.50a and 13.50b
Warning: *Before any work is performed on the fuel lines, fuel rail or injectors, the fuel system pressure must be relieved (see Section 2).*
Note: *Refer to Section 12 for the injector checking procedure.*
43   Detach the cable from the negative terminal of the battery. **Caution:** *On models equipped with a Delco Loc II or Theftlock audio system, be sure the lockout feature is turned off before performing any procedure which requires disconnecting the battery.*
44   Using a special fuel line tool to depress the seal inside the fuel line coupler **(see illustration)** and detach the fuel lines from the fuel rail.
45   Detach the vacuum line from the fuel pressure regulator on the fuel rail. **Note:** *On 1998 and later V8 engines, the fuel pressure regulator is located inside the fuel tank, therefore it will not be necessary to detach the vacuum line.*
46   Label and unplug the injector electrical connectors. Also remove the accelerator and the cruise control cable (if equipped) and any other components that would interfere with the removal of the fuel rail.

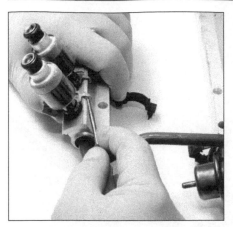

13.49a Slide the retaining clip off

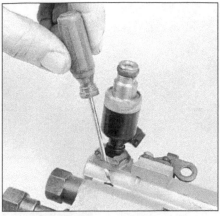

13.49b To remove an injector that is retained by a spring clip, simply pry off the clip with a small screwdriver, then pull the injector from the fuel rail

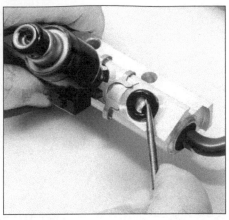

13.50a Remove the injector seal from the fuel rail. The injector seals that are positioned in the fuel rail are black while the seals that go into the engine block are brown

47 Remove the fuel rail retaining bolts **(see illustration)**.

48 Carefully remove the fuel rail with the injectors **(see illustration)**. **Caution:** *Use care when handling the fuel rail assembly to avoid damaging the injectors.* **Note:** *An eight digit identification number is stamped on the side of the fuel rail assembly. Refer to this number if servicing or parts replacement is required.*

49 To remove the fuel injectors, slide the injector retaining clip and pull the injector from the fuel rail **(see illustration)**. On 3800 models, spread open the end of the injector clip slightly and remove it from the fuel rail, then extract the injector **(see illustration)**.

50 Remove the injector O-ring seals **(see illustrations)**. **Note:** *Replace the injector seal with the correct color. Brown seals are used on the engine side of the injector while black seals are used on the fuel rail side of the injector.*

51 Install the new O-ring seal(s), as required, on the injector(s) and lubricate them with a light film of engine oil.

52 Install the injectors on the fuel rail.

53 Secure the injectors with the retainer clips.

54 Installation is the reverse of the removal procedure.

Fuel pressure regulator

Check

55 Refer to Section 3 for the fuel pressure checking procedure.

Replacement

56 Relieve the fuel system pressure (see Section 2).

57 Disconnect the cable from the negative terminal of the battery. **Caution:** *On models equipped with a Delco Loc II or Theftlock audio system, be sure the lockout feature is turned off before performing any procedure which requires disconnecting the battery.*

58 Remove the fuel rail following the procedure described earlier in this Section. **Note:** *If the vehicle is a 1998 and later 5.7L V8 engine disregard this Step and proceed to Step 60.*

5.7L V8 engine (1997 and earlier)

Refer to illustrations 13.59a and 13.59b

59 Remove the fuel line bracket from the fuel pressure regulator assembly **(see illustration)**. Remove the pressure regulator mounting screws **(see illustration)** and separate the two fuel rails from the pressure regulator assembly.

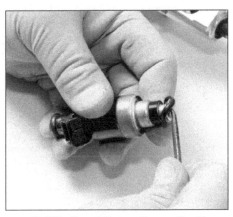

13.50b Carefully pry the seals off the injectors

5.7L V8 engine (1998 and later)

60 Remove the fuel pump/module from the fuel tank (see Section 7).

61 Detach the fuel pressure regulator retaining clip and remove the pressure regulator from the housing return line.

13.59a Remove the bracket from the fuel pressure regulator fuel lines

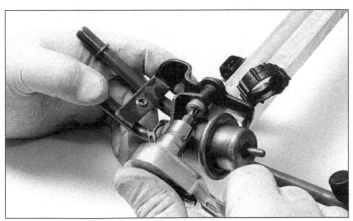

13.59b Remove the fuel pressure regulator using a Torx 27R bit

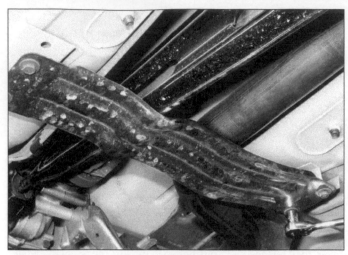

14.1a Remove the crossmember to allow removal of the exhaust system

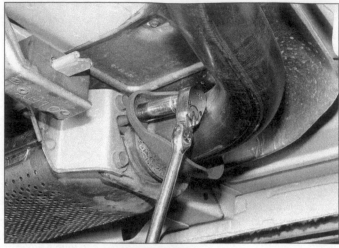

14.1b Remove the nuts that retain the flange to the catalytic converter

3800 V6 engine

62   Remove the snap-ring then twist the pressure regulator back and forth too pull it from its housing. Clean the housing and filter with carburetor cleaner. When reassembling, lightly lubricate the O-rings with clean engine oil.

3.4L V6 engine

63   The manufacturer states that the fuel pressure regulator on 3.4L V6 engines is not serviceable, requiring replacement of the entire fuel rail if malfunctioning. However, regulator rebuild kits are now available and are much less costly than an entire fuel rail assembly.

All engines

64   Reassembly is the reverse of disassembly. Be sure to replace all gaskets and seals, otherwise a dangerous fuel leak may develop. When installing the seals, lubricate them with a light film of engine oil.

---

14 Exhaust system servicing - general information

Refer to illustrations 14.1a, 14.1b and 14.1c
Warning: *The vehicle's exhaust system generates very high temperatures and must be allowed to cool down completely before any of the components are touched. Be especially*

14.1c Remove the nuts that retain the exhaust hanger

careful around the catalytic converter, where the highest temperatures are generated.

General Information

1   Replacement of exhaust system components is basically a matter of removing the heat shields, disconnecting the component and installing a new one **(see illustrations)**. The heat shields and exhaust system hangers **(see illustration)** must be reinstalled in the original locations or damage could result. Due to the high temperatures and exposed locations of the exhaust system components, rust and corrosion can seize parts together.

Penetrating oils are available to help loosen frozen fasteners. However, in some cases it may be necessary to cut the pieces apart with a hacksaw or cutting torch. The latter method should be employed only by persons experienced in this work.

Crossover pipe

2   Remove the bolts or nuts securing the crossover pipe to the exhaust manifolds. Remove the crossover pipe.
3   Installation is the reverse of removal. Tighten the fasteners evenly and securely.

Chapter 5 Engine electrical systems

Contents

Specifications

Ignition coil resistance

3.4L engine	
Primary resistance	0.35 to 1.50 ohms
Secondary resistance	5,000 to 7,000 ohms
3800 engine	
Primary resistance	0.3 to 0.6 ohms
Secondary resistance	5,000 to 6,500 ohms
5.7L engine	
1997 and earlier	
Primary resistance	0.3 to 1.3 ohms
Secondary resistance	5,000 to 7,000 ohms
1998 and later	N/A

1 General information

Warning: *Because of the very high voltage generated by the ignition system, extreme care should be taken whenever an operation involving ignition components is performed. This not only includes the distributor, coil(s), module and spark plug wires, but related items that are connected to the systems as well, such as the plug connections, tachometer and testing equipment.*

3.4L, 3800 and the 1998 and later 5.7L engines are equipped with the Distributor-less Ignition System (DIS). The 1997 and earlier 5.7L engine is equipped with the Distributor Ignition (DI) system. Although these two systems use different ways to generate the ignition signals, they share certain similar components, such as the ignition switch, battery, coil, primary (low tension) and secondary (high tension) wiring circuits and spark plugs.

The charging system consists of a belt-driven alternator with an integral voltage regulator and the battery. These components work together to supply electrical power for the ignition system, the lights and all accessories.

There are three types of CS alternators in use, the CS-130 (105 amp), CS-144 (124 amp) and the CS-130D (105 amp) alternator. All types use a conventional pulley and fan.

2 Battery - emergency jump starting

Refer to the *Booster battery (jump) starting* procedure at the front of this manual.

3 Battery - removal and installation

Refer to illustration 3.3

Warning: *Hydrogen gas is produced by the battery, so keep open flames and lighted cigarettes away from it at all times. Always wear eye protection when working around a battery. Rinse off spilled electrolyte immediately with large amounts of water.*

Removal

1 The battery is located at the left front corner or at the right front corner of the engine compartment , depending on the year the vehicle was manufactured.

2 Detach the cables from the negative and positive terminals of the battery. **Caution 1:** *To prevent arcing, disconnect the negative (-) cable first, then remove the positive (+) cable.* **Caution 2:** *On models equipped with a Delco Loc II or Theftlock audio system, be sure the lockout feature is turned off before performing any procedure which requires disconnecting the battery.*

3 Remove the hold-down clamp bolt and

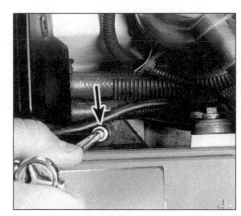

3.3 Remove the battery hold-down bolt (arrow) from the battery carrier

the clamp from the battery carrier **(see illustration)**

4 Carefully lift the battery from the carrier. **Warning:** *Always keep the battery in an upright position to reduce the likelihood of electrolyte spillage. If you spill electrolyte on your skin, rinse it off immediately with large amounts of water.*

Installation

Note: *The battery carrier and hold-down clamp should be clean and free from corrosion before installing the battery. Make certain that there are no parts in the carrier before installing the battery.*

5   Set the battery in position in its carrier. Don't tilt it.

6   Install the hold-down clamp and bolt. The bolt should be snug, but overtightening it may damage the battery case.

7   Install both battery cables, positive first, then the negative. **Note:** *The battery terminals and cable ends should be cleaned prior to connection* (see Chapter 1).

4   Battery cables - check and replacement

1   Periodically inspect the entire length of each battery cable for damage, cracked or burned insulation and corrosion. Poor battery cable connections can cause starting problems and decreased engine performance.

2   Check the cable-to-terminal connections at the ends of the cables for cracks, loose wire strands and corrosion. The presence of white, fluffy deposits under the insulation at the cable terminal connection is a sign the cable is corroded and should be replaced. Check the terminals for distortion, missing mounting bolts or nuts and corrosion.

3   If only the positive cable is to be replaced, be sure to disconnect the negative cable from the battery first. **Caution:** *On models equipped with a Delco Loc II or Theftlock audio system, be sure the lockout feature is turned off before performing any procedure which requires disconnecting the battery.*

4   Disconnect and remove the cable. Make sure the replacement cable is the same length and diameter.

5   Clean the threads of the starter or ground connection with a wire brush to remove rust and corrosion. Apply a light coat of petroleum jelly to the threads to ease installation and prevent future corrosion.

6   Attach the cable to the starter or ground connection and tighten the mounting nut securely.

7   Before connecting the new cable to the battery, make sure it reaches the terminals without having to be stretched.

8   Connect the positive cable first, followed by the negative cable. Tighten the nuts and apply a thin coat of petroleum jelly to the terminal and cable connection.

5   Ignition system - general information

Warning: *Because of the very high voltage generated by the ignition system, extreme care should be taken whenever an operation involving ignition components is performed. This not only includes the distributor, coil(s), module and spark plug wires, but related items that are connected to the systems as well, such as the plug connections, tachometer and testing equipment.*

3.4L, 3800 and 1998 and later V8 engines are equipped with the Distributorless Ignition System (DIS). The 1997 and earlier V8 engine

is equipped with the Distributor Ignition (DI) system. Although these two systems use different ways to generate the ignition signals, they share certain similar components.

Distributor Ignition (DI) system

1997 and earlier 5.7L engines are equipped with a Distributor Ignition (DI) system that incorporates the Opti Spark ignition system. The Opti Spark ignition system consists of the distributor housing, cap and rotor, optical position sensor, sensor disc, pick-up assembly, distributor drive shaft, an ignition module, coil, primary and secondary wiring, spark plugs and the necessary control circuits (wiring harness) for the entire system. Although the distributor is composed of components, the entire assembly must be replaced as a single unit from the dealer parts department.

The Opti Spark distributor is mounted on the front cover and it is driven directly by the camshaft. The cap and rotor directs the spark from the ignition coil to the proper spark plug secondary wire. The distributor cap is marked with the corresponding cylinder numbers for easy routing of the plug wires. Refer to Chapter 1 for spark plug wire routing diagrams.

The distributor contains an optical pick-up system that provides actual crankshaft position (in degrees) to the ECM/PCM. The system uses two infrared optical sensors and a flat disc with two rows of notches (slots) cut around the circumference. One row has 360 notches (one-degree widths) and the other row has eight notches in the following positions; 2, 7, 2, 12, 2, 17, 2, and 22-degrees. When the optical pick-up turns, it produces two modulated digital signals. The first row (360-degrees) produces the signal for the timing while the second row (8 notches) produces the signal for the RPM reference.

Distributorless Ignition System (DIS)

The 3.4L engine, the 3800 engine and the 1998 and later V8 engines use a distributorless ignition system called DIS. V6 engines use a "waste spark" method of spark distribution. Each cylinder is paired with its opposing cylinder in the firing order 1-4, 2-5, 3-6, so that one cylinder on compression fires simultaneously with its opposing cylinder on the exhaust stroke. Since the cylinder on the exhaust stroke requires very little of the available voltage to fire its plug, most of the voltage is used to fire the cylinder on the compression stroke. 1998 and later V8 engines use a coil over plug method of spark distribution where 8 individual coils/modules are fired sequentially by the PCM.

The DIS system includes a coil pack or coil packs, an ignition module or modules, a crankshaft reluctor ring, a magnetic crankshaft sensor and the ECM/PCM. The ignition module on V6 engines is located under the coil pack and is connected to the ECM/PCM. V8 engines have eight independent ignition modules which are part of each ignition coil assembly.

6.3 To use a calibrated ignition tester (available at most auto parts stores), simply disconnect a spark plug wire, attach the wire to the tester, clip the tester to a convenient ground and operate the starter - if there's enough power to fire the plug, sparks will be visible between the electrode tip and the tester body

The magnetic crankshaft sensor protrudes through the engine block, within about 0.050-inch of the crankshaft reluctor ring. The reluctor ring is a special disc cast into the crankshaft, which acts as a signal generator for the ignition timing.

The system uses control wires from the ECM/PCM, just like conventional distributor systems. The ECM/PCM controls timing using crankshaft position, engine rpm, engine temperature and manifold absolute pressure (MAP) sensing.

6   Ignition system - check

Warning: *Because of the very high voltage generated by the ignition system, extreme care should be taken whenever an operation is performed involving ignition components. This not only includes the coils, control module and spark plug wires, but related items connected to the system as well, such as the plug connections, tachometer and any test equipment.*

General checks

Refer to illustration 6.3

1   Check all ignition wiring connections for tightness, cuts, corrosion or any other signs of a bad connection. A faulty or poor connection at a spark plug could also result in a misfire. Also check for carbon deposits inside the spark plug boots. Remove the spark plugs, if necessary, and check for fouling.

2   Check for ignition and battery supply to the ECM/PCM. Check the ignition fuses (see Chapter 12).

3   Use a calibrated ignition tester to verify adequate available secondary voltage (25,000 volts) at the spark plug **(see illustration)**. Using an ohmmeter, check the resistance of the spark plug wires). Each wire should measure less than 30,000 ohms.

4   Check to see if the fuel pump and relay are operating properly (see Chapter 4). The fuel pump should activate for two seconds when the ignition key is cycled ON. Install an injector test light and monitor the blinks as the injector voltage signal pulses (see Chapter 4, *Fuel Injection System - check*).

Distributor Ignition (DI) systems (1997 and earlier 5.7L engines)

Refer to illustration 6.5
Note: *If the PCM sets a code 16, this indicates a low resolution pulse or a defective optical sensor in the distributor. This prevents voltage from reaching the PCM resulting in a no-start condition or a loss of engine performance. Refer to Chapter 6 for additional information on the On-Board Diagnosis (OBD) system and interpreting the coded information.*
Note: *Base timing is preset - there is NO timing adjustment possible. The PCM controls the advance and retard functions of the ignition timing.*
5   Refer to the accompanying schematic for terminal pin designations for testing the Distributor Ignition (DI) ignition system **(see illustration)**.
6   Check for spark at the spark plugs, checking at least two or more spark plug wires **(see illustration 6.3)**. On a NO START condition, check the fuel pump relay and fuel pump systems for fuel delivery problems in the event the ignition spark is available.
7   Check for spark at the coil wire with a spark tester while an assistant is cranking the engine. A spark indicates that the problem must be the distributor unit. This test separates the ignition wires from the ignition coil.
Note: *A few sparks followed by no spark is the same condition as no spark at all.*
8   If there is NO spark, check the ignition coil (see Section 7). If the coil checks are correct, disconnect the ignition module connector, turn the ignition key ON (engine not running) and check for battery voltage at the terminals A and D. Normally, there should be battery voltage at the "A" and "D" terminals. Low voltage indicates an open or high resistance circuit from the distributor to the coil or ignition switch (primary ignition circuit). Also, check the OBD self diagnosis system for codes that may indicate an ignition system failure (see Chapter 6).
9   Switch the voltmeter to the A/C scale and measure the voltage on terminal B while cranking the engine over. Have an assistant turn the engine over while observing the voltmeter. This voltage signal from the computer acts as reference voltage for ignition control. This test will eliminate the PCM as a source of the problem. If there is a voltage signal present, a no start condition will most likely be caused by a faulty module. Continue testing. If there is no voltage signal from the PCM, replace the computer (see Chapter 6).
10   Turn the ignition key OFF and install a LED type test light onto the battery positive terminal (+). Observe that the test light is ON

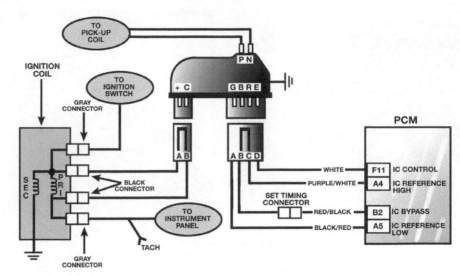

6.5 Ignition schematic of the 5.7L Distributor Ignition system

when terminal C on the ignition module connector (harness side) is probed. This checks the computer ground. This will check for a complete ground wire from the ignition module. The test light should illuminate when touched to the connector terminal C. If there is no light, check the wiring harness for an open or shorted circuit.
11   Remove the coil wire and ground it to the engine using a suitable jumper wire. Disconnect the ignition coil harness connector, connect a LED type test light to the positive (+) battery terminal and check for a pulsing signal voltage at the white/black wire while the engine is cranked over. This test will check for the pulsing signal voltage from the ignition distributor after it has been triggered by the PCM. Have an assistant crank over the engine while observing the LED test light. Be sure to remove the coil secondary wire and ground the coil to the engine using a suitable jumper wire. There should be a steady and obvious flashing LED light. Regular 12 volt test lights can be used but the bulb may not respond brightly and clearly as the LED test light.
12   If all the tests results are correct except Step 19 and there still is a NO START condition, it will be very difficult to distinguish a faulty distributor unit from a faulty ignition module. In most cases, the ignition module, when defective, will not produce a pulsing voltage signal with symptoms of a definite start or no-start condition while a faulty DI unit will produce intermittent ignition failures when the engine is running. Take the vehicle to a dealer service department or other qualified repair shop for diagnosis.
13   DI ignition systems are prone to water entering the distributor unit through the venting system. Thoroughly check the vent system for cracks, leaks or damaged components:
a) *Clean the air supply hose from the air intake duct to the distributor*
b) *Check the vacuum supply hose and check valve from the intake manifold*

c) *Check the venting system harness from the lower section of the distributor to the check valve filter*
d) *Make sure the air intake duct is not plugged or damaged*

Distributorless ignition systems

Refer to illustrations 6.14, 6.15a and 6.15b
14   Disconnect the ignition module harness connector on V6 engines or the ignition coil harness connector on 1998 and later V8 engines and check for battery voltage at the pink wire terminal from the ignition switch with the ignition key ON, engine not running **(see illustration)**. If no battery voltage is present, check and replace the ignition system fuse(s) and check for voltage again at the harness connector. Repair the circuit to the ignition system if necessary.
15   Next, check for a trigger signal from the

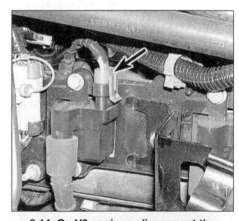

6.14 On V6 engines, disconnect the harness connector from the ignition control module - on V8 engines, disconnect the harness connector from the ignition coil (arrow) - in either case check for battery power at the pink wire

6.15a On V6 engines remove the coil pack from the ignition module and check for a trigger signal from the ignition module while an assistant cranks the engine over

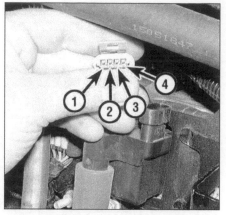

6.15b On 1998 and later V8 engines, use a test light connected to the positive battery terminal to check for a trigger signal at the ignition control terminal

1 *12-volt supply*
2 *Ignition control (from PCM)*
3 *Reference low*
4 *Ground*

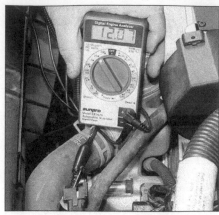

7.1 Unplug the coil electrical connector and check for voltage on the pink wire

ignition module) (V6) or the PCM (V8). On V6 engines, remove the coil pack from the ignition module to expose the module terminals. Connect a test light between each of the module terminals and have an assistant crank the engine over **(see illustration)**. On V8 engines disconnect the ignition coil harness connector and check for a trigger signal at the ignition coil **(see illustration)**. This test checks the triggering circuit in the ignition module or PCM. A blinking test light indicates the module or PCM is triggering. The test light should blink quickly and constantly as each coil pack is triggered to fire by the switching signal from the ignition module or the PCM. **Note:** *The trigger signal to each coil pack must be checked individually.*

16   If there is no blinking light on V6 engines, the ignition module is most likely the problem but not always. If there is no blinking light on V8 engines, the wiring to the ignition coil or the PCM may be faulty. Refer to the wiring diagrams and check all the related wiring from the PCM for shorts or opens.

17   A slowly blinking light, at this point, indicates the PCM is not seeing a crank sensor

signal (see Chapter 6). **Note:** *Refer to Chapter 6 for additional information and testing procedures for the camshaft sensor and the crankshaft sensors. It will be necessary to verify that the crankshaft sensors and camshaft sensor is operating correctly before changing the ignition module on V6 engines. A defective ignition module can only be diagnosed by process of elimination.*

---

7   Ignition coil - check, removal and   installation

Distributor Ignition (DI) systems

Refer to illustrations 7.1, 7.3, 7.4 and 7.7

Check

1   Check to make sure the coil is receiving battery voltage with the ignition key ON (engine not running) **(see illustration)**.

2   Detach the primary electrical connector from the coil.
3   Using the ohmmeter's low scale, hook up the ohmmeter leads to the primary terminals on the ignition coil **(see illustration)**. The ohmmeter should indicate a very low resistance value. If it doesn't, replace the coil. Refer to the Specifications listed in the beginning of this Chapter.
4   Using the high scale, hook up one lead to the ignition coil primary terminal and the other lead to the secondary terminal **(see illustration)**. Refer to the Specifications listed in the beginning of this Chapter. The ohmmeter should not indicate an infinite resistance. If it does, replace the coil.

Removal

5   Detach the cable from the negative terminal of the battery. **Caution:** *On models equipped with a Delco Loc II or Theftlock audio system, be sure the lockout feature is turned off before performing any procedure which requires disconnecting the battery.*
6   Unplug the coil high tension wire and both electrical leads from the coil.
7   Remove both mounting nuts and re-move the coil from the engine **(see illustration)**.

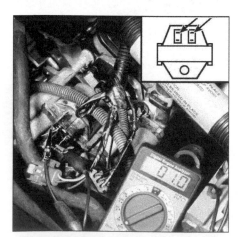

7.3 Checking the coil primary resistance on the 5.7L engine Distributor Ignition (DI)

7.4 Checking the coil secondary resistance on the 5.7L engine Distributor Ignition (DI)

7.7 Remove the mounting nuts and separate the coil/module assembly from the engine

7.10 Checking the coil pack secondary resistance on the 3800 engine

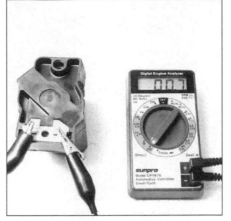

7.11 Checking the coil pack primary resistance on a 3800 engine

7.14 Original equipment coils are numbered with their corresponding cylinder numbers stamped on the top

7.15 Carefully lift the coil pack from the ignition module

Installation

8   Installation of the coil is the reverse of the removal procedure.

Distributorless Ignition System (DIS)

Check

Refer to illustrations 7.10 and 7.11

9   Refer to Section 6 and perform the simple ignition system checks.

10   Use an ohmmeter and check secondary resistance for each coil pack **(see illustration)**. Refer to the Specifications listed in this Chapter for the correct amount of resistance.

11   Next, check the primary resistance for each coil pack **(see illustration)**. Refer to the Specifications listed in this Chapter for the correct amount of resistance. **Note:** *On 1998 and V8 engines, the ignition coil primary and secondary resistance figures are not available from the manufacturer. If, after following the ignition system checks outlined in Section 6, there is trigger signal at the coil and there is no spark from the the ignition coil, then the ignition coil(s) are faulty.*

Removal and installation

12   Detach the cable from the negative terminal of the battery. **Caution:** *On models*

equipped with a Delco Loc II or Theftlock audio system, be sure the lockout feature is turned off before performing any procedure which requires disconnecting the battery.

V6 engines

Refer to illustrations 7.14 and 7.15

13   Unplug the electrical connectors from the module.

14   If the plug wires are not numbered, label them and detach the plug wires at the coil assembly **(see illustration)**.

15   Remove the module/coil assembly mounting bolts and lift the assembly from the vehicle **(see illustration)**

16   Installation is the reverse of removal. When installing the coils, make sure they are connected properly to the module and all plug wires are fully seated.

V8 engines (1998 and later)

Refer to illustrations 7.18 and 7.19

17   The ignition coils may be removed from each cylinder bank as a complete assembly or removed from the mounting bracket individually.

18   If removing the complete assembly, disconnect the ignition coils main electrical connector. Disconnect the spark plug wires from the spark plugs. Remove the ignition coil

bracket mounting nuts/bolts and remove the assembly from the engine **(see illustration)**.

19   If removing an individual coil, disconnect the spark plug wire from the coil. Remove the ignition coil mounting screws and remove the ignition coil from the bracket **(see illustration)**.

20   Installation is the reverse of removal.

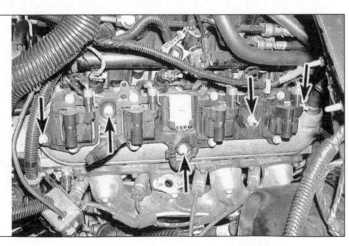

7.18 Ignition coil bracket bolts (arrows)

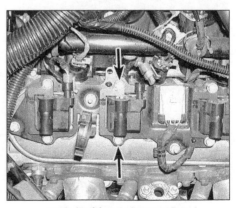

7.19 Ignition coil mounting screws (arrows)

8.5 The distributor is retained by four bolts

9.3 Remove the ignition module bolts (arrows)

9.9 Location of the module electrical connectors (arrows) on the 3.4L engine

8 Distributor (1997 and earlier 5.7L engine) - removal and installation

Removal

Refer to illustration 8.5

1   Disconnect the cable from the negative battery terminal. **Caution:** *On models equipped with a Delco Loc II or Theftlock audio system, be sure the lockout feature is turned off before performing any procedure which requires disconnecting the battery.*
2   Remove the water pump (see Chapter 3).
3   Remove the drivebelt pulley from the crankshaft (see Chapter 2C).
4   Remove the spark plug wires from the distributor (see Chapter 1).
5   Remove the distributor bolts **(see illustration)**.
6   Remove the distributor from the engine.

Installation

7   Insert the distributor unit onto the splined (or doweled, depending on year) shaft exactly the same relation to the block in which it was removed. To mesh the splines, it may be necessary to turn the distributor unit

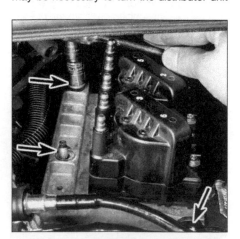

9.17 Remove the bolts (arrows) and lift the coil packs and module from the engine as a complete assembly (3800 engine)

slightly and wiggle the unit until it slides completely onto the spline and meets the engine block. **Note:** *The splined shaft can be removed for cleaning. In the event the splined shaft gets inadvertently rearranged, it can be inserted into the engine block either direction without assembly trouble.* **Caution:** *Do not install the distributor bolts and tighten the unit in such a way as to pull (press) the distributor gears onto the spline. This will damage the distributor assembly.*
8   Install the bolts securely.
9   Install the distributor ignition wires and coil wire.
10   Connect the cable to the negative terminal of the battery.

9 Electronic ignition system module - replacement

Distributor Ignition (DI) system
5.7L engine (1997 and earlier)

Refer to illustration 9.3

1   Detach the cable from the negative terminal of the battery. **Caution:** *On models equipped with a Delco Loc II or Theftlock audio system, be sure the lockout feature is turned off before performing any procedure which requires disconnecting the battery.*
2   Disconnect the ignition module electrical connector.
3   Remove both module attaching screws and lift the module up and away from its mount **(see illustration)**.
4   Install the module and tighten the screws securely.
5   Plug in the electrical connector.
6   Attach the cable to the negative terminal of the battery.

Distributorless Ignition Systems (DIS)
3.4L engine

Refer to illustration 9.9

7   Detach the cable from the negative terminal of the battery. **Caution:** *On models*

equipped with a Delco Loc II or Theftlock audio system, be sure the lockout feature is turned off before performing any procedure which requires disconnecting the battery.
8   Clearly label, then disconnect, all spark plug wires from the DIS assembly.
9   Unplug the electrical connector at the module **(see illustration)**.
10   Unbolt and remove the DIS and support bracket assembly.
11   Using a Torx screwdriver or bit (most models), remove the coil-to-module attaching screws.
12   Separate the coil and module assemblies.
13   Open the coil and module halves. Label, then detach, the wires between the module and the coil assemblies from the spade terminals on the underside of the coils.
14   Unbolt the module from the support bracket.
15   Installation is the reverse of removal. Be sure to attach the wires of the new module to the coil assembly spade terminals in exactly the same order in which they were removed.

3800 engine

Refer to illustration 9.17

16   Detach the cable from the negative terminal of the battery. **Caution:** *On models equipped with a Delco Loc II or Theftlock audio system, be sure the lockout feature is turned off before performing any procedure which requires disconnecting the battery.*
17   Remove the entire module and coil pack assembly from the engine **(see illustration)**.
18   Remove the individual coil packs from the top of the ignition module.
19   Disconnect the ignition module connector from the module.
20   Remove the bolts that retain the ignition module to the engine.
21   Installation is the reverse of removal.

5.7L engine (1998 and later)

22   The ignition module(s) on these engines are incorporated into each coil pack. Refer to Section 7 for the ignition coil replacement procedures.

11.3 Monitor the battery voltage after the engine is started - the voltage should be between 14 to 15 volts

12.2 Disconnect the electrical connectors from the backside of the alternator

12.4a Alternator mounting bolts - 1997 and earlier 5.7L engine

12.4b Alternator bolt locations (arrows) on the 3800 engine

10 Charging system - general information and precautions

Caution: *On models equipped with a Delco Loc II or Theftlock audio system, be sure the lockout feature is turned off before performing any procedure which requires disconnecting the battery.*

The charging system consists of a belt-driven alternator with an integral voltage regulator and the battery. These components work together to supply electrical power for the ignition system, the lights and all accessories.

There are three types of CS alternators in use, the CS-130 (105 amp), CS-144 (124 amp) and the CS-130D (105 amp) alternator. All types use a conventional pulley and fan.

To determine which type of alternator is fitted to your vehicle, look at the fasteners employed to attach the two halves of the alternator housing. The CS-130 and CS-130A alternators use rivets instead of screws. These alternators are rebuildable once the rivets are drilled out. However, we don't recommend this practice. For all intents and purposes, these types should be considered non-serviceable and, if found to be faulty, should be exchanged as cores for new or rebuilt units. The CS-144 alternator is bolted together with conventional bolts and locknuts. These alternators can be disassembled and repaired if necessary. Because of the expense and the limited availability of parts, no alternator overhaul information is included in this manual.

The purpose of the voltage regulator is to limit the alternator's voltage to a preset value. This prevents power surges, circuit overloads, etc., during peak voltage output. On all models with which this manual is concerned, the voltage regulator is contained within the alternator housing.

The charging system does not ordinarily require periodic maintenance. The drivebelt, electrical wiring and connections should, however, be inspected at the intervals suggested in Chapter 1.

Take extreme care when making circuit connections to a vehicle equipped with an alternator and note the following. When making connections to the alternator from a battery, always match correct polarity. Before using arc welding equipment to repair any part of the vehicle, disconnect the wires from the alternator and the battery terminal. Never start the engine with a battery charger connected. Always disconnect both battery leads before using a battery charger.

The charging indicator light on the dash lights when the ignition switch is turned on and goes out when the engine starts. If the light stays on or comes on once the engine is running, a charging system problem has occurred. See Section 11 for the proper diagnosis procedure for each type of alternator.

11 Charging system - check

Refer to illustration 11.3

1   If a malfunction occurs in the charging circuit, do not immediately assume that the alternator is causing the problem. First check the following items:

   a) *Make sure the battery cable connections at the battery are clean and tight.*
   b) *The battery electrolyte specific gravity (if possible). If it is low, charge the battery.*
   c) *Check the external alternator wiring and connections. They must be in good condition.*
   d) *Check the drivebelt condition and tension (Chapter 1).*
   e) *Make sure the alternator mounting bolts are tight.*
   f) *Run the engine and check the alternator for abnormal noise (may be caused by a loose drive pulley, loose mounting bolts, worn or dirty bearings, defective diode or defective stator).*

2   Using a voltmeter, check the battery voltage with the engine off. It should be approximately 12 volts.

3   Start the engine and check the battery voltage again. It should now be approximately 14 to 15 volts **(see illustration)**.

12 Alternator - removal and installation

Refer to illustrations 12.2, 12.4a and 12.4b

1   Detach the cable from the negative terminal of the battery. **Caution:** *On models equipped with a Delco Loc II or Theftlock audio system, be sure the lockout feature is turned off before performing any procedure which requires disconnecting the battery.*

2   Clearly label, if necessary, then unplug and unbolt the electrical connectors from the alternator **(see illustration)**. On 1998 and later V8 engines, this is best accomplished from below by raising the vehicle and supporting it securely on jackstands.

3   Remove the serpentine drivebelt (see Chapter 1). On 2000 and later V6 models, remove the EVAP canister purge solenoid.

4   Remove the alternator mounting bolts **(see illustrations)** and remove the alternator. On 1998 and later V8 engines, remove the alternator rear mounting bracket first, then remove the mounting bolts at the front of the alternator. On 2000 and later models, unsnap the transmission cooling lines from their bracket and move them aside slightly.

5   Installation is the reverse of removal.

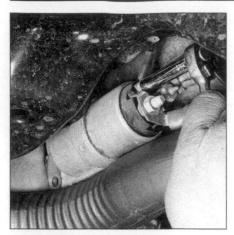

15.3 Disconnect the solenoid wire from the starter solenoid

15.4 Remove the starter bolts (arrows) on the 5.7L engine

15.5 Removing the starter from the 5.7L engine

13 Starting system - general information

Caution: *On models equipped with a Delco Loc II or Theftlock audio system, be sure the lockout feature is turned off before performing any procedure which requires disconnecting the battery.*

The function of the starting system is to crank the engine. The starting system is composed of a starter motor, solenoid and battery. The battery supplies the electrical energy to the solenoid, which then completes the circuit to the starting motor, which does the actual work of cranking the engine.

The solenoid and starter motor are mounted together at the lower front side of the engine. No periodic lubrication or maintenance is required.

The electrical circuitry of the vehicle is arranged so that the starter motor can only be operated when the clutch pedal is depressed (manual transmission) or the transmission selector lever is in Park or Neutral (automatic transmission).

Never operate the starter motor for more than 15 seconds at a time without pausing to allow it to cool for at least two minutes. Excessive cranking can cause overheating, which can seriously damage the starter.

There are three types of starters equipped in these models. The 1993 through 1995 5.7L engine is equipped with the PG 250 starter assembly which uses a planetary gear reduction system. 1993 through 1995 3.4L engines are equipped with the SD 210 starter assembly which uses a direct drive system. 1996 and later models are equipped with the PG 260 gear reduction type starter. All types are not serviceable and in the event of failure, they must be replaced as a single unit as a rebuilt (exchange) or new part.

14 Starter motor - testing in vehicle

1　If the starter motor does not turn at all when the switch is operated, make sure that the shift lever is in Neutral or Park (automatic

transmission) or that the clutch pedal is depressed (manual transmission).
2　Make sure that the battery is charged and that all cables, both at the battery and starter solenoid terminals, are secure.
3　If the starter motor spins but the engine is not cranking, the overrunning clutch in the starter motor is slipping and the motor must be removed from the engine for replacement.
4　If, when the switch is actuated, the starter motor does not operate at all but the solenoid clicks, then the problem lies with either the battery, the main solenoid contacts or the starter motor itself. **Note:** *Before diagnosing starter problems, make sure the battery is fully charged.*
5　If the solenoid plunger cannot be heard when the switch is actuated, the solenoid itself is defective or the solenoid circuit is open.
6　To check the solenoid, connect a jumper lead between the battery (+) and the "S" terminal on the solenoid. If the starter motor now operates, the solenoid is OK and the problem is in the ignition switch, neutral start switch or in the wiring.
7　If the starter motor still does not operate, remove the starter/solenoid assembly for disassembly, testing and repair.
8　If the starter motor cranks the engine at an abnormally slow speed, first make sure that the battery is charged and that all terminal connections are clean and tight. If the engine is partially seized, or has the wrong viscosity oil in it, it will crank slowly.
9　Run the engine until normal operating temperature is reached, then stop the engine, disconnect the coil wire from the distributor cap and ground it on the engine.
10　Connect a voltmeter positive lead to the starter motor terminal of the solenoid and then connect the negative lead to ground.
11　Crank the engine and take the voltmeter readings as soon as a steady figure is indicated. Do not allow the starter motor to turn for more than 15 seconds at a time. A reading of 9 volts or more, with the starter motor turning at normal cranking speed, is normal. If the reading is 9 volts or more but the cranking speed is slow, the motor is faulty. If the read-

ing is less than 9 volts and the cranking speed is slow, the solenoid contacts are probably burned.

15 Starter motor - removal and installation

Refer to illustrations 15.3, 15.4 and 15.5
1　Disconnect the negative battery cable. **Caution:** *On models equipped with a Delco Loc II or Theftlock audio system, be sure the lockout feature is turned off before performing any procedure which requires disconnecting the battery.*
2　Raise the front of the vehicle and support it securely on jackstands. On 2000 and later models, allow the engine to cool, then remove the passenger side catalytic converter.
3　From under the vehicle, disconnect the solenoid wire and battery cable from the terminals on the solenoid **(see illustration)**. On 1998 and later V8 engines, remove the front exhaust pipe from the vehicle (see Chapter 4).
4　On 2000 and later models, remove the shield over the starter motor. On V6 models the shield is secured to the left side of the engine with three bolts, while shields on V8 models are attached with clips. Unsnap the clips. Remove the starter mounting bolts **(see illustration)**.
5　Remove the starter motor **(see illustration)**. Note the location of the spacer shim(s), if equipped.
6　Installation is the reverse of removal. Be sure to install the spacer shim(s) in exactly the same location, if equipped.

16 Starter solenoid - removal and installation

Note: *The starter solenoid is not easily removed from the main starter body without complete disassembly. It is recommended that the starter/solenoid assembly be exchanged as a complete unit in the event of failure.*

Chapter 6
Emissions and engine control systems

Contents

Specifications

Torque specifications

Ft-lbs (unless otherwise indicated)

3.4L V6 engine
- Crankshaft sensor bolts .... 96 in-lbs
- Camshaft sensor bolts .... 96 in-lbs

3800 V6 engine
- Crankshaft sensor bolts 18
- Camshaft sensor bolts
  - Through 1999 .... 48 in-lbs
  - 2000 and later .... 96 in-lbs

5.7L V8 engine (1998 and later)
- Crankshaft sensor bolts 18
- Camshaft sensor bolts 18

1   General information

Refer to illustration 1.1

To prevent pollution of the atmosphere from burned and evaporating gases, a number of emissions control systems are incorporated on the vehicles covered by this manual. The combination of systems used depends on the year in which the vehicle was manufactured, the locality to which it was originally delivered and the engine type. The major systems incorporated on the vehicles with which this manual is concerned include the:

Secondary Air Injection (AIR) system
Fuel Control System
Exhaust Gas Recirculation (EGR) system
Evaporative Emissions Control (EECS) system
Transmission Converter Clutch (TCC) system
Positive Crankcase Ventilation (PCV) system
Catalytic converter

All of these systems are linked, directly or indirectly, to the On Board Diagnostic (OBD) system. The Sections in this Chapter include general descriptions, checking procedures (where possible) and component replacement procedures (where applicable) for each of the systems listed above.

Before assuming that an emissions control system is malfunctioning, check the fuel and ignition systems carefully. In some cases special tools and equipment, as well as specialized training, are required to accurately

diagnose the causes of a rough running or difficult to start engine. If checking and servicing become too difficult, or if a procedure is beyond the scope of the home mechanic, consult your dealer service department. This does not necessarily mean, however, that the emissions control systems are particularly difficult to maintain and repair. You can quickly and easily perform many checks and do most (if not all) of the regular maintenance at home with common tune-up and hand tools. **Note:** *The most frequent cause of emissions system problems is simply a loose or broken vacuum hose or wiring connection. Therefore, always check the hose and wiring connections first.*

Pay close attention to any special precautions outlined in this Chapter. It should be noted that the illustrations of the various systems may not exactly match the system installed on your particular vehicle due to changes made by the manufacturer during production or from year to year.

A Vehicle Emissions Control Information

(VECI) label is located in the engine compartment of all vehicles with which this manual is concerned **(see illustration)**. This label contains important emissions specifications and setting procedures, as well as a vacuum hose schematic with emissions components identified. When servicing the engine or emissions systems, the VECI label in your particular vehicle should always be checked for up-to-date information. **Note:** *Because of a federally mandated extended warranty which covers the emission control system components (and any components which have a primary purpose other than emission control but have significant effects on emissions), check with your dealer about warranty coverage before working on any emission related systems.*

Because of their more precise fuel/air management, fuel-injected engines use simpler emissions systems which do not use all of the systems described previously.

The number of emissions control system components on later model fuel-injected

1.5 A Vehicle Emissions Control Information (VECI) label will be found in the engine compartment of all vehicles - if it's missing, obtain a new one from a dealer parts department

vehicles has actually decreased due to the high efficiency of the new fuel injection and ignition systems. No longer needed are the dual bed catalytic converter (although a single bed or monolithic converter is still used) and many of the confusing thermal vacuum switches, valves and hoses.

2 On Board Diagnostic (OBD) system and trouble codes

Diagnostic tool information

Refer to illustrations 2.1, 2.2 and 2.4

1 A digital multimeter is a necessary tool for checking fuel injection and emission related components **(see illustration)**. A digital volt-ohmmeter is preferred over the older style analog multimeter for several reasons. The analog multimeter cannot display the volts-ohms or amps measurement in hundredths and thousandths increments. When working with electronic circuits which are often very low voltage, this accurate reading is most important. Another good reason for the digital multimeter is the high impedance circuit. The digital multimeter is equipped with a high resistance internal circuitry (10 million ohms). Because a voltmeter is hooked up in parallel with the circuit when testing, it is vital that none of the voltage being measured should be allowed to travel the parallel path set up by the meter itself. This dilemma does not show itself when measuring larger amounts of voltage (9 to 12 volt circuits) but if you are measuring a low voltage circuit such as the oxygen sensor signal voltage, a fraction of a volt may be a significant amount when diagnosing a problem.

2 Hand-held scanners are the most powerful and versatile tools for analyzing engine management systems used on later model vehicles **(see illustration)**. Early model scanners handle codes and some diagnostics for many OBD I systems. Each brand scan tool must be examined carefully to match the year, make and model of the vehicle you are working on. Often interchangeable cartridges are available to access the particular manufacturer; Ford, GM, Chrysler, etc.). Some manufacturers will specify by continent; Asia, Europe, USA, etc.

3 With the arrival of the federally mandated emission control system (OBD II), a specially designed scanner has been developed. At this time, several manufacturers have released OBD II scan tools for the home mechanic. Ask the parts salesman at a local auto parts store for additional information.

4 Another, and less expensive, type of code reader is also available at parts stores **(see illustration)**. These tools simplify the procedure for extracting codes from the engine management computer by simply "plugging in" to the diagnostic connector on the vehicle wiring harness. **Note:** *Some diagnostic connectors are located under the dash, kick panel or glovebox while others are located in the engine compartment.*

2.1 Digital multimeters can be used for testing all types of circuits; because of their high impedance, they are much more accurate than analog meters for measuring millivolts in low-voltage computer circuits

General description

5 The electronically controlled fuel and emissions system is linked with many other related engine management systems. It consists mainly of sensors, output actuators and a Powertrain Control Module (PCM). Completing the system are various other components which respond to commands from the PCM.

6 In many ways, this system can be compared to the central nervous system in the human body. The sensors (nerve endings) constantly gather information and send this data to the PCM (brain), which processes the data and, if necessary, sends out a command for some type of vehicle change (limbs).

7 Here's a specific example of how one portion of this system operates: An oxygen sensor, mounted in the exhaust manifold and protruding into the exhaust gas stream, constantly monitors the oxygen content of the exhaust gas as it travels through the exhaust pipe. If the percentage of oxygen in the exhaust gas is incorrect, an electrical signal is sent to the PCM. The PCM takes this information, processes it and then sends a command to the fuel injector(s) on the fuel injection system, telling it to change the fuel/air mixture. To be effective, all this happens in a fraction of a second, and it goes on continuously while the engine is running. The end result is a fuel/air mixture which is constantly kept at a predetermined ratio, regardless of driving conditions.

2.2 Scanners like the Actron Scantool and the AutoXray XP240 are powerful diagnostic aids - programmed with comprehensive diagnostic information, they can tell you just about anything you want to know about your engine management system

Testing

Refer to illustrations 2.11a and 2.11b

8 One might think that a system which uses exotic electrical sensors and is controlled by an on-board computer would be difficult to diagnose. This is not necessarily the case.

9 The On Board Diagnostic (OBD) system has a built-in self-diagnostic system, which indicates a problem by turning on a "SERVICE ENGINE SOON" light on the instrument panel when a fault has been detected. **Note:** *Since some of the trouble codes do not set the 'SERVICE ENGINE SOON' light, it is a good idea to access the OBD system and look for any trouble codes that may have been recorded and need tending.*

10 Perhaps more importantly, the PCM will recognize this fault, in a particular system monitored by one of the various information sensors, and store it in its memory in the form of a trouble code. Although the trouble code cannot reveal the exact cause of the malfunction, it greatly facilitates diagnosis as you or a dealer mechanic can "tap into" the PCMs memory and be directed to the problem area. **Caution:** *The material that follows in Paragraphs 11, 12 and 13 applies only to OBD I models with 12-pin connector.*

11 To retrieve this information from the PCM on OBD I models, you must use a short jumper wire to ground a diagnostic terminal. The terminal is part of an electrical connector

2.4 Trouble code tools simplify the task of extracting the trouble codes on OBD I systems

called the Assembly Line Data Link (ALDL) **(see illustrations)**. The ALDL is located under the dashboard, just below the instrument panel. To use the ALDL, remove the plastic cover (if equipped). With the electrical connector exposed to view, push one end of the jumper wire into the diagnostic TEST terminal and the other end into the GROUND terminal. **Note:** *Some 1993 through 1995 models use a 16-pin OBD II type connector, but actually use an OBD I diagnostic system. The VECI label under the hood will specify whether the sytem is OBD I or OBD II. Vehicles with 16-pin connectors and OBD I systems require a scan tool that uses a specialized connector.*

12   Turn the ignition to the ON position. **Caution:** *Do not start the engine with the TEST terminal grounded.* The "SERVICE ENGINE SOON" light should flash Trouble Code 12, indicating that the diagnostic system is working. Code 12 will consist of one flash, followed by a short pause, then two more flashes in quick succession. After a longer pause, the code will repeat itself two more times. If no other codes have been stored, Code 12 will continue to repeat itself until the jumper wire is disconnected. If additional Trouble Codes have been stored, they will follow Code 12. Again, each Trouble Code will flash three times before moving on.

13   Once the code(s) have been noted, use the Trouble Code Identification information which follows to locate the source of the fault. **Note:** *Whenever the battery cable is disconnected, all stored Trouble Codes in the PCM are erased. Be aware of this before you disconnect the battery.*

14   To retrieve this information from 1994/1995 and later models (OBD II), a SCAN tool must be connected to the Assembly Line Diagnostic Link (ALDL). The SCAN tool is a hand-held digital computer scanner that interfaces with the on-board computer. The SCAN tool is a very powerful tool; it not only reads the trouble codes but also displays the actual operating conditions of the sensors and actuators. SCAN tools are expensive, but they are necessary to accurately diagnose a modern computerized fuel injected engine. SCAN tools are available from automotive parts stores and specialty tool companies.

2.11a The 12 pin Assembly Line Data Link (ALDL) terminal identification

A   Ground
B   Diagnostic TEST terminal

15   It should be noted that the self-diagnosis feature built into this system does not detect all possible faults. If you suspect a problem with the On Board Diagnostic (OBD) system, but the SERVICE ENGINE SOON light has not come on and no trouble codes have been stored, take the vehicle to a dealer service department or other qualified repair shop for diagnosis.

16   Furthermore, when diagnosing an engine performance, fuel economy or exhaust emissions problem (which is not accompanied by a SERVICE ENGINE SOON light) do not automatically assume the fault lies in this system. Perform all standard troubleshooting procedures, as indicated elsewhere in this manual, before turning to the On Board Diagnostic (OBD) system.

17   Finally, since this is an electronic system, you should have a basic knowledge of automotive electronics before attempting any diagnosis. Damage to the PCM, Programmable Read Only Memory (PROM) calibration unit or related components can easily occur if care is not exercised.

Trouble Code Identification

18   Following is a list of the typical Trouble Codes which may be encountered while diagnosing the On Board Diagnostic system.

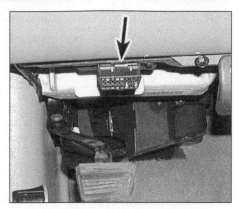

2.11b The 16 pin ALDL connector is typically located under the dash

Also included are simplified troubleshooting procedures. If the problem persists after these checks have been made, the vehicle must be diagnosed by a professional mechanic who can use specialized diagnostic tools and advanced troubleshooting methods to check the system. Procedures marked with an asterisk (*) indicate component replacements which may not cure the problem in all cases. For this reason, you may want to seek professional advice before purchasing replacement parts.

19   To clear the Trouble Code(s) from the PCM memory, momentarily remove the PCM/IGN fuse from the fuse box for 10 seconds. Clearing codes may also be accomplished by removing the fusible link (main power fuse) located near the battery positive terminal (see Chapter 12) or by disconnecting the cables from the battery. **Caution:** *On models equipped with a Delco-Loc II or Theftlock audio system, be sure the lockout feature is turned off before disconnecting the battery cable.* Disconnecting the power to the PCM to clear the memory can be an important diagnostic tool, especially on intermittent problems. **Caution:** *To prevent damage to the PCM, the ignition switch must be OFF when disconnecting or connecting power to the PCM.* **Note:** *Disconnecting the negative battery terminal will erase any radio preset codes that have been stored.*

OBD I TROUBLE CODES

Trouble Code	Circuit or system	Probable cause
13	Oxygen sensor circuit	Check the wiring and connectors from the oxygen sensor. Replace oxygen sensor (see Section 4)*
14	Coolant sensor circuit (high temperature indicated)	If the engine is experiencing overheating problems, the problem must be rectified before continuing (see Chapters 1 and 3). Check all wiring and connectors associated with the sensor. Replace the coolant sensor (see Section 4)*
15	Coolant sensor circuit (low temperature indicated)	See above. Also, check the thermostat for proper operation
16	System voltage discharge (low battery voltage) (3.4L)	Check the charging system (see Chapter 5)
16	Low resolution pulse (5.7L)	Ignition is not detecting the low resolution (4 pulses per crankshaft revolution) (see Chapter 5)
17	Camshaft Position Sensor circuit error (3.4L)	Check the camshaft position sensor circuit (refer to the wiring diagrams in Chapter 12)

OBD I TROUBLE CODES (continued)

Trouble Code	Circuit or system	Probable cause
21	TPS circuit (signal voltage high)	Check for sticking or misadjusted TPS. Check all wiring and connections at the TPS and at the ECM/PCM. Adjust or replace TPS* (see Section 4)
22	TPS circuit (signal voltage low)	See above
23	Intake Air Temperature (IAT) sensor	Low temperature indicated (see Section 4)
24	Vehicle Speed sensor (VSS)	A fault in this circuit should be indicated only while the vehicle is in motion. Disregard code 24 if set when drive wheels are not turning. Check connections at the ECM/PCM. Check the TPS setting (see Section 4)
25	Intake Air Temperature (IAT) sensor	Check the resistance of the IAT sensor. Check the wiring and connections to the sensor. Replace the IAT sensor*
26	Evaporative purge control solenoid circuit (5.7L)	EVAP purge control solenoid circuit is faulty. Check the wiring harness (see Section 6)
27	EGR vacuum control solenoid circuit (5.7L)	Check the EGR vacuum solenoid control circuit (see Section 5)
28	Transmission range (TR)	Check the transmission range control circuit
29	Secondary Air Injection	Check the Secondary Air Injection (AIR) system for an open circuit in the wiring (see Section 6)
32	Exhaust Gas Recirculation (EGR) failure (5.7L)	Check the vacuum source and all vacuum lines. Check the system electrical connectors at the ECM/PCM and EGR valve. Replace the EGR valve or ECM/PCM as necessary (see Section 5)*
33	Manifold Absolute Pressure (MAP) signal voltage high	Check vacuum hose(s) from MAP sensor. Check electrical sensor or circuit connections at the ECM/PCM. Replace MAP sensor (see Section 4)*
34	Manifold Absolute Pressure (MAP) signal voltage low	Check vacuum hose(s) from MAP sensor. Check electrical sensor or circuit connections at the ECM/PCM. Replace MAP sensor (see Section 4) *
35	Idle Air Control circuit (IAC) (3.4L)	Idle RPM too low or too high. Check minimum idle speed (see Chapter 4), check fuel pressure, check for leaking injector and obstructions in the throttle body. Replace the IAC valve*
36	Distributor Ignition system (5.7L)	Faulty high resolution pulse or multiple low resolution pulses detected in ignition system (see Chapter 5)
36	24X Signal Circuit error (3.4L)	Faulty circuit error from coil packs to module (see Chapter 5)
37	TCC brake switch error	Faulty circuit error TCC switch to ECM/PCM (see Section 4). TCC brake switch stuck ON
38	TCC brake switch error	Faulty circuit error TCC switch to ECM/PCM (see Section 4). TCC brake switch stuck OFF
39	Clutch Switch error	Fault in circuit from clutch switch to ECM/PCM
41	No ignition control circuit (5.7L)	Check DI ignition system (see Chapter 5). Open signal circuit from distributor
41	No ignition control circuit (3.4L)	Check DIS ignition system (see Chapter 5). Open signal circuit #423 from distributor
42	Ignition Control Circuit	Check the wiring and connectors between the ignition module and the ECM/PCM. Replace the ignition module.* Replace the ECM/PCM*
43	Knock Sensor (KS) circuit	Check the ECM/PCM for an open or short to ground; if necessary, reroute the harness away from other wires such as spark plugs, etc. Replace the knock sensor (see Section 4)*
44	Lean exhaust	Check the wiring and connectors from the oxygen sensor to the ECM/PCM. Check the ECM/PCM ground terminal. Check the fuel pressure (Chapter 4). Replace the oxygen sensor (see Section 4)*
45	Rich exhaust	Check the evaporative charcoal canister and its components for the presence of fuel. Check for fuel contaminated oil. Check the fuel pressure regulator. Check for a leaking fuel injector. Check for a sticking EGR valve. Replace the oxygen sensor (see Section 4)*
46	Pass Key Circuit	Check for faulty circuit from ignition switch to theft deterrent module or ECM/PCM
47	Knock Sensor Module or circuit (5.7L)	Check for faulty circuit from knock sensor module or missing knock sensor module (see Section 4)
48	MAF sensor circuit (5.7L)	Check for faulty circuit from MAF sensor to ECM/PCM (see Section 4)
50	System voltage LOW	Check the charging system (see Chapter 5)
51	PROM/EEPROM error	Faulty or incorrect PROM/EEPROM. Diagnosis should be performed by a dealer service department or other qualified repair shop
53	System voltage high	Code 53 will set if the voltage at the ECM/PCM is greater than 17.1-volts or less than 10-volts. Check the charging system (see Chapter 5)
54	Fuel pump circuit low (3.4L)	Check the fuel pump circuit for shorts or damage (see Chapter 4)

Trouble Code	Circuit or system	Probable cause
55	Fuel Lean Monitor (5.7L)	Engine running lean during power acceleration indicating a possible fuel pump failure, fuel line restriction, etc. (see Chapter 4)
58	Transmission Fluid Temp (TFT) circuit	High temperature indicated on the TFT circuit
59	Transmission Fluid Temp (TFT) circuit	Low temperature indicated on the TFT circuit
61	Air conditioning circuit performance	Possible low R134 refrigerant charge. Check air conditioning system charge pressure (see Chapter 3)
63	Right Bank Heated O2 sensor circuit	Check wiring and connections from the O2 sensor to ECM/PCM. Open circuit indicated (see Section 4)
64	Right Bank O2 sensor circuit	Check wiring and connections from the O2 sensor to the ECM/PCM. Lean exhaust indicated (see Section 4)
65	Right Bank O2 sensor circuit	Check wiring and connections from the O2 sensor to the ECM/PCM. Rich exhaust indicated (see Section 4)
66	Air conditioning refrigerant pressure sensor circuit (low pressure)	Check the sensor electrical terminal connections and possible short to ground or open circuit in the sensor wiring. Replace the Air conditioning refrigerant pressure sensor*
67	Air conditioning refrigerant pressure sensor circuit	Check the pressure sensor or air conditioning clutch circuit for possible short to ground or open circuit in the sensor wiring. Replace the air conditioning refrigerant pressure sensor*
68	Air conditioning relay circuit	Check the air conditioning relay circuit and the relay for possible short to ground or open circuit (see Chapter 3)
69	Air conditioning relay circuit	Check the air conditioning relay circuit and the relay for possible short to ground or open circuit (see Chapter 3)
70	Air conditioning refrigerant pressure sensor circuit (high pressure)	Check the pressure sensor or air conditioning clutch circuit for possible short to ground or open circuit in the sensor wiring. Replace the air conditioning refrigerant pressure sensor*
71	Air conditioning evaporator temperature sensor circuit (low temp indicated)	Check the pressure sensor or air conditioning clutch circuit for possible short to ground or open circuit in the sensor wiring. Replace the air conditioning refrigerant pressure sensor*
72	VSS Signal circuit error	Check the vehicle speed sensor and circuit for possible short to ground or open circuit in the sensor wiring. Replace the VSS (see Section 4)*
73	Air conditioning evaporator temperature sensor circuit (high temp indicated) (3.4L)	Check the pressure sensor or air conditioning clutch circuit for possible short to ground or open circuit in the sensor wiring. Replace the air conditioning refrigerant pressure sensor*
73	Pressure Control Solenoid current surge (5.7L)	Check the solenoid and transmission line pressure for possible damage
74	Traction Control system (5.7L)	Low voltage signal in Traction Control circuit
75	Transmission system low voltage (5.7L)	Possible short in ignition feed circuit to ECM/PCM. Pressure control solenoid malfunctioning
75	Digital EGR (3.4L)	Check the Digital EGR circuit for the number 1 solenoid for possible short to ground or open circuit in the solenoid wiring
76	Digital EGR (3.4L)	Check the Digital EGR circuit for the number 2 solenoid for possible short to ground or open circuit in the solenoid wiring
77	Digital EGR (3.4L)	Check the Digital EGR circuit for the number 3 solenoid for possible short to ground or open circuit in the solenoid wiring
77	Cooling fan relay control circuit (5.7L)	Check the cooling fan relay control circuit and relays for possible short to ground or open circuit in the wiring (see Chapter 3)
79	Transmission fluid temp	Transmission fluid temperature sensor defective or excessive heat in transmission
80	Transmission component harness fault	TCC problem (see Section 9) or transmission slipping
81	Transmission 2-3 shift solenoid circuit (5.7L)	2-3 shift solenoid circuit faulty
82	Transmission 1-2 shift solenoid circuit (5.7L)	1-2 shift solenoid circuit faulty
82	Ignition Control 3X Signal circuit error (3.4L)	No 3X reference pulses detected by the ECM/PCM

OBD I TROUBLE CODES (continued)

83	Reverse Inhibit system (manual transmission) (5.7L)	Vehicle must be operating under 4 mph to allow reverse selection
84	Skip shift solenoid circuit (manual transmission) (5.7L)	Skip shift solenoid circuit malfunction
84	3-2 shift control circuit (automatic transmission) (5.7L)	3-2 shift control circuit malfunction
85	Transmission TCC stuck ON (5.7L)	TCC stuck ON (see Section 9)
85	Prom Error (3.4L)	Have the ECM/PCM checked at a dealer service department
86	A/D Error (3.4L)	Have the ECM/PCM checked at a dealer service department
87	EEPROM Error (3.4L)	Have the ECM/PCM checked at a dealer service department
90	Transmission TCC solenoid circuit	TCC circuit faulty (see Section 9)
91	Skip Shift Lamp circuit (5.7L)	Skip shift lamp circuit defective. Refer to the wiring diagrams at the end of Chapter 12
93	PCS Circuit Error (3.4L)	Have the ECM/PCM checked at a dealer service department
96	Transmission System voltage low (3.4L)	Transmission circuit defective
97	VSS Output circuit voltage low (5.7L)	Check VSS and circuit (see Section 5)
98	Invalid ECM/PCM program (3.4L)	Have the ECM/PCM checked at a dealer service department
99	Invalid ECM/PCM program (3.4L)	Have the ECM/PCM checked at a dealer service department
99	TACH Output circuit (5.7L)	Have the ECM/PCM checked at a dealer service department

* Component replacement may not cure the problem in all cases. For this reason, you may want to seek professional advice before purchasing replacement parts.

OBD II TROUBLE CODES

Code	Code Identification
P0101	Mass air flow sensor circuit, range or performance problem
P0102	Mass air flow sensor circuit, low input
P0103	Mass air flow sensor circuit, high input
P0106	Manifold absolute pressure sensor circuit, range or performance problem
P0107	Manifold absolute pressure sensor circuit, low input
P0108	Manifold absolute pressure sensor circuit, high input
P0112	Intake air temperature circuit, low input
P0113	Intake air temperature circuit, high input
P0116	Engine coolant temperature circuit, out of range with IAT
P0117	Engine coolant temperature circuit, low input
P0118	Engine coolant temperature circuit, high input
P0121	Throttle position sensor circuit, range or performance problem
P0122	Throttle position sensor circuit, low input
P0123	Throttle position sensor circuit, high input
P0125	Insufficient coolant temperature for closed loop fuel control
P0128	Engine coolant temperature circuit
P0130	Heated oxygen sensor circuit, bank 1 (V6 engine)
P0131	Oxygen sensor circuit, low voltage (pre-converter sensor, left bank)
P0132	Oxygen sensor circuit, high voltage (pre-converter sensor, left bank)
P0133	Oxygen sensor circuit, slow response (pre-converter sensor, left bank)
P0134	Oxygen sensor circuit - no activity detected (pre-converter sensor, left bank)
P0135	Oxygen sensor heater circuit malfunction (pre-converter sensor, left bank)
P0137	Oxygen sensor circuit, low voltage (post-converter sensor, left bank)

Code	Code Identification
P0138	Oxygen sensor circuit, high voltage (post-converter sensor, left bank)
P0140	Oxygen sensor circuit - no activity detected (post-converter sensor, left bank)
P0141	Oxygen sensor heater circuit malfunction (post-converter sensor, left bank)
P0150	Heated oxygen sensor circuit, high-low signal bias (V6)
P0151	Oxygen sensor circuit, low voltage (pre-converter sensor, right bank)
P0152	Oxygen sensor circuit, high voltage (pre-converter sensor, right bank)
P0153	Oxygen sensor circuit, slow response (pre-converter sensor, right bank)
P0154	Oxygen sensor circuit - no activity detected (pre-converter sensor, right bank)
P0155	Oxygen sensor heater circuit malfunction (pre-converter sensor, right bank)
P0157	Oxygen sensor circuit, low voltage (post-converter sensor, right bank)
P0158	Oxygen sensor circuit, high voltage (post-converter sensor, right bank)
P0160	Oxygen sensor circuit - no activity detected (post-converter sensor, right bank)
P0161	Oxygen sensor heater circuit malfunction (post-converter sensor, right bank)
P0171	System too lean, left bank on 4 sensor systems
P0172	System too rich, left bank on 4 sensor systems
P0174	System too lean, right bank
P0175	System too rich, right bank
P0191	Injector pressure sensor system performance
P0192	Injector pressure sensor circuit low input
P0193	Injector pressure sensor circuit high input
P0200	Injector circuit malfunction
P0201	Injector 1 control circuit
P0202	Injector 2 control circuit
P0203	Injector 3 control circuit
P0204	Injector 4 control circuit
P0205	Injector 5 control circuit
P0206	Injector 6 control circuit
P0230	Fuel pump primary circuit malfunction
P0300	Engine misfire detected
P0301	Cylinder number 1 misfire detected
P0302	Cylinder number 2 misfire detected
P0303	Cylinder number 3 misfire detected
P0304	Cylinder number 4 misfire detected
P0325	Knock sensor circuit malfunction
P0326	Knock sensor circuit performance
P0327	Knock sensor circuit, low output (front knock sensor on V8 models)
P0332	Knock sensor circuit, low output (rear knock sensor on V8 models)
P0335	Crankshaft position sensor circuit malfunction
P0336	Crankshaft position sensor circuit, range or performance problem
P0341	Camshaft position sensor circuit, range or performance problem
P0342	Camshaft position sensor circuit, low output
P0343	Camshaft position sensor circuit, high output
P0351	Ignition control circuit, coil 1
P0352	Ignition control circuit, coil 2
P0353	Ignition control circuit, coil 3
P0354	Ignition control circuit, coil 4
P0355	Ignition control circuit, coil 5
P0356	Ignition control circuit, coil 6
P0357	Ignition control circuit, coil 7
P0358	Ignition control circuit, coil 8
P0400	Exhaust gas recirculation flow fault
P0401	Exhaust gas recirculation, insufficient flow detected

OBD II TROUBLE CODES (continued)

Code	Code Identification
P0402	Exhaust gas recirculation, excessive flow detected
P0403	Exhaust gas recirculation solenoid control circuit
P0404	Exhaust gas recirculation circuit, range or performance problem
P0405	Exhaust gas recirculation sensor circuit low
P0410	Secondary air injection system
P0412	Secondary air injection solenoid control circuit
P0418	Secondary air injection pump relay control circuit
P0420	Catalyst system efficiency below threshold, left bank
P0421	Catalyst system efficiency below threshold, left bank
P0430	Catalyst system efficiency below threshold, right bank
P0431	Catalyst system efficiency below threshold, right bank
P0440	Evaporative emission control system malfunction
P0441	Evaporative emission control system incorrect purge flow
P0442	Evaporative emission control system, small leak detected
P0443	Evaporative emission control system, purge control circuit malfunction
P0446	Evaporative emission control system, vent system performance
P0449	Evaporative emission control system, vent control circuit malfunction
P0452	Evaporative emission control system, pressure sensor low input
P0453	Evaporative emission control system, pressure sensor high input
P0461	Fuel level sensor circuit, range or performance problem
P0462	Fuel level sensor circuit, low input
P0463	Fuel level sensor circuit, high input
P0480	Cooling fan relay No. 1 control circuit
P0481	Cooling fan relay No. 2 and 3 control circuit
P0500	Vehicle speed sensor circuit
P0502	Vehicle speed sensor circuit low output
P0503	Vehicle speed sensor signal intermittent
P0506	Idle control system, rpm lower than expected
P0507	Idle control system, rpm higher than expected
P0530	Air conditioning refrigerant pressure sensor circuit
P0560	System voltage insufficient
P0562	System voltage low
P0563	System voltage high
P0567	Cruise control resume switch circuit
P0568	Cruise control resume switch circuit
P0571	Cruise control brake switch circuit
P0601	Powertrain Control Module, memory error
P0602	Powertrain Control module, programming error
P0604	Powertrain Control Module, memory error (RAM)
P0605	Powertrain Control Module, memory error (ROM)
P0606	Powertrain Control Module internal fault
P0608	Vehicle speed sensor output circuit
P0620	Alternator malfunction
P0645	Air conditioning clutch relay control circuit
P0650	Malfunction Indicator Light (MIL) control circuit
P0704	Clutch start switch circuit
P0705	Transmission range sensor circuit malfunction
P0801	Reverse inhibit solenoid control circuit
P0116	Throttle Position Sensor (TPS) circuit high input

Component replacement may not cure the problem in all cases. For this reason, you may want to seek professional advice before purchasing replacement parts.

3.2 Firmly tap against the ECM/PCM housing and observe any changes in the running condition of the engine

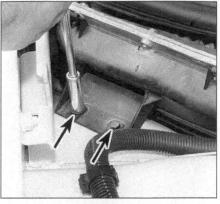

3.6 Remove the ECM/PCM mounting bolts (arrows)

3.7 Lift the ECM/PCM from the engine compartment and carefully disconnect the electrical connectors

3 Electronic Control Module/Powertrain Control Module (ECM/PCM) - check and replacement

ECM/PCM check

Refer to illustration 3.2

1   Lift the hood of the vehicle to gain access to the ECM/PCM. The ECM/PCM is located in the right rear corner of the engine compartment near the shock tower.

2   Using the tips of the fingers **(see illustration)**, tap vigorously on the side of the computer while the engine is running. If the computer is not functioning properly, the engine will stumble or stall and display glitches on the engine data stream obtained using a SCAN tool or other diagnostic equipment.

3   If the ECM/PCM fails this test, check the electrical connectors. Each connector is color coded to fit the respective slot in the computer body. If there are no obvious signs of damage, have the unit checked at a dealer service department.

ECM/PCM replacement

Refer to illustrations 3.6 and 3.7

Caution: *To prevent damage to the ECM/PCM, the ignition switch must be turned Off when disconnecting or connecting in the ECM/PCM connectors.*

4   The Electronic Control Module/Powertrain Control Module (ECM/PCM) is located in the engine compartment, on the passenger side (right side) of the vehicle.

5   Disconnect the cable from the negative battery terminal. **Caution:** *On models equipped with a Delco-Loc II or Theftlock audio system, be sure the lockout feature is turned off before disconnecting the battery cable.*

6   Remove the bolts that retain the ECM/PCM to the engine compartment body panel **(see illustration)**.

7   Carefully lift the ECM/PCM **(see illustration)** from the engine compartment without damaging the electrical connectors and wiring harness to the computer.

8   Unplug the electrical connectors from the ECM/PCM. Each connector is color coded to fit its respective receptacle in the computer.

9   Installation is the reverse of removal.

EEPROM replacement

10   These models are equipped with Electrical Erasable Programmable Read Only Memory (EEPROM) chip that is permanently soldered to the computer circuit boards. The EEPROM can be reprogrammed using a dealer's TECH 1 SCAN tool. Do not attempt to remove this component from the ECM/PCM. Have the EEPROM reprogrammed at a dealer service department.

11   The remainder of the installation is the reverse of removal.

4 Information sensors - check and replacement

Caution: *On models equipped with a Delco-Loc II or Theftlock audio system, be sure the lockout feature is turned off before disconnecting the battery cable.*

Note: *These checks and removal procedures do not include the 1995 and later models equipped with the OBD II system. These systems require a special SCAN tool in order to read out the various levels of coded information. Have the vehicle diagnosed by a dealer service department in the event of any fuel injection or emissions component failure. After performing any checking procedure to any of the information sensors, be sure to clear the ECM/PCM of all trouble codes by removing the ECM/PCM BAT fuse from the fuse box or disconnecting the cable from the negative terminal of the battery for at least ten seconds.*

Engine Coolant Temperature (ECT) sensor

General description

1   The coolant temperature sensor is a thermistor (a resistor which varies the value of its resistance in accordance with temperature changes). The change in the resistance value will directly affect the voltage signal from the coolant thermosensor. As the sensor temperature DECREASES, the resistance will IN-CREASE. As the sensor temperature IN-

CREASES, the resistance will DECREASE. A failure in the coolant sensor circuit should set either a Code 14 or a Code 15. These codes indicate a failure in the coolant temperature circuit, so the appropriate solution to the problem will be either repair of a wire or replacement of the sensor. The sensor can also be checked with an ohmmeter, by measuring its resistance when cold, then warming up the engine and taking another measurement.

Check

Refer to illustrations 4.2, 4.4 and 4.6

2   To check the sensor, check the resistance of the coolant temperature sensor while it is completely cold (50 to 80-degrees F = 5,700 to 2,200 ohms). Next, start the engine and warm it up until it reaches operating temperature. The resistance should be lower (170 to 200- degrees F = 200 to 300 ohms). **Note:** *The coolant temperature sensor on the 3.4L V6 engine is located on the left bank cylinder head directly behind the DIS coil packs and on the 3800 engines, the sensor is located under the throttle body on the engine block. The coolant temperature sensor on the 5.7L V8 engine is located on the water pump housing* **(see illustration)**. *Access to the coolant temperature sensor makes it difficult to position probes of the meter onto the terminals. It will be necessary*

4.2 Location of the Engine Coolant Temperature (ECT) sensor on the 5.7L V8 engine

to remove the sensor and perform the tests in a pan of heated water to simulate the conditions. Refer to Chapter 3 for water pump access procedures.

3   Check the reference voltage with the ignition key ON (engine not running). It should be approximately 5.0 volts. Refer to the wiring diagrams at the end of Chapter 12 for the correct wire colors.

4   Access to the coolant temperature sensor is difficult making it impossible to monitor the resistance changes without removing many extraneous components from the engine. Install the SCAN tool and switch to the ECT mode and monitor the temperature of a cold engine. The SCAN tool should indicate between 75 to 90-degrees F. Allow the engine to idle for several minutes and observe the coolant temperature increase as the engine warms up **(see illustration)**. The temperature should indicate between 180 to 210-degrees F at normal operating temperature. If there is not a definite change in temperature, remove the coolant temperature sensor and check the resistance in a pan of heated water to simulate warm-up conditions. If the sensor tests are good, check the wiring harness from the sensor to the computer (see Chapter 12 wiring diagrams).

Replacement

Warning: *Wait until the engine is completely cool before beginning this procedure.*

5   To remove the sensor, release the locking tab, unplug the electrical connector, then carefully unscrew the sensor. **Caution:** *Handle the coolant sensor with care. Damage to this sensor will affect the operation of the entire fuel injection system.*

6   Before installing the new sensor, wrap the threads with Teflon sealing tape to prevent leakage and thread corrosion **(see illustration)**.

7   Installation is the reverse of removal.

Manifold Absolute Pressure (MAP) sensor (3.4L V6 and 5.7L V8 engines)

General description

Refer to illustration 4.9

8   The Manifold Absolute Pressure (MAP) sensor monitors the intake manifold pressure changes resulting from changes in engine load and speed and converts the information into a voltage output. The ECM/PCM uses

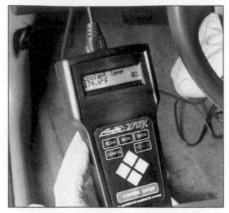

4.4 Plug the SCAN tool connector into the ALDL, switch the SCAN tool into ECT mode and observe the engine temperature through the computer data stream

the MAP sensor to control fuel delivery and ignition timing. A failure in the MAP sensor circuit should set a Code 33 or 34.

9   The MAP sensor SIGNAL voltage to the ECM/PCM varies from below 2 volts at idle (high vacuum) to above 4 volts with the ignition key ON (engine not running) or wide open throttle (WOT) (low vacuum). These values correspond with the altitude and pressure changes the vehicle experiences while driving **(see illustration)**. **Note:** *The MAP sensor is located near the throttle body on the intake manifold.*

Check

Refer to illustrations 4.10, 4.11a, 4.11b and 4.12

10   Check the REFERENCE voltage from the computer to the MAP sensor. Use pins or paper clips and backprobe the MAP sensor electrical connector terminal A (black) [-]) and terminal C (gray [+]) (two outside terminals). The voltage should be approximately 5.0 volts **(see illustration)**.

11   Check the SIGNAL voltage. Connect the positive lead of the voltmeter to terminal B (green[+]) of the sensor electrical connector and the negative lead to terminal A (black [-]). With the ignition ON (engine not running) the voltage reading should be about 4.5 to 5 volts **(see illustration)**. Start the engine and let it warm up. The voltage should be approximately 1.5 to 2.0 volts at idle **(see illustration)**. Raise the engine rpm and observe that

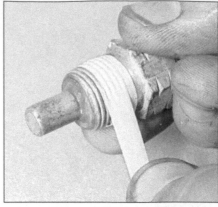

4.6 To prevent coolant leakage, be sure to wrap the temperature sensor threads with Teflon tape before installation

as vacuum decreases, voltage increases. At wide open throttle the MAP sensor SIGNAL should produce approximately 4.5 volts.

12   An alternate method of diagnosing the MAP sensor is by the use of an electronic SCAN tool. Install the SCAN tool and switch to the MAP mode and monitor the voltage signal with the engine at idle and high rpm **(see illustration)**. The SCAN tool should indicate between 1.2 to 1.6 volts at idle. Raise the engine rpm and observe that as engine rpm increases (decreasing vacuum) the MAP SIGNAL voltage increases. If the MAP sensor voltage readings are incorrect, replace the MAP sensor.

Replacement

Models through 2000

13   To replace the sensor, detach the vacuum hose, unplug the electrical connector and remove the mounting screws. Installation is the reverse of the removal procedure. On 2000 V8 models, unbolt and set aside the PCV hose assembly from the right side for MAP access.

2001 and later models

14   Remove the air intake housing (see Chapter 4).

4.10 Check the MAP sensor REFERENCE voltage on the black wire (-) and the gray wire (+). It should be approximately 5.0 volts

ALTITUDE		VOLTAGE RANGE
Meters	Feet	
Below 305	Below 1,000	3.8---5.5V
305--- 610	1,000--2,000	3.6---5.3V
610--- 914	2,000--3,000	3.5---5.1V
914--1219	3,000--4,000	3.3---5.0V
1219--1524	4,000--5,000	3.2---4.8V
1524--1829	5,000--6,000	3.0---4.6V
1829--2133	6,000--7,000	2.9---4.5V
2133--2438	7,000--8,000	2.8---4.3V
2438--2743	8,000--9,000	2.6---4.2V
2743--3048	9,000--10,000	2.5---4.0V

LOW ALTITUDE = HIGH PRESSURE = HIGH VOLTAGE

4.9 Typical Manifold Absolute Pressure (MAP) sensor altitude (pressure) vs. voltage values

4.11a Check the MAP sensor SIGNAL voltage on the green wire (+) and the black wire (-) with the ignition key ON (engine not running)

4.11b Start the engine and observe the MAP sensor voltage at idle. With the engine idling, vacuum in the intake manifold is high, therefore the voltage value will be low

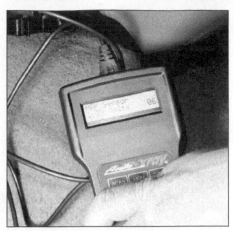

4.12 Install a SCAN tool to the ALDL and switch to MAP sensor mode. Observe the voltage values as the engine rpm goes from idle to full throttle. The voltage should increase as engine rpm increases (vacuum decreases)

15   With the vehicle raised and supported on jackstands, remove the bellhousing bolt securing the transmission dipstick tube to the bellhousing.

16   The PCV hoses and pipe must be removed from the engine's right bank for MAP sensor access. Disconnect the rubber hoses, then unbolt and remove the metal PCV pipe.

17   Disconnect the electrical connector at the MAP sensor, then twist the sensor up and out.

18   Installation is the reverse of the removal procedure. Apply a drop of engine oil to the sensor seal before installing the sensor.

Intake Air Temperature (IAT) sensor

General description

Refer to illustrations 4.20 and 4.21

19   The Intake Air Temperature (IAT) sensor is located inside the air duct directly after the air filter housing. This sensor acts as a thermistor (a resistor which changes the value of its resistance as the temperature changes). When the intake air is cold, the sensor resistance is high, therefore the ECM/PCM will read a high signal voltage. If the intake air is warm, the resistance is low giving the ECM/PCM a low voltage reading. The sensor values range from 2,700 ohms at 70-degrees F to 240 ohms at 190-degrees F. The ECM/PCM supplies approximately 5.0 volts (REFERENCE voltage) to the IAT sensor. A failure in the IAT sensor circuit should set either a Code 23 or Code 25.

Check

20   Measure the REFERENCE voltage. Disconnect the IAT electrical sensor and with the ignition key ON (engine not running), probe terminal A ground (black) and terminal B REF (tan) **(see illustration)**. The VOM should read approximately 5.0 volts.

21   Next, measure the resistance across the sensor terminals **(see illustration)**. By measuring its resistance when cold, then warming it up (a hair dryer can be used for this) and

taking another measurement. The resistance of the IAT sensor should be HIGH when the temperature is low. Next, start the engine and let it idle. Wait awhile and let the engine reach operating temperature. Turn the ignition key OFF, disconnect the IAT temperature sensor and check the resistance across the terminals. The resistance should be LOW when the air temperature is high. If the sensor does not exhibit this change in resistance, replace the sensor with a new part.

22   An alternate method of diagnosing the IAT sensor is by the use of an electronic SCAN tool. Install the SCAN tool and switch to the IAT mode and monitor the temperature with the engine cold and then completely warmed-up. The SCAN tool temperature ranges should increase while the IAT sensor resistance decreases. If the IAT sensor temperature/resistance readings are incorrect, replace the IAT sensor.

Replacement

23   To remove an IAT sensor, unplug the

electrical connector and remove the sensor from the air intake duct. Carefully twist the sensor to release it from the rubber boot.

24   Installation is the reverse of removal.

Mass Air Flow (MAF) sensor (5.7L V8 engines)

General description

25   The Mass Air Flow (MAF) sensor, which is located in a housing between the air cleaner housing and the intake duct, measures the amount of air entering the engine. A large quantity of air entering the engine indicates an acceleration or high load situation, while a small quantity of air indicates deceleration or idle. The ECM/PCM uses this information to control fuel delivery. The MAF sensor produces a frequency signal that will vary within a range of around 2,000 Hertz at idle to 10,000 Hertz at maximum load conditions. A failure in the MAF sensor circuit should set a Code 48.

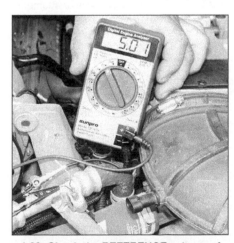

4.20 Check the REFERENCE voltage of the IAT sensor. It should be approximately 5.0 volts

4.21 Check the IAT sensor resistance and compare the value at the existing temperature

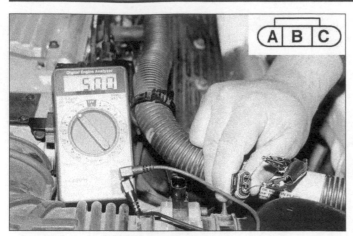

4.27 Check the MAF sensor signal voltage on terminal A (+) (yellow wire) and terminal B (-) (black/white wire). It should be approximately 5.0 volts

4.28 Check the MAF sensor supply voltage on terminal C (+) (pink wire) and terminal B (-) (black/white wire). It should be approximately 12.0 volts (battery voltage)

Check

Refer to illustrations 4.27, 4.28 and 4.29

26   A quick check of the sensor can be made by tapping the flat portion of the sensor body with a screwdriver handle as the engine is running. If the engine stumbles or stalls, the sensor is faulty. This test works only in an intermittent failure situation. If the engine does not start or fails to idle smoothly, continue the checks.

27   Check the SIGNAL voltage. Disconnect the MAF electrical connector and with the ignition key ON (engine not running), check the voltage on terminal B ground (black/white) and terminal A SIGNAL (yellow). There should be between 4 to 6 volts **(see illustration)**. If the voltage is higher or lower, check the wiring harness for shorts or check for a faulty ECM/PCM.

28   Next, check for battery voltage to the MAF sensor. With the ignition key ON (engine not running), check the voltage on terminal B ground (black/white) and terminal C (pink). There should be approximately 12 volts **(see illustration)**. If all the checks are correct, and the OBD system indicates a MAF code, have the vehicle checked at a dealer service department.

29   An alternate method of diagnosing the MAF sensor is by the use of an electronic SCAN tool. Install the SCAN tool and switch to the MAF mode and monitor the frequency signal with the engine at idle and high rpm **(see illustration)**. The SCAN tool should indicate HIGH or LOW. Raise the engine rpm and observe that as engine rpm increases the MAF signal also changes (increases). If the MAF sensor voltage readings are incorrect, replace the MAF sensor. Different SCAN tools will have different values for the MAF sensor data; some will simply read HIGH or LOW, while other will give frequency values in Hertz.

Replacement

Refer to illustration 4.30

30   To replace the MAF sensor, unplug the electrical connector, loosen the clamps and detach the sensor from the airducts **(see illustration)**.

31   Installation is the reverse of removal. **Caution:** *Make sure the arrow on the MAF sensor faces the engine when installed.*

Oxygen sensor

General description

32   The oxygen sensor, which is located in the exhaust manifold, monitors the oxygen content of the exhaust gas stream. The oxygen content in the exhaust reacts with the oxygen sensor to produce a voltage output which varies from 0.1-volt (high oxygen, lean mixture) to 0.9-volts (low oxygen, rich mixture). The ECM/PCM constantly monitors this variable voltage output to determine the ratio of oxygen to fuel in the mixture. The ECM/PCM alters the air/fuel mixture ratio by controlling the pulse width (open time) of the fuel injectors. A mixture ratio of 14.7 parts air to 1 part fuel is the ideal mixture ratio for minimizing exhaust emissions, thus allowing the catalytic converter to operate at maximum efficiency. It is this ratio of 14.7 to 1 which the ECM/PCM and the oxygen sensor attempt to maintain at all times. These engines are equipped with a left bank O_2 sensor (left exhaust manifold), a right bank O_2 sensor (right exhaust manifold) and a post-catalytic converter O_2 sensor. When checking the oxygen sensor system, it will be necessary to test all three O_2 sensors.

33   The oxygen sensor produces no voltage when it is below its normal operating temperature of about 600-degrees F. During this initial period before warm-up, the ECM/PCM operates in OPEN LOOP mode.

34   If the engine reaches normal operating temperature and/or has been running for two or more minutes, and if the computer detects a malfunction, the ECM/PCM will set a code 13, 44 or 45 for the left bank O_2 sensor or codes 63, 64 or 65 for the right bank O_2 sensor.

35   When there is a problem with the oxygen sensor or its circuit, the ECM/PCM operates in the open loop mode - that is, it controls fuel delivery in accordance with a programmed default value instead of feedback information from the oxygen sensor.

36   The proper operation of the oxygen sensor depends on four conditions:

a) *Electrical* - *The low voltages generated by the sensor depend upon good, clean connections which should be checked whenever a malfunction of the sensor is suspected or indicated.*

b) *Outside air supply* - *The sensor is designed to allow air circulation to the internal portion of the sensor. Whenever the sensor is removed and installed or replaced, make sure the air passages are not restricted.*

c) *Proper operating temperature* - *The ECM/PCM will not react to the sensor signal until the sensor reaches approximately 600 degrees-F. This factor must be taken into consideration when evaluating the performance of the sensor.*

d) *Unleaded fuel* - *The use of unleaded fuel is essential for proper operation of the sensor. Make sure the fuel you are using is of this type.*

37   In addition to observing the above conditions, special care must be taken whenever the sensor is serviced.

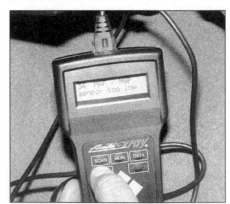

4.29 Install a SCAN tool into the ALDL and read MAF output frequency through the data stream

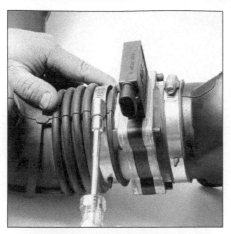

4.30 Remove the air intake duct and separate the MAF sensor clamps

a) *The oxygen sensor has a permanently attached pigtail and electrical connector which should not be removed from the sensor. Damage or removal of the pigtail or electrical connector can adversely affect operation of the sensor.*

b) *Grease, dirt and other contaminants should be kept away from the electrical connector and the louvered end of the sensor.*

c) *Do not use cleaning solvents of any kind on the oxygen sensor.*

d) *Do not drop or roughly handle the sensor.*

e) *The silicone boot must be installed in the correct position to prevent the boot from being melted and to allow the sensor to operate properly.*

Check

Refer to illustration 4.38

Note: *Access to the oxygen sensor(s) makes it very difficult but not impossible to back-probe the harness electrical connectors for testing purposes. The exhaust manifolds and pipes are extremely hot and will melt stray electrical probes and leads that touch the surface during testing. If possible, use a SCAN tool that plugs into the ALDL (diagnostic link). This tool will access the ECM/PCM data stream and indicates the millivolt changes for each individual O_2 sensor.*

38   Check the O_2 sensor millivolt signal. Locate the oxygen sensor electrical connector and carefully backprobe it using a long pin(s) into the appropriate wire terminals:

1993 through 1995 models:

Bank 1	purple/white (+) (signal
sensor (left)	wire) and black wire (-)
	(ground)
Bank 2	purple (+) (signal wire) and
sensor (right)	black (-) (ground)

1996 models:

Bank 1	purple/white (+) (signal wire)
sensor (left)	and black wire (-) (ground)
Bank 2	purple (+) (signal wire) and
sensor (right)	black wire (-) (ground)
Sensor 3	dark blue (+) (signal wire)
(before CAT)	and black wire (-) (ground)

4.38 Check the O_2 sensor millivolt signal from the single wire O_2 sensor connector (3.4L V6 engine shown)

Sensor 4	dark blue (+) (signal wire)
(after CAT)	and black wire (-) (ground)

Install the positive probe of a voltmeter onto the signal wire pin and the negative probe to the ground wire. **Note:** *Consult the wiring diagrams at the end of Chapter 12 for additional information on the oxygen sensor electrical connector wire color designations.* Monitor the voltage signal (millivolts) as the engine goes from cold to warm **(see illustration)**.

39   The oxygen sensor will produce a steady voltage signal of approximately 0.1 to 0.2 volts with the engine cold (open loop). After a period of approximately two minutes, the engine will reach operating temperature and the oxygen sensor will start to fluctuate between 0.1 to 0.9 volts (closed loop). If the oxygen sensor fails to reach the closed loop mode or there is a very long period of time until it does switch into closed loop mode, replace the oxygen sensor with a new part.

40   Also inspect the oxygen sensor heater. Disconnect the oxygen sensor electrical connector and working on the O_2 sensor side, connect an ohmmeter between the:

1993 through 1995 5.7L engines -
black wire (-) and brown wire (+)

1996 and later 5.7L engines -
black wire (-) and pink wire (+)

It should measure approximately 5 to 7 ohms. **Note:** *The 3.4L V6 engine is not equipped with heated O_2 sensors. This can be determined by the single wire. The wire colors often change from the harness to the actual O_2 sensor wires according to manufacturer's specifications. Follow the wire colors to the O_2 sensor electrical connector and determine the matching wires and their colors before testing the heater resistance.*

41   Check for proper supply voltage to the heater. Disconnect the oxygen sensor electrical connector and working on the computer side, measure the voltage between the:

1993 through 1995 5.7L engines -
black wire (-) and brown wire (+)

1996 and later 5.7L engines -
black wire (-) and pink wire (+)

4.47 Use a special slotted socket to remove the O_2 sensor from the exhaust pipe

on the oxygen sensor electrical connector. There should be battery voltage with the ignition key ON (engine not running). If there is no voltage, check the circuit between the main relay, the ECM/PCM and the sensor. **Note:** *It is important to remember that supply voltage will only last approximately 2 seconds because the system uses a relay to divert the voltage (air conditioning relay), therefore it will be necessary to have an assistant turn the ignition key ON.*

42   If the oxygen sensor fails any of these tests, replace it with a new part.

43   On the 5.7L V8 and 3.4L V6 models, access to the oxygen sensors is difficult making it impossible to monitor the SIGNAL voltage changes without removing several components to gain access to the O_2 sensor electrical connectors. Install the SCAN tool and switch to the Oxygen Sensor mode and monitor the O_2 sensor crosscounts (varying millivolt signals). The SCAN tool should indicate approximately 100 to 200 millivolts when cold (LEAN condition) and then fluctuate from 300 to 800 millivolts warm (CLOSED LOOP).

Replacement

Refer to illustration 4.47

Note: *Because it is installed in the exhaust manifold or pipe, which contracts when cool, the oxygen sensor may be very difficult to loosen when the engine is cold. Rather than risk damage to the sensor (assuming you are planning to reuse it in another manifold or pipe), start and run the engine for a minute or two, then shut it off. Be careful not to burn yourself during the following procedure.*

44   Disconnect the cable from the negative terminal of the battery. **Caution:** *On models equipped with a Delco-Loc II or Theftlock audio system, be sure the lockout feature is turned off before disconnecting the battery cable.*

45   Raise the vehicle and place it securely on jackstands.

46   Carefully disconnect the electrical connector from the sensor.

47   Carefully unscrew the sensor from the exhaust manifold **(see illustration)**.

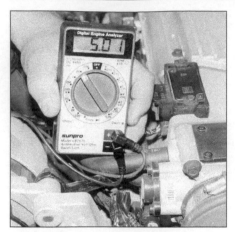

4.54 Check the TPS REFERENCE voltage on the gray (+) wire and the black (-) wire. It should be about 5.0 volts

4.55a Check the SIGNAL voltage on the dark blue (+) wire and the black (-) wire at idle and . . .

4.55b . . . then at wide open throttle. The voltage should increase smoothly to approximately 4.5 volts

48   Anti-seize compound must be used on the threads of the sensor to facilitate future removal. The threads of new sensors will already be coated with this compound, but if an old sensor is removed and reinstalled, recoat the threads. *Caution: Do not get any compound or other chemicals on the probe portion of the sensor or it will not function correctly.*
49   Install the sensor and tighten it securely.
50   Reconnect the electrical connector of the pigtail lead to the main engine wiring harness.
51   Lower the vehicle, take it on a test drive and check to see that no trouble codes set.

Throttle Position Sensor (TPS)

General description
52   The Throttle Position Sensor (TPS) is located on the end of the throttle shaft on the throttle body. By monitoring the output voltage from the TPS, the ECM/PCM can determine fuel delivery based on throttle valve angle (driver demand). A broken or loose TPS can cause intermittent bursts of fuel from the injector and an unstable idle because the ECM/PCM thinks the throttle is moving.
53   A problem in any of the TPS circuits will set a Code 21 or 22. Once a trouble code is set, the ECM/PCM will use an artificial default value for TPS and some vehicle performance will return.

Check
Refer to illustrations 4.54, 4.55a and 4.55b
54   Locate the Throttle Position Sensor (TPS) on the throttle body. Using a voltmeter, check the REFERENCE voltage from the ECM/PCM. Install the positive probe (+) onto the gray (reference wire) and negative probe (-) onto the black (ground wire) **(see illustration)**. The voltage should read approximately 5.0 volts.
55   Next, check the TPS signal voltage. With the throttle fully closed, install the positive probe (+) of the voltmeter onto the dark blue wire and the negative probe (-) onto the black wire **(see illustration)**. Gradually open the

throttle valve and observe the TPS sensor voltage. Observe a smooth change in the voltage values as the sensor travels from idle to full throttle. The voltage should increase to approximately 4.5 to 5.0 volts **(see illustration)**. If the readings are incorrect, replace the TPS sensor.
56   An alternate method of diagnosing the TPS sensor is by the use of an electronic SCAN tool. Install the SCAN tool and switch to the TPS mode and monitor the voltage signal with the engine at idle and high rpm. The SCAN tool should indicate between 1.2 to 1.6 volts at idle. Raise the engine rpm and observe that as engine rpm increases (throttle angle) the TPS voltage increases to approximately 4.5 to 5.0 volts. If the TPS sensor voltage readings are incorrect, replace the TPS sensor.

Replacement
57   Disconnect the electrical connector from the TPS.
58   Remove the Torx drive bolts from the TPS and remove the TPS from the throttle body.
59   When installing the TPS, be sure to align the socket locating tangs on the TPS with the throttle shaft in the throttle body.
60   Installation is the reverse of removal. Inspect the condition of the O-ring and replace it if necessary.

Neutral Start switch

61   The Neutral Start switch, located on the rear upper part of the automatic transmission, indicates to the ECM/PCM when the transaxle is in Park or Neutral. This information is used for Transmission Converter Clutch (TCC), Exhaust Gas Recirculation (EGR) and Idle Air Control (IAC) valve operation. **Caution:** *The vehicle should not be driven with the Neutral Start switch disconnected because idle quality will be adversely affected.*
62   For more information regarding the Neutral Start switch, which is part of the Neutral start and back-up light switch assembly, see Chapter 7B.

Air conditioning control

Air conditioning clutch control
63   During air conditioning operation, the ECM/PCM controls the application of the air conditioning compressor clutch. The ECM/PCM controls the air conditioning clutch control relay to delay clutch engagement after the air conditioning is turned ON to allow the IAC valve to adjust the idle speed of the engine to compensate for the additional load. The ECM/PCM also controls the relay to disengage the clutch on WOT (wide open throttle) to prevent excessively high rpm on the compressor. A problem in the air conditioning clutch control circuits will set a Code 61 or 66 through 71. Be sure to check the air conditioning system as detailed in Chapter 3 before attempting to diagnose the air conditioning clutch or electrical system.

Air conditioning "On" signal
Refer to illustrations 4.65 and 4.66
64   Turning on the air conditioning supplies battery voltage to the air conditioning compressor clutch of ECM/PCM electrical connector to increase idle air rate and maintain idle speed. In most cases, if the air conditioning does not function, the problem is proba-

4.65 Check for battery voltage on the air conditioning clutch relay connector

4.66 Apply battery voltage to the air conditioning clutch using a jumper wire and make sure the clutch activates

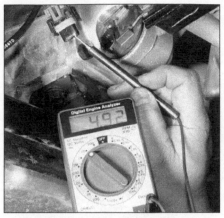

4.68 Check for REFERENCE voltage to the VSS

4.73 Measure the camshaft sensor feed voltage on the harness side of the camshaft sensor connector. It should be between 10 to 12-volts

bly related to the air conditioning system relays and switches and not the ECM/PCM.

65   Remove the air conditioning relay from the relay center and check for battery voltage to the relay **(see illustration)**. Battery voltage should exist with the ignition key ON (engine not running).

66   If battery voltage exists, install a jumper wire into the relay connector and observe that the air conditioning clutch activates **(see illustration)**. If the air conditioning clutch and relay system are working properly, check the air conditioning system pressures (see Chapter 3).

Vehicle Speed Sensor (VSS)

General description

67   The Vehicle Speed Sensor (VSS) is located in the transmission housing at the rear section near the output shaft. This permanent magnet generator sends a pulsing voltage signal to the ECM/PCM, which the ECM/PCM converts to miles per hour. The VSS is part of the Transmission Converter Clutch (TCC) system. A problem in the VSS control circuits will set a Code 24 or Code 72.

Check

Refer to illustration 4.68

68   To check the VSS, disconnect the electrical connector in the wiring harness near the sensor. Check the REFERENCE voltage from the ECM/PCM by probing the yellow (+) and the purple (-) with a voltmeter **(see illustration)**. There should be approximately 5.0 volts present. If reference voltage is present, have the VSS diagnosed by a dealer service department.

Replacement

69   To replace the VSS, detach the sensor retaining screw and bracket, unplug the sensor and remove it from the transmission.

70   Installation is the reverse of removal. Always use a new O-ring during installation.

Camshaft sensor

3.4L and 3.8L V6 engines

Note: *For additional information concerning*

the camshaft sensor circuit, refer to the wiring diagrams at the end of Chapter 12.

General description

71   The camshaft sensor is located in the top section of the timing chain cover directly in front of the intake manifold. As the camshaft turns, a magnet on the camshaft activates the Hall-effect switch in the sensor. This signal is used by the ECM/PCM to synchronize the SFI mode on the fuel injection system with the opening of the intake valves. In the event the reference signal is lost or interrupted, the ECM/PCM will continue to pulse the fuel injectors using a default value but the synchronizing effect may be slightly off (stumble or missfire) due to the lack of precision. If the cam signal is not received by the ECM/PCM, a code 17 will set.

Check

Refer to illustration 4.73 and 4.75

72   To check the cam sensor circuit, disconnect the cam sensor electrical connector. Turn the ignition key to ON but do not start the engine.

73   Measure the voltage between the pink (+) and the pink/black (-) wire terminals (3.4L) or the white/black (+) and the red/black (-) wire terminals (3800). The voltage should be approximately 10 to 12-volts **(see illustration)**.

74   If there is no voltage present, check the ignition fuse in the fuse box. If the fuse is good, check the circuit between the cam sensor and the fuse box (3.4L) or ignition module (3800). Also check the ground circuit from the sensor to the ECM/PCM or ignition module.

75   If the sensor feed and ground circuits are good, reconnect the sensor electrical connector and backprobe the signal circuit brown/white wire terminal with the positive probe of a voltmeter, connect the negative probe to a good engine ground. The signal voltage should be approximately 4 volts **(see illustration)**.

76   Using a breaker bar and socket on the crankshaft pulley bolt, rotate the crankshaft at least two full turns. Observe that the voltage drops to 0-volts as the camshaft timing

mark passes the sensor, indicating the sensor is sending a signal to the ECM/PCM.

77   If the sensor signal voltage is not correct, the sensor is probably defective. Have the cam sensor diagnosed by a dealer service department or other qualified repair shop.

78   It's also possible to observe the actual camshaft sensor signal while the engine is running. A SCAN tool is available from automotive parts stores and specialty tool companies that can be plugged into the ALDL for the purpose of monitoring the computer and the sensors. Install the SCAN tool and switch to the Camshaft sensor mode and monitor the voltage signal from the cam sensor.

Replacement

79   Disconnect the negative terminal from the battery.

80   Remove the serpentine drivebelt (see Chapter 1, Section 21).

81   Disconnect the electrical connector from the cam sensor.

82   Remove the bolt from the camshaft sensor and lift the sensor from the timing cover.

83   Installation is the reverse of removal. Be sure to torque the camshaft sensor bolt to the torque listed in this Chapter's Specifications.

4.75 Measure the camshaft sensor signal voltage. It should pulse between 4-volts and 0-volts with the engine cranking

4.96a Carefully pry the plastic protector from the timing cover studs lifting each side evenly to avoid damage

4.96b Remove the two bolts (arrows) and lift the crankshaft sensor from the timing cover (3800 engine shown)

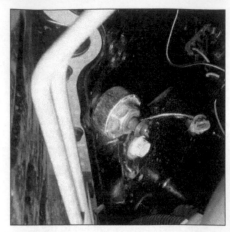

4.103 Squeeze the electrical connector to release it from the fluted portion of the knock sensor

5.7L V8 engines, 1998 and later

84 Late-model V8 engines utilize a camshaft position sensor on the back of the block, in the location formerly occupied by the distributor.
85 Disconnect the negative battery cable.
86 Refer to Chapter 2D and remove the intake manifold.
87 Disconnect the electrical connector at the CMP.
88 Remove the bolt and pull the CMP straight up to remove it.
89 Installation is the reverse of the removal procedure.

Crankshaft position sensor

3.4L and 3.8L V6 engines

Note: *For additional information concerning the crankshaft sensor circuit, refer to the wiring diagrams at the end of Chapter 12*

General description

90 The 3.4L V6 system is equipped with two crankshaft sensors. The 24X signal crankshaft position sensor is mounted on the engine front cover and partially behind the vibration damper. This sensor reads 24 positions on the crankshaft balancer assembly and is used for ignition control during cranking and idle control. The 3X crankshaft position sensor is mounted on the right side of the engine block near mid-engine near the crankshaft reluctor ring. Both sensors are Hall-effect switches, each equipped with one shared magnet mounted between them. The magnet and each Hall-effect switch are separated by an air gap. The vibration damper is equipped with specially designed blades or "interrupter rings" to create the ON OFF ON pulses necessary for the timing sequence. Although there is not an ignition timing procedure for the DIS system, the position of the crank sensor is very important. The sensor must not contact the rotating interrupter rings or the sensor will be damaged and the engine will shut down.

Check

91 To check the crankshaft sensor, it is necessary to use a special SCAN tool (Tech 1). Have the sensor diagnosed by a dealer service department or other qualified repair shop.

Replacement

Refer to illustrations 4.96a and 4.96b

92 Disconnect the negative terminal from the battery.
93 Remove the serpentine drivebelt (see Chapter 1).
94 Disconnect the electrical connector from the crankshaft sensor.
95 Remove the crankshaft vibration damper retaining bolt and damper (see Chapter 2B, Section 11).
96 Remove the protector and bolts from the crankshaft sensor and lift the sensor from the timing cover **(see illustrations)**.
97 Installation is the reverse of removal. Tighten the bolts to the torque listed in this Chapter's Specifications.

V8 engines

98 1998 and later 5.7L V8 engines have a crankshaft position sensor on the right side of the engine block.

4.106a Remove the bolts and lift the cover to expose the knock sensor module

99 Disconnect the negative battery cable, then raise the front of the vehicle and support it securely on jackstands.
100 Refer to Chapter 5 and remove the starter motor for access to the CKP.
101 Disconnect the electrical connector at the CKP, then remove the bolt and the sensor.
102 Installation is the reverse of the removal procedure.

Knock sensor

Note: *2001 and later models have two knock sensors. V6 engines have one on each side of the block, while V8 models have two under the intake manifold.*

General description

Refer to illustration 4.103

103 The knock sensor is located in the engine block near the crankshaft position sensor about mid-engine **(see illustration)**. Octane ratings vary the performance of engines and often detonation leads to "spark knock." To control spark knock, the knock control system detects abnormal vibration in the engine. This system is designed to reduce spark knock up to 10 degrees during

4.106b Pinch the tabs and lift the knock sensor module from the ECM/PCM

periods of heavy detonation. This allows the engine to use maximum spark advance to improve driveability. The knock sensor produces an AC output voltage which increases with the severity of the knock. The signal is fed into the ECM/PCM and the timing is retarded up to 10 degrees to compensate for the severe detonation. Any problems with the knock sensor circuit will set a code 43.

Check

Refer to illustrations 4.106a and 4.106b
104 Disconnect the electrical connector for the knock sensor and with the ignition key ON (engine not running), check the REFERENCE voltage from the computer using a voltmeter. It should be approximately 5.0 volts.
105 Disconnect the electrical connector for the knock sensor and using an ohmmeter, check the resistance of the sensor. It should be between 3.3K and 4.5K-ohms on the 3.4L V6 and 5.7L V8. The 3800 V6 has two knock sensors, one at each bank, and the resistance of each sensor is approximately 100K-ohms.
106 If the knock sensor resistance is correct but there is no REFERENCE voltage available, have the ECM/PCM and knock sensor module checked at a dealer service department **(see illustrations)**.

Replacement

V6 models

107 Drain the cooling system (see Chapter 1). The engine must be completely cool.
108 Disconnect the electrical connector at the sensor.
109 Unscrew the sensor. **Caution:** *Some coolant may come out of the block when the sensor is removed, so have eye protection.*
110 Installation is the reverse of the removal procedure. Do NOT use any types of sealant on the threads of the sensor.

V8 models

111 2000 and later V8 models have two knock sensors, each located in the block valley cover, under the intake manifold.
112 Refer to Chapter 2, Part D and remove the intake manifold.
113 Disconnect the electrical connectors from the sensors and remove the two sensors from the block valley cover.
114 Installation is the reverse of the removal procedure. Do NOT use any types of sealant on the threads of the sensor.

---

5   Exhaust Gas Recirculation (EGR) system

Non-digital EGR system (1997 and earlier 5.7L V8 engine)

General description

1   The system meters exhaust gases into the engine induction system through pas-

5.7 Remove the vacuum line and check for manifold vacuum to the EGR control solenoid

sages cast into the intake manifold. From there the exhaust gases pass into the fuel/air mixture for the purpose of lowering combustion temperatures, thereby reducing the amount of oxides of nitrogen (NOx) formed.
2   The amount of exhaust gas admitted is regulated by a vacuum or backpressure controlled (EGR) valve in response to engine operating conditions. The EGR valve is under the control of the EGR vacuum control solenoid valve, which, in turn, is under the control of the ECM/PCM. The vacuum signal to the EGR valve is controlled by varying the duty cycle (on-time) of the solenoid valve. The duty cycle is calculated by the computer using information from the ECT, MAF, VSS and IAT sensors. Problems with the EGR control circuit or the EGR valve will set Codes 27 or 32.
3   Common engine problems associated with the EGR system are rough idling or stalling at idle, rough engine performance during light throttle application and stalling during deceleration.

Check

Refer to illustration 5.7
4   Start the engine, warm it up to normal operating temperature and allow the engine to idle. Manually lift the EGR diaphragm. The engine should stall, or at least the idle should drop considerably. **Note:** *The EGR valve is located on the backside of the intake manifold under the hood cowl.*
5   If the EGR valve appears to be in proper operating condition, carefully check all hoses connected to the valve for breaks, leaks or kinks. Replace or repair the valve/hoses as necessary.
6   With the engine idling at normal operating temperature, disconnect the vacuum hose from the EGR valve and connect a vacuum pump. When vacuum is applied the engine should stumble or die, indicating the vacuum diaphragm is operating properly. **Note:** *Some models use a backpressure-type EGR valve. On models so equipped, backpressure must be created in the exhaust sys-*

tem before the vacuum pump will actuate the valve. To create backpressure, install a large socket into the tailpipe and clamp it to the pipe to prevent it from dropping out during the test. The 1/2 inch hole in the socket will allow the engine to idle but still create backpressure on the EGR valve. Don't restrict the exhaust system any longer than necessary to perform this test. Replace the EGR valve with a new one if the test does not affect the idle.
7   Check the operation of the EGR control solenoid. Check for vacuum to the solenoid **(see illustration)**. If vacuum exists, check for battery voltage to the solenoid.
8   Further testing of the EGR system, vacuum solenoid and the ECM/PCM will require a SCAN tool to access computer information that directly controls the EGR system. Have the vehicle checked by a dealer service department or other qualified repair facility.

Component replacement

EGR valve

9   Disconnect the vacuum hose at the EGR valve.
10   Remove the nuts or bolts which secure the valve to the intake manifold or adapter.
11   Lift the EGR valve from the engine.
12   Clean the mounting surfaces of the EGR valve. Remove all traces of gasket material.
13   Place the new EGR valve, with a new gasket, on the intake manifold or adapter and tighten the attaching nuts or bolts.
14   Connect the vacuum signal hose.

EGR vacuum solenoid

15   Remove the hoses from the EGR vacuum control solenoid, labeling them to ensure proper installation.
16   Remove the solenoid and replace it with the new one.

EGR valve cleaning

17   With the EGR valve removed, inspect the passages for excessive deposits.
18   It is a good idea to place a rag securely in the passage opening to keep debris from entering. Clean the passages by hand, using a drill bit.

Digital EGR valve system (3.4L V6 engine)

General description

19   The digital EGR valve feeds small amounts of exhaust gas back into the intake manifold and then into the combustion chamber.
20   The digital EGR valve is designed to accurately supply exhaust gas to the engine, independent of intake manifold vacuum. The valve controls EGR flow from the exhaust to the intake manifold through three orifices, which increment in size, to produce seven combinations. When a solenoid is energized, the armature, with attached shaft and swivel pintle, is lifted, opening the orifice. The flow accuracy is dependent on metering orifice size only, which results in improved control.
21   The digital EGR valve is opened by the

5.34 Remove the bolts from the base of the Linear EGR valve

6.8 Check for battery voltage to the AIR pump relay

6.9 Jump terminals E1 and E4 to activate the AIR pump

ECM/PCM, grounding each solenoid circuit. This activates the solenoid, raises the pintle, and allows exhaust gas flow into the intake manifold. The exhaust gas then moves with the air/fuel mixture into the combustion chamber. Problems with the Digital EGR control circuit or the EGR valve will set Codes 75, 76 or 77.

Check

22   Special electronic diagnostic equipment is needed to check this valve and should be left to a dealer service department or other qualified repair facility.

Replacement

23   Disconnect the electrical connector from the EGR valve.
24   Remove the two mounting bolts and remove the EGR valve from the intake manifold.
25   Remove the EGR valve and gasket.
26   Clean the mounting surface of the EGR valve. Remove all traces of gasket material from the intake manifold and from the valve, if it is to be reinstalled. Clean both mating surfaces with a cloth dipped in lacquer thinner or acetone.
27   Install a new gasket and the EGR valve and tighten the bolts securely.
28   Connect the electrical connector onto the EGR valve.

Linear EGR valve system (3800 V6 and 1998 and later 5.7L V8 engine)

General description

29   The linear EGR valve feeds small amounts of exhaust gas back into the intake manifold and then into the combustion chamber independent of intake manifold vacuum.
30   The valve controls exhaust gas flow from the exhaust to the intake manifold through a single orifice with a ECM/PCM controlled pintle. This EGR valve operates similar to the stepper motor type particular to IAC valves that control idle quality.

31   If the EGR system will not allow the ECM/PCM to control the position of the EGR valve pintle, it will set a diagnostic trouble code in the OBD II system. This system can be accessed using a special TECH 1 SCAN tool at the local dealer repair and service center.

Check

32   The vehicles that are equipped with the 3800 engine and the liner EGR valve are equipped with the OBD II system. Special electronic diagnostic equipment is needed to check this valve and should be left to a dealer service department or other qualified repair facility.

Replacement

Refer to illustration 5.34

33   Disconnect the electrical connector from the EGR valve.
34   Remove the two mounting bolts and remove the EGR valve from the intake manifold **(see illustration)**.
35   Remove the EGR valve and gasket.
36   Clean the mounting surface of the EGR valve. Remove all traces of gasket material from the intake manifold and from the valve if it is to be reinstalled. Clean both mating surfaces with a cloth dipped in lacquer thinner or acetone. On 1998 and later V8 models, it may be necessary to remove the EGR pipe from the intake and exhaust manifolds for other procedures in this manual. If so, remove the two bolts securing the EGR pipe to the exhaust manifold and the single bolt securing the EGR pipe to the top of the intake manifold. Pull the pipe from the top of the intake manifold and remove the bolts securing the the pipe bracket to the right cylinder head. Be sure to reinstall the EGR pipe assembly before installing the EGR valve on these models.
37   Install a new gasket and the EGR valve and tighten the bolts securely.
38   Connect the electrical connector onto the EGR valve.

6   Secondary Air Injection (AIR) system

Refer to illustrations 6.8, 6.9 and 6.13

General description

1   The AIR system helps reduce hydrocarbons and carbon monoxide levels in the exhaust by injecting air into the exhaust ports of each cylinder during cold engine operation, or directly into the catalytic converter during normal operation. It also helps the catalytic converter reach proper operating temperature quickly during warm-up.
2   The AIR system uses an electrically operated air pump to force the air into the exhaust stream. The AIR pump is controlled by the ECM/PCM. Battery voltage to the AIR pump is controlled by a relay. An integral stop valve prevents air flow through the pump during OFF periods. Check valves prevent backflow of exhaust gases into the pump in the event of an exhaust backfire. The ECM/PCM turns the AIR pump ON when the coolant temperature has reached 59 degrees-F or until the system enters closed loop mode.

6.13 Remove the mounting bolts from the AIR pump (top bolt visible)

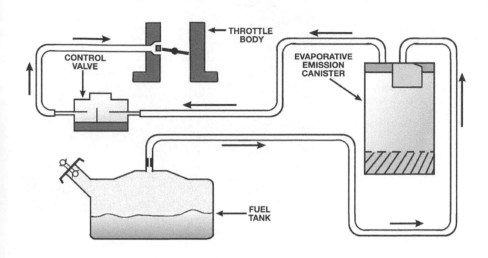

7.2 Details of a typical EVAP system

7.7a Remove the inner fender cover to expose the charcoal canister

7.7b The pressure control valve should hold at least 5 in-Hg vacuum

3   The following components are utilized in the AIR system: an electrical air pump; check valves, air injection pipes, air flow and control hoses; and a dual bed catalytic converter.

Check

AIR system

4   Because of the complexity of this system it is difficult for the home mechanic to make a proper diagnosis. If the system is suspected of not operating properly, individual components can be checked.

5   Begin any inspection by carefully checking all hoses, vacuum lines and wires. Be sure they are in good condition and that all connections are tight and clean.

6   To check the pump, allow the engine to reach normal operating temperature and then idle speed. Locate the hose running from the air pump and squeeze it to feel the pulsation. Have an assistant increase the engine speed and check for a parallel increase in airflow. If this is observed as described, the pump is functioning properly. If it is not operating in this manner, a faulty pump is indicated. If OK, raise the rpm above 2,850 rpm and check to make sure the ECM/PCM turns OFF the AIR system,

7   The check valve can be inspected by first removing it from the air line. Attempt to blow through it from both directions. Air should only pass through it in the direction of normal airflow. If it is either stuck open or stuck closed the valve should be replaced.

8   Remove the AIR system relay **(see illustration)** and with the ignition key ON (engine not running) check for power to the relay. Battery voltage should be available.

9   Place a jumper wire between E1 and E4 on the relay electrical connector **(see illustration)**. This should activate the AIR pump. If the pump activates with the jumper wire and not with the relay in place, replace the relay with a new part.

10   If the jumper wire does not activate the AIR pump, check the wiring harness (see Chapter 12).

AIR pump replacement

1997 and earlier

11   Remove the serpentine drivebelt (see Chapter 1).

12   Disconnect the ignition coil primary and secondary harness connectors (see Chapter 5).

13   Remove the mounting bolts **(see illustration)**.

14   Remove the AIR pump from the engine compartment.

15   Installation is the reverse of removal. Install the drivebelt.

1998 and later

16   Raise the vehicle and support securely on jackstands.

17   Remove the lower close out panel just forward of the left front wheel.

18   Disconnect the electrical connectors from the air pump, detach the mounting three bolts and remove the air pump from the vehicle.

19   Installation is the reverse of removal. Install the drivebelt.

7   Evaporative Emissions Control System (EECS)

General description

Refer to illustration 7.2

1   This system is designed to trap and store fuel that evaporates from the throttle body and fuel tank which would normally enter the atmosphere and contribute to hydrocarbon (HC) emissions.

2   The system consists of a charcoal-filled canister and lines running to and from the

canister. These lines include a vent line from the gas tank, a vent line from the throttle body, an EVAP pressure control valve, an EVAP purge control solenoid (electronic), a charcoal canister and the manifold **(see illustration)**. In addition, there is a purge solenoid valve in the canister. The ECM/PCM controls the vacuum to the purge solenoid valve with an electrically operated solenoid. The fuel tank cap is also an integral part of the system.

3   An indication that the system is not operating properly is a strong fuel odor.

Check

Refer to illustrations 7.7a and 7.7b

4   Maintenance and replacement of the charcoal canister filter is covered in Chapter 1.

5   Check all lines in and out of the canister for kinks, leaks and breaks along their entire lengths. Repair or replace as necessary.

6   Check the gasket in the gas cap for signs of drying, cracking or breaks. Replace the gas cap with a new one if defects are found.

7   With a hand-held vacuum pump, apply approximately 15 in-Hg to the control vacuum tube located on the pressure control valve **(see illustrations)**. After 10 seconds

there should be about 5 in-Hg vacuum remaining. **Note:** *Be sure the vacuum pump does not leak internally and there all hoses are sealed tight.* If the vacuum leaks down faster and more than the specified amount, replace the valve.

8   The computer operates the purge control solenoid by changing its frequency signal. If there is no obvious sounds from the purge control solenoid (buzzing), have it checked by a dealer service department. Remember, the purge control solenoid will not be activated by the computer until fuel tank pressure exceeds 0.7 psi.

Component replacement

9   Replacement of the canister filter is covered in Chapter 1.
10   When replacing any line running to or from the canister, make sure the replacement line is a duplicate of the one you are replacing. These lines are often color coded to denote their particular usage.

---

8   Positive Crankcase Ventilation (PCV) system

General description

1   The positive crankcase ventilation system reduces hydrocarbon emissions by circulating fresh air through the crankcase to pick-up blow-by gases, which are then rerouted through the carburetor to be burned in the engine.
2   The main components of this system are vacuum hoses and a PCV valve, which regulates the flow of gases according to engine speed and manifold vacuum.

Check and component replacement

3   Checking the system and PCV valve replacement are covered in Chapter 1.

---

9   Transmission Converter Clutch (TCC) system

General information

1   The purpose of the Torque Converter Clutch (TCC) system, equipped in automatic transmissions, is to eliminate the power loss of the torque converter stage when the vehicle is in the cruising mode (usually above 35 mph). This economizes the automatic trans-

10.1 Location of the catalytic converter

mission to the fuel economy of the manual transmission. The lock-up mode is controlled by the ECM/PCM through the activation of the TCC apply solenoid which is built into the automatic transmission. When the vehicle reaches a specified speed, the ECM/PCM energizes the solenoid and allows the torque converter to lock-up and mechanically couple the engine to the transmission, under which conditions emissions are at their minimum. However, because of other operating condition demands (deceleration, passing, idle, etc.), the transmission must also function in its normal, fluid-coupled mode. When such latter conditions exist, the solenoid de-energizes, returning the torque converter to normal operation. The converter also returns to normal operation whenever the brake pedal is depressed.

Check

2   Due to the requirement of special diagnostic equipment for the testing of this system, and the possible requirement for dismantling of the automatic transmission to replace components of this system, checking and replacing of the components should be handled by a dealer service department or other qualified repair facility.

---

10   Catalytic converter

General description

Refer to illustration 10.1
1   The catalytic converter is an emission control device added to the exhaust system to reduce pollutants from the exhaust gas stream. These systems are equipped with a

single bed monolith catalytic converter. This monolithic converter contains a honeycomb mesh which is also coated with two types of catalysts. One type is the oxidation catalyst while the other type is a three-way catalyst that contains platinum and palladium. The three-way catalyst lowers the levels of oxides of nitrogen (NOx) as well as hydrocarbons (HC) and carbon monoxide (CO) emissions. The oxidation catalyst lowers the levels of hydrocarbons and carbon monoxide **(see illustration)**.

Check

2   The test equipment for a catalytic converter is expensive and highly sophisticated. If you suspect the converter is malfunctioning, take it to a dealer service department or authorized emissions inspection facility for diagnosis and repair.
3   Whenever the vehicle is raised for service of underbody components, check the converter for leaks, corrosion and other damage. If damage is discovered, the converter should be replaced.
4   Because the converter is welded to the exhaust system, converter replacement requires removal of the exhaust pipe assembly (see Chapter 4). Take the vehicle, or the exhaust system, to a dealer service department or a muffler shop.

Chapter 7 Part A
Manual transmission

Contents

Specifications

Torque specifications

Ft-lbs (unless otherwise indicated)

Back-up light switch	28
Reverse lockout housing hold-down screw	156 in-lbs
Skip shift solenoid	30
Transmission-to-bellhousing bolts	
RPO M49 (five-speed)	55
RPO MM6 (six-speed)	
Through 1999	26
2000 and later	37

1   General information

The vehicles covered by this manual are equipped with a five- or six-speed manual transmission or a four-speed automatic. Information on the two manual transmissions is included in this part of Chapter 7. You'll find information on the automatic transmission in Part B of this Chapter.

Models with a V6 engine use an RPO M49 five-speed transmission; models with a V8 engine use an RPO MM6 six-speed transmission. Neither transmission requires shift linkage adjustment because there is no shift linkage. The shift lever, which is mounted on top of the extension housing on both transmissions, can be unbolted and replaced without removing the transmission.

Both transmissions are equipped with a back-up light switch and a speed sensor. On the M49, the back-up light switch and speed sensor are located on the left side of the transmission housing. On the MM6, the speed sensor is located on the left side and the back-up light switch is located on the right side. The procedure for checking and replacing the back-up light switch are con-tained in this Chapter. The procedure for checking and replacing the speed sensor is in Chapter 6.

The MM6 has two unique features. A reverse lockout solenoid operates a reverse lockout mechanism to prevent accidental shifting into reverse when the vehicle is moving forward at a speed of three or more mph. A "skip shift" solenoid provides better fuel economy and helps V8 models comply with Federal fuel economy standards. When you shift out of first gear, second and third gears are inhibited by the use of an electric solenoid when all three of the following con-

3.7 The back-up light switch on 5-speeds is located on the left side of the transmission (on 6-speeds it's on the right side)

ditions are present: coolant temperature is above 162 degrees Fahrenheit, vehicle speed is between 15 and 21 mph and the throttle is opened 35 percent or less. A pair of simple checks for these two solenoids are included in this Chapter.

Both the five and six-speed transmissions use Dexron II or Dexron III ATF instead of gear lubricant. Refer to Chapter 1 for information on changing transmission fluid.

2   Shift lever - removal and installation

1   Remove the console trim panel (see Chapter 11).
2   On V8 models, remove the shift lever handle retaining bolts and remove the handle.
3   Remove the shift lever boot retaining screws and remove the boot.
4   Remove the shift lever base retaining bolts.
5   Remove the shift lever assembly.
6   Remove all old sealant from the mating surface on top of the extension housing and, if you're installing the old shift lever assembly, any sealant on the mating surface of the shift lever base. Apply a 1/8-inch wide bead of RTV sealant to the mating surface of the extension housing.
7   Installation is otherwise the reverse of removal.

3   Back-up light switch - check and replacement

Check

Note: *On V6 models, the back-up light switch is located on the left side of the transmission housing; on V8 models, it's located on the right side.*
1   With the engine running and the lights

on, depress the clutch and place the shift lever in Reverse. The back-up lights should come on.
2   If the back-up lights don't come on, either a bulb is burned out, there's a short, ground or open in the circuit, or the back-up light switch is bad.
3   Check the bulbs for the back-up light circuit. If either bulb is bad, replace it (see Chapter 12). If the bulbs are good, check the circuit.
4   Using a voltmeter, verify that voltage is getting to the switch and that, when the shift lever is placed in Reverse, there is voltage to the bulb. If the bulbs are okay and the circuit is okay, check the switch.
5   Raise the vehicle and place it securely on jackstands. Using an ohmmeter, verify that there is continuity through the switch when it's closed, i.e. the shift lever is placed in Reverse. If the switch is bad, replace it.

Replacement

Refer to illustration 3.7
6   Raise the vehicle and place it securely on jackstands.
7   Unplug the electrical connector from the back-up light switch **(see illustration)**.
8   Unscrew the switch.
9   Installation is the reverse of removal. Be sure to apply a coat of thread sealant or wrap the threads with Teflon tape before installing the new switch. Tighten the switch to the torque listed in this Chapter's Specifications.

4   Reverse lockout solenoid (6-speed) - check and replacement

Check

Note: *The reverse lockout solenoid is located on the left side of the transmission extension housing.*
1   Raise the vehicle and place it securely on jackstands.
2   Locate the reverse lockout solenoid and unplug the electrical connector .
3   Apply battery voltage to the solenoid and verify that it clicks. If it clicks, the solenoid is okay. If it doesn't click, check for power to the solenoid. If there is no power to the solenoid, check the fuses and the circuit. If voltage is present, remove the solenoid and reverse lockout housing as a single assembly.
4   Apply battery voltage to the solenoid. The reverse lockout plunger should be easy to depress with your thumb.
   a) *If the plunger is easy to push in, the solenoid and the reverse lockout assembly are okay.*
   b) *If the plunger isn't easy to push in, disassemble and clean the reverse lockout assembly.*
   c) *If, after cleaning, the plunger still can't be easily depressed when the solenoid is energized, replace the reverse lockout assembly.*

Replacement

5   Unplug the electrical connector from the reverse lockout solenoid.
6   Remove the reverse lockout housing hold-down screw and remove the reverse lockout assembly.
7   Installation is the reverse of removal. Tighten the reverse lockout housing hold-down screw to the torque listed in this Chapter's Specifications.

5   Skip shift solenoid (6-speed) - check and replacement

Check

Note: *The skip shift solenoid is located on the left side of the transmission.*
1   Raise the vehicle and place it securely on jackstands.
2   Locate the skip shift solenoid and unplug the electrical connector (it looks like the back-up light switch on the 5-speed transmission, shown in **illustration 3.7**).
3   Apply battery voltage to the skip shift solenoid and verify that it clicks.
   a) *If the skip shift solenoid clicks when energized, check the skip shift solenoid circuit.*
   b) *If the skip shift solenoid doesn't click when energized, replace it.*

Replacement

4   Unplug the electrical connector.
5   Unscrew the skip shift solenoid.
6   Installation is the reverse of removal. Be sure to tighten the skip shift solenoid to the torque listed in this Chapter's Specifications.

6   Transmission - removal and installation

Removal

1   Disconnect the cable from the negative battery terminal. **Caution:** *On models equipped with a Delco Loc II or Theftlock audio system, be sure the lockout feature is turned off before performing any procedure which requires disconnecting the battery.*
2   Remove the center console trim panel (see Chapter 11).
3   On V8 models (six-speeds), remove the shift lever handle (see Section 2).
4   Raise the vehicle and place it securely on jackstands.
5   Drain the transmission lubricant (see Chapter 1).
6   Disconnect the driveshaft from the transmission (see Chapter 8).
7   Support the rear axle with a jackstand and remove the torque arm (see Chapter 10).
8   Remove the catalytic converter hanger assembly and on V6 models the converter (see Chapter 6). On five speed transmissions remove the bellhousing to oil pan support braces if equipped.

9 On all transmissions, unplug the electrical connectors from the back-up light switch (see Section 3) and the speed sensor (see Chapter 6). On V8 models (six-speeds), unplug the electrical connectors for the reverse lockout solenoid (see Section 4) and the skip shift solenoid (see Section 5). Also remove the starter motor (see Chapter 5) and any bellhousing covers.

10 On 1997 and earlier models (except 3800 V6), remove the clutch pedal-to-master cylinder pushrod, then remove the clutch release cylinder (see Chapter 8). Support the release cylinder with a piece of wire; don't let it hang by the clutch hydraulic fluid hose. On 1997 and earlier 3800 V6 models and all 1998 and later models, detach the clutch hydraulic line at the bellhousing.

11 Support the engine assembly with a floor jack (place a block of wood on the jack head to protect the engine oil pan). Support the transmission with another floor jack - preferably one equipped with a transmission holding fixture.

12 Remove the transmission crossmember and mount (see Chapter 7B, Section 9).

13 On V6 models, lower the transmission far enough to reach the shift lever base retaining screws. Reach these screws and disengage the shift lever assembly from the top of the extension housing (see Section 2). Have a helper hold the shift lever assembly to prevent it from falling when the transmission is lowered. If the shift lever assembly falls, it could be damaged. Lower the transmission far enough to remove the shift lever assembly.

14 Remove the bolts securing the bellhousing to engine. Pull the transmission and the bellhousing to the rear and separate it from the engine. Lower the transmission on the jack and wheel it out from under the vehicle.

Installation

15 Remove all old sealant from the extension housing and, on V6 models (five-speeds), from the shift lever assembly base.

16 On V6 models (five-speeds), carefully maneuver the shift lever assembly into its installed position and support it with a piece of wire or have a helper hold it there.

17 Wheel the transmission into position on a jack, raise it up and, with the aid of a helper, carefully move it forward and guide the input shaft into the clutch disc hub assembly and pilot bearing. Make sure the input shaft is fully seated into the pilot bearing. You'll know if it isn't, because the transmission case won't seat flat against the bellhousing. If you have trouble inserting the input shaft through the clutch hub or seating it into the pilot bearing, carefully wiggle the transmission. You may even have to rotate the engine or input shaft slightly to align the splines on the input shaft with the splines on the inside of the clutch hub. But do NOT try to force the input shaft through the clutch hub and into the pilot bearing, or you will damage something.

18 Install the transmission-to-bellhousing bolts and tighten them to the torque listed in this Chapter's Specifications.

19 On V6 models (five-speeds), apply a 1/8-inch wide bead of RTV sealant to the extension housing-to-shift lever base mating surface, then raise up the transmission far enough to install the shift lever assembly (see Section 2).

20 Raise the transmission all the way up and install the mount and crossmember. Tighten all fasteners securely.

21 Remove the jack stands from the engine and transmission.

22 On 1997 and earlier models (except 3800 V6), push the clutch release fork up to engage it with the release bearing. Install the actuator spacer, the actuator and the actuator mounting nuts. Tighten the nuts to the torque listed in the Chapter 8 Specifications. On 1998 and later models reattach the clutch hydraulic line to the clutch actuating cylinder

23 Reconnect all electrical connectors.

24 Reattach the catalytic converter hanger (see Chapter 6).

25 Reattach the torque arm (see Chapter 10). Remove the jack supporting the rear axle.

26 Reconnect the driveshaft (see Chapter 8).

27 Check and refill the transmission with the specified fluid (see Chapter 1).

28 Remove the jackstands and lower the vehicle.

29 Install the center console trim panel (see Chapter 11).

7 Transmission overhaul - general information

If your transmission reaches the end of its service life, you can save a great deal of money by removing and installing it yourself, but you're better off leaving the overhaul to a transmission repair shop. Better yet, buy a rebuilt unit from a dealer parts department or an auto parts store. Rebuilding a manual transmission is a difficult job involving the disassembly and reassembly of many parts. The cost in time and money to overhaul a transmission yourself will almost surely exceed the cost of a rebuilt unit.

Nevertheless, it's not impossible for an inexperienced mechanic to rebuild a transmission, if the special tools are available and the job is done in a deliberate step-by-step manner so nothing is overlooked.

The tools needed for an overhaul include internal and external snap-ring pliers, a bearing puller, a slide hammer, a set of pin punches, a dial indicator and a hydraulic press. You will also need a large, sturdy workbench and a vise or transmission stand.

During disassembly of the transmission, make careful notes of how each piece comes off, where it fits in relation to other pieces and what holds it in place. When removing parts, note how they're installed; this will make it easier to reassemble the transmission correctly.

Before taking the transmission apart for repair, it will help if you have some idea what area of the transmission is malfunctioning. Certain problems can be closely tied to specific areas in the transmission, which can make component examination and replacement easier. Refer to the *Troubleshooting* section at the front of this manual for information regarding possible sources of trouble.

Notes

Chapter 7 Part B
Automatic transmission

Contents

Specifications

Torque specifications

	Ft-lbs
Transmission-to-engine bolts	
V6	70
V8	35
Torque converter-to-driveplate bolts	46

1 General information

All vehicles covered in this manual come equipped with a four or five-speed manual transmission or an automatic transmission. All information on the automatic transmission is included in this Part of Chapter 7. Information on the manual transmission can be found in Part A of this Chapter. You'll also find two procedures common to both automatic and manual transmissions - oil seal replacement and transmission mount removal and installation - here in Part B.

All models with an automatic transmission use the same Hydra-matic 4L60-E electronic four-speed unit. This transmission is equipped with a lock-up torque converter that engages in high gear. The lock-up torque converter provides a direct connection between the engine and the drive wheels for improved efficiency and economy. The lock-up converter consists of a solenoid-controlled clutch on the torque converter that engages to lock up the converter in high gear.

Due to the complexity of the automatic transmissions covered in this manual and the need for specialized equipment to perform most service operations, this Chapter contains only general diagnosis, routine maintenance, adjustment and removal and installation procedures.

If the transmission requires major repair work, it should be left to a dealer service department or an automotive or transmission repair shop. You can, however, remove and install the transmission yourself and save the expense, even if the repair work is done by a transmission shop.

2 Diagnosis - general

Note: *Automatic transmission malfunctions may be caused by five general conditions: poor engine performance, improper adjustments, hydraulic malfunctions, mechanical malfunctions or malfunctions in the computer or its signal network. Diagnosis of these problems should always begin with a check of the easily repaired items: fluid level and condition (see Chapter 1) and shift linkage adjustment. Next, perform a road test to determine if the problem has been corrected or if more diagnosis is necessary. If the problem persists after the preliminary tests and corrections are completed, additional diagnosis should be done by a dealer service department or transmission repair shop. Refer to the* Troubleshooting *section at the front of this manual for information on symptoms of transmission problems.*

Preliminary checks

1   Drive the vehicle to warm the transmission to normal operating temperature.
2   Check the fluid level as described in Chapter 1:

 a) *If the fluid level is unusually low, add enough fluid to bring the level within the designated area of the dipstick, then check for external leaks (see below).*

 b) *If the fluid level is abnormally high, drain off the excess, then check the drained fluid for contamination by coolant. The presence of engine coolant in the automatic transmission fluid indicates that a failure has occurred in the internal radiator walls that separate the coolant from the transmission fluid (see Chapter 3).*

 c) *If the fluid is foaming, drain it and refill the transmission, then check for coolant in the fluid or a high fluid level.*

3   Check the engine idle speed. **Note:** *If the engine is malfunctioning, do not proceed with the preliminary checks until it has been repaired and runs normally.*

Fluid leak diagnosis

4   Most fluid leaks are easy to locate visually. Repair usually consists of replacing a seal or gasket. If a leak is difficult to find, the following procedure may help.
5   Identify the fluid. Make sure it's transmission fluid and not engine oil or brake fluid (automatic transmission fluid is a deep red color).
6   Try to pinpoint the source of the leak. Drive the vehicle several miles, then park it over a large sheet of cardboard. After a minute or two, you should be able to locate the leak by determining the source of the fluid dripping onto the cardboard.
7   Make a careful visual inspection of the suspected component and the area immediately around it. Pay particular attention to gasket mating surfaces. A mirror is often helpful for finding leaks in areas that are hard to see.
8   If the leak still cannot be found, clean the suspected area thoroughly with a degreaser or solvent, then dry it.
9   Drive the vehicle for several miles at normal operating temperature and varying speeds. After driving the vehicle, visually inspect the suspected component again.
10   Once the leak has been located, the cause must be determined before it can be properly repaired. If a gasket is replaced but the sealing flange is bent, the new gasket will

3.4 Remove the extension housing seal with a large screwdriver or with a seal removal tool

3.6 Drive the new seal into the extension housing with a large socket

3.8 To remove the vehicle speed sensor, unplug the electrical connector and remove the hold-down bolt (arrows)

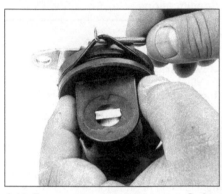

3.11 Remove the old speed sensor O-ring and install a new one

not stop the leak. The bent flange must be straightened.

11   Before attempting to repair a leak, check to make sure that the following conditions are corrected or they may cause another leak. **Note:** *Some of the following conditions cannot be fixed without highly specialized tools and expertise. Such problems must be referred to a transmission repair shop or a dealer service department.*

Gasket leaks

12   Check the pan periodically. Make sure the bolts are tight, no bolts are missing, the gasket is in good condition and the pan is flat (dents in the pan may indicate damage to the valve body inside).

13   If the pan gasket is leaking, the fluid level or the fluid pressure may be too high, the vent may be plugged, the pan bolts may be too tight, the pan sealing flange may be warped, the sealing surface of the transmission housing may be damaged, the gasket may be damaged or the transmission casting may be cracked or porous. If sealant instead of gasket material has been used to form a seal between the pan and the transmission housing, it may be the wrong type of sealant.

Seal leaks

14   If a transmission seal is leaking, the fluid level or pressure may be too high, the vent may be plugged, the seal bore may be damaged, the seal itself may be damaged or improperly installed, the surface of the shaft protruding through the seal may be damaged or a loose bearing may be causing excessive shaft movement.

15   Make sure the dipstick tube seal is in good condition and the tube is properly seated. Periodically check the area around the speedometer gear or sensor for leakage. If transmission fluid is evident, check the O-ring for damage.

Case leaks

16   If the case itself appears to be leaking, the casting is porous and will have to be repaired or replaced.

17   Make sure the oil cooler hose fittings are tight and in good condition.

Fluid comes out vent pipe or fill tube

18   If this condition occurs, the transmission is overfilled, there is coolant in the fluid, the case is porous, the dipstick is incorrect, the vent is plugged or the drain-back holes are plugged.

3   Oil seal replacement

1   Oil leaks frequently occur due to wear of the extension housing oil seal, and/or the speedometer drive gear oil seal and O-ring. Replacement of these seals is relatively easy, since the repairs can usually be performed without removing the transmission from the vehicle.

Extension housing oil seal

Refer to illustrations 3.4 and 3.6

2   The extension housing oil seal is located at the extreme rear of the transmission, where the driveshaft is attached. If leakage at the seal is suspected, raise the vehicle and support it securely on jackstands. If the seal is leaking, transmission fluid will be built up on the front of the driveshaft and may be dripping from the rear of the transmission.

3   Refer to Chapter 8 and remove the driveshaft.

4   Using a screwdriver, seal removal tool or a chisel and hammer, carefully pry the oil seal out of the rear of the transmission **(see illustration)**. Do not damage the splines on the transmission output shaft.

5   If the oil seal cannot be removed with a chisel, a special oil seal removal tool (available at auto parts stores) will be required.

6   Using a large section of pipe or a very large deep socket as a drift, install the new oil seal **(see illustration)**. Drive it into the bore squarely and make sure it's completely seated.

7   Lubricate the splines of the transmission output shaft and the outside of the driveshaft

sleeve yoke with lightweight grease, then install the driveshaft. Be careful not to damage the lip of the new seal.

Speed sensor O-ring

Refer to illustrations 3.8 and 3.11

Note: *Any time the speed sensor is removed, you MUST install a new O-ring.*

8   The speed sensor **(see illustration)** is located on the left (driver's) side of the extension housing. To determine if the O-ring is leaking, look for transmission fluid around the sensor.

9   Unplug the sensor electrical connector.

10   Remove the sensor hold-down bolt and remove the sensor.

11   Remove the old O-ring **(see illustration)** and install a new O-ring on the sensor.

12   Installation is the reverse of removal. Tighten the sensor hold-down bolt securely.

4   Shift cable - removal and installation

Warning: *The models covered by this manual are equipped with airbags. Always disable the airbag system before working in the vicinity of the impact sensors, steering column or instrument panel to avoid the possibility of accidental deployment of the airbag(s), which*

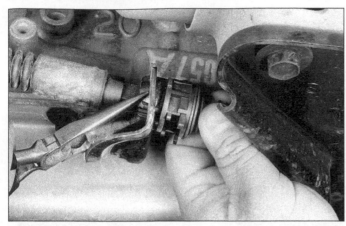

4.3 To detach the shift cable from the transmission bracket, pinch the plastic tangs together and disengage the cable from the bracket

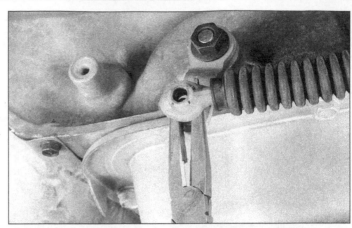

4.4 To disconnect the shift cable from the shift lever at the transmission, pop it loose from the pin on the shift lever with a screwdriver or a pair of pliers

4.6 To disconnect the shift cable from the shift lever inside the vehicle, pop it loose from the pin on the shift lever with a screwdriver or a pair of pliers

4.7 Pry the shift cable grommet out of its hole in the floor, then pull the shift cable assembly through the hole

could cause personal injury (see Chapter 12). The yellow wires and connectors routed through the console are for this system. Do not use electrical test equipment on these yellow wires or tamper with them in any way while working around the console.

Removal

Refer to illustrations 4.3, 4.4, 4.6 and 4.7

1   Disconnect the cable from the negative battery terminal. **Caution:** *On models equipped with a Delco Loc II or Theftlock audio system, be sure the lockout feature is turned off before performing any procedure which requires disconnecting the battery.*
2   Raise the vehicle and place it securely on jackstands.
3   Detach the shift cable from the transmission bracket **(see illustration)**.
4   Disconnect the shift cable from the shift lever on the transmission **(see illustration)**.
5   Remove the shift plate assembly from the front floor console (see Chapter 11).
6   Disconnect the shift cable from the shift lever **(see illustration)**.
7   Pry out the grommet **(see illustration)** from the base of the shift lever assembly and

pull out the shift cable through the hole in the floor.

Installation

8   Guide the new cable through the hole in the floor and install the grommet.
9   Connect the shift cable to the shift lever.
10   Install the center console trim panel (see Chapter 11).
11   Place the shift lever (inside the vehicle) in Neutral. Make sure it remains in the Neutral position until the shift cable is installed.
12   Attach the shift cable to the transmission bracket with the adjustment button unlocked.
13   Place the shift lever on the transmission in the Neutral position by rotating it counterclockwise from Park through Reverse into Neutral.
14   Connect the shift cable to the shift lever on the transmission.
15   Lock the adjustment button on the shift cable to the transmission bracket by turning it counterclockwise.
16   Lower the vehicle.
17   Reconnect the cable to the negative battery terminal.

5   Park/lock cable - removal and installation

Warning: *The models covered by this manual are equipped with airbags. Always disable the airbag system before working in the vicinity of the impact sensors, steering column or instrument panel to avoid the possibility of accidental deployment of the airbag(s), which could cause personal injury (see Chapter 12). The yellow wires and connectors routed through the console are for this system. Do not use electrical test equipment on these yellow wires or tamper with them in any way while working around the console.*

Removal

Refer to illustrations 5.7, 5.8, 5.10, 5.11a and 5.11b

1   Disconnect the cable from the negative battery terminal. **Caution:** *On models equipped with a Delco Loc II or Theftlock audio system, be sure the lockout feature is turned off before performing any procedure which requires disconnecting the battery.*
2   Disable the airbag system (see Chapter 12).

5.7 Pop the cable end loose from the Park/lock lever pin with a screwdriver

5.8 To detach the Park/lock cable assembly from its bracket on the shift lever base, squeeze the plastic locking tangs on the adjuster mechanism and pull the cable assembly forward

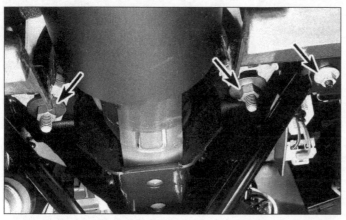

5.10 To detach the steering column from the dash, remove the four nuts (arrows - one of the nuts is not visible in this photo)

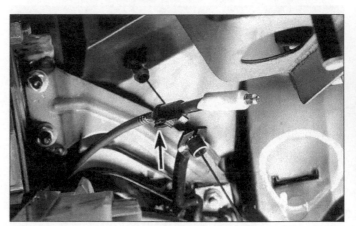

5.11a Depress this locking tab (arrow) and pull the forward end of the Park/lock cable . . .

3   Place the shift lever in the Park position.

4   Turn the ignition key to the Run position.

5   Remove the left instrument panel sound insulator assembly, the instrument panel driver knee bolster assembly and the instrument panel driver knee bolster deflector. Remove the shift plate assembly from the front floor console (see Chapter 11). On 2000 and later models, remove the front floor console (see Chapter 11).

6   Locate the Park/lock cable guide on the firewall, remove the guide retaining nut and detach the guide.

7   Unsnap the Park/lock cable from the Park/lock lever pin **(see illustration)**.

8   Detach the Park/lock cable from the bracket on the shift lever assembly **(see illustration)**.

9   Move the cable lock button to the "Up" position **(see illustration 5.14)**.

10   Lower the steering column assembly **(see illustration)**.

11   **Note:** *The ignition key must be in the Run position for this last Step.*Unsnap the Park/lock cable from the steering column **(see illustrations)**.

5.11b . . . from the front (firewall) end of this device (arrow) located on the upper left side of the steering column

Installation

Refer to illustrations 5.14, 5.15a and 5.15b

12   Make sure the ignition key is in the Run position and the shift lever is in the Park position.

13   Insert the Park/lock cable through its hole in the bracket on the shift lever base and

5.14 Before adjusting the Park/lock cable, pry up the lock button - it must be in this raised position to adjust the cable

snap the end of the cable onto the Park/Lock lever pin.

14   Raise the lock button **(see illustration)**.

15   To attach the Park/lock cable to the bracket on the shift lever base, push the cable housing forward, against the adjusting spring, to take out the slack **(see illustration)**, then push down the Park/lock cable

5.15a Slide the cable forward . . .

5.15b . . . and push down the lock button

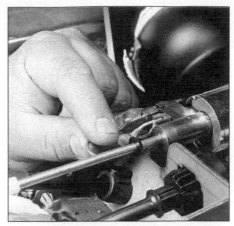

6.4a Unplug this small lock from the shift lever solenoid connector . . .

6.4b . . . then unplug the electrical connector from the shift lever solenoid

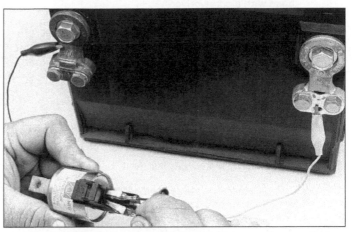

6.5 To test the solenoid, apply battery voltage to the terminals and verify that it clicks (solenoid removed from vehicle for clarity)

lock button **(see illustration)**.
16   Snap the Park/lock cable into the steering column.
17   Raise the steering column assembly and bolt it into place.
18   Install the Park/lock cable guide onto its mounting stud, install the guide retaining nut and tighten it securely.
19   Install the center console trim panel. Install the left instrument panel sound insulator, the knee bolster and the the bolster deflector (see Chapter 11).
20   Turn the ignition to the Off position.
21   Connect the cable to the negative battery terminal.
22   Enable the airbag system (see Chapter 12).
23   Check the operation of the system as follows:

a) *With the shift lever in the Park and the ignition key in the Lock position, verify that the shift lever cannot be moved to a different position. You should, however, be able to remove the ignition key from the steering column.*
b) *With the ignition key in the Run position and the shift lever in the Neutral position, verify that the key cannot be turned to the Lock position.*

24   If the system operates as described, it's working properly. If it doesn't operate as described, proceed to the next Step.
25   Put the button back to the "Up" position and readjust it as described in Step 14. Push down the button and recheck the cable as described in Step 23.
26   If you can't remove the key when the shift lever is in the Park position, move the button to its "Up" position and move the Park/lock cable connector to the rear until the key *can* be removed. Push down the button.
27   The remainder of installation is the reverse of removal.

6   Shift lever solenoid - check and replacement

Warning: *The models covered by this manual are equipped with airbags. Always disable the airbag system before working in the vicinity of the impact sensors, steering column or instrument panel to avoid the possibility of accidental deployment of the airbag(s), which could cause personal injury (see Chapter 12). The yellow wires and connectors routed through the console are for this system. Do*

not use electrical test equipment on these yellow wires or tamper with them in any way while working around the console.

Check

Refer to illustrations 6.4a, 6.4b and 6.5
1   The shift lever solenoid is part of the Brake Transmission Shift Interlock (BTSI) system. The solenoid prevents you from taking the shift lever out of Park unless the brake pedal is depressed. If you're able to do so, check the solenoid.
2   Disconnect the cable from the negative battery terminal. **Caution:** *On models equipped with a Delco Loc II or Theftlock audio system, be sure the lockout feature is turned off before performing any procedure which requires disconnecting the battery.* Disable the airbag system (see Chapter 12).
3   Remove the shift plate from the front floor console (see Chapter 11).
4   Unplug the electrical connector from the solenoid **(see illustrations)**.
5   Apply battery voltage to the solenoid **(see illustration)** and verify that it clicks. If the solenoid doesn't click, replace it and retest. If the solenoid does click, check the BTSI circuit.

6.9 Unsnap the retainer at the front of the solenoid from its pin on the shift lever base with a screwdriver

6.10 Unsnap the retainer on the rear end of the solenoid rod from its pin on the shift lever

7.9 Unplug the electrical connector from the front end of the Neutral start/back-up light switch

Replacement

Refer to illustrations 6.9 and 6.10

6   Disconnect the cable from the negative battery terminal. **Caution:** *On models equipped with a Delco Loc II or Theftlock audio system, be sure the lockout feature is turned off before performing any procedure which requires disconnecting the battery.* Disable the airbag system, if you have not already done so (see Chapter 12).

7   Remove the shift plate from the front floor console (see Chapter 11).

8   Unplug the electrical connector from the solenoid **(see illustrations 6.4a and 6.4b)**.

9   Pop loose the plastic retainer **(see illustration)** from its stud on the shift lever base.

10   Pop loose the plastic retainer on the end of the solenoid rod from its pin on the shift lever assembly **(see illustration)** and remove the solenoid.

11   Installation is the reverse of removal. Enable the airbag system (see Chapter 12).

---

7   Neutral start/back-up light switch - check, replacement and adjustment

---

Warning: *The models covered by this manual are equipped with airbags. Always disable the airbag system before working in the vicinity of the impact sensors, steering column or instrument panel to avoid the possibility of accidental deployment of the airbag(s), which could cause personal injury (see Chapter 12). The yellow wires and connectors routed through the console are for this system. Do not use electrical test equipment on these yellow wires or tamper with them in any way while working around the console.*

Check

1   The Neutral start/back-up light switch prevents the starter motor from operating when the automatic transmission is in gear by opening the starter motor circuit. When the shift lever is put into Reverse with the engine

7.10 Remove the Neutral start/back-up light switch screws (arrows)

running, the switch closes the back-up light circuit and turns on the back-up lights.

2   To check the Neutral start function of the switch, verify that the engine will start only in Neutral. If the engine starts with the transmission in gear, verify that the switch is properly adjusted (see below). If the switch still fails to work properly after it's been adjusted, it's bad. Replace it.

3   To check the back-up light function of the switch, place the shift lever in Reverse with the ignition key On. The back-up lights should come on.

4   If the back-up lights don't come on, either a fuse is blown, the bulbs are burned out, there's an open in the circuit, or the back-up light switch is bad.

5   Always begin any electrical troubleshooting procedure by checking the fuses (see Chapter 12). If a fuse is blown, replace it. If it blows again, find out why.

6   Check the bulbs for the back-up light circuit. If either bulb is bad, replace it (see Chapter 12). If the bulbs are good, check the circuit. Using a voltmeter, verify that voltage is getting to the switch and that, when the shift lever is placed in Reverse, there is voltage to the bulb. If the bulbs are okay and the circuit is okay, check the continuity of the switch.

7.15 Insert a 3/32-inch drill bit into the service adjustment hole and rotate the switch assembly until the drill bit drops to a 19/32-inch depth

Replacement

Refer to illustrations 7.9 and 7.10

7   Disconnect the cable from the negative battery terminal. **Caution:** *On models equipped with a Delco Loc II or Theftlock audio system, be sure the lockout feature is turned off before performing any procedure which requires disconnecting the battery.* Disable the airbag system (see Chapter 12).

8   Remove the center console trim panel (see Chapter 11).

9   Unplug the electrical connector **(see illustration)** from the Neutral start/back-up light switch.

10   Remove the switch retaining screws **(see illustration)** and remove the switch.

11   Be sure to adjust the switch (see below) before tightening the switch retaining screws. Installation is otherwise the reverse of removal. Enable the airbag system (see Chapter 12).

Adjustment

With old switch

Refer to illustration 7.15

12   Put the shift lever in Neutral.

8.7 Remove the shift lever base retaining bolts (arrows)

9.2 To check the transmission mount, insert a prybar as shown, between the mount and the crossmember, and note whether the mount is split or cracked when you pry it up and down

13   Insert the tang on the switch assembly into the slot on the shift lever.
14   Install the switch retaining screws loosely.
15   Insert a 3/32-inch drill bit into the service adjustment hole **(see illustration)** and rotate the switch assembly until the drill bit drops to a 19/32-inch depth.
16   Tighten the switch retaining screws securely.

With new switch

17   Put the shift lever in Neutral.
18   Insert the tang on the switch assembly into the slot on the shift lever.
19   Install the switch retaining screws and tighten them securely. **Note:** *If the screw holes will not align with the holes in the shift lever assembly, verify that the shift lever is in Neutral. Do NOT rotate the switch. If the switch has been rotated and the pin is broken, adjust the switch as described in Steps 12 through 16.*

8   Shift lever - removal and installation

Warning: *The models covered by this manual are equipped with airbags. Always disable the airbag system before working in the vicinity of the impact sensors, steering column or instrument panel to avoid the possibility of accidental deployment of the airbag(s), which could cause personal injury (see Chapter 12). The yellow wires and connectors routed through the console are for this system. Do not use electrical test equipment on these yellow wires or tamper with them in any way while working around the console.*
Refer to illustration 8.7
1   Disconnect the cable from the negative battery terminal. **Caution:** *On models equipped with a Delco Loc II or Theftlock audio system, be sure the lockout feature is turned off before performing any procedure which requires disconnecting the battery.* Disable the airbag system (see Chapter 12).
2   Remove the center console trim panel

(see Chapter 11).
3   Disconnect the shift cable from the shift lever (see Section 4).
4   Disconnect the Park/Lock cable from the shift lever (see Section 5).
5   Remove the shift lever solenoid from the shift lever base (see Section 6).
6   Remove the Park/Neutral position switch from the shift lever base (see Section 7).
7   Remove the shift lever base retaining bolts **(see illustration)**.
8   Remove the shift lever assembly.
9   Installation is the reverse of removal. Enable the airbag system (see Chapter 12).

9   Transmission mount - check and replacement

Refer to illustrations 9.2 and 9.3
1   Raise the vehicle and place it securely on jackstands.
2   Place a prybar or large screwdriver between the crossmember and the transmission **(see illustration)** and try to lever the transmission up and down. If the mount is cracked or torn, replace it.
3   Unbolt the mount from the crossmember and from the transmission **(see illustration)**.
4   Support the transmission with a floor jack and raise the transmission until the mount stud is clear of the crossmember, then remove the mount.
5   Installation is the reverse of removal.

10   Automatic transmission - removal and installation

Warning: *The models covered by this manual are equipped with airbags. Always disable the airbag system before working in the vicinity of the impact sensors, steering column or instrument panel to avoid the possibility of accidental deployment of the airbag(s), which could cause personal injury (see Chapter 12).*

9.3 To remove the transmission mount, remove the nut from the stud (lower arrow) protruding through the crossmember, then remove the mount bolts (upper arrows)

The yellow wires and connectors routed through the console are for this system. Do not use electrical test equipment on these yellow wires or tamper with them in any way while working around the console.

Removal

Refer to illustrations 10.9, 10.11 and 10.12
1   Disconnect the cable from the negative battery terminal. **Caution:** *On models equipped with a Delco Loc II or Theftlock audio system, be sure the lockout feature is turned off before performing any procedure which requires disconnecting the battery.*
2   Disable the airbag system (see Chapter 12). On 2000 and later models, remove the air intake duct (see Chapter 4).
3   Raise the vehicle and place it securely on jackstands.
4   Drain the transmission fluid (see Chapter 1).
5   Remove the driveshaft (see Chapter 8).
6   Clearly label, then unplug, all electrical connectors.
7   Support the engine with a floor jack (place a block of wood between the oil pan and the jack head). Support the transmission

10.9 To remove the crossmember, remove all four bolts (arrows) and the nut in the center for the transmission mount stud

10.11 Remove the bolts (arrows) from the torque converter access cover and remove the cover

with another jack - preferably one made for this purpose (available at most tool rental yards). Safety chains will help steady the transmission on the jack.

8   Support the rear axle with a jack and remove the torque arm (see Chapter 10).

9   Remove the transmission crossmember **(see illustration)** and mount (see Section 9). On 2002 V6 models, remove the right engine mount (see Chapter 2, Part B).

10   Remove the catalytic converter hanger (see Chapter 4).

11   On 1997 and earlier models, remove the torque converter cover **(see illustration)**. On 1998 and later models, remove the lower bellhousing to oil pan bolts.

12   Mark the relationship of the torque converter and driveplate with white paint so they can be installed in the same position. Remove the flywheel-to-torque converter bolts **(see illustration)**. Turn the crankshaft for access to each bolt. Turn the crankshaft in a clockwise direction only (as viewed from the front).

13   Remove the catalytic converter heat shield if necessary (see Chapter 6). On V6 models, remove the converter.

14   Disconnect the shift cable from the transmission (see Section 5).

15   Disconnect the oil cooler lines from their clamps on the transmission.

16   Remove the six transmission bellhousing-to-engine bolts (there are two on each side and two on the top).

17   Remove the transmission dipstick tube.

18   Lower both jacks and pull the transmission to the rear, separate it from the engine and disconnect the oil cooler lines. Use a flare-nut wrench and a back-up wrench to unscrew the cooler line fittings. As soon as you disconnect the lines, connect them together with a short section of fuel hose to prevent transmission fluid from leaking out.

19   Clamp a pair of Vise-Grip locking pliers onto the lower portion of the transmission case, just in front of the torque converter (but make sure they're not clamped in front of one of the torque converter-to-driveplate bolt locations, or the driveplate will hang up on

the pliers). The pliers will prevent the torque converter from falling out while you're removing the transmission. Move the transmission to the rear to disengage it from the engine block dowel pins and make sure the torque converter is detached from the driveplate. Lower the transmission on the jack and wheel it out from under the vehicle.

Installation

20   Remove all old sealant from the extension housing.

21   Make sure the torque converter hub is securely engaged with the pump. If you've removed the converter, spread transmission fluid on the torque converter rear hub, where the transmission front seal rides. With the front of the transmission facing up, rotate the converter back and forth. It should drop down into the transmission front pump in stages. To ensure that the converter is fully engaged, lay a straightedge across the transmission-to-engine mating surface and make sure the converter hub is at least 1/2-inch below the straightedge. Reinstall the Vise-Grips to hold the converter in this position. Lubricate the front hub of the converter with multi-purpose grease.

22   With the transmission secured to the jack, wheel it into position, raise it up into position and connect the oil cooler lines.

23   Turn the torque converter to line up the holes with the holes in the driveplate. The white paint mark made on the torque converter and driveplate in Step 12 must line up.

24   With the aid of a helper, carefully move the transmission forward until the dowel pins and the torque converter are engaged. Make sure the transmission mates with the engine with no gap. If there's a gap, make sure there are no wires or other objects pinched between the engine and transmission and also make sure the torque converter is completely engaged in the transmission front pump. Try to rotate the converter - if it doesn't rotate easily, it's probably not fully engaged in the pump. If necessary, lower the transmission and install the the converter fully. Connect the cooler lines and tighten

them securely.

25   Install the transmission-to-engine bolts and tighten them to the torque listed in this Chapter's Specifications. As you're tightening the bolts, verify that the engine and transmission mate completely at all points. If not, find out why. Never try to force the engine and transmission together with the bolts or you'll break the transmission case.

26   Raise the transmission and install the transmission mount and crossmember. Tighten all nuts and bolts securely.

27   Install the torque converter-to-driveplate bolts. Tighten them to the torque listed in this Chapter's Specifications.

28   Remove the jacks supporting the engine and transmission.

29   Install the dipstick tube.

30   Reconnect all electrical connectors.

31   Reattach the catalytic converter hanger (see Chapter 6).

32   Reattach the torque arm (see Chapter 10). Remove the jack supporting the rear axle.

33   Install the driveshaft (see Chapter 8).

34   Check and refill the transmission with the specified fluid (see Chapter 1).

35   Remove the jackstands and lower the vehicle.

10.12 Mark the relationship of the torque converter to the driveplate, then rotate the crankshaft pulley to reach each flywheel-to-torque converter bolt (arrow)

Chapter 8 Clutch and driveline

Contents

Specifications

General

Clutch hydraulic fluid type	See Chapter 1
Clutch disc lining thickness	1/16 inch (above rivet)

Torque specifications

Ft-lbs (unless otherwise indicated)

Clutch

Pressure plate-to-flywheel bolts	
Through 1999	15, then turn an additional 30-degrees
2000 and later	
Five-speed	15, then turn an additional 45-degrees
Six-speed	52
Release cylinder retaining bolts/nuts	
1997 and earlier	
All except 3800 V6 models	15
3800 V6 models	71 in-lbs
1998 and later	71 in-lbs
Master cylinder retaining nuts	20

Driveshaft

U-joint strap bolts	14
Center support bearing bolts	
Through 1999	50
2000 and later	37

Rear axle

Pinion shaft lock bolt	
Through 1999	96 in-lbs
2000 and later	27

1 General information

The information in this Chapter deals with the components from the rear of the engine to the rear wheels (except for the transmission, which is dealt with in the previous Chapter). These components are grouped into three categories: clutch, driveshaft and axles. Separate Sections within this Chapter offer general descriptions and checking procedures for components in each of the three groups.

Because nearly all the procedures covered in this Chapter involve working under the vehicle, make sure that it's securely supported on sturdy jackstands or on a hoist where the vehicle can be easily raised and lowered.

2 Clutch - description and check

1   All vehicles with a manual transmission use a single dry plate, diaphragm spring type clutch. The clutch disc has a splined hub which allows it to slide along the splines of the transmission input shaft. The clutch and pressure plate are held in contact by spring pressure exerted by the diaphragm in the pressure plate.

2   The clutch release system is operated by hydraulic pressure. The hydraulic release system consists of the clutch pedal, a master cylinder and fluid reservoir, the hydraulic line, a release (or slave) cylinder which actuates the clutch release lever (1997 and earlier models only) and the clutch release (or throwout) bearing. On all 3800 engines and 1998 and later V8 engines, the release cylinder is located inside the bellhousing and is bolted directly over the transmission input shaft to actuate the release bearing. On these models there is no release lever as on previous models.

3   When pressure is applied to the clutch pedal to release the clutch, hydraulic pres-

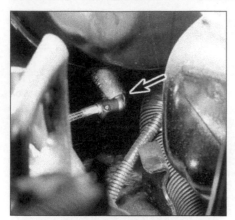

3.1 The clutch master cylinder (arrow) is mounted in the firewall, below the power brake booster (the retaining nuts must be removed from the driver's side of the firewall)

3.3 The clutch master cylinder pushrod is located at the top of the clutch pedal; to disconnect the pushrod, remove the retainer (left arrow) and pull out the pin (right arrow)

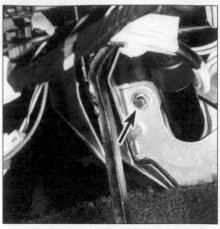

3.4 To detach the clutch master cylinder from the firewall, remove these two nuts (arrows)

sure is exerted against the outer end of the clutch release lever (except on 3800 V6 and 1998 and later models, where the release cylinder and release bearing are an integral unit). As the lever pivots the shaft fingers push against the release bearing. The bearing pushes against the fingers of the diaphragm spring of the pressure plate assembly, which in turn releases the clutch plate.

4   Terminology can be a problem when discussing the clutch components because common names are in some cases different from those used by the manufacturer. For example, the driven plate is also called the clutch plate or disc, the clutch release bearing is sometimes called a throwout bearing, the release cylinder is sometimes called the operating or slave cylinder.

5   Other than to replace components with obvious damage, some preliminary checks should be performed to diagnose clutch problems.

a) *The first check should be of the fluid level in the clutch master cylinder. If the fluid level is excessively low, add fluid as necessary and inspect the hydraulic system for leaks (fluid level will actually rise as the clutch wears).*

b) *To check "clutch spin-down time," run the engine at normal idle speed with the transmission in Neutral (clutch pedal up - engaged). Disengage the clutch (pedal down), wait several seconds and shift the transmission into Reverse. No grinding noise should be heard. A grinding noise would most likely indicate a problem in the pressure plate or the clutch disc.*

c) *To check for complete clutch release, run the engine (with the parking brake applied to prevent movement) and hold the clutch pedal approximately 1/2 inch from the floor. Shift the transmission between First gear and Reverse several times. If the shift is hard or the transmission grinds, component failure is indicated. Check the release cylinder pushrod travel. With the clutch pedal*

depressed completely, the release cylinder pushrod should extend substantially. If it doesn't, check the fluid level in the clutch master cylinder (see Chapter 1).

d) *Visually inspect the pivot bushing at the top of the clutch pedal to make sure there is no binding or excessive play.*

e) *Crawl under the vehicle and make sure the clutch release lever is solidly mounted on the ball stud.*

3   Clutch hydraulic release system - removal and installation

Note: *On 1997 and earlier models (except for 3800 V6), the clutch release system is serviced as a complete unit and has been bled of air at the factory; individual components or rebuild kits are not available separately and there are no provisions for adjustment of clutch pedal height or freeplay. On 1998 and later and all 3800 V6 models, the master cylinder and release cylinder can be removed and installed separately.*

1997 and earlier (except 3800 V6 models)

Refer to illustrations 3.1, 3.3, and 3.4

1   The clutch master cylinder **(see illustration)** is located on the firewall below the power brake booster.

2   Disconnect the cable from the negative terminal of the battery. **Caution:** *On models equipped with a Delco Loc II or Theftlock audio system, be sure the lockout feature is turned off before performing any procedure which requires disconnecting the battery.* Remove the driver knee bolster from below the instrument panel.

3   Pry off the retainer and detach the clutch master cylinder pushrod from the pin at the top of the clutch pedal **(see illustration)**.

4   Remove the clutch master cylinder retaining nuts **(see illustration)** and remove the U-bolt that attaches the clutch master cylinder to the firewall. Remove the master cylinder.

5   Remove the master cylinder reservoir retainers and remove the reservoir.

6   Raise the vehicle and place it securely on jackstands.

7   Remove the clutch release cylinder retaining nuts and remove the release cylinder.

8   Remove the clutch master cylinder and release cylinder as a single assembly.

9   Installation is the reverse of removal.

1998 and later (including all 3800 V6 models)

Master cylinder

Refer to illustration 3.12

10   Disconnect the negative battery cable. **Caution:** *On models equipped with a Delco Loc II or Theftlock audio system, be sure the lockout feature is turned off before performing any procedure which requires disconnecting the battery.*

11   Refer to Steps 1 through 5.

12   Using a hydraulic clutch line separator, disconnect the clutch fluid hydraulic line quick-connect fitting **(see illustration)** from

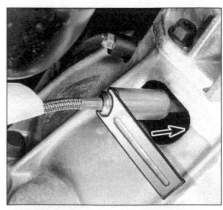

3.12 To disconnect the clutch fluid hydraulic line from the release cylinder on 3800 V6 models, put a hydraulic clutch line separator against the shoulder of the quick-connect fitting, push in on the spring-loaded fitting and pull the braided clutch line out

the release cylinder.

13   Remove the clutch master cylinder.

14   When installing the master cylinder, make sure the braided line is neither twisted nor kinked and the quick-connect fitting is properly engaged with the release cylinder. Push on the quick-connect fitting, then pull it back to verify that it's properly engaged. Don't rely on the audible clicking sound it makes. Installation is otherwise the reverse of removal.

15   Be sure to bleed the system when you're done (see Section 4).

Release cylinder

16   Disconnect the negative battery cable. **Caution:** *On models equipped with a Delco Loc II or Theftlock audio system, be sure the lockout feature is turned off before performing any procedure which requires disconnecting the battery.*

17   Remove the transmission (see Chapter 7 Part A).

18   Disconnect the clutch fluid hydraulic line (left side of the bellhousing) from the release cylinder **(see illustration 3.12)**.

19   Remove the release cylinder retaining bolts and remove the release cylinder.

20   If necessary, remove the bellhousing.

21   Installation is the reverse of removal. Be sure to tighten the release cylinder retaining bolts to the torque listed in this Chapter's Specifications.

22   Be sure to bleed the system when you're done (see Section 4).

---

4   Clutch hydraulic system - bleeding

Note: *Bleeding is necessary only when the level of fluid in the reservoir has fallen low enough to allow air to be drawn into the master cylinder.*

1997 and earlier (except 3800 V6 models)

1   Working from inside the vehicle, locate the clutch master cylinder retaining nuts and back them off until they're at the end of the threads on the U-bolt that retains the master cylinder to the firewall.

2   Clean the dirt and grease from the reservoir cap to prevent any foreign matter from entering the system, then remove the reservoir cap and diaphragm.

3   Attach a piece of wire long enough to reach the ground to the left hood strut bracket.

4   Raise the vehicle and place it securely on jackstands.

5   Remove the clutch release cylinder retaining nuts and the release cylinder. Hang the release cylinder from the piece of wire.

6   Lower the vehicle.

7   Holding the release cylinder, depress the pushrod about 25/32-inch into the release cylinder and hold it there. Install the diaphragm and reservoir cap while holding in the release cylinder pushrod.

8   Make sure the release cylinder is lower

than the master cylinder, then press the pushrod into the release cylinder bore with short strokes. Inspect the master cylinder for air bubbles. Continue stroking the pushrod until there are no more bubbles entering the reservoir.

9   Raise the vehicle and place it securely on jackstands.

10   Remove the wire suspending the release cylinder and install the cylinder and retaining nuts. Tighten the nuts to the torque listed in this Chapter's Specifications.

11   Lower the vehicle.

12   Remove the wire from the hood strut bracket.

13   Tighten the clutch master cylinder nuts to the torque listed in this Chapter's Specifications.

14   Inspect the fluid level in the clutch master cylinder reservoir. Fill the reservoir with the fluid listed in the Chapter 1 Specifications. **Caution:** *Do NOT use clutch hydraulic fluid which has been bled from a system to fill the clutch master cylinder reservoir. It may be aerated, contaminated, or contain moisture.*

1998 and later (including all 3800 V6 models)

15   Raise the vehicle and place it securely on jackstands.

16   Connect a plastic line to the bleeder screw on the clutch release cylinder and submerge the other end of the line in a container of the clutch fluid specified in the Chapter 1 Specifications.

17   Depress the clutch pedal slowly and hold it there, loosen the bleeder to purge the air, then tighten the bleeder. Repeat this procedure until all air is completely purged from the system.

18   Fill the clutch master cylinder reservoir with the fluid listed in the Chapter 1 Specifications. **Caution:** *Do NOT use clutch hydraulic fluid which has been bled from a system to fill the clutch master cylinder reservoir. It may be aerated, contaminated, or contain moisture.*

---

5   Clutch components - removal, inspection and installation

Warning: *Dust produced by clutch wear and deposited on clutch components may contain asbestos, which is hazardous to your health. DO NOT blow it out with compressed air and DO NOT inhale it. DO NOT use gasoline or petroleum-based solvents to remove the dust. Brake system cleaner should be used to flush the dust into a drain pan. After the clutch components are wiped clean with a rag, dispose of the contaminated rags and cleaner in a covered, marked container.*

Removal

Refer to illustration 5.7

1   Access to the clutch components is normally accomplished by removing the transmission, leaving the engine in the vehi-

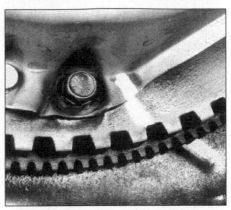

5.7 Be sure to mark the pressure plate and flywheel to insure proper alignment during installation (this won't be necessary if a new pressure plate is to be installed)

cle. If, of course, the engine is being removed for major overhaul, then check the clutch for wear and replace worn components as necessary. However, the relatively low cost of the clutch components compared to the time and trouble spent gaining access to them warrants their replacement anytime the engine or transmission is removed, unless they are new or in near perfect condition. The following procedures are based on the assumption the engine will stay in place.

2   On 1997 and earlier models (except 3800 V6), unbolt the release cylinder from the bellhousing and position it out of the way (see Section 3).

3   On 1998 and later models and all 3800 V6 models, disconnect the clutch fluid hydraulic line from the left side of the bellhousing.

4   On all 1997 and earlier models (except 3800 V6), the clutch fork and release bearing can remain attached to the bellhousing for the time being (1998 and later models and all 3800 V6 models don't have a clutch fork and the release bearing is integral with the release cylinder).

5   Referring to Chapter 7 Part A, remove the transmission from the vehicle. Support the engine while the transmission is out. Preferably, an engine hoist should be used to support it from above. However, if a jack is used underneath the engine, make sure a piece of wood is positioned between the jack and oil pan to spread the load. **Caution:** *The pick-up for the oil pump is very close to the bottom of the oil pan. If the pan is bent or distorted in any way, engine oil starvation could occur.*

6   To support the clutch disc during removal, install a clutch alignment tool through the clutch disc hub.

7   Carefully inspect the flywheel and pressure plate for indexing marks. The marks are usually an X, an O or a white letter. If they cannot be found, make scribe marks yourself so the pressure plate and the flywheel will be in the same alignment during installation **(see illustration)**.

8   Turning each bolt only 1/4-turn at a

5.10 Check the flywheel for cracks, hot spots and other obvious defects (slight imperfections can be removed by resurfacing at an automotive machine shop); this flywheel should be resurfaced

NORMAL FINGER WEAR

EXCESSIVE WEAR

EXCESSIVE FINGER WEAR

BROKEN OR BENT FINGERS

5.12 Replace the pressure plate if any of these conditions are noted

time, loosen the pressure plate-to-flywheel bolts. Work in a criss-cross pattern until all spring pressure is relieved. Then hold the pressure plate securely and remove the bolts completely, followed by the pressure plate and clutch disc.

Inspection

Refer to illustrations 5.10, 5.12 and 5.14

9   Ordinarily, when a problem occurs in the clutch, it can be attributed to wear of the clutch driven plate assembly (clutch disc). However, all components should be inspected at this time.

10   Inspect the flywheel for cracks, heat checking, grooves and other obvious defects **(see illustration)**. If the imperfections are slight, a machine shop can machine the surface flat and smooth, which is highly recommended regardless of the surface appearance. Refer to Chapter 2 for the flywheel removal and installation procedure.

11   Inspect the pilot bushing (see Section 7).

12   Inspect the lining on the clutch disc. There should be at least 1/16-inch of lining above the rivet heads. Check for loose rivets, distortion, cracks, broken springs and other

5.14 Examine the pressure plate friction surface for score marks, cracks and evidence of overheating

obvious damage **(see illustration)**. As mentioned above, ordinarily the clutch disc is routinely replaced, so if in doubt about the condition, replace it with a new one.

13   The release bearing should also be replaced along with the clutch disc (see Section 6).

14   Check the machined surfaces and the diaphragm spring fingers of the pressure plate **(see illustration)**. If the surface is grooved or otherwise damaged, replace the pressure plate. Also check for obvious damage, distortion, cracking, etc. Light glazing can be removed with medium grit emery cloth. If a new pressure plate is required, new and factory-rebuilt units are available.

Installation

Refer to illustration 5.16

Note: *The manufacturer recommends that,*

5.16 A clutch alignment tool is essential for centering the new clutch disc; if you don't have a clutch alignment tool, you can use an old transmission input shaft instead

when a clutch is replaced, the flywheel, pressure plate and disc all be replaced with new components at the same time.

15   Before installation, clean the flywheel and pressure plate machined surfaces with lacquer thinner or acetone. It's important that no oil or grease is on these surfaces or the lining of the clutch disc. Handle the parts only with clean hands.

16   Position the clutch disc and pressure plate against the flywheel with the clutch held in place with an alignment tool **(see illustration)**. Make sure it's installed properly (most replacement clutch plates will be marked "flywheel side" or something similar - if not marked, install the clutch disc with the damper springs toward the transmission).

17   Tighten the pressure plate-to-flywheel bolts only finger tight, working around the pressure plate.

18   Center the clutch disc by ensuring the alignment tool extends through the splined hub and into the pilot bushing in the crankshaft. Wiggle the tool up, down or side-to-side as needed to bottom the tool in the pilot bushing. Tighten the pressure plate-to-flywheel bolts a little at a time, working in a criss-cross pattern to prevent distorting the cover. After all of the bolts are snug, tighten them to the torque listed in this Chapter's Specifications. Remove the alignment tool.

19   On 1997 and earlier models (except 3800 V6), use high-temperature grease to lubricate the inner groove of the release bearing (see Section 6). Also place grease on the release lever contact areas and the transmission input shaft bearing retainer.

20   Install the clutch release bearing as described in Section 6.

21   Install the transmission, and all components removed previously. Tighten all fasteners to the proper torque specifications.

6.7 This is how the release fork engages the release bearing - make sure the fingers and the sleeve in which they ride are well lubricated with high temperature grease (don't overdo it, though)

6.8a Make sure the pivot ball (arrow) for the release fork is clean and well lubricated with high temperature grease

6.8b This is how the release fork and release bearing should look when they're properly installed

6 Clutch release bearing - removal, inspection and installation

Warning: *Dust produced by clutch wear and deposited on clutch components is hazardous to your health. DO NOT blow it out with compressed air and DO NOT inhale it. DO NOT use gasoline or petroleum-based solvents to remove the dust. Brake system cleaner should be used to flush the dust into a drain pan. After the clutch components are wiped clean with a rag, dispose of the contaminated rags and cleaner in a covered, marked container.*

Note: *The following procedure applies to all 1997 and earlier models (except 3800 V6). For release cylinder replacement procedures on 1998 and later models and all 3800 V6 models refer to Section 3.*

Removal

1   Disconnect the negative cable from the battery.
2   Remove the transmission (see Chapter 7 Part A).
3   Remove the bellhousing from the engine.
4   Remove the clutch release fork from the pivot stud, then remove the release bearing from the fork **(see illustration 2.1)**. On models with hydraulic clutch system, twist the bearing assembly on the actuator until the spring pressure pops the bearing off.

Inspection

5   Hold the center of the bearing and rotate the outer portion while applying pressure. If the bearing doesn't turn smoothly or if it's noisy, remove it from the sleeve and replace it with a new one. Wipe the bearing with a clean rag and inspect it for damage, wear and cracks. Don't immerse the bearing in solvent - it's sealed for life and to do so would ruin it. Also check the release fork for cracks and other damage.

Installation

Mechanical clutch release systems

Refer to illustrations 6.7, 6.8a and 6.8b

6   Lightly lubricate the clutch fork fingers and the release bearing with high-temperature grease.
7   Engage the release fork and release bearing **(see illustration)**.
8   Lubricate the ball on the pivot stud **(see illustration)** and the ball socket on the back of the release fork with high-temperature grease and push the fork onto the pivot stud until it's firmly seated **(see illustration)**.
9   Apply a light coat of high-temperature grease to the face of the release bearing, where it contacts the pressure plate diaphragm fingers.

Hydraulic clutch release systems

10   Clean the clutch actuator with a clean rag, but do NOT use any solvents or chemicals.
11   Install the new bearing by pushing it onto the actuator until the retainer tab snaps in.

All

12   Install the bellhousing and tighten the bolts to the torque listed in this Chapter's Specifications.
13   Prior to installing the transmission, apply a light coat of grease to the edge of the transmission front bearing retainer.
14   The remainder of the installation is the reverse of the removal procedure.

7 Pilot bushing - inspection and replacement

Refer to illustrations 7.8 and 7.9

1   The clutch pilot bushing is an oil-impregnated type bushing which is pressed into the rear of the crankshaft. It is greased at the factory and does not require additional lubrication. Its primary purpose is to support the front of the transmission input shaft. The pilot bushing should be inspected whenever the clutch components are removed from the

7.8 Fill the cavity behind the pilot bushing with grease . . .

engine. Due to its inaccessibility, if you are in doubt as to its condition, replace it with a new one. **Note:** *If the engine has been removed from the vehicle, disregard the following steps which don't apply.*
2   Remove the transmission (refer to Chapter 7 Part A).
3   Remove the clutch components (see Section 5).
4   Inspect for any excessive wear, scoring, lack of grease, dryness or obvious damage. If any of these conditions are noted, the bushing should be replaced. A flashlight will be helpful to direct light into the recess.
5   Removal can be accomplished with a special puller and slide hammer, but an alternative method also works very well.
6   Find a solid steel bar which is slightly smaller in diameter than the bushing. Alternatives to a solid bar would be a wood dowel or a socket with a bolt fixed in place to make it solid.
7   Check the bar for fit - it should just slip into the bushing with very little clearance.
8   Pack the bushing and the area behind it (in the crankshaft recess) with heavy grease. Pack it tightly to eliminate as much air as possible **(see illustration)**.
9   Insert the bar into the bushing bore and strike the bar sharply with a hammer which will force the grease to the back side of the

7.9 . . . then hydraulically force out the bushing with a steel rod slightly smaller than the bore in the bushing - when the hammer strikes the rod, the grease will transmit this force to the backside of the bushing and push it out

bushing and push it out **(see illustration)**. Remove the bushing and clean all grease from the crankshaft recess.

10   To install the new bushing, lightly lubricate the outside surface with lithium-based grease, then drive it into the recess with a soft-face hammer. Most new bushings come already lubricated, but if it is dry, apply a thin coat of high-temperature grease to it.

11   Install the clutch components, transmission and all other components removed previously, tightening all fasteners properly.

8   Clutch start switch - check and replacement

Check

1   The clutch start switch, which is mounted on a bracket at the top of the clutch pedal, below and to the left of the clutch anticipate switch, prevents the engine from being started unless the clutch pedal is depressed.

2   To test the operation of the switch, verify that the engine will not start unless the clutch pedal is depressed, and that it does start when the pedal is depressed.

3   If the engine starts without depressing the clutch pedal, the switch is probably bad, but check the switch circuit first. Make sure there's voltage to the switch.

4   If there's voltage to the switch, verify that there's no voltage on the other side of the switch unless the clutch pedal is depressed.

5   If there's a short or open in the circuit on either side of the switch, repair it and retest.

6   If the circuit is okay, replace the switch.

Replacement

7   Make sure the ignition key is in the Off position.

8   Remove the knee bolster from underneath the left side of the dash.

9   Unplug the electrical connector from the clutch start switch.

10   Remove the switch from its bracket.

11   Installation is the reverse of removal.

9   Clutch anticipate switch - check and replacement

Check

1   The clutch anticipate switch, which is mounted on a bracket at the top of the clutch pedal, to the right of the clutch start switch, deactivates the cruise control system when the clutch pedal is depressed.

2   To test the operation of the clutch anticipate switch, activate the cruise control system while driving the vehicle, then depress the clutch and verify that the cruise control system is deactivated.

3   If the cruise control system is not deactivated when the clutch pedal is depressed, unplug the electrical connector from the clutch anticipate switch and, using a continuity tester or an ohmmeter, verify that there is continuity through the switch when the clutch pedal is released, and no continuity through the switch when the clutch pedal is depressed.

4   If the clutch anticipate switch is faulty, replace it. If the switch is okay, the problem is in the cruise control circuit. Checking this circuit should be left to a dealer service department.

Replacement

5   Make sure the ignition key is in the Off position.

6   Remove the knee bolster from underneath the left side of the dash (see Chapter 11).

7   Unplug the electrical connector from the clutch anticipate switch.

8   Remove the switch from its bracket.

9   Installation is the reverse of removal.

10   Driveshaft and universal joints - general information

1   A driveshaft is a tube, or a pair of tubes, that transmits power between the transmission and the differential. Universal joints are located at either end of the driveshaft. All V8 models utilize a one-piece driveshaft; all V6 models use a two-piece design. A center support bearing is employed at the junction between the two driveshaft halves on two-piece units.

2   All driveshafts employ a splined yoke at the front, which engages the splined output shaft inside the end of the transmission extension housing. This setup allows the driveshaft to slide in and out of the extension housing during operation. An oil seal in the end of the extension housing prevents transmission fluid from leaking out and dirt from entering the transmission. If leakage is evident at the front of the driveshaft, replace the

oil seal (see Chapter 7, Part B).

3   Center bearings support the driveline when two-piece driveshafts are used. The center bearing is a ball-type bearing mounted in a rubber cushion attached to a frame crossmember. The bearing is pre-lubricated and sealed at the factory. The center support bearing is rebuildable on 1993 and 1994 models. If the center support bearing wears out on a 1995 or later model, it must be replaced.

4   Generally speaking, the driveshaft assembly requires very little service. The universal joints are lubricated for life. If, however, problems develop, the driveshaft must be removed from the vehicle and the U-joints replaced.

5   Since the driveshaft is a balanced unit, it's important that no undercoating, mud, etc. be allowed to remain on it. When the vehicle is raised for service it's a good idea to clean the driveshaft and inspect it for any obvious damage. Also, make sure the small weights used to originally balance the driveshaft are in place and securely attached. Whenever the driveshaft is removed it must be reinstalled in the same relative position to preserve the balance.

6   Problems with the driveshaft are usually indicated by a noise or vibration while driving the vehicle. A road test should verify if the problem is the driveshaft or another vehicle component. Refer to the Troubleshooting Section at the front of this manual. If you suspect trouble, inspect the driveline (see the next Section).

11   Driveline inspection

1   Raise the rear of the vehicle and support it securely on jackstands. Block the front wheels to keep the vehicle from rolling off the stands.

2   Crawl under the vehicle and visually inspect the driveshaft. Look for any dents or cracks in the tubing. If any are found, the driveshaft must be replaced.

3   Check for oil leakage at the front and rear of the driveshaft. Leakage where the driveshaft enters the transmission indicates a defective transmission extension housing seal (see Chapter 7). Leakage where the driveshaft enters the differential indicates a defective pinion seal (see Section 19).

4   While under the vehicle, have an assistant rotate a rear wheel so the driveshaft will rotate. As it does, make sure the universal joints are operating properly without binding, noise or looseness. Listen for any noise from the center bearing (if equipped), indicating it's worn or damaged. Also check the rubber portion of the center bearing for cracking or separation, which will necessitate replacement.

5   The universal joint can also be checked with the driveshaft motionless, by gripping your hands on either side of the joint and attempting to twist the joint. Any movement at all in the joint is a sign of considerable wear. Lifting up on the shaft will also indicate

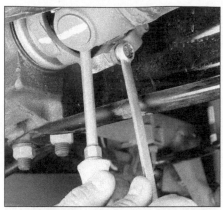

12.3 Before disconnecting the rear end of the driveshaft from the differential, mark the pinion flange and the U-joint to ensure proper balance after reassembly; hold the driveshaft with a prybar or large screwdriver while breaking loose the clamp bolts

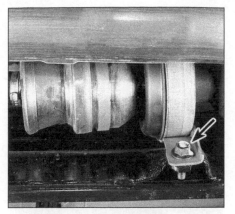

12.5 To detach the center support bearing from the torque arm on V6 models, remove the upper bolt (not visible in this photo) and the lower bolt (arrow)

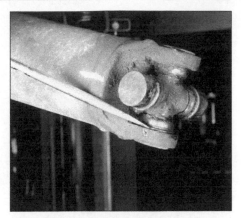

13.2 Remove the inner retainers from the U-joint by tapping them off with a screwdriver and hammer

movement in the universal joints.

6   Finally, check the driveshaft mounting bolts at the ends to make sure they're tight.

12 Driveshaft - removal and installation

Removal

Refer to illustrations 12.3 and 12.5

1   Raise the vehicle and support it securely on jackstands. Block the front wheels to prevent the vehicle from rolling.

2   Place the transmission in Park or Neutral with the parking brake off.

3   Make reference marks on the driveshaft and the pinion flange in line with each other **(see illustration)**. This is to make sure the driveshaft is reinstalled in the same position to preserve the balance.

4   Remove the rear universal joint bolts and clamps **(see illustration 12.3)**. Turn the driveshaft (or wheels) as necessary to bring the bolts into the most accessible position.

5   On vehicles with a two-piece driveshaft, unbolt the center support bearing **(see illustration)**.

6   Tape the bearing caps to the spider to prevent the caps from coming off during removal.

7   Lower the rear of the driveshaft. Slide the front of the driveshaft out of the transmission.

8   Wrap a plastic bag over the transmission extension housing and secure it in place with a rubber band. This will prevent loss of fluid and protect against contamination while the driveshaft is out.

Installation

9   Remove the plastic bag from the transmission and wipe the area clean. Inspect the oil seal carefully. Procedures for replacement of this seal can be found in Chapter 7.

10   Slide the splined front end of the driveshaft into the transmission.

11   If the driveshaft is a two-piece unit, raise the center support bearing into place and bolt it to the vehicle. Tighten the bolts securely.

12   Raise the rear of the driveshaft into position, checking to be sure the marks are in alignment. If not, turn the rear wheels to match the pinion flange and the driveshaft.

13   Remove the tape securing the bearing caps and install the clamps and bolts. Tighten all bolts to the torque listed in this Chapter's Specifications. Lower the vehicle.

13 Universal joints - replacement

Refer to illustrations 13.2, 13.3, 13.4a, 13.4b, 13.11 and 13.13

Note: *Always purchase a universal joint service kit(s) for your model vehicle before starting the procedure which follows. Also, read through the entire procedure before beginning work.*

1   Remove the driveshaft (refer to Section 10).

2   Place the shaft on a workbench equipped with a vise. If equipped, remove the snap-rings **(see illustration)**. **Note:** *If no snap-rings are present, the bearing cups are retained by injected-nylon rings which will break as the cups are pressed out. Snap-rings will be used on reassembly.*

3   Place the universal joint in the vise with a 1-1/8 inch socket against one ear of the shaft yoke and a 5/8-inch socket placed on the opposite bearing cup **(see illustration)**. **Caution:** *Never clamp the driveshaft tubing itself in a vise, as the tube may be bent.*

4   Press the bearing cup out of the yoke ear, shearing the plastic retaining ring on the bearing, if equipped **(see illustrations)**. **Note:** *If the cup does not come all the way out of the yoke, it may be pulled free with adjustable pliers, then removed.*

5   Turn the driveshaft 180-degrees and

13.3 Press out the bearing cups as shown

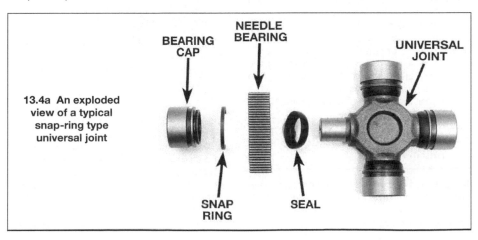

13.4a An exploded view of a typical snap-ring type universal joint

BEARING CAP

NEEDLE BEARING

UNIVERSAL JOINT

SNAP RING

SEAL

13.4b After it has been pressed out, remove the bearing cup from the yoke ear

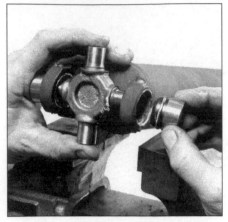

13.11 Assemble the new cross and bearing cups

13.13 Install snap-rings in the snap-ring grooves

press the opposing bearing cup out of the yoke.

6   Disengage the cross from the yoke and remove the cross. If you're replacing the front universal joint, repeat Steps 3 through 5 to remove the remaining cups.

7   If the remaining universal joint is being replaced, press the bearing cups from the slip yoke as detailed above.

8   When reassembling the driveshaft, always install all parts included in the U-joint service kit.

9   If necessary, remove all remnants of the plastic bearing retainers from the grooves in the yokes. Failure to do so may keep the bearing cups from being pressed into place and prevent the bearing retainers from seating properly.

10   Using multi-purpose grease to retain the needle bearings, assemble the bearings, cups and washers. Make sure the bearings do not become dislodged during the assembly and installation procedures.

11   In the vise, assemble the cross and cups in the yoke, installing the cups as far as possible by hand **(see illustration)**.

12   Move the cross back and forth horizontally to assure alignment, then press the cups into place a little at a time, continuing to center the cross to keep the proper alignment.

13   As soon as one snap-ring groove clears the inside of the yoke, stop pressing and install the snap-ring **(see illustration)**.

14   Continue to press on the bearing cup until the opposite snap-ring can be installed. If difficulty is encountered, strike the yoke sharply with a hammer. This will spring the yoke ears slightly and allow the snap-ring groove to move into position.

15   Install the driveshaft (see Section 12).

14   Center support bearing - removal and installation

Note 1: *Replacement parts may not be available for this model. Check with a dealer parts department or driveline repair facility if replacement is required.*

Note 2: *On 2000 and later models, the manufacturer recommends that the driveshaft and center-bearing assembly on V6 models be replaced as a unit only.*

1   Remove the driveshaft (see Section 12).

2   To replace the old center support bearing and seal, you need to press the two driveshaft halves apart, press off the old center support bearing and press on the new bearing. Instead, take the driveshaft assembly, along with a new center support bearing and a new seal, to an automotive machine shop and have the driveshaft halves separated, the old bearing and seal pressed off, the new parts pressed on, and the driveshafts pressed together.

3   Install the driveshaft (see Section 12).

15   Rear axle - description, check and identification

Description

1   The rear axle assembly is a hypoid, semi-floating type (the centerline of the pinion gear is below the centerline of the ring gear). When the vehicle goes around a corner, the differential allows the outer rear tire to turn more quickly than the inner tire. The axleshafts are splined to the differential side gears, so when the vehicle goes around a corner, the inner tire, which turns more slowly than the outer tire, turns its side gear more slowly than the outer tire turns its side gear. The differential pinion (or "spider") gears roll around the slower side gear, driving the outer side gear - and tire - more quickly. The differential is housed within a casting with a pressed steel cover, known as the "carrier." The steel axle tubes are pressed into and welded to the carrier.

2   An optional locking limited-slip rear axle is also available. This differential allows for normal operation until one wheel loses traction. A limited-slip unit is similar in design to a conventional differential, except for the addition of a pair of clutch "cones" which slow the rotation of the differential case when one

wheel is on a firm surface and the other on a slippery one. The difference in wheel rotational speed produced by this condition applies additional force to the pinion gears and through the cone, which is splined to the axleshafts, equalizes the rotation speed of the axleshaft driving the wheel with traction.

Check

3   Often, a suspected "axle" problem lies elsewhere. Do a thorough check of other possible causes before assuming the axle is the problem.

4   The following noises are those commonly associated with axle diagnosis procedures:

a)   *Road noise is often mistaken for mechanical faults. Driving the vehicle on different surfaces will show whether the road surface is the cause of the noise. Road noise will remain the same if the vehicle is under power or coasting.*

b)   *Tire noise is sometimes mistaken for mechanical problems. Tires which are worn or low on pressure are particularly susceptible to emitting vibrations and noises. Tire noise will remain about the same during varying driving situations, where axle noise will change during coasting, acceleration, etc.*

c)   *Engine and transmission noise can be deceiving because it will travel along the driveline. To isolate engine and transmission noises, make a note of the engine speed at which the noise is most pronounced. Stop the vehicle and place the transmission in Neutral and run the engine to the same speed. If the noise is the same, the axle is not at fault.*

5   Because of the special tools needed, overhauling the differential isn't cost effective for a do-it-yourselfer. The procedures included in this Chapter describe axleshaft removal and installation, axleshaft oil seal replacement, axleshaft bearing replacement and removal of the entire unit for repair or replacement. Any further work should be left to a dealer service department or other qualified repair facility.

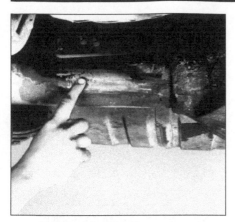

15.6 The identification code and manufacturer's code are stamped on the right rear axle tube on the front side

16.3 Remove the lock bolt and pull out the differential pinion gear shaft (arrow)

16.4 Remove the C-lock from the axleshaft

17.2 You can pry out the old seal with the axleshaft

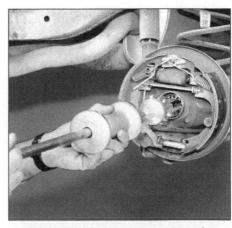

17.3 Install the new axleshaft seal with a seal driver

Identification

Refer to illustration 15.6

6   If the rear axle must be replaced, refer to the identification code and manufacturer's code stamped on the right rear axle tube on the front side of the rear axle tube **(see illustration)**. This number contains information on the rear axle ratio, differential type, manufacturer and build date information, all of which are necessary to ensure that you get the right axle.

16 Rear axleshaft - removal and installation

Refer to illustrations 16.3 and 16.4
Caution: *On later models with traction control, remove the axle very carefully to avoid damaging the speed sensor reluctor ring. The axle and reluctor are an assembly, and if the ring is damaged, the axle and reluctor must be replaced as an assembly only.*
1   Raise the rear of the vehicle, support it securely and remove the wheel and brake drum (refer to Chapter 9). On models with rear disc brakes, remove the caliper and rotor (see Chapter 9).
2   Remove the cover from the differential

carrier and allow the oil to drain into a container.
3   Remove the lock bolt from the differential pinion shaft. Remove the pinion shaft **(see illustration)**.
4   Push the outer (flanged) end of the axleshaft in and remove the C-lock from the inner end of the shaft **(see illustration)**.
5   Withdraw the axleshaft, taking care not to damage the oil seal in the end of the axle housing as the splined end of the axleshaft passes through it.
6   Installation is the reverse of removal. Tighten the lock bolt to the torque listed in this Chapter's Specifications.
7   Always use a new cover gasket and tighten the cover bolts to the torque listed in the Chapter 1 Specifications.
8   Refill the axle with the correct quantity and grade of lubricant (see Chapter 1).

17 Rear axleshaft oil seal - replacement

Refer to illustrations 17.2 and 17.3
1   Remove the axleshaft (see Section 16).
2   Pry the oil seal out of the end of the axle

housing with a large screwdriver or the inner end of the axleshaft **(see illustration)**.
3   Apply multi-purpose grease to the oil seal lips and recess and tap the new seal evenly into place with a hammer and seal installation tool **(see illustration)**, large socket or piece of pipe so the lips are facing in and the metal face is visible from the end of the axle housing. When correctly installed, the face of the oil seal should be flush with the end of the axle housing.

18 Rear axleshaft bearing - replacement

Refer to illustrations 18.3 and 18.4
1   Remove the axleshaft (see Section 16) and the oil seal (see Section 17).
2   A bearing puller which grips the bearing from behind will be required for this job.
3   Attach a slide hammer to the puller and extract the bearing from the axle housing **(see illustration)**.
4   Clean out the bearing recess and drive in the new bearing with a bearing installer or a piece of pipe positioned against the outer

18.3 Use a slide hammer and bearing removal attachment to remove the old axleshaft bearing

18.4 Install the new axleshaft bearing with a bearing driver, large socket or piece of pipe

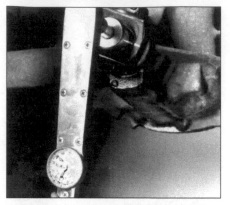

19.3 Use an inch-pound torque wrench to check the torque necessary to rotate the pinion shaft

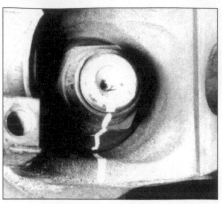

19.4 Mark the relative positions of the pinion, nut and flange (arrows) before removing the nut

bearing race **(see illustration)**. Make sure the bearing is tapped in to the full depth of the recess and the numbers on the bearing are visible from the outer end of the axle housing.
5   Install a new oil seal (see Section 17), then install the axleshaft.

19   Pinion oil seal - replacement

Refer to illustrations 19.3, 19.4 and 19.6
1   Raise the rear of the vehicle and support it securely on jackstands. Block the front wheels to keep the vehicle from rolling off the stands.
2   Disconnect the driveshaft and fasten it out of the way.
3   Use an inch-pound torque wrench to check the torque required to rotate the pinion. Record the torque value for use later **(see illustration)**.
4   Scribe or punch alignment marks on the pinion shaft, nut and flange **(see illustration)**.
5   Count the number of threads visible between the end of the nut and the end of the pinion shaft and record the number for use later.
6   A special tool can be used to keep the companion flange from moving while the self-locking pinion nut is loosened. If the special tool isn't available, try using a large pair of pliers **(see illustration)**.
7   Remove the pinion nut.
8   Pull off the companion flange. It may be necessary to use a two or three-jaw puller engaged behind the flange to draw it out. Do Not attempt to pry behind the flange or hammer on the end of the pinion shaft.
9   Pry out the old seal and discard it.
10   Lubricate the lips of the new seal with multi-purpose grease and tap it evenly into position with a seal installation tool or a large socket. Make sure it enters the housing squarely and is tapped in to its full depth.
11   Align the mating marks made before disassembly and install the companion flange. If necessary, tighten the pinion nut to draw the flange into place. Do not try to hammer the flange into position.
12   Apply non-hardening sealant to the

ends of the splines visible in the center of the flange so oil will be sealed in.
13   Install the washer (if equipped) and pinion nut. Tighten the nut carefully until the original number of threads are exposed.
14   Measure the torque required to rotate the pinion and tighten the nut in small increments until it matches the figure recorded in Step 4. In order to compensate for the drag of the new oil seal, the nut should be tightened more until the rotational torque of the pinion exceeds the earlier recording by 10 to 15 in-lbs. Be very careful and do not overtighten the pinion nut and do not back-off the pinion nut after it has been tightened.
15   Connect the driveshaft and lower the vehicle.

20   Rear axle assembly - removal and installation

Warning: *Do not attempt this procedure if your vehicle is equipped with the Traction Control System (TCS), as the brake system cannot be bled without the use of a Tech-1 or T-100 (CAMS) scan tool. Have the following procedure performed at a dealer service department or other qualified repair shop equipped with this special tool.*

Removal

1   Loosen the rear wheel lug nuts, raise the rear of the vehicle and support it securely on jackstands placed under the frame rails. Block the front wheels and remove the rear wheels.
2   Mark the driveshaft and companion flange, unbolt the driveshaft and support it out of the way on a wire hanger (see Section 12).
3   Position a floor jack under the differential housing. Raise the jack just enough to take up the weight, but not far enough to take the weight of the vehicle off the jackstands.
4   Remove the rear brake assemblies (see Chapter 9).
5   Remove the cover from the differential and drain the lubricant into a container.
6   Remove the axleshafts (see Section 16).

7   Disconnect the brake lines from the axle housing clips.
8   Remove the brake backing plates. If equipped with traction control, disconnect the electrical connector from the rear wheel speed sensor.
9   Disconnect both shock absorbers (see Chapter 10).
10   Remove the bolt that attaches the left side of the Panhard rod to the axle (see Chapter 10).
11   Remove the brake line junction block bolt at the axle housing, then disconnect the brake lines at the junction block.
12   Lower the jack under the differential housing enough to remove the springs.
13   Disconnect the lower control arms from the axle housing (see Chapter 10).
14   Disconnect the torque arm from the axle housing (see Chapter 10).
15   Remove the rear axle assembly from under the vehicle.

Installation

16   Installation is the reverse of the removal procedure.
17   When installation is complete, fill the differential with the recommended oil and bleed the brake system.

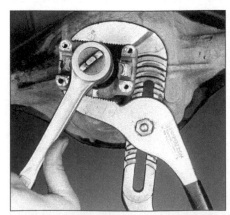

19.6 If you don't have the factory tool or one like it, try holding the flange with a large pair of pliers while you loosen the locknut

Chapter 9   Brakes

Contents

Specifications

Disc brakes

Disc thickness after resurfacing (minimum)	
Front	
1997 and earlier	1.250 inches
1998 and later	1.223 inches
Rear	
1997 and earlier	0.733 inch
1998 and later	0.985 inch
Discard thickness*	
Front	1.209 inches
Rear	
1997 and earlier	0.724 inch
1998 and later	0.965 inch
Disc runout (maximum)	
Front	0.005 inch
Rear	
Disc (installed on vehicle)	0.006 inch
Rear axle shaft flange	0.002 inch
Disc thickness variation (maximum)	0.0005 inch

Refer to the marks stamped on the disc (they supersede information printed here)

Rear drum brakes

Drum diameter	
Standard	9.50 inches
Service limit*	9.59 inches
Drum taper (maximum)	0.003 inch
Out-of-round (maximum)	0.002 inch

Refer to marks stamped on the drum (they supersede information printed here)

Torque specifications

	Ft-lbs (unless otherwise indicated)
Brake booster retaining nuts	15
Caliper bleeder valve	
Front	106 to 115 in-lbs
Rear	97 in-lbs
Caliper mounting bolts	
Front caliper guide pins	
1997 and earlier	38
1998 and later	23
Rear caliper guide pins	27
Caliper mounting plate bolts	74
Front brake hose-to-caliper banjo bolt	30 to32
Master cylinder mounting nuts	21
Rear brake hose-to-caliper banjo bolt	22
Rear wheel cylinder bolts	
1993 and 1994	115 in-lbs
1995 and later	156 in-lbs

1 General information

All vehicles covered by this manual are equipped with hydraulically operated front and rear brake systems. All front brake systems are disc type, while rear brake systems are drum brakes on some models and disc type on others. A quick visual inspection will tell you what type of brakes you have at the rear of your vehicle. All brakes are self-adjusting.

The hydraulic system consists of two separate front and rear circuits. The master cylinder has separate reservoirs for the two circuits and in the event of a leak or failure in one hydraulic circuit, the other circuit will remain operative. A visual warning of circuit failure or air in the system is given by a warning light activated by displacement of the piston in the brake distribution (pressure differential warning) switch from its normal "in balance" position.

The parking brake mechanically operates the rear brakes only. It is activated by a pull-handle in the center console between the front seats.

A combination valve, located in the engine compartment, consists of three sections providing the following functions: The metering section limits pressure to the front brakes until a predetermined front input pressure is reached and until the rear brakes are activated. There is no restriction at inlet pressures below 3 psi, allowing pressure equalization during non-braking periods. The proportioning section proportions outlet pressure to the rear brakes after a predetermined rear input pressure has been reached, preventing early rear wheel lock-up under heavy brake loads. The valve is also designed to assure full pressure to the rear brakes should the front brakes fail, and vice-versa.

The pressure differential warning switch is designed to continuously compare the front and rear brake pressure from the master cylinder and energize the dash warning light in the event of either front or rear brake system failure. The design of the switch and valve are such that the switch will stay in the "warning" position once a failure has occurred. The only way to turn the light off is to repair the cause of the failure and apply a brake pedal force of 450 psi.

The power brake booster, utilizing engine manifold vacuum and atmospheric pressure to provide assistance to the hydraulically operated brakes, is located in the engine compartment, mounted on the driver's side of the firewall.

After completing any operation involving the disassembly of any part of the brake system, always test drive the vehicle to check for proper braking performance before resuming normal driving. When testing the brakes, perform the tests on a clean, dry, flat surface. Conditions other than these can lead to inaccurate test results. Test the brakes at various speeds with both light and heavy pedal pressure. The vehicle should stop evenly without pulling to one side or the other. Avoid locking the brakes because this slides the tires and diminishes braking efficiency and control.

Tires, vehicle load and front-end alignment are factors which also affect braking performance.

Torque values given in the Specifications Section are for dry, unlubricated fasteners.

2 Anti-lock Brake System (ABS) - general information

A four-wheel Anti-lock Brake System (ABS) maintains vehicle maneuverability, directional stability, and optimum deceleration under severe braking conditions on most road surfaces. It does so by monitoring the rotational speed of the wheels and controlling the brake line pressure to the wheels during braking. This prevents the wheels from locking up on slippery roads or during hard braking.

Hydraulic modulator/motor pack assembly

The hydraulic modulator/motor pack assembly, mounted in the left front corner of the engine compartment, controls hydraulic pressure to the front calipers and rear wheel cylinders or calipers by modulating hydraulic pressure to prevent wheel lock-up.

The Electronic Brake Control Module (EBCM) is located above the left (driver's side) kick panel. The EBCM monitors the ABS system and controls the anti-lock valve solenoids. It accepts and processes information received from the brake switch and wheel speed sensors to control the hydraulic line pressure and avoid wheel lock up. It also monitors the system and stores fault codes which indicate specific problems.

Each sensor assembly consists of a variable reluctance sensor mounted adjacent to a "toothed ring" with an air gap between them. A wheel speed sensor and toothed ring are mounted in the hub/bearing unit of each front wheel. A third sensor is mounted on the rear differential housing; its toothed ring is located within the rear differential. The air gap between the sensors and the rings is not adjustable. And the sensors themselves are not rebuildable. If a front sensor or ring malfunctions, the entire hub and bearing assembly must be replaced. The rear sensor and ring can be replaced separately.

A wheel speed sensor measures wheel speed by monitoring the rotation of the toothed ring. As the teeth of the ring move through the magnetic field of the sensor, an AC voltage is generated. This signal frequency increases or decreases in proportion to the speed of the wheel. The ECBM monitors these three signals for changes in wheel speed; if it detects the sudden deceleration of a wheel, i.e. wheel lockup, the EBCM activates the ABS system.

The ABS system has self-diagnostic capabilities. Each time the vehicle is started, the EBCM runs a self-test. The red BRAKE warning light should come on briefly then go out. The EBCM also monitors the ABS system continuously during vehicle operation. If the ABS INOP light on the dash comes on while you're driving, there is a fault somewhere in the ABS system and the ABS system may be inoperative, but the brakes should still function in their non-ABS mode. Take the vehicle to a dealer service department or other qualified repair shop immediately and have the ABS serviced.

If the ABS INOP light *flashes*, however, there is a more serious fault in the ABS system which may have affected the regular braking system. Pull over immediately and have the vehicle towed to a dealer or other qualified repair shop for service.

Although a special electronic tester is necessary to properly diagnose the system, the home mechanic can perform a few preliminary checks before taking the vehicle to a dealer service department which is equipped with this tester.

a) *Make sure the brake calipers are in good condition.*
b) *Check the electrical connector at the controller.*
c) *Check the fuses.*
d) *Follow the wiring harness to the speed sensors and brake light switch and make sure all connections are secure and the wiring isn't damaged.*

If the above preliminary checks don't rectify the problem, the vehicle should be diagnosed by a dealer service department.

3 Disc brake pads - replacement

Warning: *Disc brake pads must be replaced on both front or rear wheels at the same time - never replace the pads on only one wheel. Also, the dust created by the brake system is harmful to your health. Never blow it out with compressed air and don't inhale any of it. An approved filtering mask should be worn when working on the brakes. Do not, under any circumstances, use petroleum-based solvents to clean brake parts. Use brake system cleaner only!*

1997 and earlier models

Refer to illustrations 3.3, 3.4a through 3.4h and 3.4i through 3.4p

1 Loosen the front or rear wheel lug nuts, raise the front or rear of the vehicle and support it securely on jackstands, then remove the wheels. Be sure to block the wheels at the opposite end.

2 Remove about two-thirds of the fluid from the master cylinder reservoir and discard it. **Caution:** *Do not spill brake fluid on painted surfaces.* Position a drain pan under the brake assembly and clean the caliper and surrounding area with brake system cleaner.

3 Position a large C-clamp over the caliper as shown and squeeze the piston back into in its bore **(see illustration)** to provide room for the new brake pads. As the pis-

ton is depressed to the bottom of its caliper bore, the fluid in the master cylinder will rise. Make sure it doesn't overflow. If necessary, siphon off some of the fluid.

4   Work on one brake assembly at a time. If you're replacing the *front brake pads*, follow the accompanying photos, beginning with **illustrations 3.4a through 3.4h**; if you're replacing the *rear pads*, start at **illustration 3.4i**. Be sure to stay in order and read the caption under each illustration.

5   While the pads are removed, inspect the caliper for seal leaks (evidenced by moisture around the cavity) and for any damage to the piston dust boots. If excessive moisture is evident, rebuild the caliper (see Section 4). When you're done, make sure you tighten the caliper bolts to the torque listed in this Chapter's Specifications.

3.3 Depress the piston into the caliper with a large C-clamp as shown

Front brake pads (1997 and earlier)

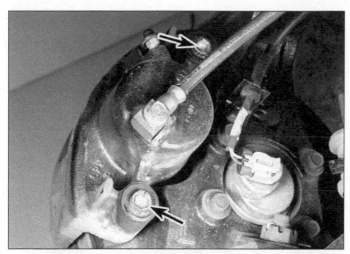

3.4a To detach the front caliper from the steering knuckle, remove these two bolts (upper and lower arrows); don't remove the banjo bolt for the brake hose unless you intend to overhaul the caliper

3.4b While you're replacing the pads, hang the caliper with a piece of wire

3.4c Remove the outer brake pad

3.4d Remove the inner brake pad

3.4e Pull the two caliper guide pins out of their sleeves, clean them, inspect them for wear, replacing as necessary . . .

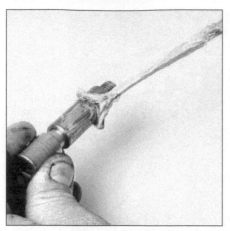

3.4f . . . then lubricate them with high-
temperature grease and install them in
the caliper

3.4g Install the new inner pad

3.4h Install the new outer pad, then install
the caliper and tighten the caliper guide
pins to the torque listed in this
Chapter's Specifications

Rear brake pads (1997 and earlier)

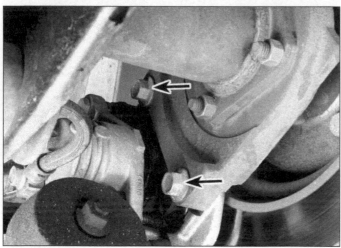

3.4i To detach the rear caliper from the rear axle, remove these
two bolts (arrows) (don't remove the brake hose banjo bolt unless
you intend to overhaul the caliper)

3.4j Suspend the rear caliper with a piece of wire while you're
replacing the pads

3.4k Remove the outer brake pad from
the caliper

3.4l Remove the inner brake pad
from the caliper

3.4m Remove the rear caliper
anchor bracket

3.4n Depress the piston back into the caliper with a C-clamp

3.4o Install the new inner brake pad

3.4p Install the new outer brake pad, install the caliper assembly over the rear disc and tighten the caliper mounting bolts to the torque listed in this Chapter's Specifications

Front and rear brake pads (1998 and later)

3.13a Before disassembling the brake, wash it thoroughly with brake system cleaner and allow it to dry - position a drain pan under the brake to catch the residue - DO NOT use compressed air to blow off brake dust!

3.13b To make room for the new pads, use a C-clamp to depress the piston into the caliper before removing the caliper and pads - do this a little at a time, keeping an eye on the fluid level in the master cylinder to make sure it doesn't overflow

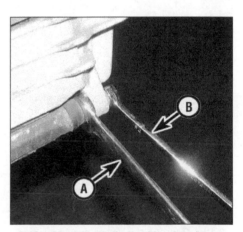

3.13c Hold the caliper slide pin with an open-end wrench (A) and loosen the lower mounting bolt with another wrench (B)

6   Install the brake pads on the opposite wheel, then install the wheels and lower the vehicle. Tighten the lug nuts to the torque listed in the Chapter 1 Specifications. Add brake fluid to the reservoir until it's full (see Chapter 1).

7   Pump the brakes several times to seat the pads against the disc, then check the fluid level again.

8   Check the operation of the brakes before driving the vehicle in traffic. Try to avoid heavy brake applications until the brakes have been applied lightly several times to seat the pads.

1998 and later models (front or rear)

Refer to illustrations 3.13a through 3.13m
Warning: *Disc brake pads must be replaced on both front or both rear wheels at the same time - never replace the pads on only one*

wheel. Also, the dust created by the brake system is harmful to your health. Never blow it out with compressed air and don't inhale any of it. An approved filtering mask should be worn when working on the brakes. Do not, under any circumstances, use petroleum-based solvents to clean brake parts. Use brake system cleaner only!
Note: *This procedure applies to the front **and** rear brake pads.*

9   Remove the cap from the brake fluid reservoir. Remove about two-thirds of the fluid from the reservoir. **Caution:** *Brake fluid will damage paint. If any fluid is spilled, wash it off immediately with plenty of clean, cold water.*

10   Loosen the front or rear wheel lug nuts, raise the front or rear of the vehicle and support it securely on jackstands. Block the wheels at the opposite end.

11   Remove the wheels. Work on one brake assembly at a time, using the assembled brake for reference, if necessary.

12   Inspect the brake disc carefully as outlined in Section 5. If machining is necessary, follow the information in that Section to remove the disc.

13   Follow the accompanying photo sequence for the actual pad replacement procedure **(see illustrations 3.13a through 3.13m)**. Be sure to stay in order and read the caption under each illustration.

14   When reinstalling the caliper, be sure to tighten the mounting bolts to the torque listed in this Chapter's Specifications. Tighten the wheel lug nuts to the torque listed in the Chapter 1 Specifications.

15   After the job has been completed, firmly depress the brake pedal a few times to bring the pads into contact with the disc. Check the level of the brake fluid, adding some if necessary (see Chapter 1). Check the operation of the brakes carefully before placing the vehicle into normal service.

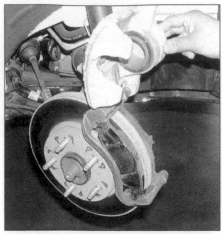

3.13d Pivot the caliper up and support it in this position for access to the brake pads

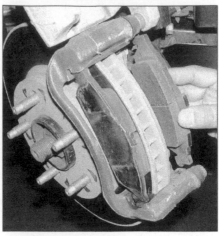

3.13e Remove the inner brake pad

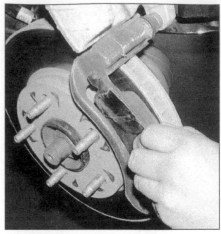

3.13f Remove the outer brake pad

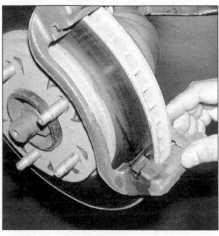

3.13g Remove the upper and lower pad retainers from the caliper mounting bracket and inspect them for wear and damage

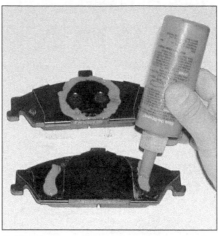

3.13h Apply anti-squeal compound to the back of both pads (let the compound "set up" a few minutes before installing them)

3.13i Install the upper and lower pad retainers

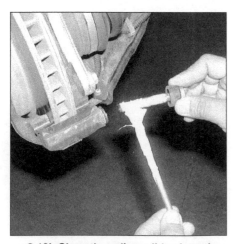

3.13j Clean the caliper slide pin and inspect it for scoring and corrosion; coat the pin with high-temperature grease

3.13k Install the inner brake pad

3.13l Install the outer brake pad

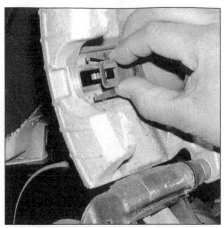

3.13m Check the condition of the anti-rattle spring in the center of the caliper, replacing it if necessary. Swing the caliper down over the pads and install the lower mounting bolt, tightening it to the torque listed in this Chapter's Specifications. Note: *If the caliper won't fit over the pads, use a C-clamp to push the piston into the caliper a little further*

4.10 Using a wood block as a cushion, ease the piston out of the caliper bore with compressed air

4.11 Pry the dust boot out of the caliper bore with a screwdriver - make sure you don't nick or gouge anything

---

4   Disc brake caliper - removal, overhaul and installation

Warning 1: *Do not attempt this procedure if your vehicle is equipped with the Traction Control System (TCS), as the brake system cannot be bled without the use of a Tech-1 or T-100 (CAMS) scan tool. Have the following procedure performed at a dealer service department or other qualified repair shop equipped with this special tool.*
Warning 2: *The dust created by the brake system is harmful to your health. Never blow it out with compressed air and don't inhale any of it. An approved filtering mask should be worn when working on the brakes. Do not, under any circumstances, use petroleum-based solvents to clean brake parts. Use brake system cleaner only!*
Note: *If an overhaul is indicated (usually because of fluid leaks, a stuck piston or broken bleeder screw) explore all options before beginning this procedure. New and factory rebuilt calipers are available on an exchange basis, which makes this job quite easy. If you decide to rebuild the calipers, make sure rebuild kits are available before proceeding. Always rebuild or replace the calipers in pairs - never rebuild just one of them.*

Removal

1   Remove the cover from the brake fluid reservoir and siphon off two-thirds of the fluid into a container and discard it.
2   Loosen the wheel lug nuts, raise the front or rear of the vehicle and place it securely on jackstands. Remove the front or rear wheel.
3   Reinstall one wheel lug (flat side toward the disc) to hold the disc in place. Don't remove both calipers at the same time.

Instead, remove only one caliper at a time so you can use the other assembled unit for reference.
4   Push the caliper piston back into its bore with a C-clamp **(see illustration 3.3)**. As the piston is depressed to the bottom of the caliper bore, the fluid in the master cylinder will rise. Make sure that it does not overflow.
5   Remove the banjo fitting bolt holding the brake hose, then remove and discard the sealing washers found on either side of the banjo fitting. Always use new sealing washers when reinstalling the brake hose.
6   To prevent brake fluid leakage and contamination, plug the openings in the caliper and brake hose. **Note:** *If you're just removing the caliper for access to other components, don't disconnect the hose.*
7   Remove the caliper and separate the caliper from the disc.. On 1997 and earlier models, refer to **illustration 3.4a** for the front caliper and **illustration 3.4i** for the rear caliper. On 1998 and later models refer to **illustration 3.13c** for both the front and the rear caliper mounting bolts.
8   On 1997 and earlier models, remove the brake pads from the caliper (see Section 3).

Overhaul (1997 and earlier models only)
Front caliper
Refer to illustrations 4.10, 4.11, 4.12, 4.16, 4.17, 4.18 and 4.19
Note: *Purchase a brake caliper overhaul kit for your particular vehicle before beginning this procedure.*
9   Clean the exterior of the brake caliper with brake system cleaner (never use gasoline, kerosene or any petroleum-based solvents), then place the caliper on a clean workbench.
10   Place a wooden block or shop rag in the caliper as a cushion, then use compressed air to remove the piston from the caliper **(see illustration)**. Use only enough air pressure to ease the piston out of the bore. If the piston is blown out, even with the cushion in place,

it may be damaged. **Warning:** *Never place your fingers in front of the piston in an attempt to catch or protect it when applying compressed air - serious injury could occur.*
11   Carefully pry the dust boot out of the caliper bore **(see illustration)**.
12   Using a wooden or plastic tool, remove the piston seal from the groove in the caliper bore **(see illustration)**. Metal tools may cause bore damage.
13   Remove the caliper bleeder valve, then remove and discard the sleeves and bushings from the caliper ears. Also discard all rubber parts.
14   Clean the remaining parts with brake fluid or brake system cleaner. Allow them to drain and then shake them vigorously to remove as much fluid as possible.
15   Carefully examine the piston for nicks and burrs and loss of plating. If surface defects are present, parts must be replaced. Check the caliper bore in a similar way, but light polishing with crocus cloth is permissible to remove light corrosion and stains. Discard the mounting bolts if they are corroded or damaged.

4.12 To remove the seal from the caliper bore, use a plastic or wooden tool, such as a pencil

4.16 Lubricate the piston bores and seal with clean brake fluid before placing the seal in the caliper bore groove

4.17 Install a new dust boot in the piston groove - note that the folds (the accordion-like pleats) are facing out

4.18 Lubricate the piston with clean brake fluid, insert it squarely in the bore and carefully push it into the caliper

4.19 Use a hammer and driver to seat the boot in the caliper housing counterbore

4.23 Remove the anchor bracket

4.24 Remove the springs from each end of the parking brake collar

16   When assembling, lubricate the piston bores and seal with clean brake fluid; position the seal in the caliper bore groove. Make sure the seal seats properly and isn't twisted **(see illustration).**

17   Lubricate the piston with clean brake fluid, then install a new boot in the piston groove with the fold toward the open end of the piston **(see illustration).**

18   Insert the piston squarely into the caliper bore, then apply force to bottom the piston in the bore **(see illustration).**

19   Position the dust boot in the caliper counterbore, then use a drift to drive it into position **(see illustration)**. Make sure that the boot is evenly installed below the caliper face.

20   Install the bleeder valve.

21   Proceed to Step 43.

Rear caliper

Refer to illustrations 4.23, 4.24, 4.25, 4.26, 4.27, 4.28, 4.29, 4.34, 4.35, 4.36, 4.38, 4.39 and 4.40

Note: *Purchase a brake caliper overhaul kit for your particular vehicle before beginning this procedure.*

22   Clean the exterior of the brake caliper with brake system cleaner (never use gasoline, kerosene or any petroleum-based cleaning solvents), then place the caliper on a clean workbench.

23   Remove the anchor bracket **(see illustration).**

24   Remove the parking brake collar return springs **(see illustration).**

25   Place a wood block or shop rag in the caliper as a cushion, then use compressed air to remove the piston from the caliper **(see illustration)**. Use only enough air pressure to ease the piston out of the bore. If the piston is blown out, even with the cushion in place, it may be damaged. **Warning:** *Never place your fingers in front of the piston in an attempt to catch or protect it when applying compressed air - serious injury could occur.*

4.25 Place a wood block or shop rag in the caliper as a cushion, then use compressed air to remove the piston from the caliper (use no more air pressure than necessary to ease the piston out of the bore)

4.26 Remove the clamp rod, actuating collar boot retainers, actuating collar boots, actuating collar and piston as a single assembly

4.27 Slide the piston off the clamp rod bushing

4.28 Separate the clamp rod from the actuating collar boot retainers and actuating collar boots

4.29 Use a wood or plastic tool to remove the piston seal from the bore

4.34 Lubricate the clamp rod bushing with the supplied lubricant, then slide the piston onto the clamp rod bushing

4.35 Lubricate the piston bore with brake fluid, then place the assembled piston, actuating collar retainers and dust boots and clamp rod into the caliper assembly

26   Remove the clamp rod, actuating collar boot retainers, actuating collar boots, actuating collar and piston as a single assembly **(see illustration)**.
27   Slide the piston off the clamp rod bushing **(see illustration)**.
28   Separate the clamp rod from the actuating collar boot retainers and actuating collar boots **(see illustration)**.
29   Use a wooden or plastic tool to remove the piston seal from the bore **(see illustration)**.
30   Carefully check the caliper bore for score marks, nicks, corrosion and excessive wear. Light corrosion may be removed with crocus cloth; otherwise, replace the caliper housing with a new one.
31   Clean all parts not included in the caliper repair kit with clean brake fluid or brake system cleaner. Do not, under any circumstances, use petroleum-based solvents.
32   Use compressed air to dry the parts and blow out all the passages in the caliper housing and bleeder valve.
33   Lubricate the new piston seal with clean brake fluid and install it in the caliper bore groove. Make sure that the seal is not twisted.
34   Lubricate the clamp rod bushing and clamp rod with the liquid lubricant supplied in the overhaul kit, then slide the piston onto the clamp rod bushing **(see illustration)**.

35   Also lubricate the bead of the actuating collar, actuating collar boot and the boot's grove in the caliper with the same liquid lubricant. Lubricate the piston bore with brake fluid, then place the assembled piston, actuating collar retainers and dust boots and clamp rod into the caliper assembly **(see illustration)**.
36   With the piston, actuating collar retain-

ers and dust boots and clamp rod in position, square to the bore, use your thumbs to push the piston into the caliper **(see illustration)**.
37   Install the parking brake collar return springs **(see illustration 4.24)**.
38   Lubricate the caliper guide pins with high temperature grease **(see illustration)**.

4.36 With the piston, actuating collar retainers and dust boots and clamp rod in position, square to the bore, use your thumbs to push the piston into the caliper

4.38 Lubricate the caliper guide pins with high-temperature grease

4.39 Inspect the guide pin boots for cracks and replace as necessary

4.40 Install the caliper anchor bracket

5.5 Check the runout of the brake disc with a dial indicator

39   Inspect the guide pin boots for cracks. Replace as necessary **(see illustration)**.
40   Install the caliper anchor bracket **(see illustration)**.
41   Install the brake pads (see Section 3).
42   Install the bleeder valve.
43   Installation is the reverse of removal. Always use new sealing washers when connecting the brake hose. Tighten the caliper guide pins or mounting bolts to the torque listed in this Chapter's Specifications. Be sure to fill the master cylinder and bleed the brakes (see Section 10).

5   Brake disc - inspection, removal and installation

Refer to illustrations 5.5, 5.6 and 5.7
1   Loosen the wheel lug nuts, raise the vehicle and place it securely on jackstands. Remove the wheel.
2   Remove the caliper assembly (see Section 4). It is not necessary to disconnect the brake hose. After removing the caliper mounting bolts, hang the caliper out of the way on a piece of wire. Never hang the caliper by the brake hose because damage to the hose will occur.
3   Inspect the disc surfaces. Light scoring or grooving is normal, but deep grooves or severe erosion is not. If pulsating has been noticed during application of the brakes, suspect disc runout.
4   Reinstall the wheel lug nuts - flat side toward the disc - to hold the disc in a flat, vertical plane during the following inspection.
5   Attach a dial indicator to the caliper mounting bracket, turn the disc and note the amount of runout **(see illustration)**. Check both inner and outer surfaces. If the runout is more than the specified allowable maximum, the disc must be removed from the vehicle and taken to an automotive machine shop for resurfacing.
6   Using a micrometer, measure the thickness of the disc **(see illustration)**. If it is less than the specified minimum, replace the disc. Also measure the disc thickness at several points to determine variations in the surface.

Any variation over 0.0005-inch may cause pedal pulsations during brake application. If this condition exists and the disc thickness is not below the minimum, the disc can be removed and taken to an automotive machine shop for resurfacing.
7   If the disc needs to be removed for repair or replacement on it can be pulled off after the lug nuts are removed. On 1998 and later models, be sure to remove the caliper mounting bracket first **(see illustration)**.

6   Drum brake shoes - replacement

Refer to illustrations 6.5a through 6.5ss and 6.6
Warning: *Drum brake shoes must be replaced on both wheels at the same time - never replace the pads on only one wheel. Also, the dust created by the brake system may contain asbestos, which is harmful to your health. Never blow it out with compressed air and don't inhale any of it. An approved filtering mask should be worn when working on the brakes. Do not, under any circumstances, use petroleum-based solvents to clean brake parts. Use brake system cleaner only!*
Caution: *Whenever the brake shoes are replaced, the return and hold-down springs should be replaced. Due to the continuous*

5.6 Use a micrometer to check the thickness of the disc and compare this measurement to the minimum allowable thickness stamped onto the disc

heating/cooling cycle the springs are subjected to, they lose their tension over a period of time and may allow the shoes to drag on the drum and wear at a much faster rate than normal.
1   Release the parking brake handle.
2   Loosen the rear wheel lug nuts, raise the vehicle and place it securely on jackstands. Remove the wheel.
3   All four rear shoes should be replaced at the same time, but to avoid mixing up parts, work on only one brake assembly at a time.
4   Wash down the brake assembly with brake cleaner. Do NOT blow it out with compressed air.
5   To replace the brake shoes, refer to the accompanying photographs, beginning with **illustration 6.5a**. If the brake drum cannot be easily pulled off the axle and shoe assembly, make sure that the parking brake is completely released, then squirt some penetrating oil around the center hub area. Allow the oil to soak in and try to pull the drum off again. If the drum still cannot be pulled off, the brake shoes will have to be retracted. Remove the lanced cutout in the backing plate with a hammer and chisel. With this lanced area punched in, pull the lever off the adjusting screw wheel with one small screwdriver while turning the adjusting wheel with another small screwdriver, moving the shoes away from the drum. The drum may now be pulled off.

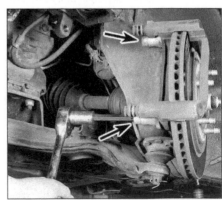

5.7 On 1998 and later models, it will be necessary to remove the caliper mounting bracket and bolts (arrows) before removing the brake disc

6.5a Pry off and discard the drum retaining washers, if present

6.5b Mark the relationship of the axle and brake drum

6.5c Remove the brake drum (if it cannot be pulled off, refer to text)

6.5d Typical drum brake assembly (left side shown, right side is the reverse)

1　Primary shoe return spring
2　Wheel cylinder
3　Shoe guide
4　Anchor pin
5　Secondary shoe return spring
6　Actuator link
7　Parking brake cable
8　Secondary brake shoe
9　Actuator lever pivot and hold-down spring
10　Actuator lever
11　Lever return spring
12　Pawl
13　Adjusting screw wheel
14　Adjusting screw
15　Adjusting screw spring
16　Primary brake shoe
17　Hold-down pin
18　Hold-down spring

6.5e Use a brake spring tool to remove the primary shoe return spring from the anchor pin

6.5f Remove the spring after unhooking it from the primary brake shoe

6.5g Push on the bottom of the automatic adjuster actuator lever (bottom hand) and remove the actuator link from the top of the actuator lever

6.5h Remove the actuator link and secondary shoe return spring from the anchor pin pivot with a brake spring tool

6.5i The actuator link and secondary shoe return spring after removal

6.5j Remove the shoe guide from the anchor pin

6.5k Remove the primary shoe hold-down spring and pin

6.5l The primary shoe hold-down spring and pin after removal

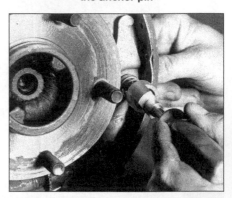

6.5m Remove the secondary shoe hold-down spring and pin (note that this spring is shorter than the primary shoe hold-down spring)

6.5n Remove the actuator lever, pawl and lever return spring (note L mark on the actuator lever denoting left side brake assembly)

6.5o Turn the adjuster wheel all the way in

6.5p Pivot the secondary shoe to the rear and down

6.5q Remove the secondary shoe from the adjusting screw spring

6.5r Remove the adjusting screw and spring

6.5s Note that the longer hook goes toward the rear

6.5t Pull the parking brake strut and spring assembly from behind the axle flange

6.5u Pull out on the primary shoe and pivot it down until it is free of the backing plate

6.5v Remove the spring from the parking brake lever

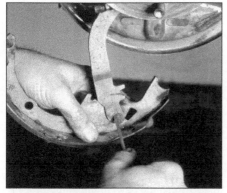

6.5w Remove the E-clip retaining the parking brake lever to the secondary shoe

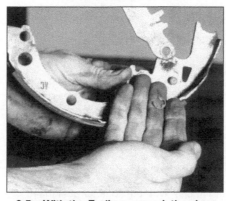

6.5x With the E-clip removed, the shoe may be separated from the parking brake lever

6.5y Check the shoe guide pads on the backing plate for wear, making sure a ridge has not formed which could hang up the brake shoe (if a sharp ridge is present, it may be removed with 150-grit emery cloth). This step completes the shoe removal procedure. At this time, check the wheel cylinder for leakage. If leakage is noted, remove the wheel cylinder for rebuilding or replacement (refer to Section 10)

6.5z Lubricate the shoe guide pads and wheel cylinder ends with multi-purpose grease (application on the wheel cylinder ends should be light)

6.5aa Lightly lubricate the anchor pin with multi-purpose grease

6.5bb Attach the parking brake lever to the primary brake shoe and the parking brake lever spring to the lever (note that the primary shoe has the shorter lining)

6.5cc Install the primary shoe in position on the backing plate and install the hold-down spring

6.5dd Place the parking brake strut into position in the primary shoe slot

6.5ee Lubricate both ends of the adjuster screw

6.5ff Engage the short hook end of the adjusting spring in the hole in the primary shoe, with the hook inserted from the front

6.5gg Engage the long hook end of the spring in the hole in the secondary shoe, with the hook inserted from the rear

6.5hh With the adjuster positioned between the shoes, pivot the secondary shoe into place on the backing plate

6.5ii Attach the pawl to the actuator lever and install the actuator assembly in place inside the secondary shoe

6.5jj Attach the bushing to the actuator and shoe

6.5kk Install the secondary shoe hold-down spring

6.5ll Install the shoe guide, with the rounded side facing in

6.5mm Install the rear hook of the secondary shoe return spring in the shoe

6.5nn Assemble the actuating link to the spring

6.5oo Attach the spring and actuating link to the anchor pin

6.5pp Attach the actuating link to the parking brake actuating lever

6.5qq Attach the primary shoe return spring to the shoe . . .

6.5rr . . . and to the anchor pin

6.5ss With all parts installed, rock the assembly back and forth with your hands to ensure that all parts are seated (to complete the job, refer to Steps 6 through 9)

6.6 The maximum allowable diameter is cast inside the brake drum

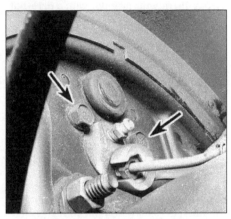

7.2 Unscrew the brake line fitting from the rear of the wheel cylinder with a flare nut wrench to prevent rounding off the corners of the nut, then remove the wheel cylinder retaining bolts (arrows)

6 Before reinstalling the drum, it should be checked for cracks, score marks, deep scratches and "hard spots," which will appear as small discolored areas. If the hard spots cannot be removed with fine emery cloth and/or if any of the other conditions listed above exist, the drum must be taken to an automotive machine shop to have it resurfaced. If the drum will not "clean up" before the maximum drum diameter is reached in the machining operation, the drum will have to be replaced with a new one. **Note:** *The maximum diameter is cast into each brake drum* **(see illustration)**.

7 Install the brake drum, lining up the marks made before removal if the old brake drum is used. Turn the adjuster wheel until the brake shoes just drag on the drum when the drum is rotated, then back-off the adjuster wheel until the shoes don't drag.

8 Mount the wheel and tire, install the wheel lug nuts and tighten them to the torque listed in the Chapter 1 Specifications, then lower the vehicle.

9 Make a number of forward and reverse stops to adjust the brakes until a satisfactory pedal action is obtained.

7 Wheel cylinder - removal, overhaul and installation

Refer to illustrations 7.2 and 7.5
Note: *Obtain a wheel cylinder rebuild kit before beginning this procedure.*

1 Remove the brake shoes (see Section 6).

2 Remove the brake line fitting from the rear of the wheel cylinder **(see illustration)**. Cap the brake line to prevent contamination

and excessive fluid loss.

3 Remove the wheel cylinder retaining bolts from the rear of the backing plate.

4 Remove the cylinder and place it on a clean workbench.

5 Remove the bleeder valve, seals, pistons, boots and spring assembly from the cylinder body **(see illustration)**.

6 Clean the wheel cylinder with brake fluid

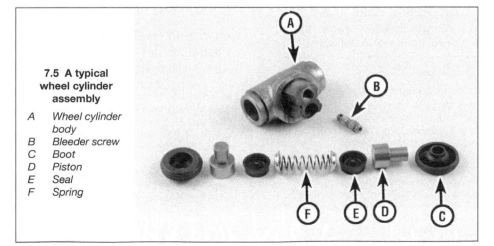

7.5 A typical wheel cylinder assembly

A   Wheel cylinder body
B   Bleeder screw
C   Boot
D   Piston
E   Seal
F   Spring

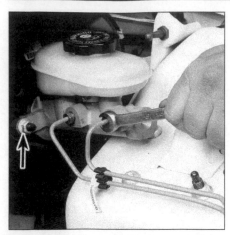

8.3 Remove the two brake line fittings from the master cylinder with a flare-nut wrench; to detach the master cylinder from the power brake booster, remove the nuts (arrow) (the other nut isn't visible in this photo)

8.10 The secondary seals must be installed with the lips facing out

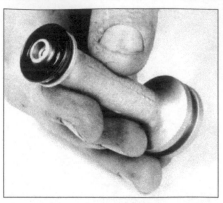

8.13 The primary seal must be installed with the lip facing away from the piston

or brake system cleaner. Do not, under any circumstances, use petroleum-based solvents to clean brake parts.

7   Use filtered, unlubricated compressed air to remove excess fluid from the wheel cylinder and to blow out the passages.

8   Check the cylinder bore for corrosion and scoring. Crocus cloth may be used to remove light corrosion and stains, but the cylinder must be replaced with a new one if the defects cannot be removed easily, or if the bore is scored.

9   Lubricate the new seals with clean brake fluid.

10   Assemble the brake cylinder, making sure the boots are properly seated.

11   Place the wheel cylinder in position on the backing plate. Thread the brake line fitting into the cylinder, being careful not to cross-thread it. Don't tighten the fitting yet.

12   Install the wheel cylinder retaining bolts and tighten them to the torque listed in this Chapter's Specifications. Now tighten the brake line fitting securely.

13   Bleed the brake system (see Section 10).

---

8   Master cylinder - removal, overhaul and installation

Removal

Refer to illustration 8.3

Warning: *Do not attempt this procedure if your 1999 or earlier vehicle is equipped with the Traction Control System (TCS), as the brake system cannot be bled without the use of a Tech-1 or T-100 (CAMS) scan tool. Have the following procedure performed at a dealer service department or other qualified repair shop equipped with this special tool.*

Caution: *Have some plugs ready to cap the metal lines that connect the master cylinder to the combination valve. Failure to do so will*

allow air into the ABS modulator and can allow dirt and moisture to enter the system.

Note: *A master cylinder overhaul kit should be purchased before beginning this procedure. The kit will include all the replacement parts necessary for the overhaul procedure. The rubber replacement parts, particularly the seals, are the key to fluid control within the master cylinder. As such, it's very important that they be installed securely and facing in the proper direction. Be careful during the rebuild procedure that no grease or mineral-based solvents come in contact with the rubber parts.*

1   Completely cover the front fender and cowling area of the vehicle; brake fluid can ruin painted surfaces if it is spilled.

2   Remove the brake fluid from the master cylinder reservoir and discard it. **Caution:** *Do not spill brake fluid on painted surfaces.* Place a drain pan under the master cylinder assembly and clean the area around the brake line fittings with brake system cleaner.

3   Disconnect the brake line fittings **(see illustration)**. Rags or newspapers should be placed under the master cylinder to soak up the fluid that will drain out.

4   Remove the two master cylinder mounting nuts, and remove the master cylinder from the vehicle. Do not to bend the hydraulic lines running to the combination valve. Plug the ends of both of these lines immediately to prevent air from entering the ABS modulator and to protect the system from moisture and dirt.

Overhaul (1997 and earlier models only)

Refer to illustrations 8.10 and 8.13

5   Remove the reservoir cover and reservoir diaphragm, then discard any remaining fluid in the reservoir. Knock out the two roll pins that secure the reservoir to the master cylinder body, remove the reservoir and the old O-rings between the reservoir and the master cylinder.

6   Remove the primary piston lock ring by depressing the piston and prying out the ring with a screwdriver.

7   Remove the primary piston assembly with a wire hook, being careful not to scratch the bore surface.

8   Remove the secondary piston assembly in the same manner.

9   Inspect the cylinder bore for corrosion and damage. If any corrosion or damage is found, replace the master cylinder body with a new one, as abrasives cannot be used on the bore.

10   Remove the old seals from the secondary piston assembly and install the new seals so that the cups face out **(see illustration)**.

11   Attach the spring retainer to the secondary piston assembly.

12   Lubricate the cylinder bore with clean brake fluid and install the spring and secondary piston assembly in the cylinder.

13   Disassemble the primary piston assembly, noting the position of the parts, then lubricate the new seals with clean brake fluid and install them on the piston **(see illustration)**.

14   Install the primary piston assembly in the cylinder bore, depress it and install the lock ring.

15   Lubricate the new O-rings with silicone brake lube and place them in position.

16   Install the reservoir on the master cylinder body and secure it with the two roll pins.

17   Inspect the reservoir cover and diaphragm for cracks and deformation. Replace any damaged parts with new ones and attach the diaphragm to the cover.

18   Anytime the master cylinder is removed, the brake hydraulic system must be bled. It's much easier to bleed the rest of the system quickly and effectively if you "bench bleed" the master cylinder before installing it on the vehicle. Bench bleed the master cylinder as follows.

19   Insert threaded plugs of the correct size into the cylinder outlet holes and fill the reservoirs with brake fluid (the master cylinder should be supported in such a manner that brake fluid will not spill out of it during the bench bleeding procedure).

20   Loosen one plug at a time and push the piston assembly into the bore to force air from the master cylinder. To prevent air from being drawn back into the cylinder, the appropriate plug must be tightened before allowing the piston to return to its original position.

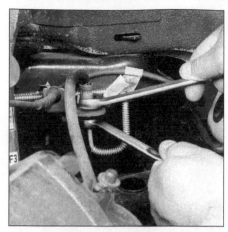

9.2 Using a back-up wrench, unscrew the brake line from the hose fitting, being careful not to bend the frame bracket or brake line (front brake hose/line connection shown, rear setup the same)

9.3 Use pliers to remove the U-clip from the female fitting at the bracket, then remove the hose from the bracket (front brake hose/line connection shown, rear setup the same)

21   Stroke the piston three or four times for each outlet to assure that all air has been expelled.
22   Refill the master cylinder reservoirs and install the diaphragm and cap assembly. **Note:** *The reservoir should only be filled to the top of the reservoir divider to prevent overflowing when the cover and diaphragm are installed.*

Installation

23   The remainder of installation is the reverse of removal. Make sure you bleed the rest of the system when you're done (see Section 10). **Warning:** *If the ABS INOP light stays on or flashes after the engine is started, the system is not fully bled. Brake bleeding must be completed with a Tech 1 scan tool. Have the vehicle towed to a dealer (or other qualified repair shop) immediately and have the service department finish the job.*

9   Brake hoses and lines - inspection and replacement

Inspection

1   About every six months, loosen the wheel lug nuts, raise the vehicle, place it securely on jackstands, remove the wheels, and inspect the flexible hoses which connect the steel brake lines with the front and rear brake assemblies. Look for cracks, chafing, leaks, blisters and any other damage. These are important and vulnerable parts of the brake system, so your inspection should be thorough. You'll need a flashlight and mirror to do the job right. If a hose exhibits any of the above conditions, replace it as follows. **Warning:** *Do not attempt this procedure if your vehicle is equipped with the Traction Control System (TCS), as the brake system cannot be bled without the use of a Tech-1 or T-100 (CAMS) scan tool. Have the following procedure performed at a dealer service*

department or other qualified repair shop equipped with this special tool.

Brake hose

Refer to illustrations 9.2 and 9.3

2   Using a back-up wrench, disconnect the brake line from the hose fitting, being careful not to bend the frame bracket or brake line **(see illustration)**.
3   Use pliers to remove the U-clip from the female fitting at the bracket **(see illustration)**, then remove the hose from the bracket.
4   At the caliper end of the hose, remove the banjo bolt from the fitting block, then remove the hose and the sealing washers on either side of the fitting block.
5   When installing the hose, always use new sealing washers on either side of the fitting block and lubricate all bolt threads with clean brake fluid before installing them.
6   With the fitting flange engaged with the caliper locating ledge, attach the hose to the caliper and tighten it to the torque listed in this Chapter's Specifications.
7   Without twisting the hose, install the female fitting in the hose bracket (it will fit the bracket in only one position).
8   Install the U-clip retaining the female fitting to the frame bracket.
9   Using a back-up wrench, attach the brake line to the hose fitting and tighten it to the torque listed in this Chapter's Specifications.
10   When the brake hose installation is complete, there should be no kinks in the hose. Also make sure that the hose does not contact any part of the suspension. If you're replacing a front brake hose, verify this by turning the wheels to the extreme left and right positions. If the hose contacts anything, disconnect it and correct the installation as necessary.
11   Fill the master cylinder reservoir and bleed the system (see Section 10).

Steel brake lines

12   When it becomes necessary to replace steel lines, use only double-walled steel tubing. Never substitute copper tubing because copper is subject to fatigue cracking and corrosion. The outside diameter of the tubing is used for sizing.
13   Auto parts stores and brake supply houses carry various lengths of prefabricated brake line. Depending on the type of tubing used, these sections can either be bent by hand into the desired shape or must be bent in a tubing bender.
14   When installing the brake line, leave at least 3/4-inch clearance between the line and any moving parts.

10   Brake hydraulic system - bleeding

Refer to illustration 10.16
Warning 1: *Do not attempt this procedure if your vehicle is equipped with the Traction Control System (TCS), as the brake system cannot be bled without the use of a Tech-1 or T-100 (CAMS) scan tool. Have the following procedure performed at a dealer service department or other qualified repair shop equipped with this special tool.*
Warning 2: *This procedure can only be performed if the amber ABS warning light on the dash isn't illuminated. If the light is On, have the vehicle towed to a dealer service department or other repair shop equipped with a Tech-1 or T-100 (CAMS) scan tool.*
Warning 3: *Use only DOT 3 brake fluid.*
Warning 4: *Wear eye protection when bleeding the brake system. If the fluid comes in contact with your eyes, immediately rinse them with water and seek medical attention.*
Note: *Bleeding the brake system is necessary to remove any air that's trapped in the system when it's opened during removal and installation of a hose, line, caliper, wheel cylinder or master cylinder.*

1   Bleeding of the hydraulic system is necessary to remove air whenever it is introduced into the brake system.
2   The manufacturer specifies that it will be necessary to bleed the *entire* system whenever air is allowed into *any* part of the system.
3   Have an assistant on hand, as well as a supply of new brake fluid, an empty clear container, a length of 3/16-inch clear plastic tubing to fit over the bleeder valve and a wrench to open and close the bleeder valve.
4   Apply the parking brake and remove your foot from the service brake pedal. Start the engine and let it run for a minimum of 10 seconds. Check to see if the amber ABS light is still On after approximately 10 seconds.
5   If the light remained on after 10 seconds, have the vehicle towed to a dealer service department or other repair shop equipped with a Tech-1 or T-100 (CAMS) scan tool and have the ABS system diagnosed.
6   If the light came On for about three seconds and then turned off (and remained Off),

10.16 When bleeding the brakes, a hose is connected to the bleeder screw and then submerged in brake fluid - air will be seen as bubbles in the tube and container (all air must be expelled before moving to the next brake or component)

11.7a To detach the power brake booster from the firewall, remove these two nuts (arrows) on the left side of the pushrod (barely visible in this photo) . . .

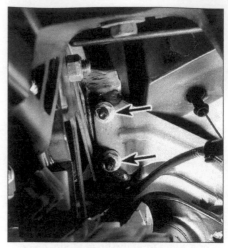

11.7b . . . and these two (arrows) to the right of the pushrod

turn off the ignition and repeat Steps 4 and 5. If the light again goes out after about three seconds, continue with the bleeding procedure.

7   Remove the cap from the brake fluid reservoir and add fluid, if necessary (see Chapter 1). Don't allow the fluid level to drop too low during this procedure - check it frequently. Reinstall the cap.

8   Connect a clear plastic hose to the rear bleeder screw on the ABS modulator. Place the other end of the hose into a container partially filled with clean brake fluid. Make sure the end of the hose is submerged in the fluid.

9   Open the bleeder screw, slowly, about 1/2-to-3/4-turn. Have an assistant push on the brake pedal until fluid flows from the bleeder screw. If air is present in the ABS hydraulic modulator, bubbles will be present.

10   Tighten the bleeder screw and have your assistant release the brake pedal.

11   Repeat Steps 9 and 10 until the fluid flowing from the bleeder screw is free of bubbles.

12   Move the bleeder hose to the front bleeder screw on the modulator and repeat Steps 9 through 11.

13   Check the fluid level again, adding fluid as necessary.

14   Raise the vehicle and support it securely on jackstands.

15   Beginning at the right rear brake, loosen the bleeder screw slightly, then tighten it to a point where it's snug but can still be loosened quickly and easily.

16   Place one end of the tubing over the bleeder screw and submerge the other end in brake fluid in the container **(see illustration)**.

17   Open the bleeder screw and have your assistant slowly depress the brake pedal and hold the pedal firmly depressed. Watch for air bubbles to exit the submerged end of the

tube. When the fluid flow slows, tighten the screw, then have your assistant slowly release the pedal. Wait five seconds before proceeding.

18   Repeat Step 17 until no more air is seen leaving the tube, then tighten the bleeder screw and proceed to the left rear brake, right front brake and left front brake, in that order. Be sure to check the fluid in the master cylinder reservoir frequently.

19   After bleeding all of the wheel brakes in the proper order, repeat Steps 16 through 18 on the left rear brake, right front brake and the left front brake, in that order.

20   Lower the vehicle and check the fluid level in the brake fluid reservoir, adding fluid as necessary.

21   Once again attach the bleeder hose to the rear bleeder screw on the ABS hydraulic modulator. Submerge the other end of the hose in the container partially filled with brake fluid.

22   Have your assistant firmly depress the brake pedal, then slowly open the bleeder screw until fluid flows out. When the flow of fluid slows, tighten the bleeder screw.

23   Pause for five seconds, then repeat Step 22 until the flow of fluid from the bleeder screw is free of air bubbles.

24   Move the bleeder hose to the front bleeder screw on the ABS modulator and repeat Steps 22 and 23.

25   Never use old brake fluid. It contains moisture which will allow the fluid to boil, rendering the brakes useless. When bleeding, make sure the fluid coming out of the bleeder is not only free of bubbles, but clean also.

26   Refill the fluid reservoir with new fluid at the end of the operation.

27   Check the operation of the brakes. The pedal should feel solid when depressed, with no sponginess. If necessary, repeat the entire process. **Warning:** *Do not operate the vehicle if you are in doubt about the effectiveness of the brake system.*

11   Power brake booster - inspection, removal and installation

Refer to illustrations 11.7a and 11.7b

Operating check

1   Depress the pedal and start the engine. If the pedal goes down slightly, operation is normal.

2   Depress the brake pedal several times with the engine running and make sure that there is no change in the pedal reserve distance.

Airtightness check

3   Start the engine and turn it off after one or two minutes. Depress the brake pedal several times slowly. If the pedal goes down farther the first time but gradually rises after the second or third depression, the booster is airtight.

4   Depress the brake pedal while the engine is running, then stop the engine with the pedal depressed. If there is no change in the pedal reserve travel after holding the pedal for 30 seconds, the booster is airtight.

Removal

Note: *Dismantling of the power brake unit requires special tools. If a problem develops, it is recommended that a new or factory-exchange unit be installed rather than trying to overhaul the original booster.*

5   Remove the mounting nuts which hold the master cylinder to the power brake unit (see Section 8). Move the master cylinder forward, but be careful not to bend or kink the lines leading to the master cylinder. If there is any strain on the lines, disconnect them at the master cylinder and plug the ends (refer to the **Warnings** at the beginning of Section 8).

6   Disconnect the vacuum hose leading to the front of the power brake booster. Cover the end of the hose. **Note:** *On 2000 and later models, remove the left side cowl cover for access (see Chapter 11).*

12.4a Pull the cable end off the parking brake actuator lever

12.4b Squeeze the ferrule and pull the parking brake cable through the bracket

and master cylinder. If the brake lines were disconnected from the master cylinder, bleed the brake system to eliminate any air which has entered the system (see Section 10).

12   Parking brake cables - replacement

Rear cable

Refer to illustrations 12.4a, 12.4b and 12.5

1   Fully release the parking brake lever.
2   Loosen the rear wheel lug nuts, raise the rear of the vehicle and place it securely on jackstands. Remove the rear wheel(s).
3   To detach a rear cable from the rear brake assembly on a vehicle with rear drum brakes, remove the rear drum, disassemble the brake assembly (see Section 6), detach the cable from the parking brake lever, squeeze the ferrule at the backing plate with a pair of pliers and pull the cable through the backing plate.
4   To detach a rear cable from the rear brake assembly on a vehicle with rear disc brakes, pull the cable end off the parking brake actuator lever **(see illustration)**, squeeze the ferrule with a pair of pliers and pull the cable through the bracket **(see illustration)**.
5   At the forward end of the cable, separate the cable from the equalizer **(see illustration)**, squeeze the ferrule and pull the cable through the bracket.
6   Installation is the reverse of removal.

Front cable

Refer to illustrations 12.7, 12.8, 12.9a, 12.9b, 12.10, 12.15, 12.17 and 12.18

7   Pull the parking brake lever up and remove the pre-tension spring **(see illustration)**.
8   Release the parking brake lever and rotate the adjuster arm toward the front of the vehicle until a 3mm drill bit can be inserted into the hole **(see illustration)**.

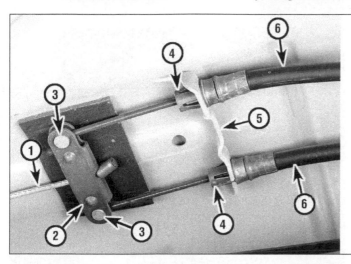

12.5 The equalizer assembly

1   *Front parking brake cable*
2   *Equalizer*
3   *Rear cable barrel plugs*
4   *Ferrules for rear cables*
5   *Rear cable bracket*
6   *Rear parking brake cables*

7   Inside the vehicle, loosen the four nuts that secure the booster to the firewall **(see illustrations)**. Do not remove these nuts at this time.
8   Disconnect the power brake pushrod from the brake pedal. Do not force the pushrod to the side when disconnecting it.

9   Now remove the four booster mounting nuts and carefully lift the unit out of the engine compartment.
10   When installing, loosely install the four mounting nuts, then connect the pushrod to the brake pedal. Tighten the nuts to the specified torque and reconnect the vacuum hose

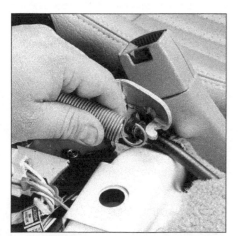

12.7 Pull the parking brake lever up and remove the pre-tension spring

12.8 Release the parking brake lever and rotate the adjuster arm toward the front of the vehicle until a 3mm drill bit can be inserted into the holes (arrow) of the adjuster and the arm

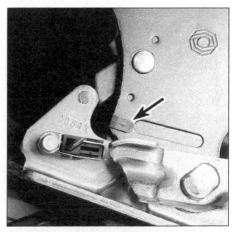

12.9a The rest of the plastic shear pin (arrow) . . .

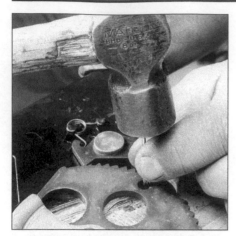

12.9b . . . can be knocked out with the drill bit

12.10 Insert the 3mm drill bit through the two holes; the self adjuster is now locked out

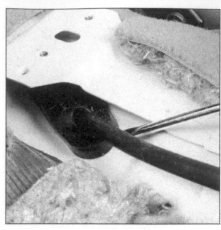

12.15 Pry the grommet from the body pan with a screwdriver

12.17 Bend back the cable retainer tab (arrow) on the lower edge of the pulley (parking lever assembly unbolted from floor and tilted to one side for clarity)

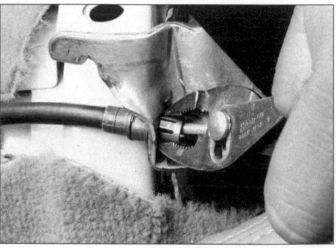

12.18 Compress the ferrule with a pair of pliers and disengage the cable from the parking brake lever bracket

9   Remove the remaining piece of plastic shear pin from the hole **(see illustrations)**.

10   Insert the 3mm drill bit into the hole **(see illustration)**. The self adjuster is now locked out.

11   Pull the parking brake lever back as far as it will go.

12   Raise the vehicle and place it securely on jackstands.

13   Disconnect the left and right rear cables from the equalizer (see Steps 1 through 5).

14   Disengage the front cable from the equalizer **(see illustration 12.5)**.

15   Remove the grommet from the body pan **(see illustration)**.

16   Lower the vehicle.

17   Bend back the cable retainer tab on the pulley **(see illustration)**. Disengage the plug-shaped forward end of the cable from the pulley on the parking brake lever.

18   Pinch the ferrule at the rear of the parking brake lever bracket **(see illustration)** and pull the cable through the bracket.

19   Pull out the cable through the hole in the floor.

20   Installation is the reverse of removal.

13   Parking brake - adjustment

1   The vehicles covered by this manual are equipped with a self-adjusting parking brake lever assembly that automatically takes up slack in the cables as they stretch. There is no need to adjust the parking brake cables manually during their service life. However, whenever new rear brake shoes are installed on models with rear drum brakes, make sure the new shoes are properly adjusted; the automatic cable adjuster mechanism can't do its job if the rear shoes are out of adjustment. On models with rear disc brakes, the parking brake lever on the caliper must be checked and, if necessary, readjusted when new pads are installed. It's also a good idea to check cable adjustment whenever a rear drum brake is disassembled or a rear caliper is removed.

Drum brakes

2   Raise the rear of the vehicle and support it securely on jackstands.

3   Insert a screwdriver through the adjustment hole and lift the actuator off the adjuster wheel. Insert a brake adjusting tool (or another screwdriver) into the hole and turn the adjuster wheel until the shoes drag on the drum as the drum is turned. Now back off the adjuster wheel until the drums rotate freely. Apply the parking brake and verify that the shoes hold the drums. Release the parking brake and make sure that both rear wheels turn freely and that there is no brake drag in either direction. **Note:** *If the brake shoes have never been adjusted before, there may be a small plate where the adjustment hole should be located. Simply knock out this plate with a small punch. If you do this, you'll have to remove the brake drum and retrieve the plate (see Section 6).*

4   Remove the jackstands and lower the vehicle.

Rear disc brakes

1997 and earlier

Refer to illustrations 13.7, 13.9a and 13.9b
Caution: *The following adjustment procedure for parking brake cable free travel assumes that either new pads have just been installed,*

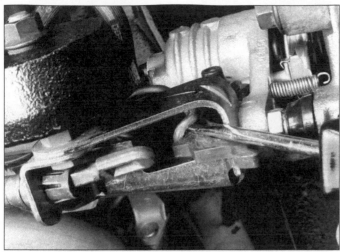

13.7 To remove the actuator lever return spring, pry it loose with a screwdriver

13.9a To remove freeplay from the lever itself, exert light pressure on the parking brake lever and measure the clearance between the lever and the caliper as shown; you should be able to insert a 0.024 to 0.028-inch feeler gauge between the lever and the caliper at the indicated point

13.9b If the gap is too small (insufficient cable free travel), turn the adjustment screw clockwise; if the gap is too big (too much cable free travel), turn the adjustment screw counterclockwise (caliper removed for clarity)

13.16 Parking brake shoe adjusting screw - 1998 and later models

or that the current pads are parallel within 0.006-inch. Doing the following adjustment with heavily tapered pads will not produce the correct amount of cable free travel, and may cause the pads and caliper to bind.

Note: *Parking brake cable free travel should be adjusted only after new pads are installed, or after the caliper has been overhauled.*

5   Make sure the parking brake lever is completely released.

6   Raise the vehicle and place it securely on jackstands.

7   Disconnect the parking brake cable from the caliper (see Section 12). Remove the actuator lever return spring **(see illustration)**.

8   Spin the wheel and have an assistant apply light pressure to the brake pedal, just enough to stop the disc. This aligns the pads with the disc and takes up the clearance between the pads and the disc. Your assistant must maintain this same light pressure - just enough to prevent the disc from turning - while you check and, if necessary, adjust the cable free travel.

9   To remove freeplay from the lever itself, exert light pressure on the parking brake lever as shown **(see illustration)**. With freeplay removed from the lever, you should be able to insert a 0.024 to 0.028-inch feeler gauge between the lever and the caliper at the indicated point **(see illustration)**. If the gap is incorrect, remove the adjuster screw, remove the old thread adhesive residue from the screw, coat the threads with new adhesive and install the screw. If the gap is too small (insufficient cable free travel), turn the adjustment screw clockwise; if the gap is too big (too much cable free travel), turn the adjustment screw counterclockwise.

10   Have your assistant release pressure on the brake pedal and apply firm brake pressure three times.

11   Recheck cable free travel as described in Steps 9 and 10 and, if necessary, readjust it.

12   Install the actuator lever return spring and reconnect the parking brake cable to the lever (see Section 12).

13   Remove the jackstands, lower the vehicle

to the ground and check parking brake operation. Repeat the above steps if necessary.

1998 and later

Refer to illustration 13.16

Note: *On 2001 and later models, the park brake system is self-adjusting. If components have been replaced, just apply and release the parking brake three times to let it self-adjust.*

14   Raise the vehicle and place it securely on jackstands.

15   Remove the rear brake discs (see Section 5)

16   Turn the parking brake adjuster screw outward until the shoe lining just drags on the braking surface inside the disc **(see illustration)**. Then remove the disc and back-off the adjuster screw until the shoe lining doesn't drag when the disc is installed and turned. The actual clearance between the lining surface of the shoe and the braking surface inside the disc should be 0.026-inch.

17   Install the brake discs and the wheels

and check the parking brake operation.
18   Remove the jackstands and lower the vehicle to the ground.

14   Brake light switch - check, replacement and adjustment

Check

Refer to illustration 14.1

1   The brake light switch **(see illustration)** is located on the brake pedal bracket. You'll need to remove the knee bolster (the trim panel beneath the steering column) to get to the switch and connector (see Chapter 11).
2   With the brake pedal in the fully released position, the switch plunger is pressed into the switch housing. When the brake pedal is depressed, the plunger protrudes from the switch, which closes the circuit and sends current to the brake lights.
3   If the brake lights are inoperative, check the fuse (see Chapter 12).
4   If the fuse is okay, verify that voltage is available at the switch.
5   If there's no voltage to the switch, use a test light to find the open circuit condition between the fuse panel and the switch. If there is voltage to the switch, close the switch (depress the brake pedal) and verify that there's voltage on the other side of the switch.
6   If there's no voltage on the other side of the switch with the brake pedal depressed, replace the switch (see Step 7). If voltage is

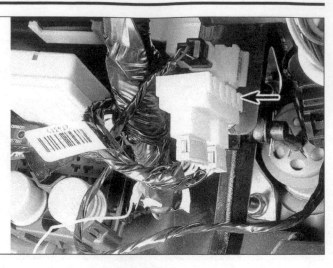

14.1 The brake light switch (arrow) is located at the upper end of the brake pedal

available, check for voltage at the brake lights. If the no power is present, look for an open circuit condition between the switch and the brake lights. Also check the brake light bulbs, even though it isn't likely that both of them would fail at the same time.

Replacement

7   Remove the knee bolster (see Chapter 11).
8   Unplug the electrical connector from the switch **(see illustration 14.1)**.
9   Remove the switch from the bracket.
10   The switch must be adjusted as it's installed (see below).

Adjustment

11   Depress the brake pedal, insert the switch into its bracket and push it in until it's fully seated.
12   Slowly pull the brake pedal to the rear until you no longer hear any "clicking" sounds. The switch should now be adjusted.
13   You can check your work with an ohm-meter or continuity tester by verifying that the switch contacts are open at one inch or less of brake pedal travel, and closed thereafter.
14   Installation is otherwise the reverse of removal.

Chapter 10
Suspension and steering systems

Contents

Specifications

General

Power steering fluid type	See Chapter 1

Torque specifications

Ft-lbs (unless otherwise indicated)

Front suspension

Balljoints	
Lower balljoint nuts	81
Upper balljoint nuts	39
Lower control arms	
Front bolts/nuts	74
Rear bolts/nuts	85
Shock absorber/coil spring assemblies	
Upper nuts	32
Upper bolts	37
Lower bolts/nuts	48
Stabilizer bar	
Clamp bolts	41
Link bolts/nuts	17
Upper control arms	
Control arm-to-support bolts/nuts	72
Upper control arm supports	
Nuts	32
Bolts	37
Hub and bearing assembly (front) mounting bolts	63

Torque specifications

Ft-lbs (unless otherwise indicated)

Rear suspension

Lower control arm bolts	
Through 2001	80
2002	87
Lower control arm nuts	60
Panhard rod	
Models through 1999	
Upper bolt/nut	61
Lower bolt/nut	61
2000 and later	
Bolts	87
Nuts	55

Torque specifications (continued)

	Ft-lbs (unless otherwise indicated)
Shock absorbers	
Upper nuts	156 in-lbs
Lower nuts	66
Stabilizer bar	
Clamp bolts	18
Link bolts/nuts	16
Torque arm	
At transmission	
Nuts	30
Long bolts	37
Short bolt	20
At rear axle	
Bolts	96
Nuts	97

Steering

Steering gear bolts/nuts	63
Tie-rod end castle nuts	35, plus up to 1/6 additional turn, or 52 ft-lbs maximum, to allow cotter pin to be inserted*
Steering shaft U-joint pinch bolt/nut	35
Steering wheel nut	32

Do not back off castle nut for cotter pin insertion.

---

1   General information

Suspension

Refer to illustrations 1.1 and 1.2

The fully-independent front suspension **(see illustration)** allows each wheel to compensate for road surface irregularities without any appreciable effect on the other wheel. The suspension at each front wheel consists of a shock absorber/coil spring assembly situated between the upper control arm supports and the lower control arms. A steering knuckle is located between the upper and lower arms by a pair of balljoints, one in each control arm. The lower balljoints are pressed into the lower control arms; the upper balljoints are riveted to the upper control arms. Both upper and lower balljoints are replaceable. A stabilizer bar controls vehicle roll during cornering. The stabilizer bar is attached to the frame by a pair of steel clamps and to the lower control arms by link bolts.

The front hub assembly contains sealed bearings that do not require maintenance. In the event of bearing failure, the hubs must be removed and new bearings pressed in by an automotive machine shop. The semi-independent rear suspension **(see illustration)** consists of a solid rear axle housing suspended by a pair of shock absorbers and coil springs and is located by a pair of lower control arms, a Panhard rod and a torque arm.

Steering

The steering system **(see illustration 1.1)** consists of a power-assisted, rack-and-pinion type steering gear connected to the steering knuckles by adjustable tie-rods.

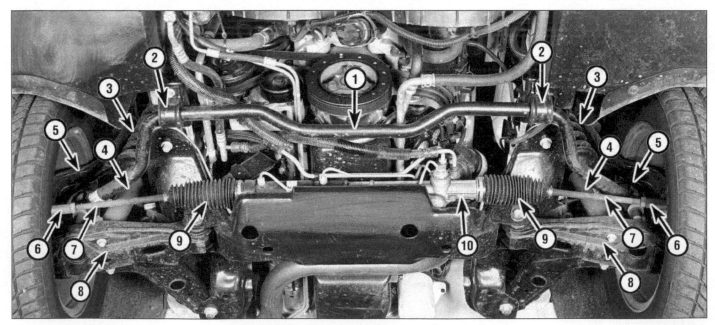

1.1 Typical front suspension and steering layout

1	Stabilizer bar	4	Shock absorber	6	Tie-rod end	8	Lower control arm
2	Stabilizer clamp	5	Steering knuckle	7	Tie-rod	9	Steering gear boot
3	Coil spring					10	Steering gear assembly

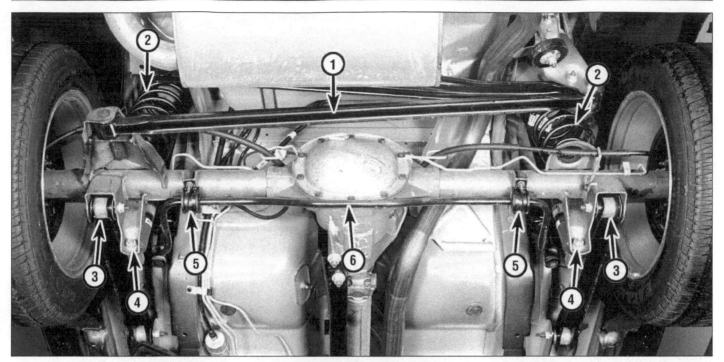

1.2 Typical rear suspension layout

1	Panhard rod	3	Lower control arm	5	Stabilizer bar clamp
2	Coil spring	4	Shock absorber lower mounting nut	6	Stabilizer bar

2.2a To disconnect the stabilizer bar link bolt from each lower control arm, hold the bolt (right arrow) with a socket (the other two bolts on the left attach the lower end of the shock absorber/coil spring assembly to the lower control arm) . . .

2.2b . . . remove the nut at the upper end of the bolt and pull out the bolt from the bottom of the lower control arm; note the order in which the parts are installed to ensure proper reassembly

2.3 To detach the stabilizer bar from the frame, remove the bolts and nuts (arrows) from both bushing clamps - make sure the slit in the bushing faces forward when you reinstall the stabilizer bar

2 Stabilizer bar (front) - removal and installation

Refer to illustrations 2.2a, 2.2b and 2.3

1   Raise the vehicle and support it securely on jackstands.

2   Remove the nut, bushings and sleeve from the link bolts that attach each end of the stabilizer bar to the lower control arms **(see illustrations)**. Note the order in which the bushings and sleeves are removed to ensure proper reassembly.

3   Remove the stabilizer bar clamps and rubber bushings **(see illustration)**.

4   Remove the stabilizer bar.

5   Inspect the rubber bushings and link bolt grommets for cracks and tears. Replace all damaged bushings and grommets.

6   Apply multi-purpose grease to the bushing installation areas on the stabilizer bar before installing the bushings and clamps.

7   Make sure the slits in the bushings are facing toward the front of the vehicle.

8   Make sure the stabilizer bar is centered in the bushings and clamps before tightening the mounting bolts.

9   Installation is otherwise the reverse of removal procedure. Be sure to tighten all bolts to the torque listed in this Chapter's Specifications.

3 Lower control arm (front) - removal and installation

Refer to illustrations 3.5a, 3.5b, 3.5c, 3.6a and 3.6b

1   Loosen the wheel lug nuts, raise the vehicle and place it securely on jackstands.

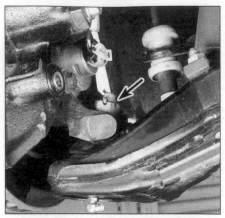

3.5a Remove the cotter pin from the lower control arm balljoint stud and loosen the castle nut (arrow) but don't remove it until the stud has been disengaged from the steering knuckle

3.5b To separate the lower control arm from the steering knuckle, insert a picklefork between the control arm and the knuckle as shown and drive it in with a hammer until the balljoint stud pops loose from the knuckle, then remove the stud nut

3.5c If you don't have a picklefork, knock the balljoint stud loose with direct hammer blows to the knuckle (this method is effective on a newer vehicle but not necessarily on an older model, especially one which has been driven in a wet or snowy climate)

3.6a To detach the front of the lower control arm from the crossmember, remove this nut and bolt (arrow)

3.6b To detach the rear of the lower control arm from the crossmember, remove this nut and bolt (arrow)

Remove the wheel.

2   Disconnect the stabilizer bar link bolt from the lower control arm **(see illustrations 2.2a and 2.2b).**

3   Disconnect the tie-rod end from the steering knuckle (see Section 16).

4   Remove the pair of nuts and bolts which attach the lower end of the shock absorber/coil spring assembly to the lower control arm **(see illustration 2.2a).**

5   Remove the cotter pin, loosen (but don't remove) the castle nut from the balljoint stud and separate the lower control arm balljoint from the steering knuckle with a "picklefork" type balljoint separator or a hammer **(see illustrations).** Note: *The use of a picklefork balljoint separator will probably result in damage to the balljoint boot.*

6   Remove the front and rear nuts and bolts that attach the lower control arm to the crossmember **(see illustrations).**

7   Remove the lower control arm.

8   Inspect the front and rear bushings in the lower control arm. If either bushing is torn or cracked, take the control arm to an automotive machine shop and have the old bushing pressed out and a new bushing pressed in.

9   Installation is the reverse of the removal procedure. Be sure to tighten all fasteners to the torque listed in this Chapter's Specifications.

4   Shock absorber/coil spring (front) - removal and installation

Refer to illustrations 4.2 and 4.5

1   If you're removing the shock absorber/coil spring assembly from the left (driver's side), separate the master cylinder from the power brake booster (see Chapter 9) and move it aside so that you can remove the outer of the two upper rear shock absorber mounting bolts.

2   Remove the upper shock absorber-to-body nuts and bolts **(see illustration).**

3   Loosen the wheel lug nuts, raise the front of the vehicle and support it securely on jackstands, then remove the wheel.

4   Disconnect the stabilizer bar from the lower control arm (see Section 2).

5   Remove both pairs of bolts and nuts

that attach the lower end of the shock absorber to the lower control arm **(see illustration).**

6   Separate the lower control arm from the steering knuckle (see Section 3).

7   Remove the shock absorber/coil spring assembly.

8   Installation is the reverse of removal. Be sure to tighten all fasteners to the torque listed in this Chapter's Specifications.

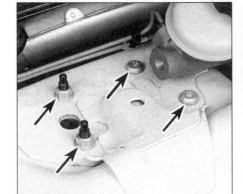

4.2 To detach the upper end of the shock absorber from the vehicle, remove these two nuts and two bolts (arrows) (the rear bolts on some models, such as this one, have Torx heads)

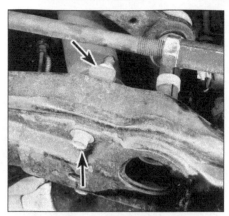

4.5 To detach the lower end of the shock absorber from the lower control arm, remove the nuts and bolts (arrows)

5.3a Mark the coil spring to the upper spring seat, and a L or R indicating on which side of the vehicle the spring seat belongs

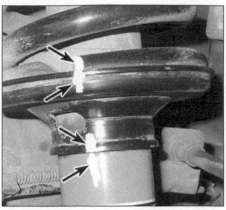

5.3b Also mark the relationship between the end of the coil spring, lower seat and shock absorber body

5 Shock absorber/coil spring (front) - replacement

1   If the shocks or coil springs exhibit the telltale signs of wear (leaking fluid, loss of damping capability, chipped, sagging or cracked coil springs), explore all options before beginning any work. The shock absorber assemblies are not serviceable and must be replaced if a problem develops. However, shock assemblies complete with springs may be available on an exchange basis, which eliminates much time and work. Whichever route you choose to take, check on the cost and availability of parts before beginning disassembly. **Warning:** *Disassembling a coil-over shock absorber is potentially dangerous and utmost attention must be directed to the job, or serious injury may result. Use only a high-quality spring compressor and carefully follow the manufacturer's instructions furnished with the tool. After removing the coil spring, set it aside in a safe, isolated area.*

Disassembly

Refer to illustrations 5.3a, 5.3b and 5.4

2   Remove the shock/coil spring assembly (see Section 4). Mount the shock/spring assembly in a vise. Line the vise jaws with wood or rags to prevent damage to the unit and don't tighten the vise excessively.
3   Mark the relationship of the upper spring seat to the spring, the spring to the lower spring seat and the lower spring seat to the shock absorber body **(see illustrations)**. **Note:** *This is necessary to ensure that the upper spring seat and shock absorber lower mount are oriented properly when the shock absorber and coil spring are reassembled. Also make a mark on the upper mount indicating which side of the vehicle it is installed on (they are not interchangeable).*
4   Install a spring compressor in accordance with the tool manufacturer's instructions **(see illustration)**. Compress the spring enough to relieve all pressure from the upper spring seat. When you can wiggle the spring, it's compressed enough. (You can buy a suit-

able spring compressor at most auto parts stores or rent one from a tool rental yard.)
5   Loosen the damper shaft nut with a box-end wrench. Use another wrench to hold the damper shaft. Do not allow the shaft to turn while loosening the nut.
6   Remove the nut and the upper shock mount. Discard the nut - the manufacturer states that a new nut must be installed during reassembly.
7   Carefully lift the compressed spring from the assembly and set it in a safe place, such as a steel cabinet. **Warning:** *Never place your head near the end of the spring!*

Reassembly

8   Reassembly is the reverse of disassembly, noting the following points:
a) *Install the lower spring seat onto the shock absorber body, aligning the marks made in step 3. If you're installing a new shock absorber, make a mark on the new shock absorber body in the same place as the mark you made on the old one (make sure the shock absorber lower mounting cross bar is angled the same way; when installed, it must slant upwards towards the center of the vehicle).*
b) *Install the coil spring, aligning the marks made in Step 3. If you're installing a new coil spring, align the spring end with the mark on the lower seat. Note: Aftermarket coil springs may or may not have the ends of the springs arranged in the same orientation to each other as factory coil springs. Compare the new spring carefully to the one being replaced. If the ends of the springs are not arranged in a like manner, keep in mind that the important thing is to properly set the relationship of the upper mount and the shock absorber lower mounting cross bar, so the shock absorber will fit when you go to reinstall it.*
c) *Install the upper spring seat, aligning the marks made in Step 3. If you're installing a new coil spring, make a mark on the new spring in the same place as the mark you made on the old one.*

5.4 Install the spring compressor in accordance with the manufacturer's instructions and compress the spring until all pressure is relieved from the upper mount

d) *Install a new damper shaft nut, tightening it to the torque listed in this Chapter's Specifications. Don't allow the shaft to turn while tightening the nut (use the method described in Step 5 to prevent that from happening).*
9   Install the shock absorber/coil spring assembly (see Section 4).

6 Upper control arm - removal and installation

Refer to illustrations 6.7a and 6.7b

1   If you're removing the upper control arm from the left (driver's side), separate the master cylinder from the power brake booster (see Chapter 9) and move it aside so that you can remove the outer of the two upper rear shock absorber mounting bolts.
2   Remove the upper shock absorber mounting nuts and bolts **(see illustration 4.2)**.
3   Loosen the wheel lug nuts, raise the front of the vehicle and support it securely on jackstands, then remove the wheel.
4   Disconnect the stabilizer bar from the lower control arm (see Section 2).

6.7a To detach the upper control arm from the upper end of the steering knuckle, remove the cotter pin and loosen (but don't remove) the castle nut . . .

6.7b . . . then install a small puller and separate the upper balljoint stud from the knuckle

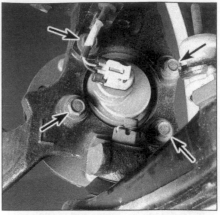

7.3 To detach the hub and bearing assembly from the steering knuckle, unplug the ABS electrical connector and remove the four bolts (arrows)

5   Remove the nuts and bolts that attach the lower end of the shock absorber to the lower control arm **(see illustrations 4.5a and 4.5b)**.

6   Support the steering knuckle and lower control arm with a jackstand or a floor jack.

7   Remove the cotter pin, loosen (but don't remove) the castle nut on the upper balljoint stud, install a small puller **(see illustrations)** and separate the upper balljoint stud from the steering knuckle. Now remove the nut from the balljoint stud.

8   Remove the upper control arm, the control arm support bracket and the shock absorber/coil spring assembly from the vehicle as a single assembly.

9   Remove the front and rear nuts and bolts that attach the upper control arm to the control arm support bracket. Separate the upper control arm from the support.

10   Installation is the reverse of removal. Be sure to tighten all fasteners to the torque listed in this Chapter's Specifications.

7   Hub and bearing assembly - removal and installation

Refer to illustration 7.3

1   Loosen the wheel lug nuts, raise the vehicle and place it securely on jackstands. Remove the wheel.

2   Remove the brake caliper and disc (see Chapter 9).

3   Unplug the electrical connector for the wheel speed sensor **(see illustration)**.

4   Remove the four bolts that attach the hub to the steering knuckle.

5   Remove the hub assembly.

6   Installation is the reverse of removal. Tighten the hub retaining bolts to the torque listed in this Chapter's Specifications. Make sure the grommet for the ABS speed sensor electrical leads is properly seated in its bracket on the knuckle. The wires could be damaged if they're not secured to this bracket.

8   Steering knuckle - removal and installation

1   Loosen the wheel lug nuts, raise the front of the vehicle and support it securely on jackstands. Remove the wheel.

2   Remove the brake caliper and disc (see Chapter 9).

3   Remove the hub assembly (see Section 7).

4   Disconnect the tie-rod end from the steering knuckle (see Section 17).

5   Disconnect the stabilizer bar from the lower control arm (see Section 2).

6   Disconnect the lower end of the shock absorber/coil spring assembly from the lower control arm (see Section 4).

7   Separate the lower control arm from the steering knuckle (see Section 3).

8   Separate the upper control arm from the steering knuckle (see Section 6).

9   Remove the steering knuckle.

10   Installation is the reverse of removal. Be sure to tighten all suspension fasteners to the torque listed in this Chapter's Specifications.

9   Balljoints - check and replacement

Check

1   Inspect the control arm balljoints for looseness whenever either of them is separated from the steering knuckle. See if you can turn the ballstud in its socket with your fingers.

2   If the balljoint is loose, or if the ballstud can be turned, replace the balljoint. You can also check the balljoints with the suspension assembled as follows.

3   Raise the front of the vehicle and support it securely on jackstands placed under the lower control arms. Position the stands as close to each balljoint as possible. Make sure the vehicle is stable. It should not rock on the stands.

Lower balljoints

Refer to illustration 9.5

4   Wipe each balljoint clean and inspect the seal for cuts and tears. If the seal is damaged, replace the balljoint.

5   The lower balljoints on 1993 and 1994 models employ a visual wear indicator **(see illustration)** that allows easy diagnosis. The shoulder at the base of the grease fitting protrudes about 0.050-inch (3/64-inch) from the lower surface of the balljoint's lower cover when the balljoint is new. As the balljoint wears, this shoulder slowly moves up into the balljoint. If the shoulder is flush with or inside the surface of the lower cover, replace the lower balljoint.

6   To check the lower balljoint on 1995 and later models, position a dial indicator against the wheel rim and insert a prybar between the lower control arm and the steering knuckle. As you lever the prybar, the needle should not deflect more than 0.046-inch. If it does, replace the lower balljoint.

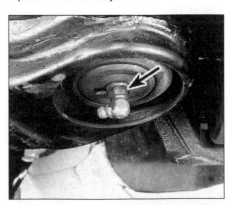

9.5 On 1993 and 1994 models, it's easy to check the lower control arm balljoint for wear: The shoulder (arrow) for the grease fitting protrudes 0.050-inch from the surface of the cover; as the balljoint wears, this shoulder moves up into the balljoint assembly; if the shoulder is flush with the cover, or is up inside the hole, replace the balljoint

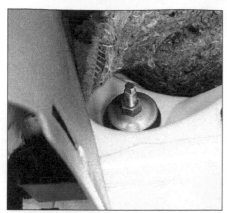

10.1 To detach the upper end of a rear shock absorber from the body, remove the quarter trim panel, peel back the carpet and remove this nut and the retainer (large washer) and insulator underneath it

Upper balljoints

7   Position a dial indicator against the wheel rim, grasp the top and bottom of the tire and "rock" the tire, alternately pushing the top and pulling the bottom, and vice-versa. The dial indicator should indicate no more than 0.125-inch deflection. If the indicated reading exceeds this figure, replace the upper balljoint.

Replacement

8   Remove the control arm (see Section 3 or 6).
9   To replace the lower balljoint, take the control arm and a new balljoint to an automotive machine shop. The machine shop will press out the old balljoint and press in the new unit. You cannot do this at home unless you have a hydraulic press.
10   To replace the upper balljoint, secure the upper control arm in a bench vise, drill out the four rivets as follows: Using a 1/8-inch drill bit, drill a 1/4-inch deep hole in the center of each rivet. Then switch to a 1/2-inch drill bit and finish the job; drill just deep enough to remove the rivet head. Insert a punch through the rivet holes, and knock out the old balljoint. Install the new balljoint with four bolts and nuts (included with the new balljoint). Tighten the nuts to the torque specified in the instructions that come with the kit.
11   Install the lower or upper control arm.
12   Have the front end alignment checked by a dealer service department or alignment shop.

10   Shock absorber (rear) - removal and installation

Refer to illustrations 10.1 and 10.5
1   With the rear seat in the folded-down position, remove the quarter trim panel from the side of the cargo area, lift the edge of the carpeting and fold it back to expose the rear shock upper mounting nut **(see illustration)**.

10.5 To detach the lower end of the shock absorber from the rear axle, remove this nut

2   Raise the rear of the vehicle and support it on jackstands.
3   Support the rear axle assembly with a floor jack under the side being worked on. **Warning:** *Failure to support the rear axle could allow a rear coil spring to fly out and cause injury. It could also result in damage to the brake hose, the driveshaft U-joint or the Panhard rod.*
4   Remove the upper mounting nut, the retainer (large washer) and the insulator.
5   Remove the shock absorber lower mounting nut **(see illustration)** from its bracket on the rear axle.
6   Remove the shock absorber. Check the damper rod for another retainer and insulator. If they're not present, check the underside of the floorpan - they may have stuck there.
7   Installation is the reverse of the removal procedure. Be sure to use new retainers and insulators and tighten all nuts to the torque listed in this Chapter's Specifications.

11   Coil spring (rear) - removal and installation

1   Raise the rear of the vehicle and support it on jackstands. Place the jackstands under the vehicle jacking points, not under the rear axle.
2   Support the rear axle assembly with a floor jack. **Warning:** *Failure to support the rear axle could allow a rear coil spring to fly out and cause injury. It could also result in damage to the brake hose, the driveshaft U-joint or the Panhard rod.*
3   It's a good idea to chain the coil spring(s) to the rear axle to prevent a spring from popping out when the axle is lowered.
4   Remove the shock absorber lower mounting nut and disconnect the shock from the rear axle bracket (see Section 10).
5   Carefully lower the jack until all tension is removed from the coil spring, then remove the spring, upper insulator and seat.
6   Inspect the coil spring for chips in the corrosion protection coating. If the coating has been chipped or damaged, replace the spring.
7   Installation is the reverse of removal.

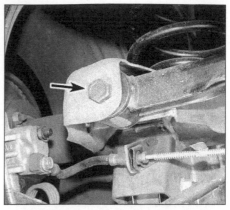

12.3 To detach the lower end of the Panhard rod from the rear axle, remove the nut (not visible in this photo) and this bolt (arrow)

12.4 To detach the upper end of the Panhard rod from its bracket, remove the nut and bolt with a pair of wrenches as shown

Don't forget to install the seat (on top) and the upper insulator (between the seat and the coil spring). The lower end of the spring should face toward the front of the vehicle, and the pigtail at the upper end of the spring should be positioned 1/4-inch from the end of the recessed portion of the spring seat. Be sure to tighten all fasteners to the torque listed in this Chapter's Specifications.

12   Panhard rod - removal and installation

Refer to illustrations 12.3 and 12.4
1   Raise the rear of the vehicle and support it with jackstands. Place the jackstands under the vehicle jacking points, not under the rear axle.
2   Support the rear axle assembly with a floor jack. **Caution:** *Failure to support the rear axle could result in damage to the brake hose or the driveshaft U-joint.*
3   Remove the Panhard rod lower mounting nut and bolt at the axle **(see illustration)**.
4   Remove the Panhard rod upper mounting nut and bolt at the underbody brace bracket **(see illustration)**.
5   Remove the Panhard rod.

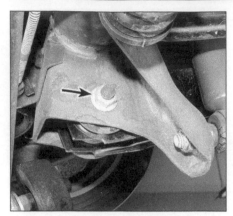

13.3 To detach the lower control arm from the rear axle, remove this nut (arrow) and bolt

13.4 To detach the lower control arm from the body, remove this nut (arrow) and bolt

14.4 To detach the forward end of the torque arm from the extension housing, remove these nuts and bolts

14.5 To detach the rear end of the torque arm from the rear axle, remove these two nuts (arrows) and their two large through-bolts; note that washers are installed both above and below the torque arm on each bolt

15.2 To disconnect the stabilizer bar from the rear axle, remove the nut, bolt, grommets and spacer from each end of the stabilizer bar . . .

6   A brace for the Panhard rod, located right above the Panhard rod, is bolted to a bracket at either end. The left end of the brace is attached to the left bracket by three bolts; the right end of the brace is attached to the right bracket (the same bracket to which the upper end of the Panhard rod is attached). Unless you're removing the fuel tank, there is no reason to remove this brace. If you are servicing the tank, photos depicting removal of the brace are contained in *"Fuel tank - removal and installation"* in Chapter 4.

7   Installation is the reverse of the removal procedure. Tighten all fasteners to the torque listed in this Chapter's Specifications.

13 Lower control arm (rear) - removal and installation

Refer to illustrations 13.3 and 13.4
Note: *If both lower control arms are to be removed, remove and install them one at a time to prevent the axle assembly from slipping sideways or rocking forward, complicating reassembly.*

1   Raise the rear of the vehicle and support it securely on jackstands. Place the jackstands under the vehicle jacking points, not under the rear axle.

2   Support the rear axle assembly with a floor jack. **Caution:** *Failure to support the rear axle could result in damage to the brake hose, the driveshaft U-joint or the Panhard rod.*

3   Remove the lower control arm-to-axle bracket nut and bolt **(see illustration)**.

4   Remove the lower control arm-to-body bracket nut and bolt **(see illustration)**.

5   Remove the control arm.

6   Inspect the bushings at either end of the control arm. If either bushing is cracked or torn, take the lower control arm to a machine shop to have the bushing replaced.

7   Installation is the reverse of the removal procedure. Tighten all fasteners to the torque listed in this Chapter's Specifications.

14 Torque arm - removal and installation

Refer to illustrations 14.4 and 14.5

1   Raise the rear of the vehicle and place it securely on jackstands.

2   Support the rear axle with a floor jack.

3   On V6 models, detach the center support bearing assembly from the torque arm (see Chapter 8).

4   Remove the nuts and bolts that attach the forward end of the torque arm to the transmission extension housing **(see illustration)**. Note how the two brackets (inner and outer) on the left side of the extension housing and the catalytic converter hanger/torque arm bracket on the right side of the extension housing are oriented to ensure proper reassembly.

5   Remove the nuts and bolts that attach the rear end of the torque arm to the rear axle **(see illustration)**.

6   To install the torque arm, raise it into position and install all the nuts and bolts. Do NOT tighten anything until all fasteners at both ends have been completely installed. When all bolts and nuts have been installed, gradually tighten them until they're all snug, then tighten everything to the torque listed in this Chapter's Specifications.

7   Installation is otherwise the reverse of removal.

15 Stabilizer bar (rear) - removal and installation

Refer to illustrations 15.2 and 15.3

1   Raise the vehicle and support it securely on jackstands.

2   To disconnect the stabilizer bar from the rear axle, remove the link bolt **(see illustration)**, nut, grommets, spacer and retainers from each end of the stabilizer bar. Note the sequence in which they're installed to ensure proper reassembly.

3   Remove the two clamps that attach the stabilizer bar to the rear axle **(see illustration)**.

4   Remove the stabilizer bar.

5   Before installing the stabilizer bar, apply multi-purpose grease to the bushing mounting points on the bar, then install the bushings and clamps. Make sure the slits in the bushings face toward the front of the vehicle. Make sure the stabilizer is centered in the clamps before tightening the mounting bolts.

6   Installation is otherwise the reverse of

15.3 . . . then remove the nuts (arrows) from each bushing clamp

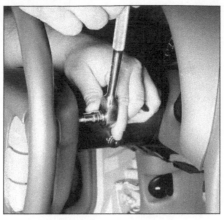

16.3 Remove the four airbag module retaining screws

16.4 Remove the airbag module and unplug the two electrical connectors (arrows)

removal. Tighten all fasteners to the torque listed in this Chapter's Specifications.

16 Steering wheel - removal and installation

Refer to illustrations 16.3, 16.4, 16.6 and 16.7
Warning: *These models have airbags. Always turn the steering wheel to the straight ahead position, place the ignition switch in the Lock position, remove the Air Bag fuse and unplug the yellow Connector Position Assurance (CPA) connectors at the base of the steering column and under the right side of the instrument panel before working in the vicinity of the impact sensors, steering column or instrument panel to avoid the possibility of accidental deployment of the airbag, which could cause personal injury* (see Chapter 12).
Caution: *On models equipped with a Delco Loc II or Theftlock audio system, be sure the lockout feature is turned off before performing any procedure which requires disconnecting the battery.*
1   Disconnect the cable from the negative battery terminal.
2   Disable the airbag system (see Chapter 12 and **Warning** above).

3   Remove the airbag module retaining screws **(see illustration)**.
4   Remove the airbag module and unplug the electrical connectors **(see illustration)**. Set the airbag module in an isolated area with the trim side facing up.
5   Remove the steering wheel retaining nut.
6   Mark the relationship of the steering wheel to the steering shaft **(see illustration)**.
7   Remove the steering wheel with a puller **(see illustration)**. **Warning:** *Don't allow the steering shaft to turn with the steering wheel removed. If for some reason the steering shaft does turn, refer to Chapter 12, Section 8, for the airbag coil centering procedure.*
8   Installation is the reverse of removal. Be sure to tighten the steering wheel nut to the torque listed in this Chapter's Specifications.
9   To enable the airbag system, refer to Chapter 12.

17 Tie-rod ends - removal and installation

Refer to illustrations 17.2, 17.3 and 17.4
1   Loosen the wheel lug nuts, raise the vehicle and place it securely on jackstands.

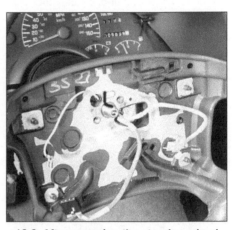

16.6 After removing the steering wheel nut, mark the relationship of the steering wheel to the steering shaft to ensure proper reassembly

Remove the wheel.
2   Loosen the tie-rod end locknut **(see illustration)** and mark the position of the tie-rod end on the threaded portion of the tie-rod.
3   Remove the cotter pin and loosen the castle nut from the tie-rod end balljoint stud,

16.7 Use a steering wheel puller to remove the steering wheel

17.2 Back off the tie-rod end locknut

17.3 Separate the tie-rod end from the steering knuckle with a small puller (don't use a pickle fork unless you plan to replace the tie-rod end; a pickle fork will damage the boot)

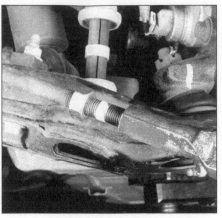

17.4 To ensure proper wheel alignment, make sure the tie-rod end is screwed onto the threaded portion of the tie-rod right up to the paint mark you made before disassembly

18.4 To remove a boot from the power steering gear, remove the tie-rod end, remove these two boot clamps (arrows) and slide off the old boot; use new clamps with the new boot

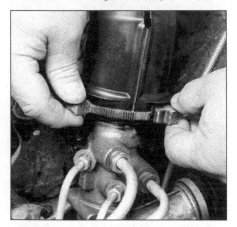

19.4a To remove the plastic protective sheath from the steering shaft-to-steering gear coupling, pop loose this locking strap (arrow) . . .

19.4b . . . then open the sheath halves and pull it off (engine removed for clarity)

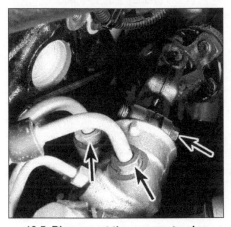

19.5 Disconnect the power steering pressure and return line fittings (arrows) from the steering gear, and remove the pinch bolt (upper arrow) from the U-joint coupling

then install a small puller **(see illustration)** and separate the tie-rod end from the steering knuckle. Remove the nut and detach the tie-rod end from the steering knuckle arm.

4   Unscrew the old tie-rod end and install the new one. Make sure the new tie-rod end is aligned with the mark you made on the threads of the tie-rod **(see illustration)**.

5   Installation is the reverse of removal. Be sure to tighten the tie-rod end balljoint nut to the torque listed in this Chapter's Specifications. Tighten the locknut securely.

18 Steering gear boots - replacement

Refer to illustration 18.4

1   If a steering gear boot is torn, dirt and moisture can damage the steering gear. Replace it.

2   Loosen the wheel lug nuts, raise the vehicle and place it securely on jackstands. Remove the front wheels.

3   Disconnect the tie-rod ends from the

steering knuckles and remove them from the tie-rods (see Section 17).

4   Remove the boot clamps **(see illustration)** and slide the boots off the tie-rods.

5   Installation is the reverse of removal. Be sure to use new clamps on the boots.

19 Steering gear - removal and installation

Refer to illustrations 19.4a, 19.4b, 19.5, 19.7a and 19.7b

Warning: *Make sure the steering shaft is not turned while the steering gear is removed or you could damage the airbag system. To prevent the shaft from turning, turn the ignition key to the lock position before beginning work or run the seat belt through the steering wheel and clip the seat belt into place.*

1   Park the vehicle with the wheels pointing straight ahead. Disconnect the cable from the negative battery terminal. **Caution:** *On models equipped with a Delco Loc II or Theft-lock audio system, be sure the lockout feature is turned off before performing any procedure which requires disconnecting the battery.*

2   Loosen the wheel lug nuts, raise the vehicle and place it securely on jackstands. Remove the front wheels.

3   Disconnect the tie-rod ends from the steering knuckles (see Section 17).

4   Remove the plastic protective shroud from the steering gear coupling **(see illustrations)**.

5   Place a drain pan under the left (driver's side) of the steering gear, then disconnect the pressure and return hoses attached to the power steering gear assembly **(see illustration)**. Plug the ends of the disconnected hoses and the holes in the power steering housing to prevent contamination.

6   Remove the lower pinch bolt from the steering shaft U-joint. On 2000 and later models, remove the serpentine drivebelt. On 2001 and later models, remove the two suspension support braces.

7   Remove the nuts and bolts securing the steering gear to the crossmember **(see illustrations)**.

8   Remove the steering gear.

19.7a To detach the steering gear from the crossmember, remove the nuts inside these holes (arrows) . . .

19.7b . . . while holding the bolt heads (arrows), which can be accessed from the upper side of the crossmember, on either side of the oil pan (engine removed for clarity)

9   Installation is the reverse of removal. Be sure to tighten all fasteners to the torque listed in this Chapter's Specifications. **Note:** *If you're installing a new steering gear, be sure to center the rack in the steering gear housing before installation.*

10   Top up the power steering pump reservoir when you're done and bleed the power steering system (see Section 21).

20   Power steering pump - removal and installation

Removal

V6 models

1   Disconnect the cable from the negative battery terminal. **Caution:** *On models equipped with a Delco Loc II or Theftlock audio system, be sure the lockout feature is turned off before performing any procedure which requires disconnecting the battery.*

2   Remove the serpentine drivebelt (see Chapter 1).

3   Remove the air intake duct (see Chapter 4).

4   Detach the inlet and outlet hoses from the power steering pump.

5   Remove the pulley from the power steering pump with a suitable pulley removal tool (available at most auto parts stores).

6   Remove the pump support bracket.

7   Remove the power steering pump bolts.

8   Remove the pump.

V8 models

Refer to illustration 20.17

9   Disconnect the cable from the negative battery terminal.

10   Drain the coolant (see Chapter 1).

11   Remove the serpentine drivebelt (see Chapter 1).

12   Remove the air intake duct (see Chapter 4).

13   On 1997 and earlier models, remove the alternator (see Chapter 5) and the radiator outlet hose. Also detach the heater inlet and outlet hoses from the water pump and the

throttle body heater return hose from the heater outlet hose (see Chapter 3).

14   On 1998 and later models, remove the power steering pump pulley. this requires a special puller commonly available at most auto parts stores.

15   On 1998 and later models, remove the return hose from the reservoir, then remove the inlet line from the pump body.

16   Be sure to place a drain pan underneath the pump to catch the residual fluid in the lines.

17   Remove the power steering pump bolts **(see illustration)**.

18   On 1997 and earlier models detach the inlet and outlet hoses from the power steering pump after the pump has been removed from the engine.

19   Remove the pump from the engine compartment.

Installation

20   Installation is the reverse of removal. Tighten the pump bolts securely. To install the pulley, a pulley installation tool will be required (these are available at most auto parts stores).

21   Prime the pump by turning the pulley in the reverse direction to that of normal rotation (counterclockwise as viewed from the front) until air bubbles cease to emerge from the fluid when observed through the reservoir filler cap.

22   Install the serpentine drivebelt (see Chapter 1).

23   Bleed the power steering system (see Section 21).

21   Power steering system - bleeding

1   This is not a routine operation and normally will only be required when the system has been dismantled and reassembled.

2   Fill the reservoir to the correct level with fluid of the recommended type and allow it to remain undisturbed for at least two (2) minutes.

3   Start the engine and run it for two or three seconds only. Check the reservoir and

20.17 To remove the power steering pump 1997 and earlier V8 models, rotate the pulley so that the holes line up with each of the two mounting bolts (arrow points to upper bolt; lower bolt, which is located at about 7 o'clock, is not visible in this photo)

add more fluid as necessary.

4   Repeat the operations described in the preceding paragraph until the fluid level remains constant.

5   Raise the front of the vehicle until the wheels are clear of the ground.

6   Start the engine and increase the speed to about 1500 rpm. Now turn the steering wheel gently from stop-to-stop. Check the reservoir fluid level.

7   Lower the vehicle to the ground and, with the engine still running, move the vehicle forward sufficiently to obtain full right lock followed by full left lock. Recheck the fluid level. If the fluid in the reservoir is extremely foamy, allow the vehicle to stand for a few minutes with the engine switched off and then repeat the previous operations. At the same time, check the belt tightness and check for a bent or loose pulley. Check also to make sure the power steering hoses are not touching any other part of the vehicle, especially sheet metal or the exhaust manifold.

8   The procedures above will normally remedy an extreme foam condition and/or an objectionably noisy pump (low fluid level

METRIC TIRE SIZES

P 185 / 80 R 13

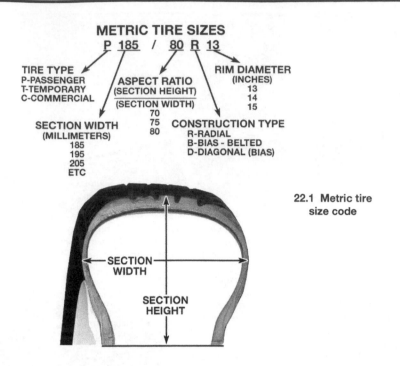

TIRE TYPE
P-PASSENGER
T-TEMPORARY
C-COMMERCIAL

ASPECT RATIO
(SECTION HEIGHT)
──────────────
(SECTION WIDTH)
70
75
80

RIM DIAMETER
(INCHES)
13
14
15

SECTION WIDTH
(MILLIMETERS)
185
195
205
ETC

CONSTRUCTION TYPE
R-RADIAL
B-BIAS - BELTED
D-DIAGONAL (BIAS)

SECTION WIDTH

SECTION HEIGHT

22.1 Metric tire size code

and/or air in the power steering fluid are the leading causes of this condition). If, however, either or both conditions persist after a few trials, the power steering system will have to be thoroughly checked. Do not drive the vehicle until the condition(s) have been remedied.

22   Wheels and tires - general information

Refer to illustration 22.1

All vehicles covered by this manual are equipped with metric-sized fiberglass or steel-belted radial tires (see illustration). Use of other size or type of tires may affect the ride and handling of the vehicle. Don't mix different types of tires, such as radials and bias belted, on the same vehicle as handling may be seriously affected. It's recommended that tires be replaced in pairs on the same axle, but if only one tire is being replaced, be sure it's the same size, structure and tread design as the other. Because tire pressure has a substantial effect on handling and wear, the pressure on all tires should be checked at least once a month or before any extended trips (see Chapter 1).

Wheels must be replaced if they are bent, dented, leak air, have elongated bolt holes, are heavily rusted, out of vertical symmetry or if the lug nuts won't stay tight. Wheel repairs that use welding or peening are not recommended.

Tire and wheel balance is important to the overall handling, braking and performance of the vehicle. Unbalanced wheels can adversely affect handling and ride characteristics as well as tire life. Whenever a tire is installed on a wheel, the tire and wheel

should be balanced by a shop with the proper equipment.

23   Front end alignment - general information

Refer to illustration 23.3

A front end alignment refers to the adjustments made to the front wheels so they

are in proper angular relationship to the suspension and the ground (see illustration). Front wheels that are out of proper alignment not only affect steering control, but also increase tire wear. Camber, caster and toe-in can be adjusted on the vehicles covered by this manual.

Getting the proper front wheel alignment is a very exacting process, one in which complicated and expensive machines are necessary to perform the job properly. Because of this, you should have a technician with the proper equipment perform these tasks. We will, however, use this space to give you a basic idea of what is involved with front end alignment so you can better understand the process and deal intelligently with the shop that does the work.

Camber is the tilting of the front wheels from vertical when viewed from the front of the vehicle. Caster is the tilting of the top of the front steering axis from the vertical. A tilt toward the rear is positive caster and a tilt toward the front is negative caster. These two angles are adjusted by altering the relationship of the front lower control arms to the crossmember.

Toe-in is the turning in of the front wheels. The purpose of a toe specification is to ensure parallel rolling of the front wheels. In a vehicle with zero toe-in, the distance between the front edges of the wheels will be the same as the distance between the rear edges of the wheels. The actual amount of toe-in is normally only a fraction of an inch. Toe-in adjustment is controlled by the position of the tie-rod end on the tie-rod. Incorrect toe-in will cause the tires to wear improperly by making them scrub against the road surface.

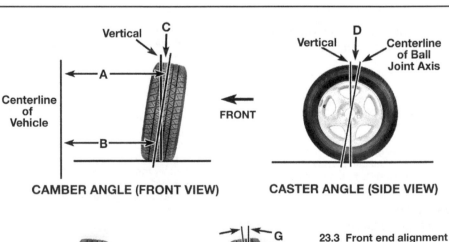

CAMBER ANGLE (FRONT VIEW)   CASTER ANGLE (SIDE VIEW)

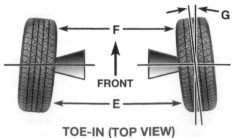

TOE-IN (TOP VIEW)

23.3 Front end alignment details

1   ***A minus B = C***
    (degrees camber)

2   ***E minus F = toe-in***
    (measured in inches)

3   ***G - toe-in***
    (expressed in degrees)

Chapter 11   Body

Contents

1   General information

Warning: *The models covered by this manual are equipped with airbags. Impact sensors for the airbag system on 1993 through 1995 models are located in the instrument panel and just in front of the radiator. On 1996 and later models a single impact sensor is located under the center console. The airbag(s) could accidentally deploy if these sensors are disturbed, so be extremely careful when working in these areas. Airbag system components are also located in the steering wheel, steering column, the base of the steering column and the passenger side of the instrument panel, so be extremely careful when working in these areas and don't disturb any airbag system components or wiring. You could be injured if an airbag accidentally deploys, and the airbag might not deploy correctly in a collision if any of the components or wiring in the system have been disturbed.*

Caution: *On models equipped with a Delco Loc II or Theftlock audio system, be sure the lockout feature is turned off before performing any procedure which requires disconnecting the battery.*

These models are of unitized construction. The frame consists of a floorpan with front and rear frame side rails which support the body components, front and rear suspension systems and other mechanical components.

Except for the hood and rear quarter panels, the exterior body panels of these models are made from various types of plastic, such as Sheet Molded Compound (SMC) or Thermo-Plastic Olefin (TPO). All these materials are rustproof and can withstand minor impacts without damage.

Only general body maintenance practices and body panel repair procedures within the scope of the average home mechanic are included in this Chapter.

2   Maintenance - body

1   The condition of the body is very important, because the value of the vehicle is dependent on it. It's much more difficult to repair a neglected or damaged body than it is to repair mechanical components. The hidden areas of the body, such as the fender wells and the engine compartment, are equally important, although they obviously don't require as frequent attention as the rest of the body.

2   Once a year, or every 12,000 miles, it's a good idea to have the underside of the body steam cleaned. All traces of dirt and oil will be removed and the underside can then be inspected carefully for damaged brake lines, frayed electrical wiring, damaged cables and other problems.

3   At the same time, clean the engine and the engine compartment with a water soluble degreaser.

4   The body should be washed as needed. Wet the vehicle thoroughly to soften the dirt, then wash it down with a soft sponge and plenty of clean soapy water. If the surplus dirt isn't washed off very carefully, it will in time wear down the paint.

5   Spots of tar or asphalt coating thrown up from the road should be removed with a cloth soaked in solvent.

6   Once every six months, wax the body thoroughly.

3   Maintenance - upholstery and carpets

1   Every three months remove the floormats and clean the interior of the vehicle (more frequently if necessary). Use a stiff whisk broom to brush the carpeting and loosen dirt and dust, then vacuum the upholstery and carpets thoroughly, especially along seams and crevices.

2   Dirt and stains can be removed from carpeting with basic household or automotive carpet shampoos available in spray cans. Follow the directions and vacuum again, then use a stiff brush to bring back the "nap" of the carpet.

3   Most interiors have cloth or vinyl upholstery, either of which can be cleaned and maintained with a number of material-specific cleaners or shampoos available in auto supply stores. Follow the directions on the product for usage, and always spot-test any upholstery cleaner on an inconspicuous area (bottom edge of a back seat cushion) to ensure that it doesn't cause a color shift in the material.

4   After cleaning, vinyl upholstery should be treated with a protectant. **Note:** *Make sure the protectant container indicates the product can be used on seats - some products may make a seat too slippery.* **Caution:** *Do not use protectant on vinyl-covered steering wheels.*

5   Leather upholstery requires special care. It should be cleaned regularly with saddlesoap or leather cleaner. Never use alcohol, gasoline, nail polish remover or thinner to clean leather upholstery.

6   After cleaning, regularly treat leather upholstery with a leather conditioner, rubbed in with a soft cotton cloth. Never use car wax on leather upholstery.

7   In areas where the interior of the vehicle is subject to bright sunlight, cover leather seating areas of the seats with a sheet if the vehicle is to be left out for any length of time.

4   Maintenance - vinyl trim

Vinyl trim should not be cleaned with detergents, caustic soaps or petroleum-based cleaners. Plain soap and water or a mild vinyl cleaner is best for stains. Test a small area for color fastness. Bubbles under the vinyl can be eliminated by piercing them with a pin and then working the air out.

5   Maintenance - hinges and locks

Every 3000 miles or three months, the door, hood and trunk lid hinges should be lubricated with a few drops of oil. The door striker plates should also be given a thin coat of white lithium-based grease to reduce wear and ensure free movement.

6   Body repair - minor damage

Plastic body panels

The following repair procedures are for minor scratches and gouges. Repair of more serious damage should be left to a dealer service department or qualified auto body shop. Below is a list of the equipment and materials necessary to perform the following repair procedures on plastic body panels. Although a specific brand of material may be mentioned, it should be noted that equivalent products from other manufacturers may be used instead.

Wax, grease and silicone removing
 solvent
Cloth-backed body tape
Sanding discs
Drill motor with three-inch disc holder
Hand sanding block
Rubber squeegees
Sandpaper
Non-porous mixing palette
Wood paddle or putty knife
Curved tooth body file
Compoxy repair material (for flexible
 panels)
Structural bonding epoxy (for rigid panels)

Flexible panels (front fenders, front and rear bumper covers and rocker panel covers)

1   Remove the damaged panel, if necessary or desirable. In most cases, repairs can be carried out with the panel installed.
2   Clean the area(s) to be repaired with a wax, grease and silicone removing solvent applied with a water-dampened cloth.
3   If the damage is structural, that is, if it extends through the panel, clean the backside of the panel area to be repaired as well. Wipe dry.
4   Sand the rear surface about 1-1/2 inches beyond the break.
5   Cut two pieces of fiberglass cloth large enough to overlap the break by about 1-1/2 inches. Cut only to the required length.
6   Mix the adhesive from the repair kit according to the instructions included with the kit, and apply a layer of the mixture approximately 1/8-inch thick on the backside of the panel. Overlap the break by at least 1-1/2 inches.
7   Apply one piece of fiberglass cloth to

the adhesive and cover the cloth with additional adhesive. Apply a second piece of fiberglass cloth to the adhesive and immediately cover the cloth with additional adhesive insufficient quantity to fill the weave.
8   Allow the repair to cure for 20 to 30 minutes at 60-degrees to 80-degrees F.
9   If necessary, trim the excess repair material at the edge.
10   Remove all of the paint film over and around the area(s) to be repaired. The repair material should not overlap the painted surface.
11   With a drill motor and a sanding disc (or a rotary file), cut a "V" along the break line approximately 1/2-inch wide. Remove all dust and loose particles from the repair area.
12   Mix and apply the repair material. Apply a light coat first over the damaged area; then continue applying material until it reaches a level slightly higher than the surrounding finish.
13   Cure the mixture for 20 to 30 minutes at 60-degrees to 80-degrees F.
14   Roughly establish the contour of the area being repaired with a body file. If low areas or pits remain, mix and apply additional adhesive.
15   Block sand the damaged area with sandpaper to establish the actual contour of the surrounding surface.
16   If desired, the repaired area can be temporarily protected with several light coats of primer. Because of the special paints and techniques required for flexible body panels, it is recommended that the vehicle be taken to a paint shop for completion of the body repair.

Rigid plastic panels (doors, rear compartment lid and roof panels)

17   Clean the repair area. Start with soap and water, then finish with isopropyl alcohol.
18   Using 80-grit sandpaper, taper the scratch or gouge about 1-1/2 inches around the scratch or gouge. Be sure to penetrate well into the plastic material; a D-A-type power sander and/or coarser-grit sandpaper may help.
19   Using 220-grit sandpaper, feather-edge the paint around the repair area. Sand the repair area with 180-grit sandpaper.
20   Wipe the repair area clean with a clean tack rag or other clean, dry rag. **Note:** *The filler will adhere better to a smooth surface than a rough surface.*
21   According to label instructions, mix the epoxy repair material and apply it to the repair area with a squeegee or plastic spreader. Build the material slightly higher than the surrounding undamaged surface. Allow the material to cure approximately 20 to 30 minutes.
22   Using a sanding block, sand the repair area with 180-grit sandpaper, followed by 240-grit sandpaper.
23   Fill scratches and pinholes in the repair area with a light coat of the repair material and sand lightly, as described in the previous Step.

24   Finish by contour sanding the area with 320-grit sandpaper on a sanding block, then wipe with a dry cloth. Because of the special paints used on these panels, it's recommended that you allow a dealer service department or qualified auto body shop to apply the final paint.

Steel body panels (quarter panels and hood)

See photo sequence

Repair of minor scratches

25   If the scratch is superficial and does not penetrate to the metal of the body, repair is very simple. Lightly rub the scratched area with a fine rubbing compound to remove loose paint and built-up wax. Rinse the area with clean water.
26   Apply touch-up paint to the scratch, using a small brush. Continue to apply thin layers of paint until the surface of the paint in the scratch is level with the surrounding paint. Allow the new paint at least two weeks to harden, then blend it into the surrounding paint by rubbing with a very fine rubbing compound. Finally, apply a coat of wax to the scratch area.
27   If the scratch has penetrated the paint and exposed the metal of the body, causing the metal to rust, a different repair technique is required. Remove all loose rust from the bottom of the scratch with a pocket knife, then apply rust inhibiting paint to prevent the formation of rust in the future. Using a rubber or nylon applicator, coat the scratched area with glaze-type filler. If required, the filler can be mixed with thinner to provide a very thin paste, which is ideal for filling narrow scratches. Before the glaze filler in the scratch hardens, wrap a piece of smooth cotton cloth around the tip of a finger. Dip the cloth in thinner and then quickly wipe it along the surface of the scratch. This will ensure that the surface of the filler is slightly hollow. The scratch can now be painted over as described earlier in this section.

Repair of dents

28   When repairing dents, the first job is to pull the dent out until the affected area is as close as possible to its original shape. There is no point in trying to restore the original shape completely as the metal in the damaged area will have stretched on impact and cannot be restored to its original contours. It is better to bring the level of the dent up to a point which is about 1/8-inch below the level of the surrounding metal. In cases where the dent is very shallow, it is not worth trying to pull it out at all.
29   If the back side of the dent is accessible, it can be hammered out gently from behind using a soft-face hammer. While doing this, hold a block of wood firmly against the opposite side of the metal to absorb the hammer blows and prevent the metal from being stretched.
30   If the dent is in a section of the body

which has double layers, or some other factor makes it inaccessible from behind, a different technique is required. Drill several small holes through the metal inside the damaged area, particularly in the deeper sections. Screw long, self-tapping screws into the holes just enough for them to get a good grip in the metal. Now the dent can be pulled out by pulling on the protruding heads of the screws with locking pliers.

31   The next stage of repair is the removal of paint from the damaged area and from an inch or so of the surrounding metal. This is done with a wire brush or sanding disk in a drill motor, although it can be done just as effectively by hand with sandpaper. To complete the preparation for filling, score the surface of the bare metal with a screwdriver or the tang of a file, or drill small holes in the affected area. This will provide a good grip for the filler material. To complete the repair, see the subsection on filling and painting later in this Section.

Repair of rust holes or gashes

32   Remove all paint from the affected area and from an inch or so of the surrounding metal using a sanding disk or wire brush mounted in a drill motor. If these are not available, a few sheets of sandpaper will do the job just as effectively.

33   With the paint removed, you will be able to determine the severity of the corrosion and decide whether to replace the whole panel, if possible, or repair the affected area. New body panels are not as expensive as most people think and it is often quicker to install a new panel than to repair large areas of rust.

34   Remove all trim pieces from the affected area except those which will act as a guide to the original shape of the damaged body, such as headlight shells, etc. Using metal snips or a hacksaw blade, remove all loose metal and any other metal that is badly affected by rust. Hammer the edges of the hole in to create a slight depression for the filler material.

35   Wire brush the affected area to remove the powdery rust from the surface of the metal. If the back of the rusted area is accessible, treat it with rust inhibiting paint.

36   Before filling is done, block the hole in some way. This can be done with sheet metal riveted or screwed into place, or by stuffing the hole with wire mesh.

37   Once the hole is blocked off, the affected area can be filled and painted. See the following subsection on filling and painting.

Filling and painting

38   Many types of body fillers are available, but generally speaking, body repair kits which contain filler paste and a tube of resin hardener are best for this type of repair work. A wide, flexible plastic or nylon applicator will be necessary for imparting a smooth and contoured finish to the surface of the filler material. Mix up a small amount of filler on a clean piece of wood or cardboard (use the hardener sparingly). Follow the manufacturer's instructions on the package, otherwise the filler will set incorrectly.

39   Using the applicator, apply the filler paste to the prepared area. Draw the applicator across the surface of the filler to achieve the desired contour and to level the filler surface. As soon as a contour that approximates the original one is achieved, stop working the paste. If you continue, the paste will begin to stick to the applicator. Continue to add thin layers of paste at 20-minute intervals until the level of the filler is just above the surrounding metal.

40   Once the filler has hardened, the excess can be removed with a body file. From then on, progressively finer grades of sandpaper should be used, starting with a 180-grit paper and finishing with 600-grit wet-or-dry paper. Always wrap the sandpaper around a flat rubber or wooden block, otherwise the surface of the filler will not be completely flat. During the sanding of the filler surface, the wet-or-dry paper should be periodically rinsed in water. This will ensure that a very smooth finish is produced in the final stage.

41   At this point, the repair area should be surrounded by a ring of bare metal, which in turn should be encircled by the finely feathered edge of good paint. Rinse the repair area with clean water until all of the dust produced by the sanding operation is gone.

42   Spray the entire area with a light coat of primer. This will reveal any imperfections in the surface of the filler. Repair the imperfections with fresh filler paste or glaze filler and once more smooth the surface with sandpaper. Repeat this spray-and-repair procedure until you are satisfied that the surface of the filler and the feathered edge of the paint are perfect. Rinse the area with clean water and allow it to dry completely.

43   The repair area is now ready for painting. Spray painting must be carried out in a warm, dry, windless and dust free atmosphere. These conditions can be created if you have access to a large indoor work area, but if you are forced to work in the open, you will have to pick the day very carefully. If you are working indoors, dousing the floor in the work area with water will help settle the dust which would otherwise be in the air. If the repair area is confined to one body panel, mask off the surrounding panels. This will help minimize the effects of a slight mismatch in paint color. Trim pieces such as chrome strips, door handles, etc., will also need to be masked off or removed. Use masking tape and several thickness of newspaper for the masking operations.

44   Before spraying, shake the paint can thoroughly, then spray a test area until the spray painting technique is mastered. Cover the repair area with a thick coat of primer. The thickness should be built up using several thin layers of primer rather than one thick one. Using 600-grit wet-or-dry sandpaper, rub down the surface of the primer until it is very smooth. While doing this, the work area should be thoroughly rinsed with water and the wet-or-dry sandpaper periodically rinsed as well. Allow the primer to dry before spraying additional coats.

45   Spray on the top coat, again building up the thickness by using several thin layers of paint. Begin spraying in the center of the repair area and then, using a circular motion, work out until the whole repair area and about two inches of the surrounding original paint is covered. Remove all masking material 10 to 15 minutes after spraying on the final coat of paint. Allow the new paint at least two weeks to harden, then use a very fine rubbing compound to blend the edges of the new paint into the existing paint. Finally, apply a coat of wax.

7   Body repair - major damage

1   Major damage must be repaired by an auto body/frame repair shop with the necessary welding and hydraulic straightening equipment.

2   If the damage has been serious, it is vital that the structure be checked for proper alignment or the vehicle's handling characteristics may be adversely affected. Other problems, such as excessive tire wear and wear in the driveline and steering may occur.

3   Due to the fact that all of the major body components (hood, fenders, etc.) are separate and replaceable units, any seriously damaged components should be replaced rather than repaired. Sometimes these components can be found in a wrecking yard that specializes in used vehicle components, often at considerable savings over the cost of new parts.

8   Windshield and fixed glass - replacement

1   Replacement of the windshield and fixed glass requires the use of special fast setting adhesive/caulk materials. These operations should be left to a dealer or a shop specializing in glass work.

2   Windshield-mounted rear view mirror support removal is also best left to experts, as the bond to the glass also requires special tools and adhesives.

9   Hood - removal, installation and adjustment

Caution: *On models equipped with a Delco Loc II or Theftlock audio system, be sure the lockout feature is turned off before performing any procedure which requires disconnecting the battery.*

Note: *The hood is somewhat awkward to remove and install, at least two people should perform this procedure.*

These photos illustrate a method of repairing simple dents. They are intended to supplement *Body repair - minor damage* in this Chapter and should not be used as the sole instructions for body repair on these vehicles.

1 If you can't access the backside of the body panel to hammer out the dent, pull it out with a slide-hammer-type dent puller. In the deepest portion of the dent or along the crease line, drill or punch hole(s) at least one inch apart . . .

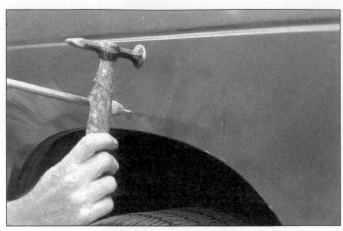

2 . . . then screw the slide-hammer into the hole and operate it. Tap with a hammer near the edge of the dent to help 'pop' the metal back to its original shape. When you're finished, the dent area should be close to its original contour and about 1/8-inch below the surface of the surrounding metal

3 Using coarse-grit sandpaper, remove the paint down to the bare metal. Hand sanding works fine, but the disc sander shown here makes the job faster. Use finer (about 320-grit) sandpaper to feather-edge the paint at least one inch around the dent area

4 When the paint is removed, touch will probably be more helpful than sight for telling if the metal is straight. Hammer down the high spots or raise the low spots as necessary. Clean the repair area with wax/silicone remover

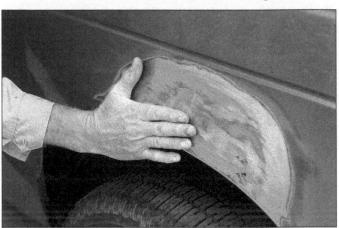

5 Following label instructions, mix up a batch of plastic filler and hardener. The ratio of filler to hardener is critical, and, if you mix it incorrectly, it will either not cure properly or cure too quickly (you won't have time to file and sand it into shape)

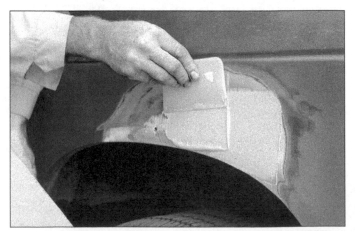

6 Working quickly so the filler doesn't harden, use a plastic applicator to press the body filler firmly into the metal, assuring it bonds completely. Work the filler until it matches the original contour and is slightly above the surrounding metal

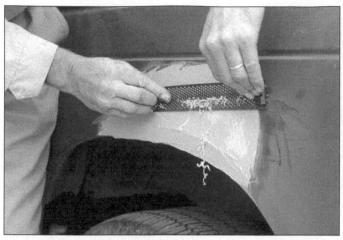

7 Let the filler harden until you can just dent it with your fingernail. Use a body file or Surform tool (shown here) to rough-shape the filler

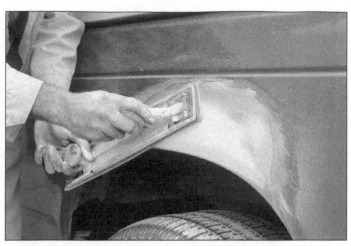

8 Use coarse-grit sandpaper and a sanding board or block to work the filler down until it's smooth and even. Work down to finer grits of sandpaper - always using a board or block - ending up with 360 or 400 grit

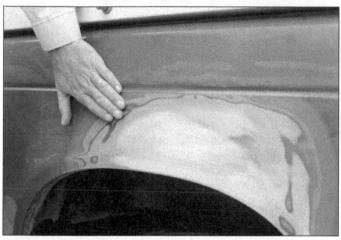

9 You shouldn't be able to feel any ridge at the transition from the filler to the bare metal or from the bare metal to the old paint. As soon as the repair is flat and uniform, remove the dust and mask off the adjacent panels or trim pieces

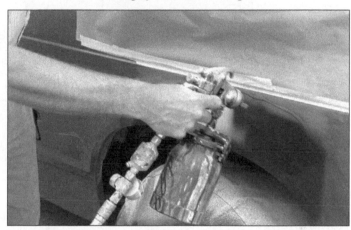

10 Apply several layers of primer to the area. Don't spray the primer on too heavy, so it sags or runs, and make sure each coat is dry before you spray on the next one. A professional-type spray gun is being used here, but aerosol spray primer is available inexpensively from auto parts stores

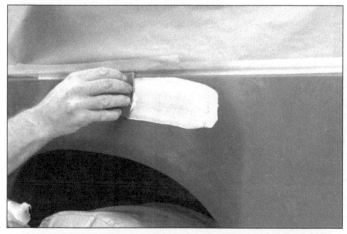

11 The primer will help reveal imperfections or scratches. Fill these with glazing compound. Follow the label instructions and sand it with 360 or 400-grit sandpaper until it's smooth. Repeat the glazing, sanding and respraying until the primer reveals a perfectly smooth surface

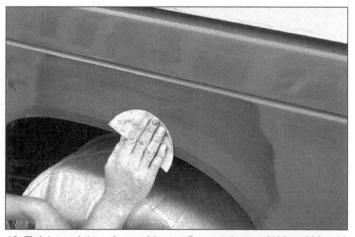

12 Finish sand the primer with very fine sandpaper (400 or 600-grit) to remove the primer overspray. Clean the area with water and allow it to dry. Use a tack rag to remove any dust, then apply the finish coat. Don't attempt to rub out or wax the repair area until the paint has dried completely (at least two weeks)

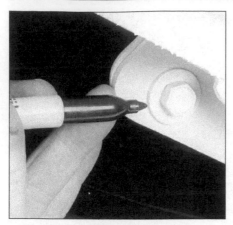

9.3 **Clearly mark around the bolt heads with a marking pen**

9.10 **Loosen the bolts (arrows) and move the hood latch to adjust the hood closed position**

9.11 **Screw the hood bumpers in or out to adjust the hood flush with the fenders**

Removal and installation

Refer to illustration 9.3

1   Disconnect the negative cable from the battery.

2   Open the hood and place rags or covers over the fenders to protect them during the removal procedure.

3   Draw around the hinge bolt washers with a marking pen **(see illustration)**.

4   Have an assistant support the weight of the hood and detach the support strut (see Section 16).

5   Remove the hinge-to-hood bolts and lift off the hood.

6   Installation is the reverse of removal. Align the hinge bolts with the marks made in step 3.

Adjustment

Refer to illustrations 9.10 and 9.11

7   Fore-and-aft and side-to-side adjustment of the hood is done by moving the hood in relation to the hinge plate after loosening the hinge-to-hood bolts or nuts.

8   Mark around the entire hinge plate so you can judge the amount of movement.

9   Loosen the bolts and move the hood into correct alignment. Move it only a little at a time. Tighten the hinge bolts and carefully lower the hood to check the alignment.

10   If necessary after installation, the entire hood latch assembly can be adjusted up-and-down as well as from side-to-side on the upper radiator support so the hood closes securely and is flush with the fenders **(see illustration)**. To do this, scribe a line around the hood latch mounting bolts to provide a reference point. Then loosen the bolts and reposition the latch assembly as necessary. Following adjustment, retighten the mounting bolts.

11   Finally, adjust the hood bumpers on the radiator support so the hood, when closed, is flush with the fenders **(see illustration)**.

12   The hood latch assembly, as well as the hinges, should be periodically lubricated with white lithium-based grease to prevent sticking and wear.

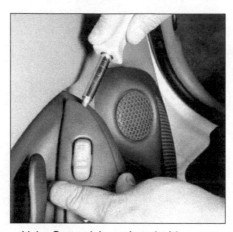

11.1a **On models equipped with power windows and door locks, gently pry off the armrest control panel . . .**

10 Hood release latch and cable - removal and installation

Removal

1   Remove the cable from the latch assembly by uncoupling the release mechanism retaining clip.

2   Attach a piece of string or thin wire to the end of the cable and unclip all cable retaining clips.

3   Working in the passenger compartment, remove two bolts and detach the hood release lever. On 2000 and later models, remove the driver's side kick panel.

4   Remove the cable and grommet rearward, pulling them into the passenger compartment.

Installation

5   Ensure that the new cable has a grommet attached and then connect the string or wire to the new cable and pull it forward into the engine compartment.

6   Connect the cable and secure it with the retaining clips.

7   Ensure that the grommet is in place and install the hood release lever.

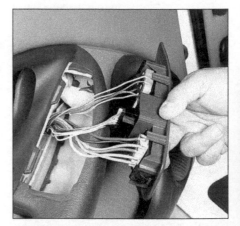

11.1b **. . . and disconnect the electrical connectors from the backside**

11 Door trim panel - removal and installation

Caution: *On models equipped with a Delco Loc II or Theftlock audio system, be sure the lockout feature is turned off before performing any procedure which requires disconnecting the battery.*

Removal

Refer to illustrations 11.1a, 11.1b, 11.1c, 11.5a, 11.5b and 11.6

1   On power window equipped models, disconnect the negative cable at the battery. Pry out the control switch assembly, disconnect the electrical connectors and remove the screw inside the opening **(see illustrations)**.

2   On manual window equipped models, remove the window regulator handle by pressing in on the bearing plate and door trim panel and, with a piece of hooked wire, pull off the spring clip. A special tool is available for this purpose but its use is not essential. With the clip removed, remove the handle and the bearing plate.

3   Remove the door lock knob by sliding it

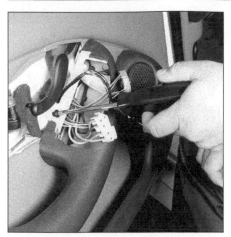

11.1c Remove the screw located in the switch opening

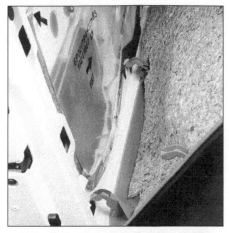

11.5a Pull up on the door trim panel to detach it, then rotate it back and out of the door

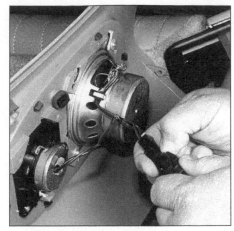

11.5b Unplug the speaker connectors

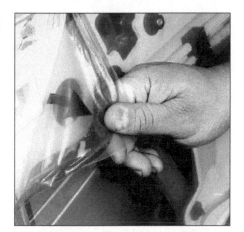

11.6 For access to the inner door components, peel back the plastic water shield - if you're careful, it can be reused

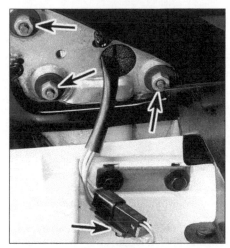

12.3 Remove the outside mirror retaining nuts and on power models, unplug the electrical connector (arrows)

forward and using a small screwdriver to detach the rod lock.

4   Remove the screw and pry off the outside mirror cover from the inner door.

5   Remove any remaining door trim panel retaining screws, then carefully pull the trim panel up and away from the door. Do not pull straight out from the door or you might break the trim panel retaining hooks. Unplug any wire harness connectors and remove the panel **(see illustrations)**.

6   For access to the door inside handle or the door window regulator, raise the window fully, remove the power window control unit (if equipped), the door panel bracket and the speaker assembly (see Chapter 12), then carefully peel back the plastic water shield **(see illustration)**.

Installation

7   Prior to installation of the door trim panel, be sure to reinstall any clips in the panel which may have come out when you removed the panel.

8   Plug in the wire harness connectors for the power door lock switch and the power window switch, if equipped, and place the

panel in position in the door. Press the door panel into place and push down until the hooks are seated. Install the trim panel retaining screws. Install the power door lock switch assembly, if equipped. Install the manual regulator crank handle or power window switch assembly.

12 Outside mirror - removal and installation

Refer to illustration 12.3
Caution: *On models equipped with a Delco Loc II or Theftlock audio system, be sure the lockout feature is turned off before performing any procedure which requires disconnecting the battery.*
1   Disconnect the negative cable from the battery.
2   Remove the retaining screw and detach the mirror cover.
3   Remove the mirror-to-door retaining nuts and, on power mirrors, unplug the electrical connector **(see illustration)**.

4   Lift off the mirror assembly.
5   Installation is the reverse of removal.

13 Door - removal and installation

Caution: *On models equipped with a Delco Loc II or Theftlock audio system, be sure the lockout feature is turned off before performing any procedure which requires disconnecting the battery.*
1   On doors with power components (power windows and door locks), disconnect the negative cable from the battery.
2   Remove the door trim panel (see Section 11). On doors with power components, unplug all electrical connections from the door (it is a good idea to label all connections to aid the reassembly process) and remove the electrical harness from the door.
3   Open the door all the way and support it on jacks or blocks covered with cloth or pads to prevent damaging the paint.
4   Mark around the door hinges to facilitate realignment during reassembly.
5   Have an assistant hold the door, remove the hinge nuts and lift the door off.
6   Install the door by reversing the removal procedure. Tighten the nuts securely.

14 Front fender - removal and installation

Refer to illustrations 14.5, 14.6, 14.7, 14.8 and 14.9
1   Raise the vehicle, support it securely on jackstands and remove the front wheel.
2   Disconnect all lighting assembly wiring harness connectors and other components that would interfere with fender removal.
3   On models through 1998, remove the hood (see Section 9).
4   Remove the bolts and plastic clips and detach the wheelhouse cover. Remove the cowl grille (see Section 24).
5   Remove the two nuts holding the fender

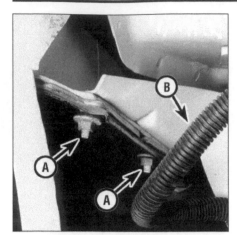

14.5 Remove the two nuts (A) holding the bumper cover to the fender bracket, then unbolt the bracket (B) from the fender

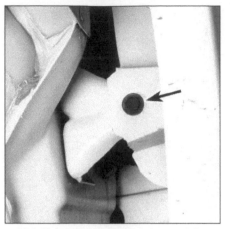

14.6 Remove the rear fender retaining bolt (arrow)

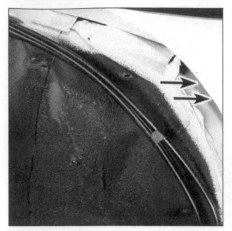

14.7 There are two more bolts retaining the rear edge of the fender accessible through this opening (arrows)

14.8 Remove the two bolts retaining the lower rear edge of the fender to the chassis (arrows)

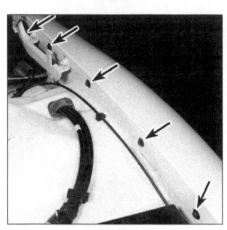

14.9 Remove the five bolts along the top edge of the fender (arrows) - a sixth bolt is located out of view behind the hinge bracket

15.5 Draw around the rear compartment lid hinge plates with a marking pen

bracket to the bumper cover, then unbolt the bracket from the fender **(see illustration)**.

6   Working in the rear of the fender well, remove the center fender-to-body bracket bolt **(see illustration)**.

7   Remove the two remaining fender-to-bracket mounting bolts **(see illustration)**.

8   Remove the fender-to-chassis mounting bolts at the rear lower edge of the fender **(see illustration)**.

9   Remove the six fender mounting bolts along the top edge of the fender **(see illustration)**.

10   Detach the fender. It is a good idea to have an assistant support the fender while it's being moved away from the vehicle to prevent damage to the surrounding body panels.

11   Installation is the reverse of removal. Tighten all nuts, bolts and screws securely.

---

15 Rear compartment lid - removal, installation and adjustment

Refer to illustrations 15.5 and 15.9
Caution: *On models equipped with a Delco Loc II or Theftlock audio system, be sure the*

lockout feature is turned off before performing any procedure which requires disconnecting the battery.
Note: *The liftgate is heavy and somewhat awkward to remove and install - at least two people should perform this procedure.*

1   Disconnect the negative cable from the battery.

2   Open the rear compartment lid and cover the upper body area around the opening with pads or blankets to protect the painted surfaces when the rear compartment lid is removed.

3   Unplug all electrical connectors and pull the wire harness out of the rear compartment lid (tie string or wire to the cables so they can be pulled back into the body when the rear compartment lid is reinstalled).

4   While an assistant supports the rear compartment lid, detach the support struts (see Section 16).

5   Draw around the hinge plates with a marking pen **(see illustration)**.

6   Remove the hinge nuts and detach the rear compartment lid from the vehicle.

7   Installation is the reverse of removal. Be extremely careful with the wire harness when

threading it back into the rear compartment lid or you could cut it on a metal edge.

8   After installation, close the rear compartment lid and make sure it's in proper alignment with the surrounding body panels.

9   If the rear compartment lid needs to be adjusted, loosen the bolts and slightly reposition the latch, repeating the procedure as necessary **(see illustration)**.

15.9 Loosen the bolts and move the rear compartment lid latch to adjust the closed position

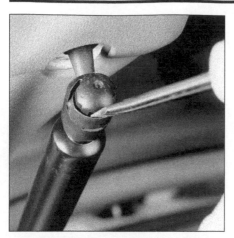

16.2 Use a small screwdriver to pry off the clip, then detach the end of the strut from the ballstud

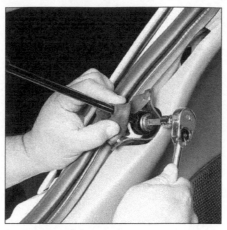

16.3 Remove the Torx screw securing the top of the strut, then remove the strut assembly

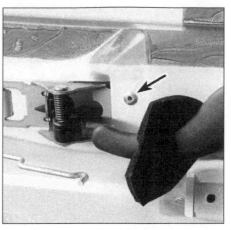

17.3 Drill out the rivet (arrow) and rotate the handle out - on installation, install a new rivet with a hand riveting tool

10   You can also adjust the rear compartment lid position in relation to the adjacent fender panels by screwing the rear compartment lid edge cushions in (to depress the rear compartment lid) or out (to raise it).

16   Support struts- replacement

Refer to illustrations 16.2 and 16.3
Note: *The hood and liftgate are heavy and somewhat awkward to hold - at least two people should perform this procedure.*
1   Open the hood or rear compartment lid and support it securely.
2   On both ends or the hood strut and the lower end of the rear compartment lid strut, use a small screwdriver to detach the retaining clip and use a prybar or grasp the strut end and pull sharply to detach it from the vehicle **(see illustration)**. **Note:** *On 2000 and later models, the upper end of the liftgate strut is attached with a Torx bolt, not a clip.*
3   On the upper end of the liftgate strut, pull the rubber cover out of the way and unscrew the Torx-head mounting bolt stud from the rear compartment lid **(see illustration)**.
4   Installation is the reverse of removal.

17   Door latch, lock cylinder and handles - removal and installation

Inside handle

Refer to illustration 17.3
1   With the window glass in the full up position, remove the door trim panel and plastic water shield (see Section 11).
2   Disconnect the lock rod from the handle.
3   Drill out the rivet securing the handle to the door frame and remove the handle **(see illustration)**.

4   Installation is the reverse of the removal procedure.

Outside handle

5   Remove the door trim panel and plastic water shield (see Section 11).
6   Remove the handle-to-door nuts.
7   Disengage the handle from the lock rod and pull the handle free.
8   Installation is the reverse of the removal procedure.

Lock cylinder

9   Remove the door trim panel and water shield (see Section 11).
10   Remove the lock rod from the cylinder.
11   Disengage the cylinder retainer and pull the lock cylinder from the door.
12   Installation is the reverse of the removal procedure.

Latch

Refer to illustration 17.15
13   Remove the door trim panel and water shield (see Section 11).
14   From inside the door, disconnect the locking knob rod, inside handle rod and inside handle lock rod from the lock assembly.
15   Remove the latch assembly retaining screws and pull the assembly free **(see illustration)**.
16   Installation is the reverse of the removal procedure.

18   Door window glass - removal and installation

Warning: *Safety glasses and gloves should be worn when performing this procedure.*
1   If the glass being replaced is cracked, criss-cross it with masking tape to help hold the glass together.
2   Lower the glass half way and use rubber door stops wedged between the door opening and the glass or duct tape to hold the window in position.

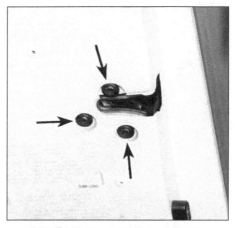

17.15 To remove the lock assembly, unscrew the retaining screws (arrows) at the rear edge of the door

3   Remove the door trim panel and plastic water shield (see Section 11).
4   Working through the access holes in the door, punch out the center pins of the three rivets that secure the glass to the window channel. Using a 1/4-inch drill bit, drill out the rivets and pull the window up and clear of the door.
5   Installation is the reverse of removal.

19   Front door window glass regulator - removal and installation

Warning: *Do not remove the electric motor from the regulator assembly. The lift arm is under tension from the counterbalance spring and could cause personal injury if allowed to retract.*
Caution: *On models equipped with a Delco Loc II or Theftlock audio system, be sure the lockout feature is turned off before performing any procedure which requires disconnecting the battery.*

19.4a Loosen the two upper (arrows) . . .

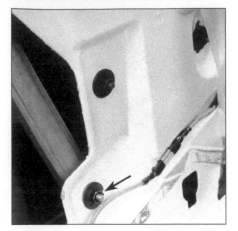

19.4b . . . and one lower guide channel nuts (arrow)

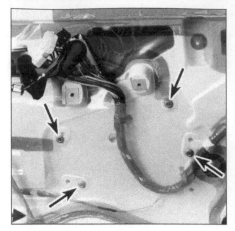

19.5 Drill out the door glass regulator rivets (arrows) - power regulator shown

Removal

Refer to illustrations 19.4a, 19.4b and 19.5

1   On power window equipped models, disconnect the negative cable at the battery.

2   With the window glass in the full up position, remove the door trim panel and water shield (see Section 11).

3   Secure the window glass in the half way "Up" position with door stops wedged

20.3a Remove the retaining screw, lift up on the rear edge of the shift plate and detach it from the console

between the door opening and the glass. Do not proceed until you're sure the weight of the window is supported.

4   Remove the rear guide channel nuts so the assembly can be moved out of the way **(see illustrations)**.

5   Punch out the center pins of the rivets that secure the window regulator to the door frame and drill the rivets out with a 3/16-inch (manual) or 1/4-inch (power) drill bit **(see illustration)**.

6   On power window equipped models, unplug the electrical connector.

7   Move the regulator until it is disengaged from the guide channel assemblies and lift it from the door.

Installation

8   Place the regulator in position in the door and engage to the guide channel assemblies.

9   Secure the regulator to the door using 3/16-inch (manual) or 1/4-inch (power) rivets and a rivet tool.

10   Plug in the electrical connector (if equipped).

11   Tighten the channel nuts securely.

12   Install the door trim panel. Connect the negative battery cable.

20   Console - removal and installation

Refer to illustrations 20.3a, 20.3b, 20.4 and 20.5

Warning: *The models covered by this manual are equipped with airbags. Always disable the airbag system before working in the vicinity of the impact sensors, steering column or instrument panel to avoid the possibility of accidental deployment of the airbag(s), which could cause personal injury (see Chapter 12). The yellow wires and connectors routed through the console are for this system. Do not use electrical test equipment on these yellow wires or tamper with them in any way while working around the console.*

Caution: *On models equipped with a Delco Loc II or Theftlock audio system, be sure the lockout feature is turned off before performing any procedure which requires disconnecting the battery.*

1   Remove the shift knob or lever. On models with an automatic transmission, pry out the clip and pull off the knob. On models with a manual transmission, simply remove the bolts and detach shifter knob and lever assembly.

20.3b Unplug the electrical connectors and remove the shift plate

20.4 Detach the cigarette lighter from the console and any additional switch connectors

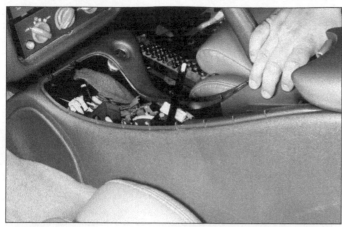

20.5 Remove the screws and pull the console back, then lift it out

21.4 Detach the instrument cluster bezel by pulling it straight out

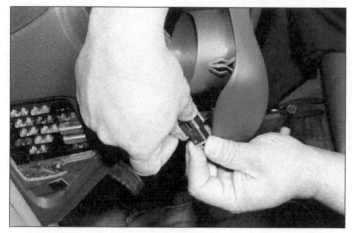

21.5 Unplug the electrical connectors and remove the bezel

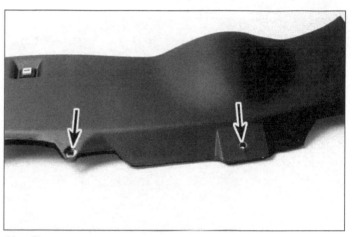

22.1 The knee bolster is held in place by bolts at the bottom edge (arrows) and Velcro fasteners or clips at the top edge

On 2000 and later models, block the vehicle's wheels and with the key in the Run position, move the shift lever to the Neutral position for console cover removal.

2   Apply the parking brake lever.

3   Open the console door for access, remove the retaining screw and lift up the rear edge of the shift plate, then unplug the electrical connectors and remove the shift plate **(see illustrations)**.

4   Detach the cigarette lighter and remove the console retaining screws **(see illustration)**.

5   Lift the console up and out of the vehicle **(see illustration)**.

6   Installation is the reverse of removal.

---

21   Instrument cluster bezel - removal and installation

Warning: *The models covered by this manual are equipped with airbags. Always disable the airbag system before working in the vicinity of the impact sensors, steering column or instrument panel to avoid the possibility of accidental deployment of the airbag(s), which could cause personal injury (see Chapter 12). The yellow wires and connectors under the instrument panel are for this system. Do not*

use electrical test equipment on these yellow wires or tamper with them in any way while working around the instrument panel.

Caution: *On models equipped with a Delco Loc II or Theftlock audio system, be sure the lockout feature is turned off before performing any procedure which requires disconnecting the battery.*

Camaro

1   Remove the four or five bezel-to-instrument panel retaining screws. On models so equipped, remove the fog light switch from the bezel.

2   Pull the bezel off the instrument panel and unplug the electrical connectors.

3   Installation is the reverse of removal.

Firebird

Refer to illustrations 21.4 and 21.5

4   On models so equipped, remove the fog light switch from the bezel. Grasp the bezel securely and pull back sharply to detach the clips from the instrument panel **(see illustration)**.

5   Unplug the electrical connectors and remove the bezel **(see illustration)**.

6   Installation is the reverse of body.

---

22   Instrument panel trim panels - removal and installation

Warning: *The models covered by this manual are equipped with airbags. Always disable the air bag system before working in the vicinity of the impact sensors, steering column or instrument panel to avoid the possibility of accidental deployment of the airbag(s), which could cause personal injury (see Chapter 12). The yellow wires and connectors routed through the instrument panel are for this system. Do not use electrical test equipment on these yellow wires or tamper with them in any way while working around the instrument panel.*

Caution: *On models equipped with a Delco Loc II or Theftlock audio system, be sure the lockout feature is turned off before performing any procedure which requires disconnecting the battery.*

Knee bolster

Refer to illustration 22.1

1   Remove the two bolts at the bottom edge of the knee bolster, then detach the clips or fasteners at the upper edge **(see illustration)**. On later models, there are four screws instead of clips.

22.7 Detach the center trim panel and pull it straight back

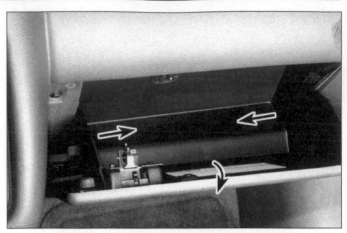

22.9 Open the glove compartment, squeeze the sides in and rotate it out of the instrument panel

22.10 Remove the three screws (arrows) that attach the glove box hinge to the instrument panel

23.2 Remove the bumper cover plastic retainers (arrows)

2   Unplug any electrical connectors, then lower the knee bolster panel from the instrument panel.

3   Installation is the reverse of the removal procedure.

Sound insulator panel

4   Remove the knee bolster.

5   Pry out the plastic clips, detach the sound insulator panel and lower it from the instrument panel.

6   Installation is the reverse of the removal procedure.

Center trim panel

Refer to illustration 22.7

7   Remove the retaining screw (if equipped), grasp the panel securely and detach it from the instrument panel by pulling it straight back **(see illustration)**.

8   Installation is the reverse of the removal procedure.

Glove box

Refer to illustrations 22.9 and 22.10

9   Open the glove box door, squeeze the plastic sides in and lower the glove box from the instrument panel **(see illustration)**.

10   Remove the three screws from the hinge

and lower the assembly from the instrument panel **(see illustration)**.

11   Installation is the reverse of the removal procedure.

---

23 Bumper covers - removal and installation

Warning: *The models covered by this manual are equipped with airbags. Always disable the airbag system before working in the vicinity of the impact sensors, steering column or instrument panel to avoid the possibility of accidental deployment of the airbag(s), which could cause personal injury* (see Chapter 12).
Caution: *On models equipped with a Delco Loc II or Theftlock audio system, be sure the lockout feature is turned off before performing any procedure which requires disconnecting the battery.*

Front bumper cover

Refer to illustrations 23.2, 23.5a, 23.5b and 23.6

1   Open the hood and disconnect the negative battery cable.

2   Remove the bumper cover plastic retainers and retaining nuts **(see illustration)**.

23.5a Remove the screws (arrows) . . .

3   Raise the front of the vehicle and support it securely on jackstands.

4   Disconnect the front turn signal lights and fog lights, if equipped (see Chapter 12).

5   From under the vehicle, remove the bumper cover plastic retainers and retaining nuts, screws and clips **(see illustrations)**. On 2000 and later models, remove the one screw and detach the clips to remove the (roughly) triangular air deflector panels at the left and right corners below the front bumper cover.

6   Remove the four retaining nuts (two on

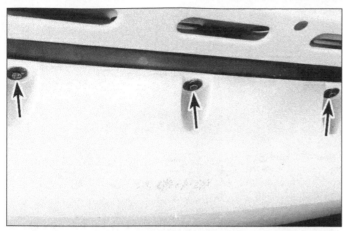

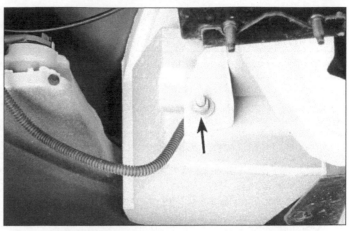

23.5b . . . and plastic retainers (arrows) along the bottom of the bumper cover

23.6 Remove the bumper cover retaining nuts (arrow)

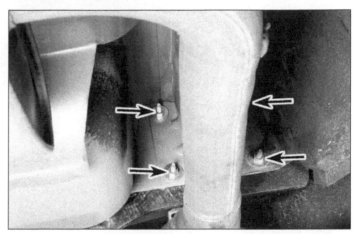

23.10 The rear bumper cover is retained by bolts and plastic retainers (arrows)

23.12 Remove the rear bumper cover retaining nuts (arrows)

each side, located inside the turn signal recesses) **(see illustration)**.
7   Lift the bumper cover, slide it forward and remove it from the vehicle.
8   Installation is the reverse of removal.

Rear bumper cover

Refer to illustrations 23.10 and 23.12
9   Remove the two bolts retaining the forward ends of the bumper cover to the fender wells. On 2000 and later models, remove the nuts inside the fenderwell securing the bumper cover.
10 Remove the plastic screws retaining the bumper cover to the body **(see illustration)**. On 2000 and later models, remove the spare tire, the jack and the finish panel above the bumper fascia to access the pushpins securing the upper/center of the bumper cover.
11 Remove the side marker lights (see Chapter 12).
12 From under the vehicle, remove the eight bumper cover retaining nuts (four on each side) **(see illustration)**. On 2000 and later models, some of the bumper cover fasteners are plastic pushpins.
13 Slide the bumper cover to the rear and remove it.
14 Installation is the reverse of removal.

24 Cowl grille - removal and installation

Refer to illustrations 24.2 and 24.4
1   Mark the position of the windshield wiper blade on the windshield with a wax marking pen.

24.2 Flip up the plastic cover, remove the nut and detach the wiper arm

2   Remove the wiper arms **(see illustration)**.
3   Remove the cowl weatherstripping.
4   Remove the plastic cowl grille retainers, disconnect the windshield washer hoses and detach the cowl grilles from the vehicle **(see illustration)**.
5   Installation is the reverse of removal.

24.4 Use wire cutters to gently pull up the centers of the plastic retainers, then remove the retainers and the cowl grilles

Make sure to align the wiper blades with the marks made during removal.

25 Seats - removal and installation

Refer to illustration 25.6
Caution: *On models equipped with a Delco Loc II or Theftlock audio system, be sure the lockout feature is turned off before performing any procedure which requires disconnecting the battery.*

Front seats

1   Disconnect the negative cable from the battery.
2   Disconnect the electrical connectors.

3   Remove the rear nuts from the seat anchor plate.
4   Remove the front securing nuts and remove the seat.
5   Installation is the reverse of the removal procedure. Tighten the seat securing nuts securely.

Rear seat

6   Remove the bolt at the base of the seat cushion **(see illustration)**.
7   Lift the seat cushion out of the vehicle.
8   Release the seat back, push it forward and remove the two retaining nuts, then lift the seat back out through the rear hatch opening.
9   Installation is the reverse of the removal procedure.

25.6 Remove each rear seat cushion bolt (arrow)

Chapter 12
Chassis electrical system

Contents

1   General information

The electrical system is a 12-volt, negative ground type. Power for the lights and all electrical accessories is supplied by a lead/acid-type battery which is charged by the alternator.

This Chapter covers repair and service procedures for the various electrical components not associated with the engine. Information on the battery, alternator, ignition system and starter motor can be found in Chapter 5. It should be noted that when portions of the electrical system are serviced, the negative battery cable should be disconnected from the battery to prevent electrical shorts and/or fires. **Caution:** *On models equipped with a Delco Loc II or Theftlock audio system, be sure the lockout feature is turned off before performing any procedure which requires disconnecting the battery.*

2   Electrical troubleshooting - general information

A typical electrical circuit consists of an electrical component, any switches, relays, motors, fuses, fusible links or circuit breakers related to that component and the wiring and electrical connectors that link the component to both the battery and the chassis. To help you pinpoint an electrical circuit problem, wiring diagrams are included at the end of this Chapter.

Before tackling any troublesome electrical circuit, first study the appropriate wiring diagrams to get a complete understanding of what makes up that individual circuit. Trouble spots, for instance, can often be narrowed down by noting if other components related to the circuit are operating properly. If several components or circuits fail at one time, chances are the problem is in a fuse or ground connection, because several circuits are often routed through the same fuse and ground connections.

3.1a The passenger compartment fuse block is accessible after removing the cover

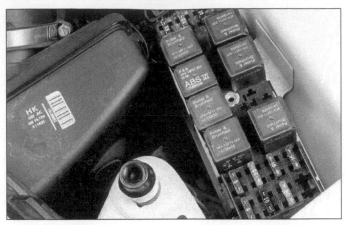

3.1b The engine compartment fuse block also contains relays

Electrical problems usually stem from simple causes, such as loose or corroded connections, a blown fuse, a melted fusible link or a bad relay. Visually inspect the condition of all fuses, wires and connections in a problem circuit before troubleshooting it.

If testing instruments are going to be utilized, use the diagrams to plan ahead of time where you will make the necessary connections in order to accurately pinpoint the trouble spot.

The basic tools needed for electrical troubleshooting include a circuit tester or voltmeter (a 12-volt bulb with a set of test leads can also be used), a continuity tester, which includes a bulb, battery and set of test leads, and a jumper wire, preferably with a circuit breaker incorporated, which can be used to bypass electrical components. Before attempting to locate a problem with test instruments, use the wiring diagram(s) to decide where to make the connections.

Voltage checks

Voltage checks should be performed if a circuit is not functioning properly. Connect one lead of a circuit tester to either the negative battery terminal or a known good ground. Connect the other lead to an electrical connector in the circuit being tested, preferably nearest to the battery or fuse. If the bulb of the tester lights, voltage is present, which means that the part of the circuit between the electrical connector and the battery is problem free. Continue checking the rest of the circuit in the same fashion. When you reach a point at which no voltage is present, the problem lies between that point and the last test point with voltage. Most of the time the problem can be traced to a loose connection. **Note:** *Keep in mind that some circuits receive voltage only when the ignition key is in the Accessory or Run position.*

Finding a short

One method of finding shorts in a circuit is to remove the fuse and connect a test light or voltmeter in its place to the fuse terminals. There should be no voltage present in the cir-

cuit. Move the wiring harness from side-to-side while watching the test light. If the bulb goes on, there is a short to ground somewhere in that area, probably where the insulation has rubbed through. The same test can be performed on each component in the circuit, even a switch.

Ground check

Perform a ground test to check whether a component is properly grounded. Disconnect the battery and connect one lead of a self-powered test light, known as a continuity tester, to a known good ground. Connect the other lead to the wire or ground connection being tested. If the bulb goes on, the ground is good. If the bulb does not go on, the ground is not good.

Continuity check

A continuity check is done to determine if there are any breaks in a circuit - if it is passing electricity properly. With the circuit off (no power in the circuit), a self-powered continuity tester can be used to check the circuit. Connect the test leads to both ends of the circuit (or to the "power" end and a good ground), and if the test light comes on the circuit is passing current properly. If the light doesn't come on, there is a break somewhere in the circuit. The same procedure can be used to test a switch, by connecting the continuity tester to the switch terminals. With the switch turned On, the test light should come on.

Finding an open circuit

When diagnosing for possible open circuits, it is often difficult to locate them by sight because oxidation or terminal misalignment are hidden by the electrical connectors. Merely wiggling an electrical connector on a sensor or in the wiring harness may correct the open circuit condition. Remember this when an open circuit is indicated when troubleshooting a circuit. Intermittent problems may also be caused by oxidized or loose connections.

Electrical troubleshooting is simple if

you keep in mind that all electrical circuits are basically electricity running from the battery, through the wires, switches, relays, fuses and fusible links to each electrical component (light bulb, motor, etc.) and to ground, from which it is passed back to the battery. Any electrical problem is an interruption in the flow of electricity to and from the battery.

3   Fuses - general information

Refer to illustrations 3.1a, 3.1b and 3.3

1   The electrical circuits of the vehicle are protected by a combination of fuses, circuit breakers and fusible links. The two fuse blocks are located under a cover on the left end of the instrument panel and in the left side of the engine compartment **(see illustrations)**.

2   Each of the fuses is designed to protect a specific circuit, and the various circuits are identified on the fuse panel itself.

3   Miniaturized fuses are employed in the fuse block. These compact fuses, with blade terminal design, allow fingertip removal and replacement. If an electrical component fails, always check the fuse first. A blown fuse is easily identified through the clear plastic body. Visually inspect the element for evidence of damage **(see illustration)**. If a continuity check is called for, the blade terminal tips are exposed in the fuse body.

4   Be sure to replace blown fuses with the correct type. Fuses of different ratings are physically interchangeable, but only fuses of the proper rating should be used. Replacing a fuse with one of a higher or lower value than specified is not recommended. Each electrical circuit needs a specific amount of protection. The amperage value of each fuse is molded into the fuse body.

If the replacement fuse immediately fails, don't replace it again until the cause of the problem is isolated and corrected. In most cases, the cause will be a short circuit in the wiring caused by a broken or deteriorated wire.

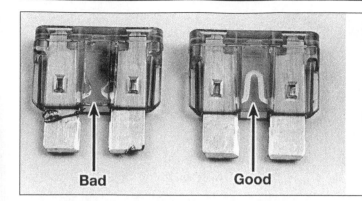

3.3 When a fuse blows, the element between the terminals melts - the fuse on the left is blown, the fuse on the right is good

Bad   Good

4 Fusible links - general information

Some circuits are protected by fusible links. Fusible links are used in circuits which are not ordinarily fused, such as the ignition circuit.

Although the fusible links appear to be a heavier gauge than the wire they are protecting, the appearance is due to the thick insulation. All fusible links are four wire gauges smaller than the wire they are designed to protect.

Fusible links cannot be repaired, but a new link of the same size wire can be put in its place. The procedure is as follows:

a) *Disconnect the negative cable from the battery.*
b) *Disconnect the fusible link from the wiring harness.*
c) *Cut the damaged fusible link out of the wiring just behind the electrical connector.*
d) *Strip the insulation back approximately 1/2-inch.*
e) *Position the electrical connector on the new fusible link and crimp it into place.*
f) *Use rosin core solder at each end of the new link to obtain a good solder joint.*
g) *Use plenty of electrical tape around the soldered joint. No wires should be exposed.*
h) *Connect the battery ground cable. Test the circuit for proper operation.*

5 Circuit breakers - general information

Circuit breakers protect components such as power windows, power door locks and headlights. On some models the circuit breaker resets itself automatically, so an electrical overload in a circuit breaker protected system will cause the circuit to fail momentarily, then come back on. If the circuit doesn't come back on, check it immediately. Once the condition is corrected, the circuit breaker will resume its normal function. Some circuit breakers must be reset manually.

6 Relays - general information

Several electrical accessories in the vehicle use relays to transmit the electrical signal to the component. If the relay is defective, that component will not operate properly.

The various relays are grouped together in several locations. Some relays are grouped together in the engine compartment **(see illustration 3.1b)**.

If a faulty relay is suspected, it can be removed and tested by a dealer service department or a repair shop. Defective relays must be replaced as a unit.

7 Turn signal/hazard flasher - check and replacement

Warning: *The models covered by this manual are equipped with airbags. Always disable the airbag system before working in the vicinity of the impact sensors, steering column or instrument panel to avoid the possibility of accidental deployment of the airbag(s), which could cause personal injury (see Section 26). The yellow wires and connectors routed through the instrument panel are for this system. Do not use electrical test equipment on these yellow wires or tamper with them in any way while working under the instrument panel.*
Caution: *On models equipped with a Delco Loc II or Theftlock audio system, be sure the lockout feature is turned off before performing any procedure which requires disconnecting the battery.*

Turn signal flasher

1   The turn signal flasher, a small canister-shaped unit located under the dash, in a clip to the left of the steering column, flashes the turn signals.
2   When the flasher unit is functioning properly, an audible click can be heard during its operation. If the turn signals fail on one side or the other and the flasher unit does not make its characteristic clicking sound, a faulty turn signal bulb is indicated.
3   If both turn signals fail to blink, the problem may be due to a blown fuse, a faulty flasher unit, a broken switch or a loose or

open connection. If a quick check of the fuse box indicates that the turn signal fuse has blown, check the wiring for a short before installing a new fuse.
4   To replace the flasher, simply unplug it and pull it out of the clip.
5   Make sure that the replacement unit is identical to the original. Compare the old one to the new one before installing it.
6   Installation is the reverse of removal.

Hazard flasher

Refer to illustration 7.9
7   The hazard flasher, a small canister-shaped unit located in the convenience center, flashes all four turn signals simultaneously when activated.
8   The hazard flasher is checked in a fashion similar to the turn signal flasher (see Steps 2 and 3).
9   To replace the hazard flasher, remove the knee bolster for access (see Chapter 11), then pull the flasher from the convenience center **(see illustration)**.
10   Make sure the replacement unit is identical to the one it replaces. Compare the old one to the new one before installing it.
11   Installation is the reverse of removal.

8 Turn signal switch assembly - removal and installation

Warning: *The models covered by this manual are equipped with airbags. Always disable the airbag system before working in the vicinity of the impact sensors, steering column or instrument panel to avoid the possibility of accidental deployment of the airbag(s), which could cause personal injury (see Section 26). The yellow wires and connectors routed through the instrument panel are for this system. Do not use electrical test equipment on these yellow wires or tamper with them in any way while working under the instrument panel.*
Caution: *On models equipped with a Delco Loc II or Theftlock audio system, be sure the lockout feature is turned off before performing any procedure which requires disconnecting the battery.*

7.9 Grasp the hazard flasher unit (arrow) and pull it straight off the clip

8.3a Remove the snap-ring and lift the airbag coil off

8.3b Use a special tool (available at most auto parts stores) to compress the lock plate for access to the retaining ring

8.4 Remove the hazard warning knob screw and the knob

8.5 Remove the cancel cam assembly

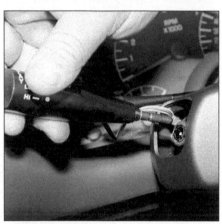

8.6a Grasp the turn signal lever securely and pull it straight out to detach it

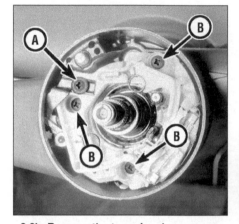

8.6b Remove the turn signal arm screw (A), followed by the switch mounting screws (B) - place the switch in the right turn position to access all the screws

Removal

Refer to illustrations 8.3a, 8.3b 8.4, 8.5, 8.6a, 8.6b and 8.8

1   Detach the cable from the negative battery terminal and disable the airbag system (see Section 26).

2   Remove the steering wheel (see Chapter 10).

3   Remove the airbag coil retaining snap-ring and remove the coil assembly; let the coil hang by the wiring harness. Remove the wave washer. Using a lock plate removal tool, depress the lock plate for access to the retaining ring **(see illustrations)**. Use a small screwdriver to pry the retaining ring out of the groove in the steering column and remove the lock plate.

4   Use a small Phillips head screwdriver to remove the hazard warning knob **(see illustration)**.

5   Remove the turn signal cancel cam **(see illustration)**.

6   Pull the multi-function lever straight out to detach the lever, then remove the turn signal switch mounting screws **(see illustrations)**.

7   Remove the knee bolster panel below the steering column.

8   Locate the turn signal switch electrical connector and unplug it **(see illustration)**.

Remove the wiring protector.

9   Pull the wiring harness and electrical connector up through the steering column and remove the switch assembly.

Installation

Refer to illustrations 8.14 and 8.15

10   Feed the turn signal switch connector and wiring harness down through the col-

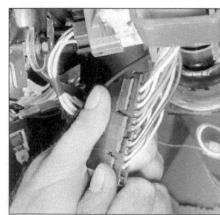

8.8 The turn signal switch electrical connector is located under the dash near the steering column - unplug it as shown

umn. Use a section of mechanics wire to pull it through, if necessary.

11   Plug in the connector and replace the wiring protector.

12   Seat the turn signal switch on the column and install the turn signal switch mounting screws and lever arm. Install the hazard knob and multi-function lever. Press the multi-function lever straight in until it snaps in place.

13   Install the cancel cam and the lock plate. Depress the lock plate and install the retaining ring.

14   If necessary center the airbag coil as follows (it will only become uncentered if the spring lock is depressed and the hub rotated with the coil off the column) **(see illustration)**:

a) *Turn the coil over and depress the spring lock.*

b) *Rotate the hub in the direction of the arrow until it stops.*

c) *Rotate the hub in the opposite direction 2-1/2 turns and release the spring lock.*

15   Install the wave washer and the airbag coil **(see illustration)**. Pull the slack out of airbag coil lower wiring harness to keep it tight through the steering column, or it may

8.14 To center the airbag coil, depress the spring lock (arrow); rotate the hub in the direction of the arrow; back the hub off 2-1/2 turns and release the spring lock

8.15 When properly installed, the airbag coil will be centered with the marks aligned (circle) and the tab fitted between the projections on the top of steering column (arrow)

9.5 Lift off the buzzer switch and clip with needle-nose pliers

be cut when the steering wheel is turned. Install the airbag coil retaining snap-ring.
16 The remainder of installation is the reverse of removal.
17 On all later models, the turn signal, wipe/wash and cruise control switches are all part of the lever/switch assembly on the left of the steering column, called the multi-function switch.
18 Diagnosis of the multi-function switch is best performed with a scan tool.
10 To replace the multi-function switch, remove the small plastic access cover at the column, just below the switch lever.
20 Disconnect the electrical connector to the multi-function lever and pull the lever straight out.
21 Installation is the reverse of the removal procedure.

9 Ignition key lock cylinder - replacement

Refer to illustrations 9.5 and 9.7
Warning: *The models covered by this manual are equipped with airbags. Always disable the airbag system before working in the vicinity of the impact sensors, steering column or instrument panel to avoid the possibility of accidental deployment of the airbag(s), which could cause personal injury (see Section 26). The yellow wires and connectors routed through the instrument panel are for this system. Do not use electrical test equipment on these yellow wires or tamper with them in any way while working under the instrument panel.*
Caution: *On models equipped with a Delco Loc II or Theftlock audio system, be sure the lockout feature is turned off before performing any procedure which requires disconnecting the battery.*
1 Detach the cable from the negative battery terminal and disable the airbag system (see Section 26).
2 Remove the steering wheel (see Chapter 10).

3 Remove the turn signal switch assembly (see Section 8).
4 Make sure the key is removed from the lock cylinder.
5 Using needle nose pliers, remove the buzzer switch **(see illustration)**.
6 Insert the key and place the lock cylinder in the Lock position.
7 Remove the lock cylinder retaining screw **(see illustration)**.
8 Remove the lock cylinder, with key, by pulling the assembly straight out.
9 Disconnect the Pass-key electrical connector from the bulkhead connector and remove the retaining clip from the housing cover.
10 Attach a section of mechanics wire to the connector to aid in installation and pull the harness through the column.
11 Installation is the reverse of removal.

10 Headlight switch - removal and installation

Refer to illustrations 10.2 and 10.4
Warning: *The models covered by this manual*

9.7 The lock cylinder is held in place by a Torx-head screw (arrow)

are equipped with airbags. Always disable the airbag system before working in the vicinity of the impact sensors, steering column or instrument panel to avoid the possibility of accidental deployment of the airbag(s), which could cause personal injury (see Section 26). The yellow wires and connectors routed through the instrument panel are for this system. Do not use electrical test equipment on these yellow wires or tamper with them in any way while working under the instrument panel.*
Caution: *On models equipped with a Delco Loc II or Theftlock audio system, be sure the lockout feature is turned off before performing any procedure which requires disconnecting the battery.*
1 Detach the cable from the negative battery terminal.
2 On models through 1999, remove the instrument cluster bezel (see Chapter 11). Using a screwdriver, pry the retaining clips in and rotate the switch out of the instrument panel **(see illustration)**.
3 On 2000 and later models, pull out on the left instrument panel trim plate until the clips are released. Remove the two bolts securing the switch.
4 Withdraw the switch from the opening

10.2 Use a screwdriver to pry the retaining clips in

10.4 **Unplug the electrical connectors**

11.2 **On Firebird models, the headlight bezel is held in place by screws (arrows) on each side of the housing**

11.3 **Remove the Phillips-head retaining screws - don't confuse them with the Torx-head adjustment screws**

11.4 **Pull the headlight forward and unplug the connector**

12.1 **The headlight vertical adjustment screw is located at the top of the headlight and the horizontal screw is on the side of the headlight (arrows) - a Torx-head tool will be required for making headlight adjustments**

and disconnect the electrical connector **(see illustration).**

5   Plug in the electrical connector, rotate the switch into position, then press in until the clips engage.

6   Install the instrument cluster bezel.

11   Headlights - replacement

Refer to illustrations 11.2, 11.3 and 11.4

Caution 1: *On models equipped with a Delco Loc II or Theftlock audio system, be sure the lockout feature is turned off before performing any procedure which requires disconnecting the battery.*

Caution 2: *Halogen-gas-filled bulbs are under pressure and may shatter if the surface is scratched or the bulb is dropped. Wear eye protection and handle the bulbs carefully, grasping only the base. Do not touch the surface of the bulb with your fingers because the oil from your skin could cause it to overheat and fail prematurely. If you do touch the bulb surface, clean it with rubbing alcohol.*

1   Disconnect the negative cable from the battery. On 1998 and later Camaro models, remove the bolt securing the headlight housing to the headlight housing bracket, then slide the headlight housing outward to disen-

gage the housing tabs from the body. Unplug the electrical connectors from the bulb and twist headlight bulbs one quarter turn counter clockwise to remove the bulbs. On 2000 and later models, the bulbs can be replaced without removing the headlight housings. Just open the hood and there is access behind the housings to reach in and twist the bulbs out for replacement.

2   On all other models, remove the retaining screws and detach the headlight bezel **(see illustration).**

3   Remove the retainer screws, taking care not to disturb the adjusting screws **(see illustration).**

4   Remove the retainer and pull the headlight out enough to allow the connector to be unplugged **(see illustration).**

5   Remove the headlight.

6   To install the headlight, plug the connector in, place the headlight in position and install the retainer and screws. Tighten the screws securely.

7   Place the bezel in position and install the retaining screws.

12   Headlights - adjustment

Refer to illustrations 12.1 and 12.5

Caution: *The headlights must be aimed cor-*

rectly. If adjusted incorrectly they could blind the driver of an oncoming vehicle and cause a serious accident or seriously reduce your ability to see the road. The headlights should be checked for proper aim every 12 months and any time a new headlight is installed or front end body work is performed. It should be emphasized that the following procedure is only an interim step which will provide temporary adjustment until the headlights can be adjusted by a properly equipped shop.

1   Headlights have two spring loaded adjusting screws, one on the top controlling up-and-down movement and one on the side controlling left-and-right movement **(see illustration).**

2   There are several methods of adjusting the headlights. The simplest method requires a blank wall, masking tape and a level floor.

3   Position masking tape vertically on the wall in reference to the vehicle centerline and the centerlines of both headlights.

4   Position a horizontal tape line in reference to the centerline of all the headlights. **Note:** *It may be easier to position the tape on the wall with the vehicle parked only a few inches away.*

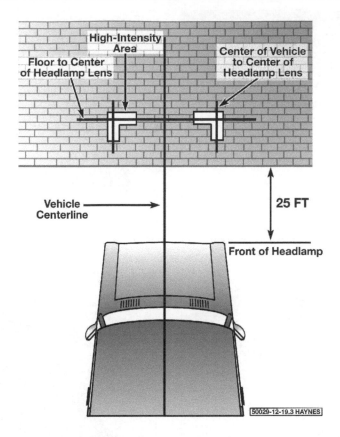

12.5 Headlight aiming details

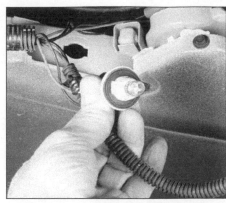

13.2a Rotate the turn signal bulb housing and lift it out - the bulb pulls straight out

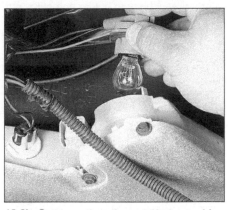

13.2b Squeeze the clip and lift the parking light bulb holder up - push in and rotate the bulb to remove it

5   Adjustment should be made with the vehicle parked 25 feet from the wall, sitting level, the gas tank half-full and no unusually heavy load in the vehicle **(see illustration)**.
6   Starting with the low beam adjustment, position the high intensity zone so it's two inches below the horizontal line and two inches to the right of the headlight vertical line. Adjustment is made by turning the top adjusting screw clockwise to raise the beam and counterclockwise to lower the beam. The adjusting screw on the side should be used in the same manner to move the beam left or right.
7   With the high beams on, the high intensity zone should be vertically centered with the exact center just below the horizontal line. **Note:** *It may not be possible to position the headlight aim exactly for both high and low beams. If a compromise must be made, keep in mind that the low beams are the most used and have the greatest effect on driver safety.*
8   Have the headlights adjusted by a dealer service department or service station at the earliest opportunity.

13 Bulb replacement

Front

1   Remove the plastic retainers and detach the air deflector panel from under the front bumper cover.

Parking/signal

Refer to illustrations 13.2a and 13.2b

2   Reach up under the front bumper and remove the bulb holders from the housing **(see illustrations)**. On 2000 and later models, lower the access panels in the air deflector under the front bumper cover to access the turn/park bulbs.
3   Replace the bulbs.
4   Installation is the reverse of removal.

13.5 Remove the fog light bolts and lower the housing

Fog light

Refer to illustrations 13.5 and 13.6

5   On models through 1999, remove the bolts and lower the fog light housing for access to the bulb **(see illustration)**. On 2000 and later models, lower the access panels in the air deflector under the front bumper cover to access the fog light bulbs.
6   Rotate the bulb housing counterclockwise to remove it **(see illustration)**.
7   Installation is the reverse of removal.

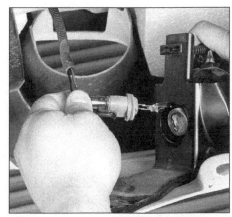

13.6 Rotate the fog light bulb holder, then pull it out

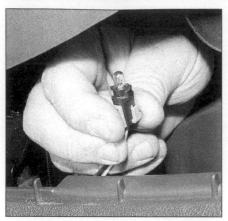

13.11 Reach up under the console shift plate, grasp the bulb holder and pull it out - the bulb pulls straight out of the holder

13.13 Remove the instrument cluster cover

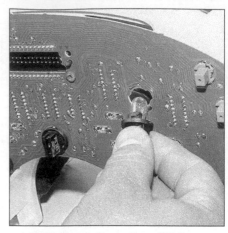

13.14 Rotate the bulb and lift it out of the cluster

13.16 Remove the screw and rotate the rear side marker light out of the panel

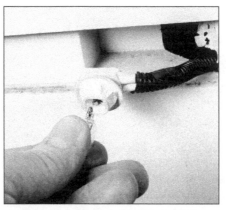

13.17 Detach the holder and pull the bulb out

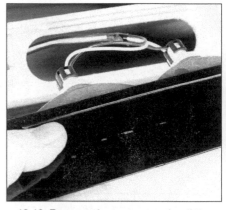

13.19 Remove the screws and pull the high mounted light assembly out

Interior

Dome light

8   Loosen the mounting screws, detach the lamp housing and unplug the electrical connector.

9   Detach the bulb by using a small screwdriver to pry back one of the contacts.

Console lamp

Refer to illustration 13.11

10   Separate the shift plate from the console (see Chapter 11).

11   Reach up under the shift plate, grasp the bulb holder and detach it. Remove the bulb by pulling it straight out of the holder **(see illustration)**.

12   Installation is the reverse of removal.

Instrument panel light

Refer to illustrations 13.13 and 13.14

13   To gain access to the instrument panel lights, the instrument cluster will have to be removed first (see Section 19). Remove the screws and detach the rear cover **(see illustration)**.

14   Rotate the bulb holder counterclockwise and pull it out of the cluster **(see illustration)**. To remove the bulb from the holder, simply pull it straight out.

15   Installation is the reverse of removal.

Rear

Side marker light

Refer to illustrations 13.16 and 13.17

16   Remove the screw and rotate the housing out of the body **(see illustration)**.

17   Rotate the bulb holder counterclockwise to remove it, then pull the bulb out of the housing **(see illustration)**.

18   Installation is the reverse of removal.

13.20 Rotate the holder and pull it out, then remove the bulb

High-mounted brake light

Refer to illustrations 13.19 and 13.20

19   Remove the Torx-head screws and detach the light housing **(see illustration)**.

20   Twist the bulb holder counterclockwise to remove it, then pull the bulb straight out of the holder **(see illustration)**.

21   Installation is the reverse of removal.

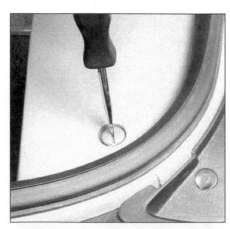

13.22 Turn the plastic screws until the slots all face front-to-back and detach the panel

13.23a Remove the plastic tail light housing nuts (arrows)

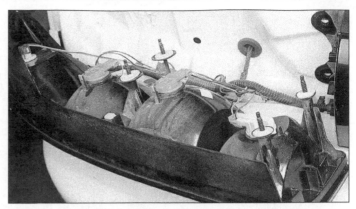

13.23b Rotate the housing out for access - squeeze the gray plastic tabs to remove the bulb holders (on holders without tabs, rotate counterclockwise)

Tail/backup lights

Refer to illustrations 13.22, 13.23a, 13.23b and 13.24

22   Open the rear compartment lid and remove the rear interior trim panels **(see illustration)**.

23   Remove the five plastic nuts and rotate the housing out for access to the bulb holders **(see illustrations)**.

24   Depress the clip, turn the bulb holder counterclockwise and withdraw it from the housing, then remove the bulb by pulling it straight out **(see illustration)**.

25   Installation is the reverse of removal.

License plate light

Refer to illustration 13.26

26   Reach up behind the license plate light housing, rotate the bulb holder counterclockwise, then withdraw the holder from the housing and pull the bulb out **(see illustration)**.

27   Installation is the reverse of removal.

---

14   Radio and speakers - general information, removal and installation

Warning: *The models covered by this manual are equipped with airbags. Always disable the*

13.24 Remove the bulbs by pulling straight out

airbag system before working in the vicinity of the impact sensors, steering column or instrument panel to avoid the possibility of accidental deployment of the airbag(s), which could cause personal injury (see Section 26). The yellow wires and connectors routed through the instrument panel are for this system. Do not use electrical test equipment on these yellow wires or tamper with them in any way while working under the instrument panel.
Caution: *On models equipped with a Delco Loc II or Theftlock audio system, be sure the lockout feature is turned off before perform-*

13.26 Detach the holder, then pull the bulb out

ing any procedure which requires disconnecting the battery.

Radio

Refer to illustrations 14.3a and 14.3b

1   Detach the cable from the negative terminal of the battery.

2   Remove the instrument panel center trim panel (see Chapter 11).

3   Remove the bolts, and pull the radio out and disconnect the electrical connection and antenna lead **(see illustrations)**.

14.3a Remove the bolts (arrows) securing the radio

14.3b Pull the radio out, support it and unplug the connectors

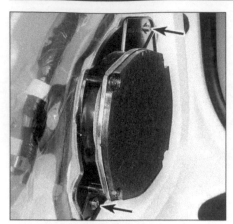

14.7 Remove the screws (arrows), detach the speaker and unplug the electrical connector

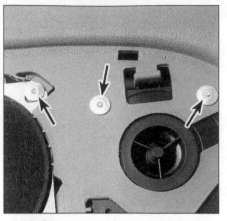

14.9 On some speakers, you'll have to drill out the rivets to remove them from the trim panel

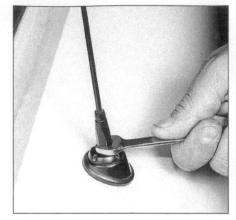

15.2 Use a small wrench to unscrew the antenna mast

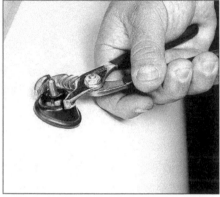

15.3 Carefully unscrew the antenna cap

16.4 Remove the bolts (A), the stud nut (B) and detach the wiper linkage. Unplug the electrical connector (C) and lift the windshield wiper motor assembly out

4   Remove the radio from the instrument panel.

5   Installation is the reverse of removal.

Speakers

Refer to illustrations 14.7 and 14.9

6   On door mounted speakers, remove the door trim panel (see Chapter 11).

7   Remove the screws, detach the speaker and unplug the electrical connector **(see illustration)**.

8   Installation is the reverse of removal.

9   Some speakers are riveted to the back side of the trim panel. On these models it will be necessary to remove the panel and drill out the rivets to remove the speaker **(see illustration)**. A small hand-operated rivet gun will be necessary when installing the new speaker.

15   Antenna - removal and installation

Refer to illustrations 15.2 and 15.3

Caution: *On models equipped with a Delco Loc II or Theftlock audio system, be sure the lockout feature is turned off before performing any procedure which requires disconnecting the battery.*

1   Disconnect the negative battery cable.

2   Use a small open-end wrench to unscrew the antenna mast **(see illustration)**.

3   When replacing the antenna cable, remove the cap nut and lift off the upper adapter and gasket **(see illustration)**.

4   Open the rear compartment lid and remove the spare tire for access to the antenna mount.

5   Detach the antenna mount from the tire bracket and remove the cable from the fender opening. Always replace the cable with one using the same barbless-type connector to ensure good reception.

6   Installation is the reverse of removal.

16   Windshield wiper motor - removal and installation

Refer to illustration 16.4

Caution: *On models equipped with a Delco Loc II or Theftlock audio system, be sure the lockout feature is turned off before performing any procedure which requires disconnecting the battery.*

1   Raise the hood.

2   Detach the cable from the negative terminal of the battery.

3   Remove the left side windshield wiper arm and cowl grille (Chapter 11).

4   Remove the wiper linkage assembly bolt and stud nut. Detach the wiper link and unplug the electrical connector. Remove the motor assembly-to-chassis bolt, detach the motor from the bracket and lift it from the cowl opening **(see illustration)**.

5   Installation is the reverse of removal.

17   Rear window defogger switch - removal and installation

Warning: *The models covered by this manual are equipped with airbags. Always disable the airbag system before working in the vicinity of the impact sensors, steering column or instrument panel to avoid the possibility of accidental deployment of the airbag(s), which could cause personal injury (see Section 26). The yellow wires and connectors routed through the instrument panel are for this system. Do not use electrical test equipment on these yellow wires or tamper with them in any way while working under the instrument panel.*

Caution: *On models equipped with a Delco Loc II or Theftlock audio system, be sure the lockout feature is turned off before performing any procedure which requires disconnecting the battery.*

1   On models through 1999, remove the

18.5 When measuring the voltage at the rear window defogger grid, wrap a piece of aluminum foil around the positive probe of the voltmeter and press the foil against the wire with your finger

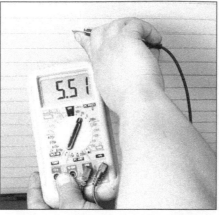

18.6 To determine if a heating element has broken, check the voltage at the center of each element - if the voltage is 6-volts, the element is unbroken

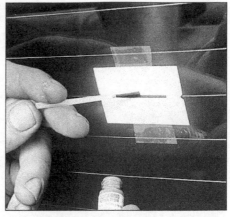

18.13 To use a defogger repair kit, apply masking tape to the inside of the window at the damaged area, then brush on the special conductive coating

dashboard center trim panel (see Chapter 11). On 2000 and later models, the rear window defroster switch is located in the HVAC control panel. Refer to Chapter 3 for removal of the control panel to replace switches.

2   Unplug the electrical connector and detach the defogger switch from the panel.

3   Installation is the reverse of removal.

---

18 Rear window defogger - check and repair

---

1   The rear window defogger consists of a number of horizontal elements baked onto the glass surface.

2   Small breaks in the element can be repaired without removing the rear window.

Check

Refer to illustrations 18.5 and 18.6

3   Turn the ignition switch and defogger system switches to the ON position.

4   Using a voltmeter, place the positive probe against the battery feed terminal and the negative probe against the negative (ground) bus bar. The positive terminal is located on the drivers side and the negative terminal is located on the passengers side. If battery voltage is not indicated, check the fuse, defogger switch and related wiring.

5   When measuring voltage during the next two tests, wrap a piece of aluminum foil around the tip of the voltmeter positive probe and press the foil against the heating element with your finger **(see illustration)**.

6   Place the negative lead against the negative (ground) bus bar. Check the voltage at the center of each heating element **(see illustration)**. If the voltage is 6-volts, the element is okay (there is no break). If the voltage is 10-volts or more, the element is broken somewhere between the mid-point and ground. If the voltage is 0-volts the element is broken between the mid-point and the positive side.

7   To find the break, slide the probe toward the positive side. The point at which the voltmeter deflects from zero to several volts is the point at which the heating element is broken. **Note:** *If the heating element is not broken, the voltmeter will indicate 12-volts at the positive side and gradually decrease to 0-volts as you slide the positive probe toward the ground side.*

Repair

Refer to illustration 18.13

8   Repair the break in the element using a repair kit specifically recommended for this purpose, such as Dupont paste No. 4817 (or equivalent). Included in this kit is plastic conductive epoxy.

9   Prior to repairing a break, turn off the system and allow it to cool off for a few minutes.

10   Lightly buff the element area with fine steel wool, then clean it thoroughly with rubbing alcohol.

11   Use masking tape to mask off the area being repaired.

12   Thoroughly mix the epoxy, following the instructions provided with the repair kit.

13   Apply the epoxy material to the slit in the masking tape, overlapping the undamaged area about 3/4-inch on either end **(see illustration)**.

14   Allow the repair to cure for 24 hours

before removing the tape and using the system.

---

19 Instrument cluster - removal and installation

Refer to illustrations 19.3a and 19.3b

Warning: *The models covered by this manual are equipped with airbags. Always disable the airbag system before working in the vicinity of the impact sensors, steering column or instrument panel to avoid the possibility of accidental deployment of the airbag(s), which could cause personal injury (see Section 26). The yellow wires and connectors routed through the instrument panel are for this system. Do not use electrical test equipment on these yellow wires or tamper with them in any way while working under the instrument panel.*

Caution: *On models equipped with a Delco Loc II or Theftlock audio system, be sure the lockout feature is turned off before performing any procedure which requires disconnecting the battery.*

1   Detach the cable from the negative battery terminal.

2   Remove instrument cluster bezel (Chapter 11).

3   Remove the four bolts, detach the instrument cluster from the electrical wiring harness connector and lift it from the dash **(see illustrations)**.

4   Installation is the reverse of removal.

19.3a Remove the upper cluster bolts (arrows)

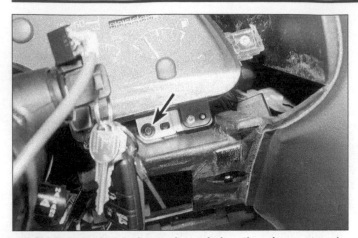

19.3b Remove the two lower cluster bolts - there is one at each corner (arrow)

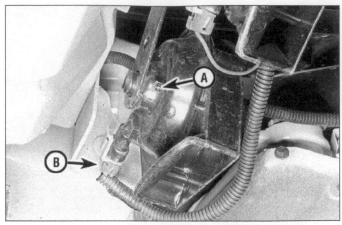

20.2 The horn can be checked after unplugging the electrical connector (B) and adjusted by turning the adjustment screw (A)

20 Horn - check and replacement

Refer to illustration 20.2

1   Remove the air deflector from bumper cover.
2   Unplug the electrical connector from the horn **(see illustration)**.
3   To test the horn, connect battery voltage to the two terminals with a pair of jumper wires. If the horn doesn't sound, check the current draw with an ammeter - it should be around five amps at battery voltage.
4   If the amperage is over 20 amps, replace the horn.
5   If the reading is around 18 amps, the contact points are not open. Turn the adjustment screw counterclockwise a quarter turn at a time to open the points and lower the amperage until the proper level is reached **(see illustration 20.2)**.
6   If the horn doesn't sound, it could mean the problem lies in the switch, relay, the wiring between the components or open points. Turn the adjustment screw clockwise a quarter turn at a time to close the points until the amperage is around five amps.
7   To replace the horn, unplug the electrical connector and remove the bracket bolt.
8   Installation is the reverse of removal.
9   Install the air deflector.

21 Electric rear view mirrors - description and check

Refer to illustration 21.9

1   Electric rear view mirrors use two motors to move the glass; one for up-and-down adjustments and one for left-to-right adjustments.
2   The control switch has a selector portion which sends voltage to the left or right side mirror. With the ignition ON, engine OFF, roll down the windows and operate the mirror control switch through all functions (left-right and up-down) for both the left and right side mirrors.

3   Listen carefully for the sound of the electric motors running in the mirrors.
4   If the motors can be heard but the mirror glass doesn't move, there's probably a problem with the drive mechanism inside the mirror. Remove and disassemble the mirror to locate the problem.
5   If the mirrors don't operate and no sound comes from the mirrors, check the fuse (see Section 3).
6   If the fuse is OK, remove the mirror control switch from its mounting without disconnecting the wires attached to it. Turn the ignition ON and check for voltage at the switch. There should be voltage at one terminal. If there's no voltage at the switch, check for an opening or short in the wiring between the fuse panel and the switch.
7   If there's voltage at the switch, disconnect it. Check the switch for continuity in all its operating positions. If the switch does not have continuity, replace it.
8   Re-connect the switch. Locate the wire going from the switch to ground. Leaving the switch connected, connect a jumper wire between this wire and ground. If the mirror works normally with this wire in place, repair

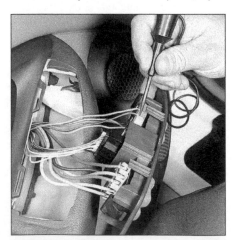

21.9 Use a test light to make sure there is voltage with the mirror switch in each operating position

the faulty ground connection.
9   If the mirror still doesn't work, remove the cover and check the wires at the mirror for voltage with a test light **(see illustration)**. Check with ignition ON and the mirror selector switch on the appropriate side. Operate the mirror switch in all its positions. There should be voltage at one of the switch-to-mirror wires in each switch position (except the neutral position).
10   If there's not voltage in each switch position, check the wiring between the mirror and control switch for opens and shorts.
11   If there's voltage, remove the mirror and test it off the vehicle with jumper wires. Replace the mirror if it fails this test (see Chapter 11).

22 Cruise control system - description and check

Refer to illustration 22.5 and 22.7

1   The cruise control system maintains vehicle speed with an electronic servo motor located in the engine compartment, which is connected to the throttle linkage by a cable. The system consists of the electronic control module, brake switch, control switches, a relay, the vehicle speed sensor and associated wiring. Listed below are some general procedures that may be used to locate common cruise control problems.
2   Locate and check the fuse (see Section 3).
3   Have an assistant operate the brake lights while you check their operation (voltage from the brake light switch deactivates the cruise control).
4   If the brake lights don't come on or don't shut off, correct the problem and retest the cruise control.
5   Inspect the cable linkage between the cruise control actuator and the throttle linkage. The cruise control module is located on the frame rail under the left (drivers) corner of the vehicle **(see illustration)**.
6   Visually inspect the wires connected to the cruise control actuator and check for

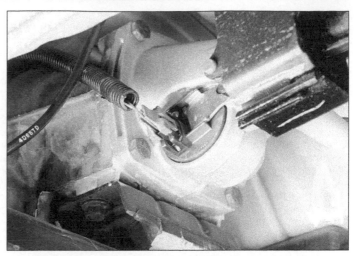

22.5 The cruise control module (arrow) is located under the front of the vehicle - check for damage to the connectors

22.7 The speed sensor in mounted in the transmission

damage and broken wires.

7   The vehicle speed sensor is located on the transmission **(see illustration)**. Raise the front of the vehicle and support it on jack stands. Unplug the electrical connector and touch one probe of a digital voltmeter to the orange wire of the connector and the other to a good ground. With the vehicle in Neutral and key On, measure the voltage while rotating one wheel with the other one blocked. If the voltage doesn't vary as the wheel rotates, the sensor is defective.

8   Test drive the vehicle to determine if the cruise control is now working. If it isn't, take it to a dealer service department or an automotive electrical specialist for further diagnosis and repair.

23 Power door lock system - description and check

Caution: *On models equipped with a Delco Loc II or Theftlock audio system, be sure the lockout feature is turned off before performing any procedure which requires disconnecting the battery.*

1   Power door lock systems are operated by bi-directional solenoids located in the doors. The lock switches have two operating positions: Lock and Unlock. These switches activate a relay which in turn connects voltage to the door lock solenoids. Depending on which way the relay is activated, it reverses polarity, allowing the two sides of the circuit to be used alternately as the feed (positive) and ground side.

2   Always check the circuit protection first. Some vehicles use a combination of circuit breakers and fuses.

3   Operate the door lock switches in both directions (Lock and Unlock) with the engine off. Listen for the faint click of the relay operating.

4   If there's no click, check for voltage at the switches. If no voltage is present, check

the wiring between the fuse panel and the switches for shorts and opens.

5   If voltage is present but no click is heard, test the switch for continuity. Replace it if there's not continuity in both switch positions.

6   If the switch has continuity but the relay doesn't click, check the wiring between the switch and relay for continuity. Repair the wiring if there's no continuity.

7   If the relay is receiving voltage from the switch but is not sending voltage to the solenoids, check for a bad ground at the relay case. If the relay case is grounding properly, replace the relay.

8   If all but one lock solenoid operates, remove the trim panel from the affected door (see Chapter 11). and check for voltage at the solenoid while the lock switch is operated. One of the wires should have voltage in the Lock position; the other should have voltage in the unlock position.

9   If the inoperative solenoid is receiving voltage, replace the solenoid.

10   If the inoperative solenoid isn't receiving voltage, check for an open or short in the wire between the lock solenoid and the relay. **Note:** *It's common for wires to break in the portion of the harness between the body and door (opening and closing the door fatigues and eventually breaks the wires).*

24 Power window system - description and check

Caution: *On models equipped with a Delco Loc II or Theftlock audio system, be sure the lockout feature is turned off before performing any procedure which requires disconnecting the battery.*

1   The power window system consists of the control switches, the motors, glass mechanisms (regulators), and associated wiring.

2   Power windows are wired so they can be lowered and raised from the master con-

trol switch by the driver or by remote switches located at the individual windows. Each window has a separate motor which is reversible. The position of the control switch determines the polarity and therefore the direction of operation. The system is equipped with a relay that controls current flow to the motors.

3   The power window system operates when the ignition switch is ON. In addition, these models have a window lockout switch at the master control switch which, when activated, disables the switches at the rear windows and, sometimes, the switch at the passenger's window also. Always check these items before troubleshooting a window problem.

4   These procedures are general in nature, so if you can't find the problem using them, take the vehicle to a dealer service department or other qualified repair shop.

5   If the power windows don't work at all, check the fuse or circuit breaker.

6   If only the rear windows are inoperative, or if the windows only operate from the master control switch, check the rear window lockout switch for continuity in the unlocked position. Replace it if it doesn't have continuity.

7   Check the wiring between the switches and fuse panel for continuity. Repair the wiring, if necessary.

8   If only one window is inoperative from the master control switch, try the other control switch at the window. **Note:** *This doesn't apply to the drivers door window.*

9   If the same window works from one switch, but not the other, check the switch for continuity.

10   If the switch tests OK, check for a short or open in the wiring between the affected switch and the window motor.

11   If one window is inoperative from both switches, remove the trim panel from the affected door and check for voltage at the motor while the switch is operated.

12   If voltage is reaching the motor, discon-

nect the glass from the regulator (see Chapter 11). Move the window up and down by hand while checking for binding and damage. Also check for binding and damage to the regulator. If the regulator is not damaged and the window moves up and down smoothly, replace the motor (see Chapter 11). If there's binding or damage, lubricate, repair or replace parts, as necessary.

13   If voltage isn't reaching the motor, check the wiring in the circuit for continuity between the switches and motors. Check that the relay is grounded properly and receiving voltage from the switches. Also check that the relay sends voltage to the motor when the switch is turned on. If it doesn't, replace the relay.

14   Test the windows after you are done to confirm proper repairs.

25   Power seats - description and check

1   Power seats allow you to adjust the position of the seat with little effort. The optional power seats on these models adjust forward and backward, up and down and tilt forward and backward.

2   The power seat system consists of a motor, a switch on the seat and a relay and fuse in the engine compartment fuse block.

3   Look under the seat for any objects which may be preventing the seat from moving.

4   If the seat won't work at all, check the fuse.

5   With the engine off to reduce the noise level, operate the seat controls in all directions and listen for sound coming from the seat motor(s).

6   If the motor runs or clicks but the seat doesn't move, the integral the seat drive mechanism is damaged and the motor assembly must be replaced.

7   If the motor doesn't work or make noise, check for voltage at the motor while an assistant operates the switch.

8   If the motor is getting voltage but doesn't run, test it off the vehicle with jumper wires. If it still doesn't work, replace it.

9   If the motor isn't getting voltage, check for voltage at the switch. If there's no voltage at the switch, check the wiring between the fuse panel and the switch. If there's voltage at the switch, obtain the wiring diagrams for the vehicle and check the switch for continuity in all its operating positions. Replace the switch if there's no continuity.

10   If the switch is OK, check for a short or open in the wiring between the switch and motor. If there's a relay between the switch and motor, check that it's grounded properly and there's voltage to the relay. Also check that there's voltage going from the relay to the motor when the when the switch is operated. If there's not, and the relay is grounded properly, replace the relay.

11   Test the completed repairs.

26   Airbag - general information

Warning: *The models covered by this manual are equipped with airbags. Airbag system components are located in the steering wheel, steering column, instrument panel and center console. The airbag(s) could accidentally deploy if any of the system components or wiring harnesses are disturbed, so be extremely careful when working in these areas and don't disturb any airbag system components or wiring. You could be injured if an airbag accidentally deploys, or the airbag might not deploy correctly in a collision if any components or wiring in the system have been disturbed. The yellow wires and connectors routed through the instrument panel and center console are for this system. Do not use electrical test equipment on these yellow wires or tamper with them in any way while working in their vicinity.*

Caution: *On models equipped with a Delco Loc II or Theftlock audio system, be sure the lockout feature is turned off before performing any procedure which requires disconnecting the battery.*

Description

1   The models covered by this manual are equipped with a Supplemental Inflatable Restraint (SIR) system, more commonly known as an airbag system. The SIR system is designed to protect the driver and passenger from serious injury in the event of a head-on or frontal collision.

2   The SIR system consists of two airbags; one located in the center of the steering wheel, and another located in the top of the dashboard, above the glove box. 1993 through 1995 models utilize two impact sensors; one located in the instrument panel and another located just in front of the radiator; an arming sensor located under the center console; and a diagnostic/energy reserve module located at the right end of the instrument panel. On 1996 and later models, the sensors and the diagnostic/energy reserve module have been incorporated into one unit and is located under the center console.

Sensors

3   The 1993 through 1995 system has three separate sensors; two impact sensors and an arming sensor. The sensors are basically pressure sensitive switches that complete an electrical circuit during an impact of sufficient G force. The electrical signal from the crash sensors is sent to the diagnostic module, that then completes circuit and inflates the airbags.

4   On the 1996 and later system, the sensing circuitry is contained in the sensing diagnostic/energy reserve module. If a frontal crash of sufficient force is detected, the circuitry allows current to flow to the airbags, inflating them.

Diagnostic/energy reserve module

5   The diagnostic/energy reserve module

contains an on-board microprocessor which monitors the operation of the system. It performs a diagnostic check of the system every time the vehicle is started. If the system is operating properly, the AIRBAG warning light will blink on and off seven times. If there is a fault in the system, the light will remain on and the airbag control module will store fault codes indicating the nature of the fault. If the AIRBAG warning light remains on after staring, or comes on while driving, the vehicle should be taken to your dealer immediately for service. The diagnostic/energy reserve module also contains a back-up power supply to deploy the airbags in the event battery power is lost during a collision.

Operation

6   For the airbag(s) to deploy, an impact of sufficient G force must occur within 30-degrees of the vehicle centerline. When this condition occurs, the circuit to the airbag inflator is closed and the airbag inflates. If the battery is destroyed by the impact, or is too low to power the inflators, a back-up power supply inside the diagnostic/energy reserve module supplies current to the airbags.

Self-diagnosis system

7   A self-diagnosis circuit in the module displays a light when the ignition switch is turned to the On position. If the system is operating normally, the light should go out after seven flashes. If the light doesn't come on, or doesn't go out after seven flashes, or if it comes on while you're driving the vehicle, there's a malfunction in the SIR system. Have it inspected and repaired as soon as possible. Do not attempt to troubleshoot or service the SIR system yourself. Even a small mistake could cause the SIR system to malfunction when you need it.

Servicing components near the SIR system

8   Nevertheless, there are times when you need to remove the steering wheel, radio or service other components on or near the instrument panel. At these times, you'll be working around components and wiring harnesses for the SIR system. SIR system wiring is easy to identify; they're all covered by a bright yellow conduit. Do not unplug the connectors for the SIR system wiring, except to disable the system. And do not use electrical test equipment on the SIR system wiring. ***ALWAYS DISABLE THE SIR SYSTEM BEFORE WORKING NEAR THE SIR SYSTEM COMPONENTS OR RELATED WIRING.***

Disabling the SIR system

Refer to illustrations 26.11 and 26.13

9   Turn the steering wheel to the straight ahead position, place the ignition switch in Lock and remove the key. Remove the airbag fuse from the fuse block (see Section 3).

10   Unplug the yellow Connector Position Assurance (CPA) connectors at the base of the steering column and under the right side

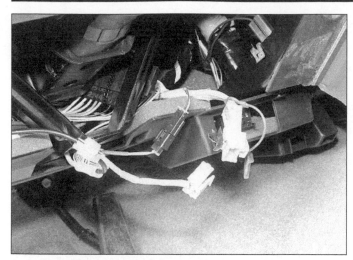

26.11 The driver's side airbag Connector Position Assurance (CPA) connector is found at the base of the steering column

26.13 The passenger airbag Connector Position Assurance (CPA) connector is located under the instrument panel and is accessible through the glove box opening

of the instrument panel as described in the following steps.

Driver's side airbag

11   Remove the knee bolster and sound insulator panel below the instrument panel (see Chapter 11) and unplug the yellow Connector Position Assurance (CPA) steering column harness connector **(see illustration)**.

Passenger's side airbag

12   Remove the glove box (see Chapter 11).
13   Unplug the CPA electrical connector from the passenger inflator module under the right side of the dash **(see illustration)**.

Enabling the SIR system

14   After you've disabled the airbag and performed the necessary service, plug in the steering column (driver's side) and passenger side CPA connectors. Reinstall the knee bolster, sound insulator panel and the glove box.
15   Install the airbag fuse.

27   Wiring diagrams - general information

Since it isn't possible to include all wiring diagrams for every year covered by this manual, the following diagrams are those that are typical and most commonly needed.

Prior to troubleshooting any circuit, check the fuse and circuit breakers (if equipped) to make sure they're in good condition. Make sure the battery is properly charged and check the cable connections (see Chapter 1, Section 9).

When checking a circuit, make sure that all electrical connectors are clean, with no broken or loose terminals. When unplugging an electrical connector, do not pull on the wires. Pull only on the connector housings themselves.

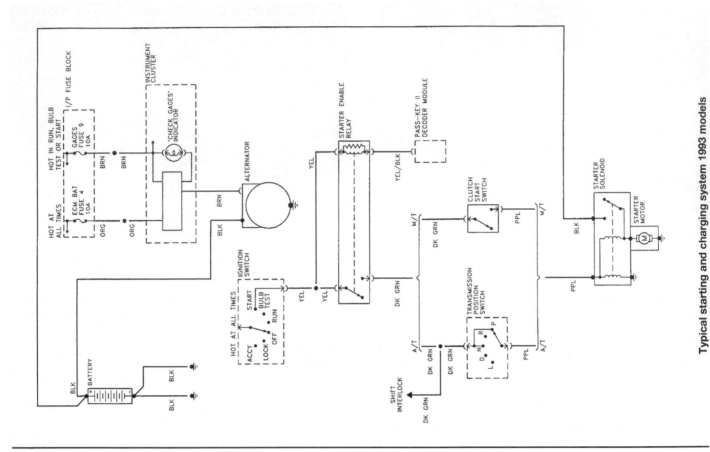

Typical starting and charging system 1993 models

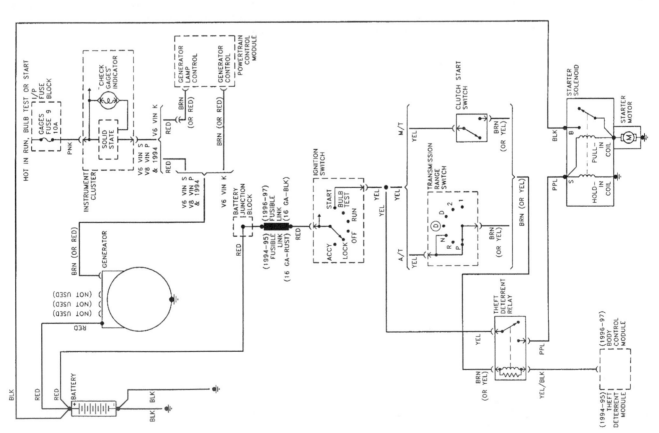

Typical starting and charging system 1994 and later models

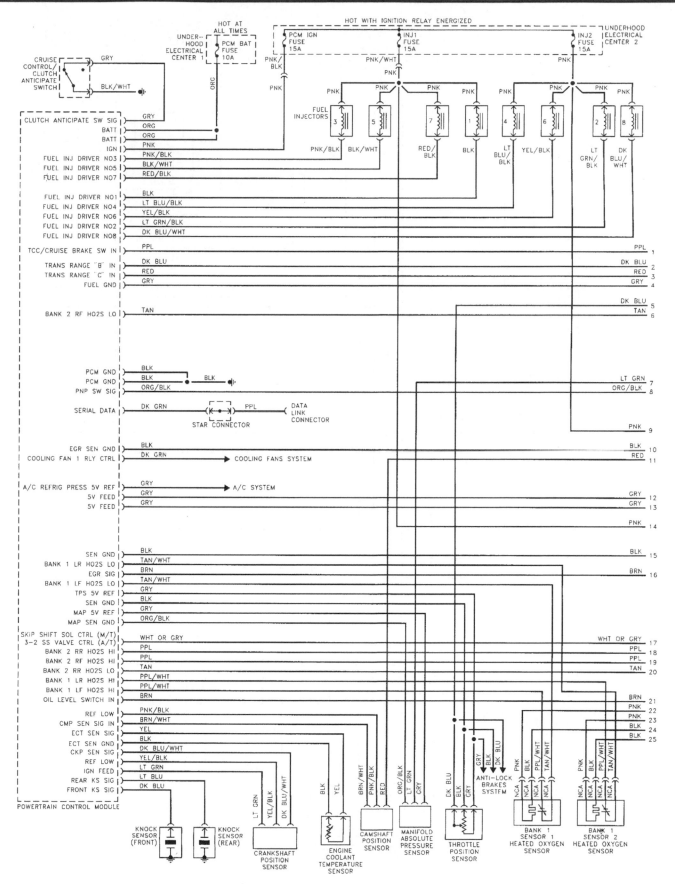

Typical 5.7L V8 engine controls 1998 (1 of 4)

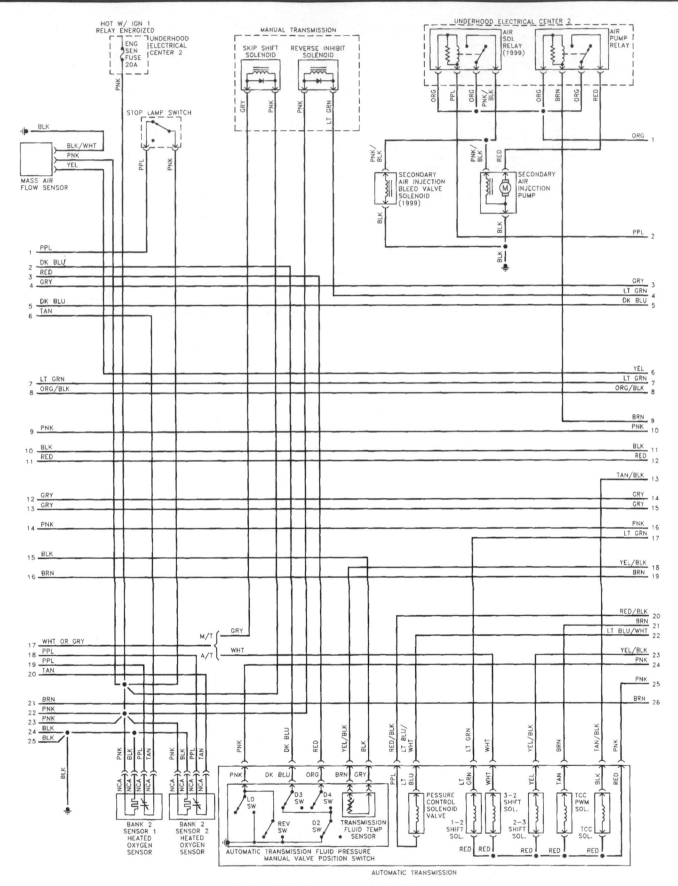

Typical 5.7L V8 engine controls 1998 (2 of 4)

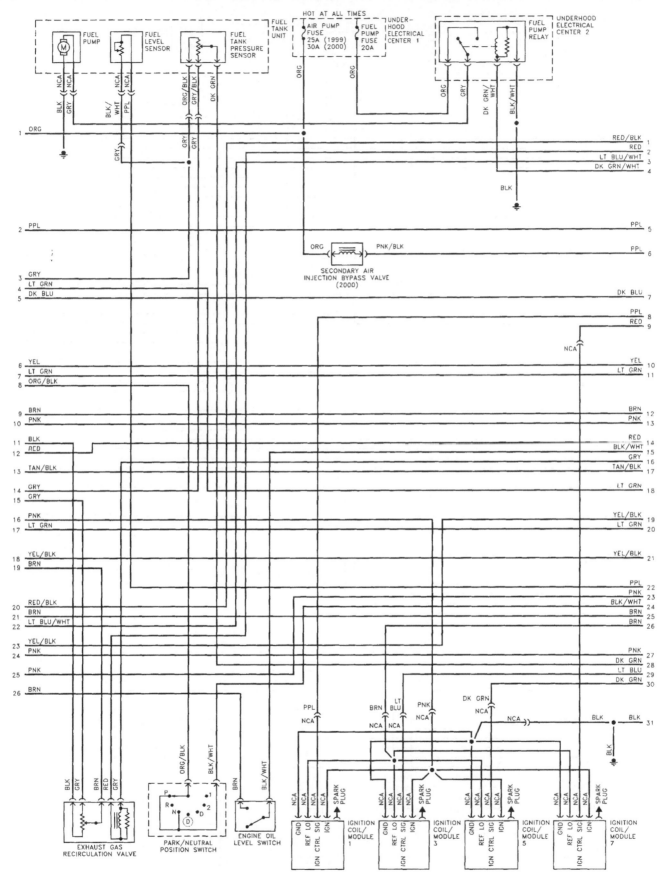

Typical 5.7L V8 engine controls 1998 (3 of 4)

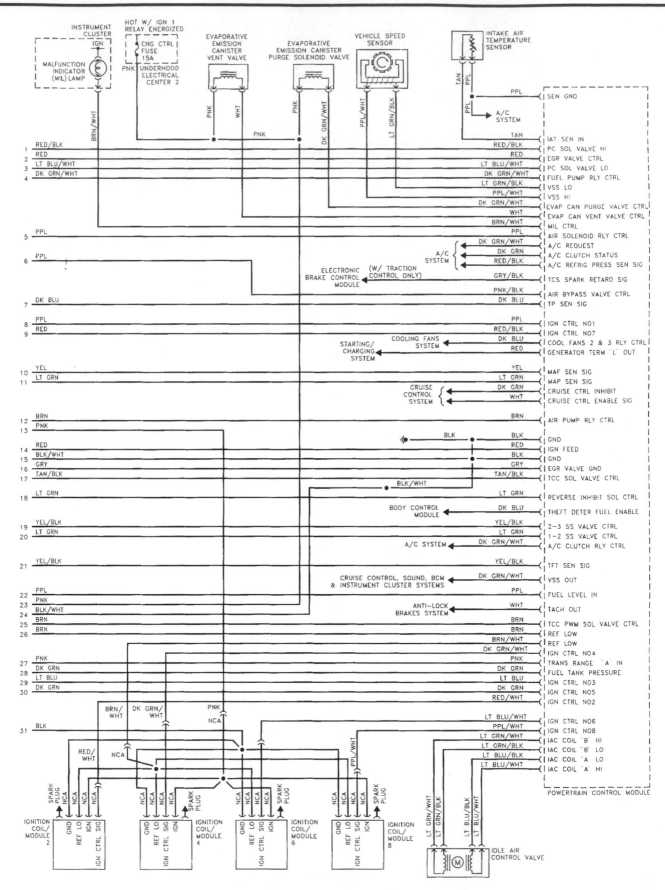

Typical 5.7L V8 engine controls 1998 (4 of 4)

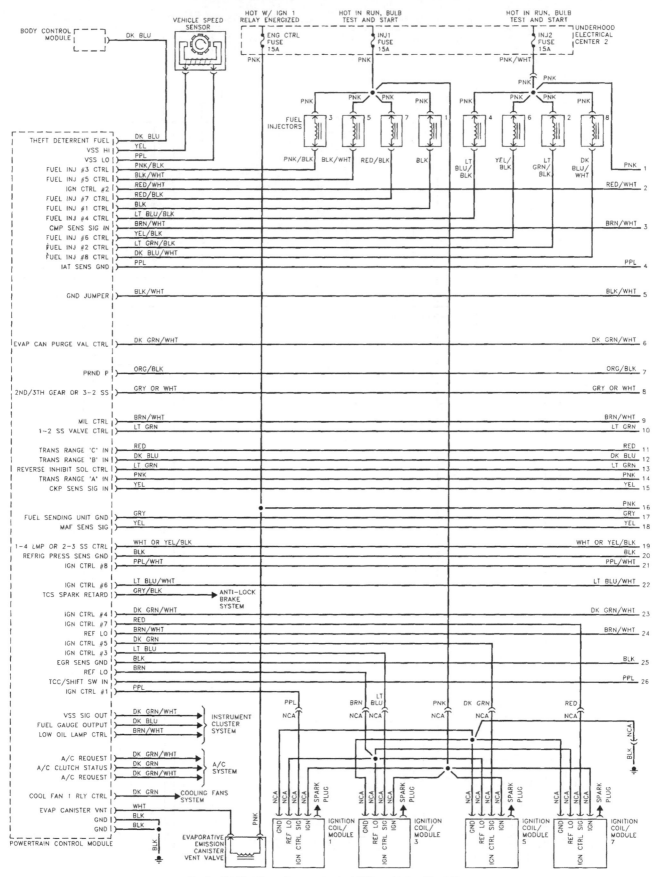

Typical 5.7L V8 engine controls 1999 and later (1 of 4)

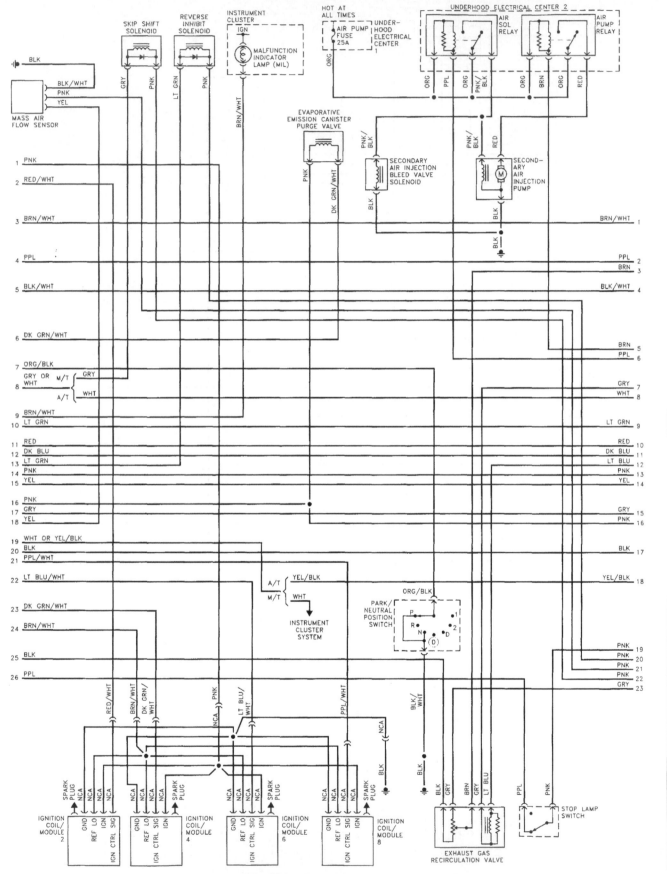

Typical 5.7L V8 engine controls 1999 and later (2 of 4)

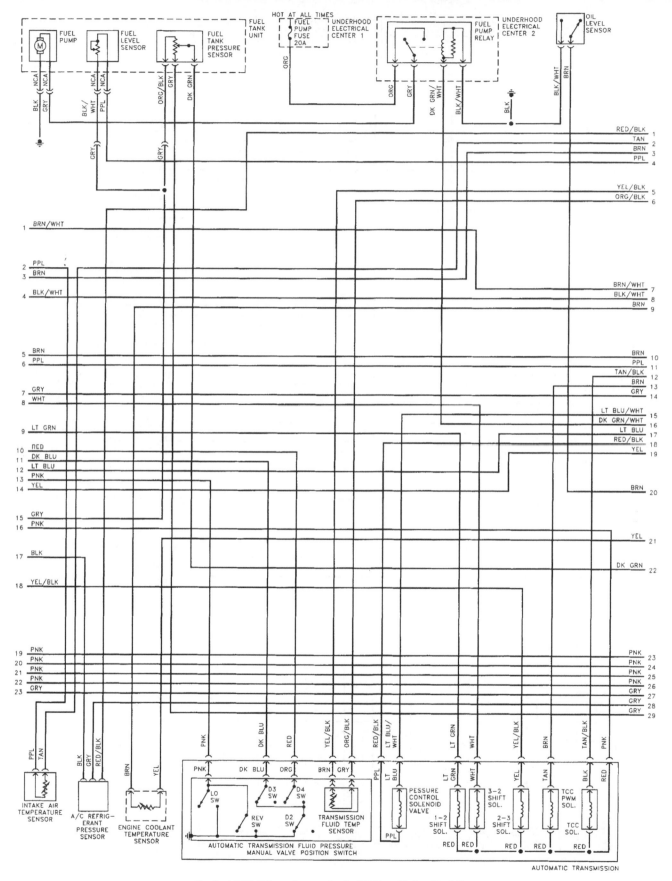

Typical 5.7L V8 engine controls 1999 and later (3 of 4)

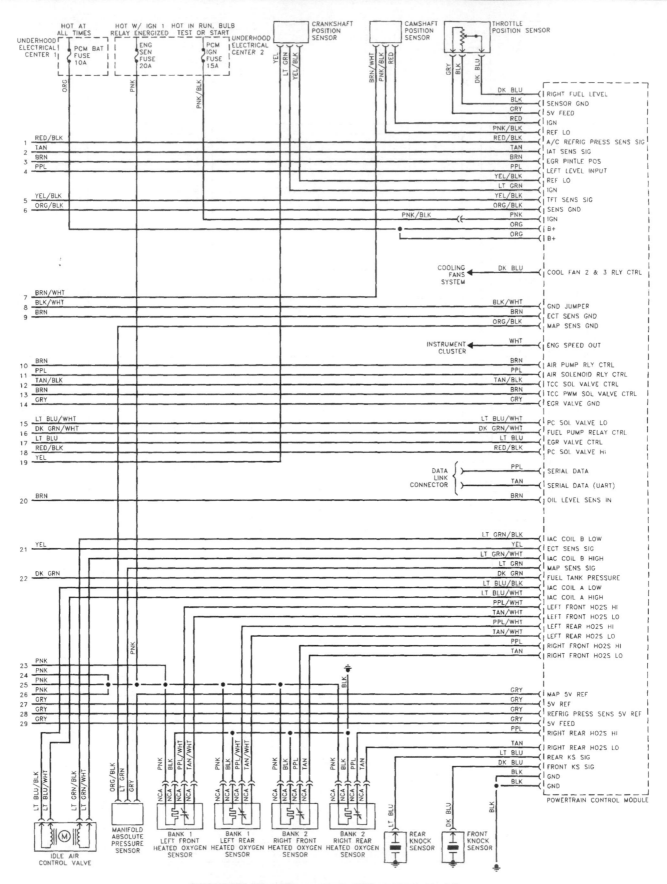

Typical 5.7L V8 engine controls 1999 and later (4 of 4)

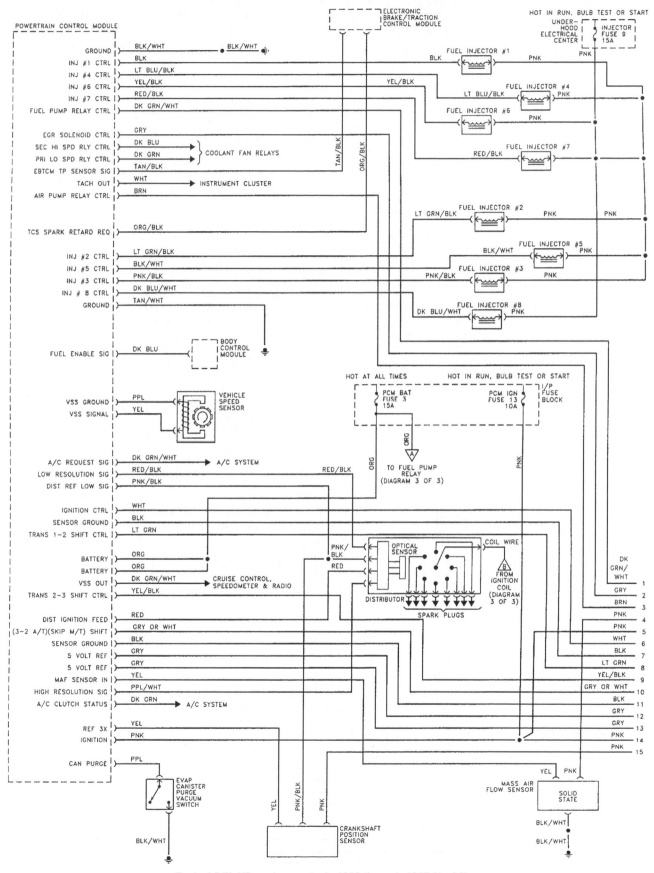

Typical 5.7L V8 engine controls 1996 through 1997 (1 of 3)

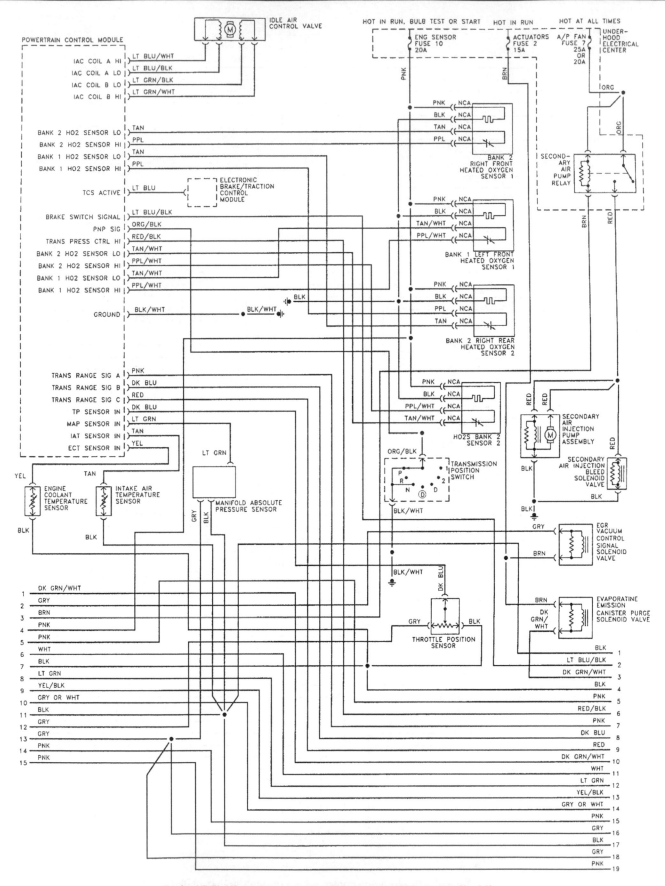

Typical 5.7L V8 engine controls 1996 through 1997 models (2 of 3)

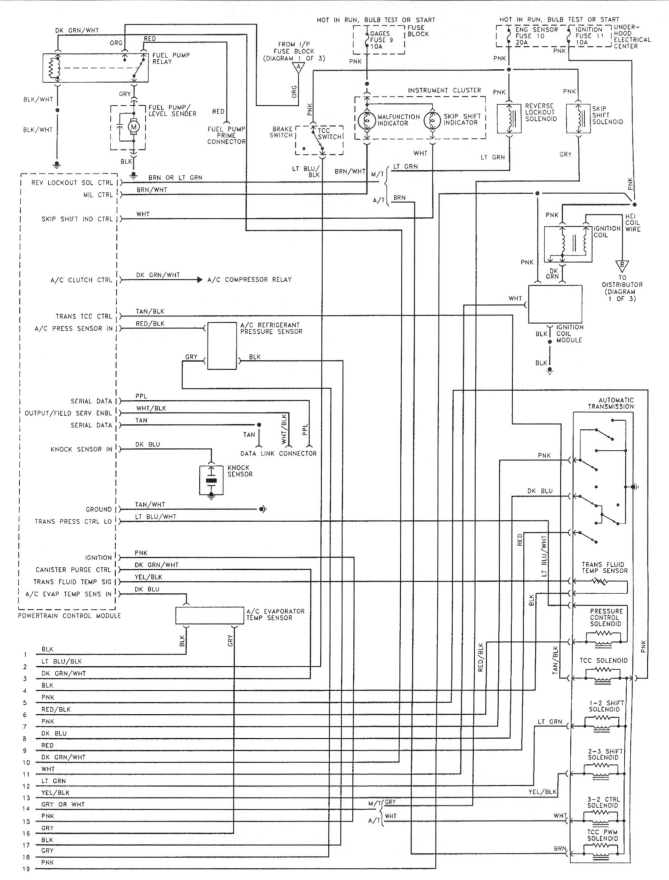

Typical 5.7L V8 engine controls 1996 through 1997 models (3 of 3)

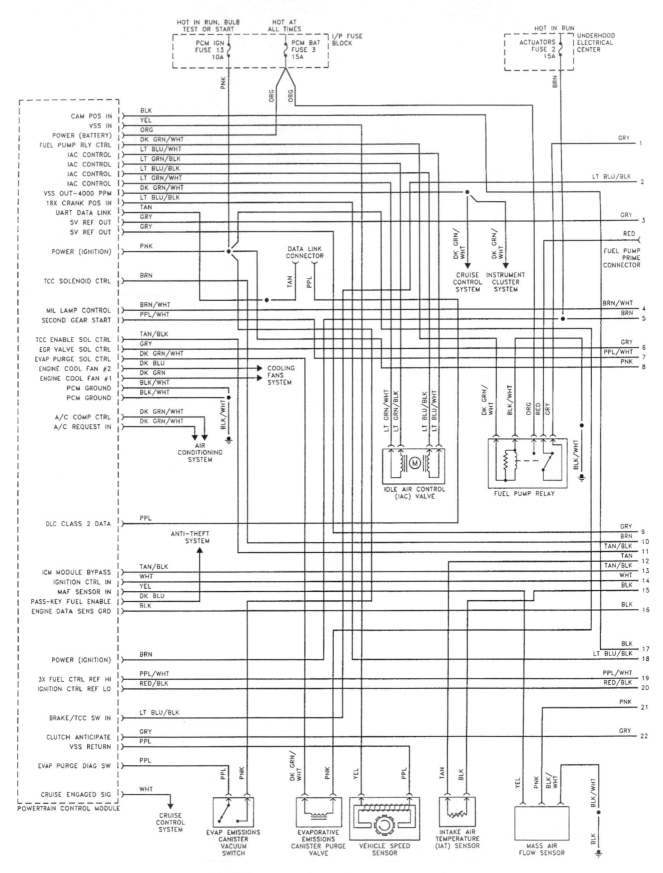

Typical 3.8L V6 engine controls 1996 and later models (1 of 3)

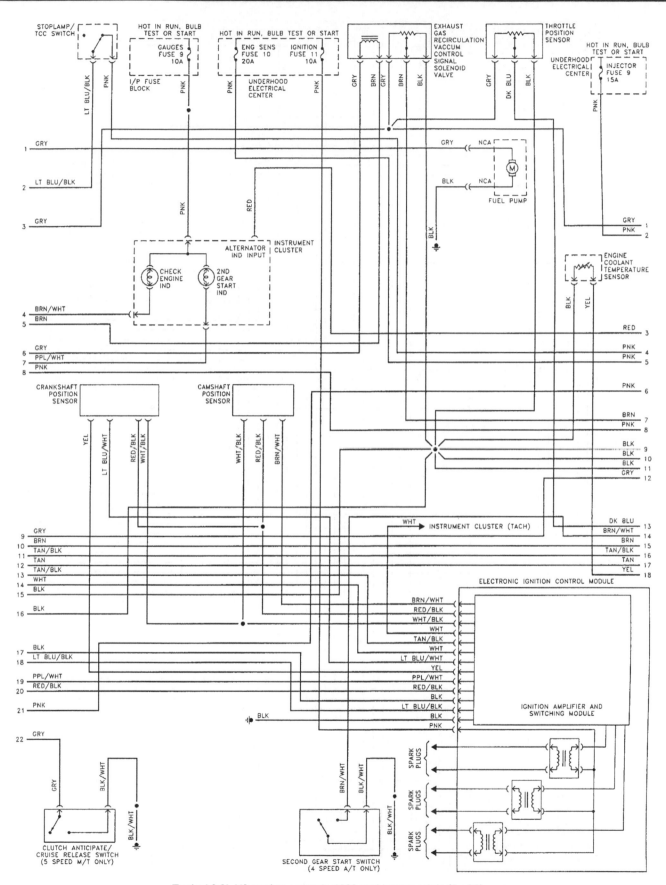

Typical 3.8L V6 engine controls 1996 and later models (2 of 3)

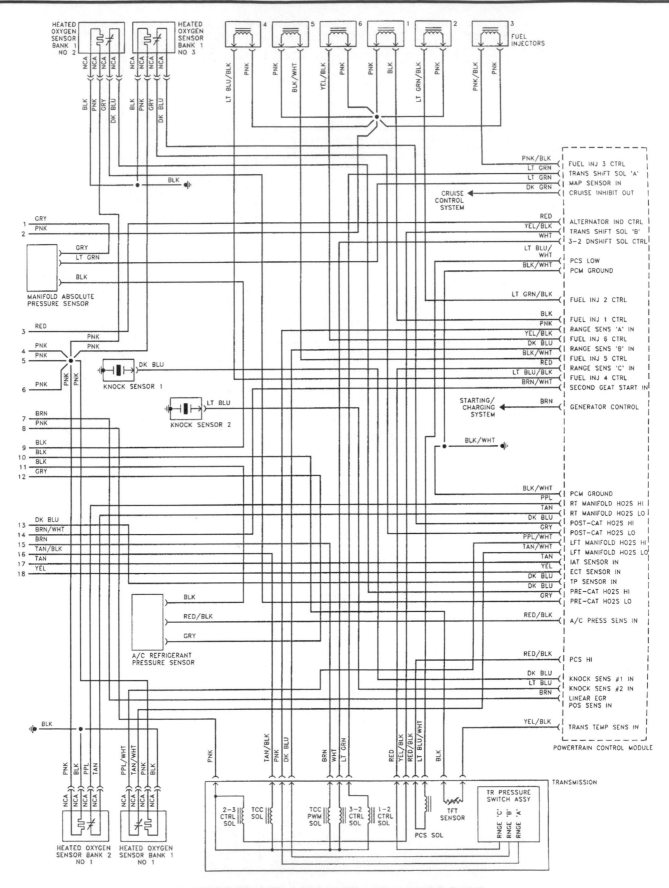

Typical 3.8L V6 engine controls 1996 and later models (3 of 3)

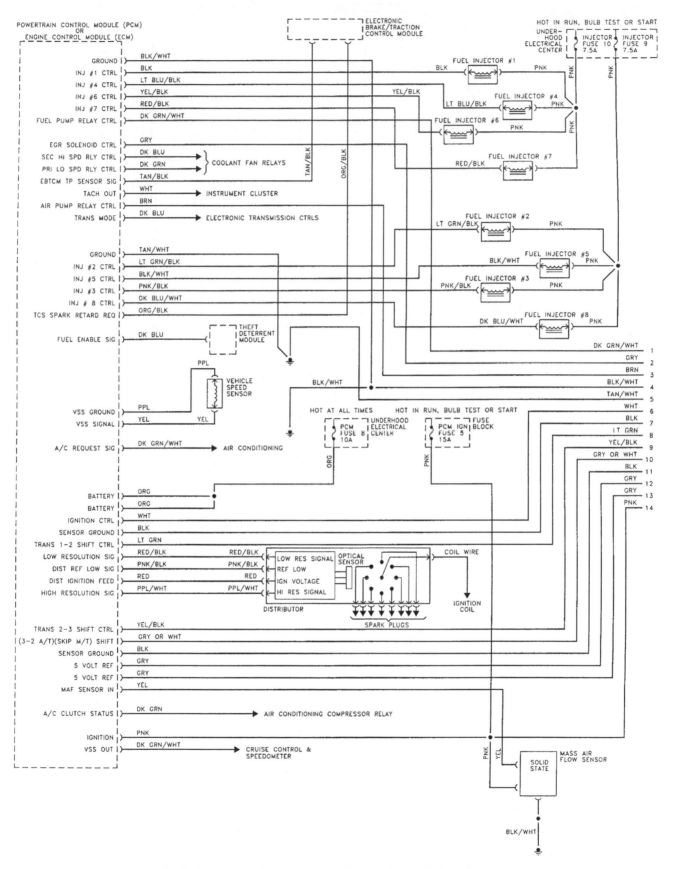

Typical 5.7L V8 engine controls 1995 and earlier models (1 of 3)

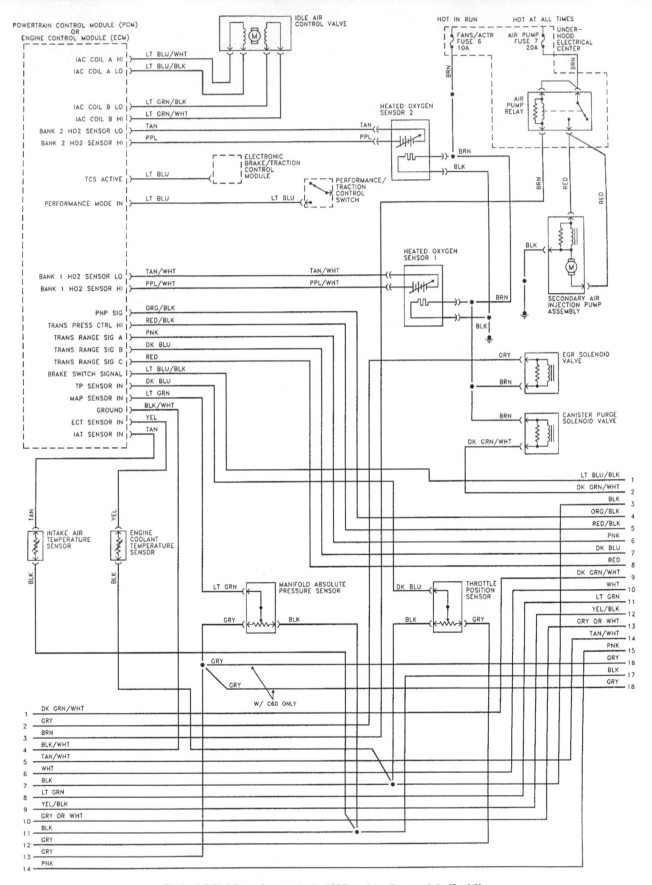

Typical 5.7L V8 engine controls 1995 and earlier models (2 of 3)

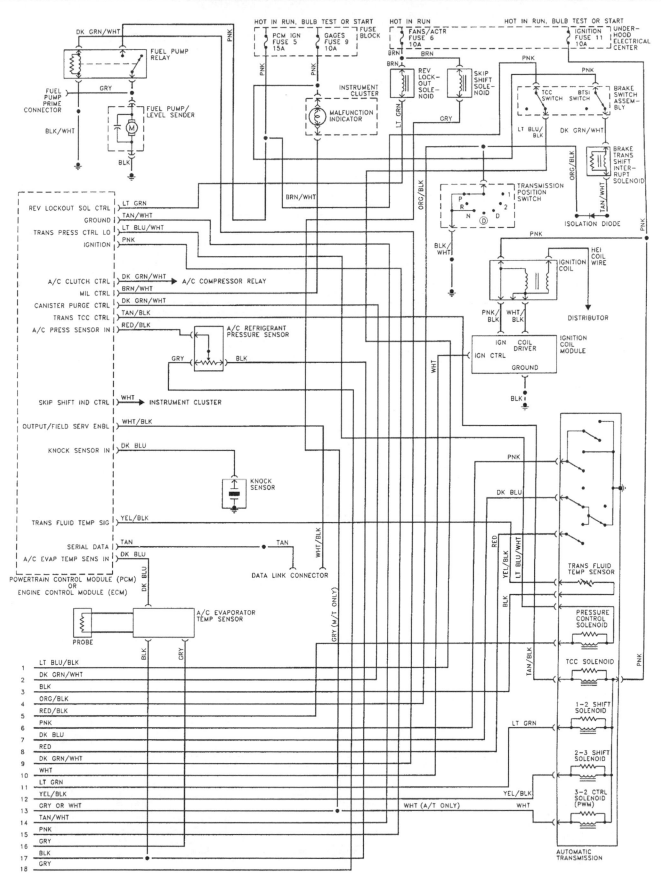

Typical 5.7L V8 engine controls 1995 and earlier models (3 of 3)

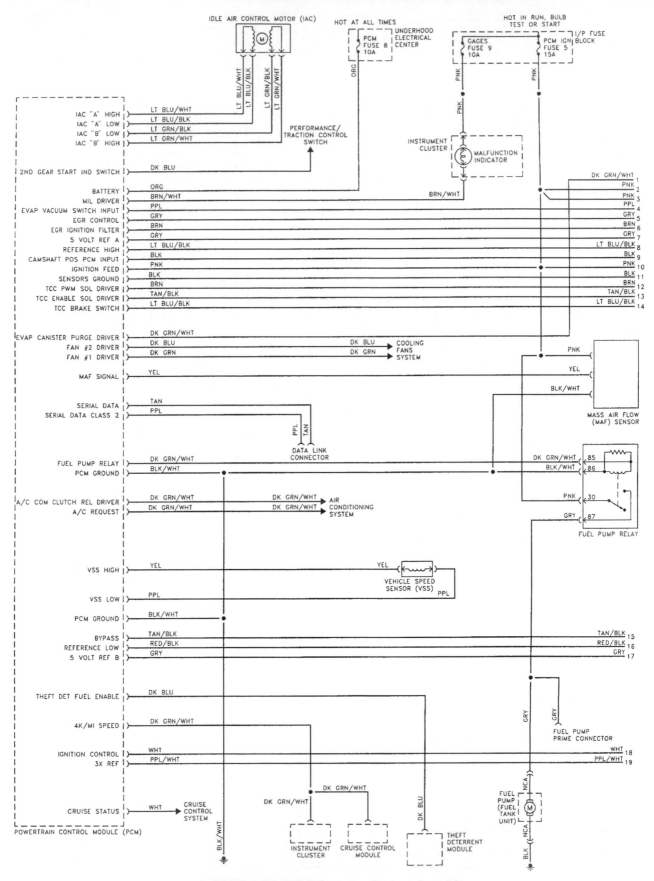

Typical 3.8L V6 engine controls 1995 models (1 of 3)

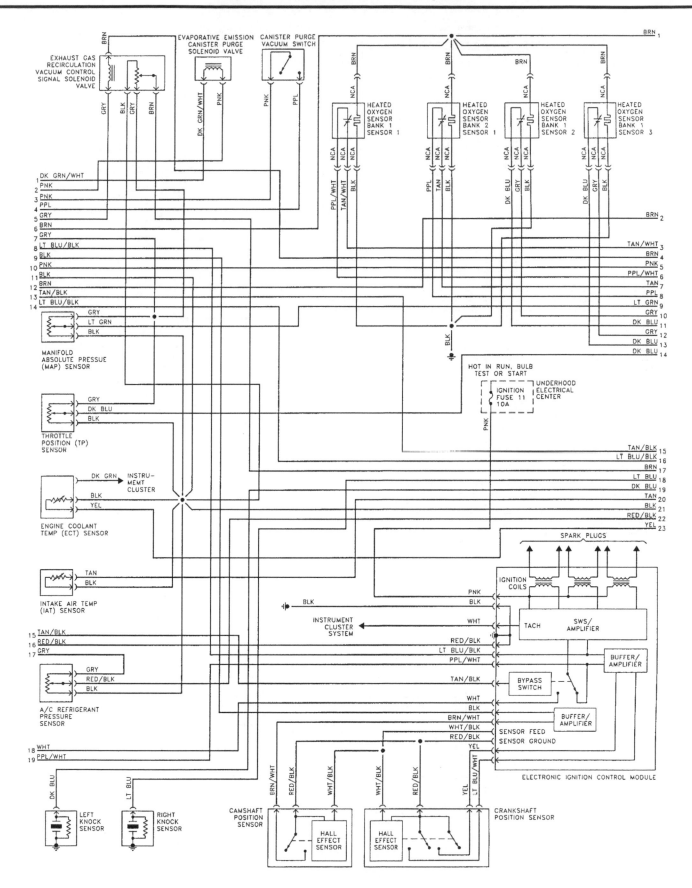

Typical 3.8L V6 engine controls 1995 models (2 of 3)

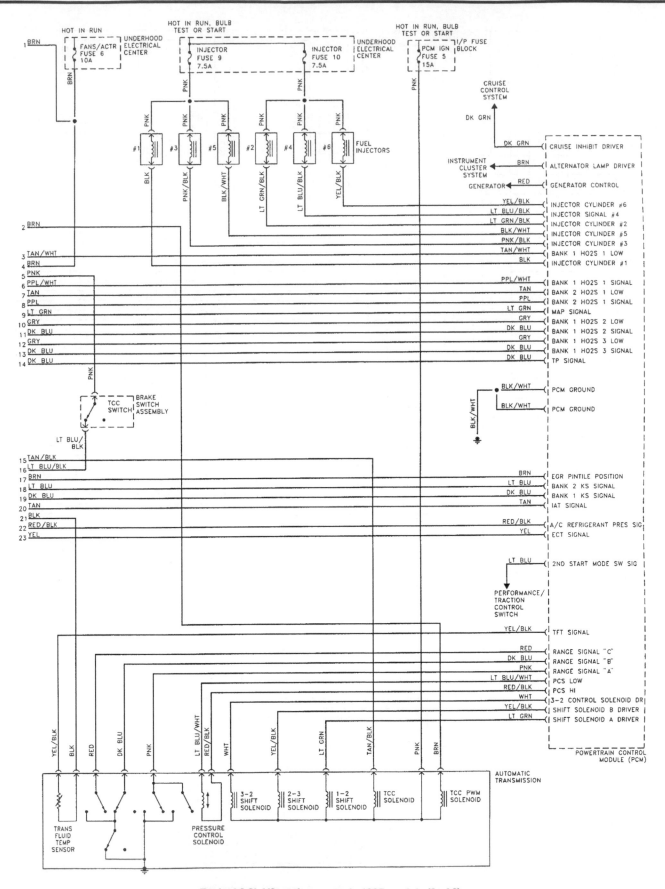

Typical 3.8L V6 engine controls 1995 models (3 of 3)

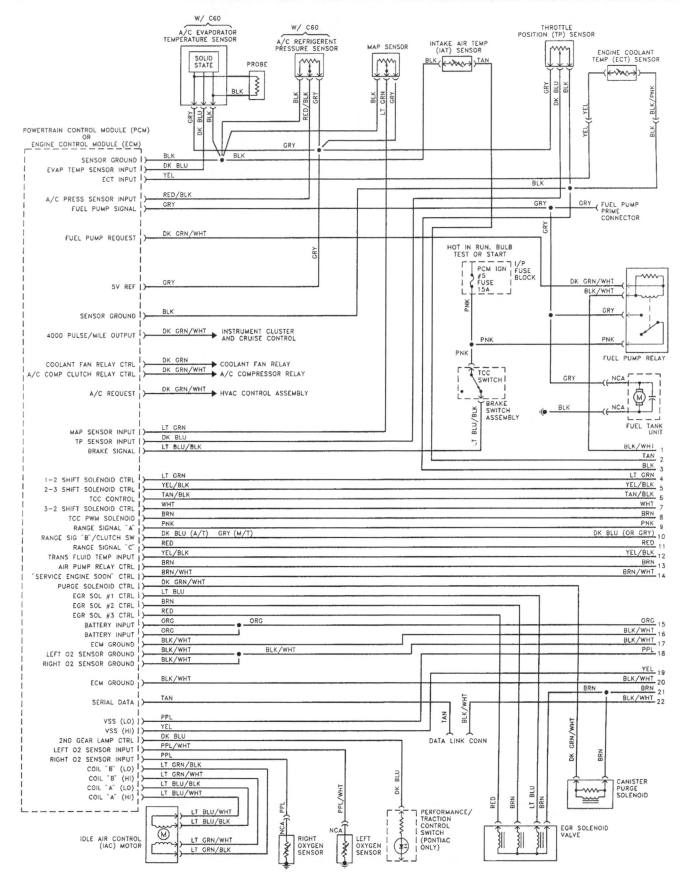

Typical 3.4L V6 engine controls 1995 and earlier models (1 of 3)

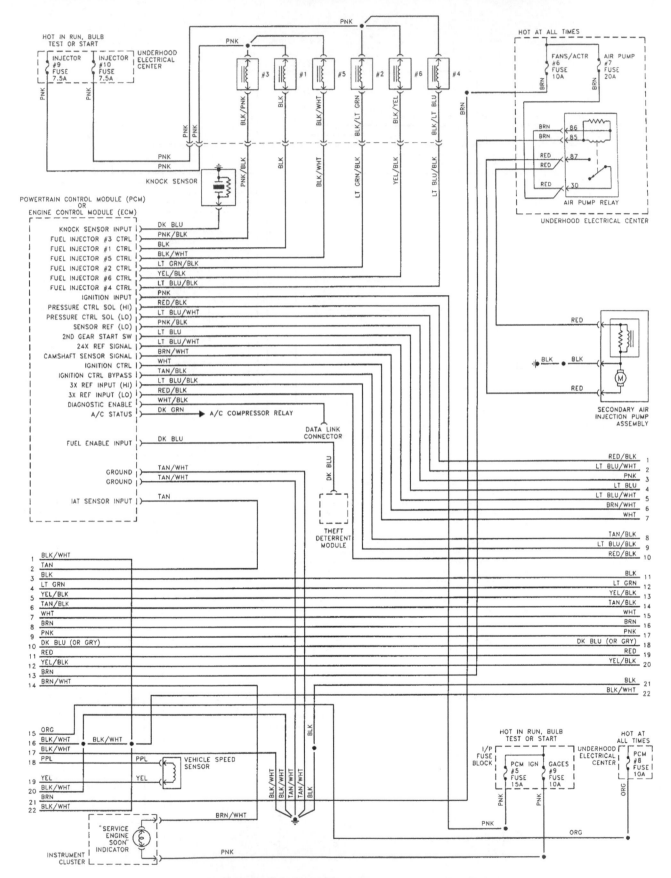

Typical 3.4L V6 engine controls 1995 and earlier models (2 of 3)

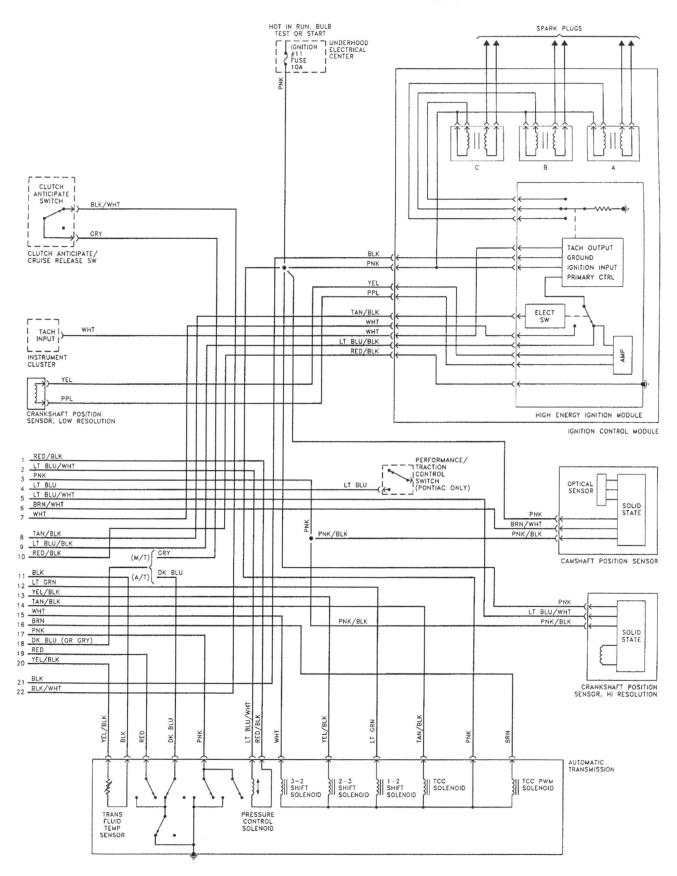

Typical 3.4L V6 engine controls 1995 and earlier models (3 of 3)

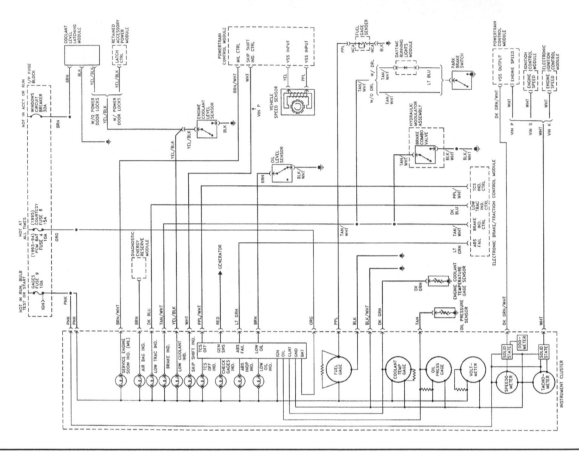

Typical engine warning system 1995 and earlier models

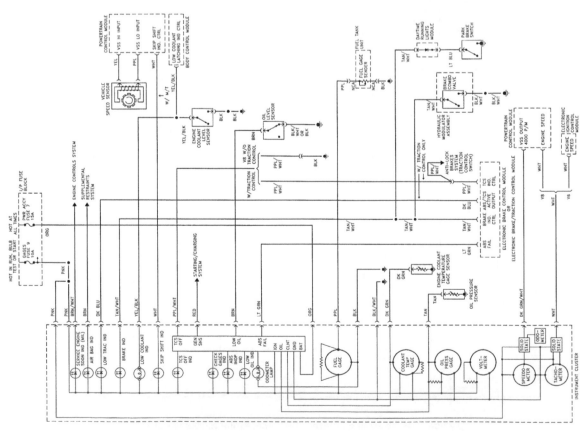

Typical engine warning system 1996 and later models

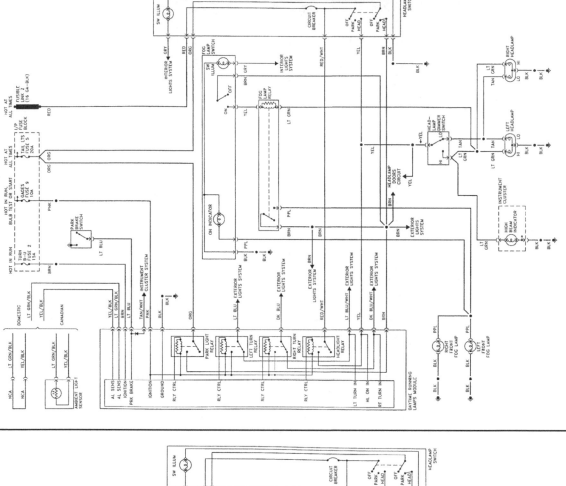

Typical Firebird headlight and fog light system 1996 and later models

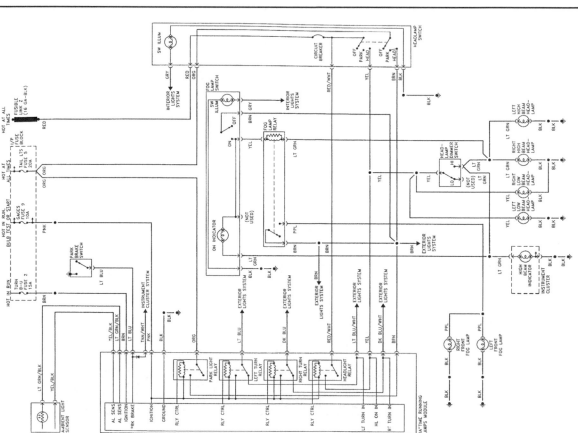

Typical Camaro headlight and fog light system 1996 and later models

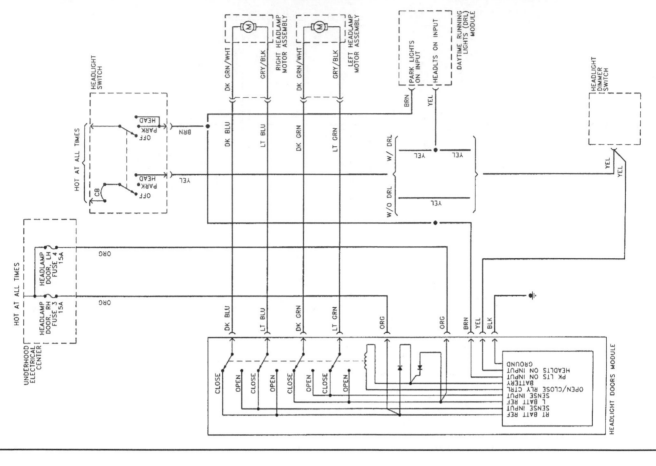

Typical Firebird headlamp door system

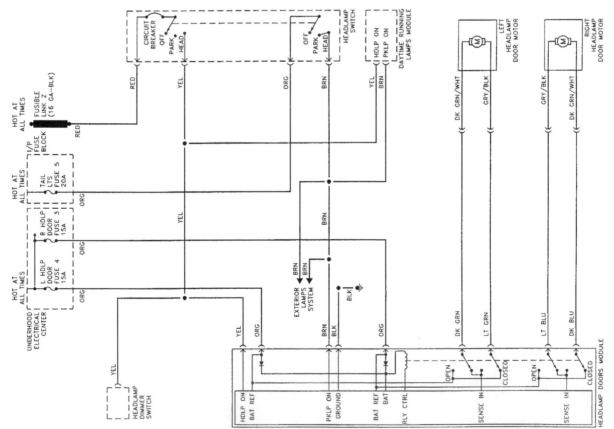

Typical Firebird headlamp door system 1996 and later models

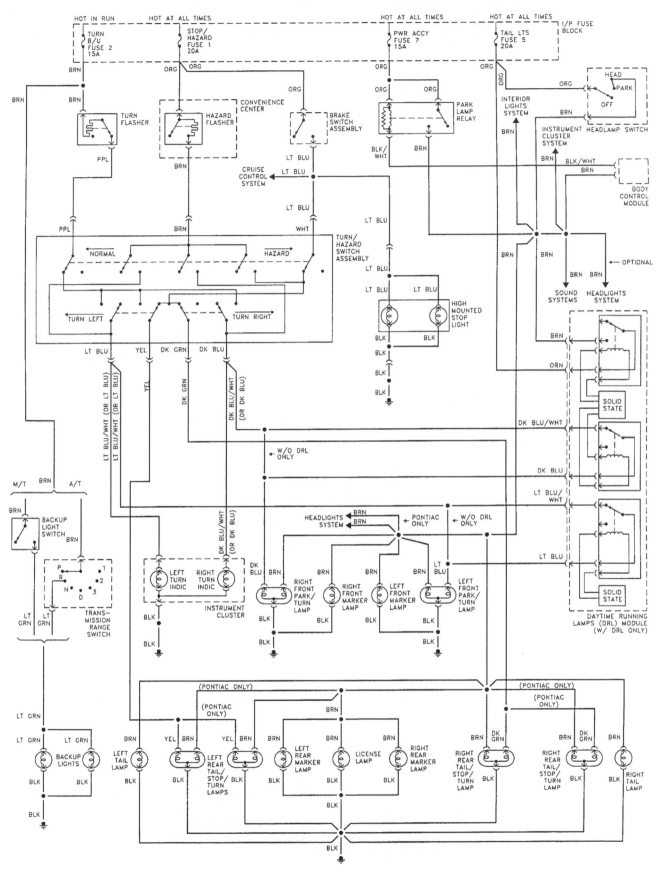

Typical exterior light system 1996 and later models

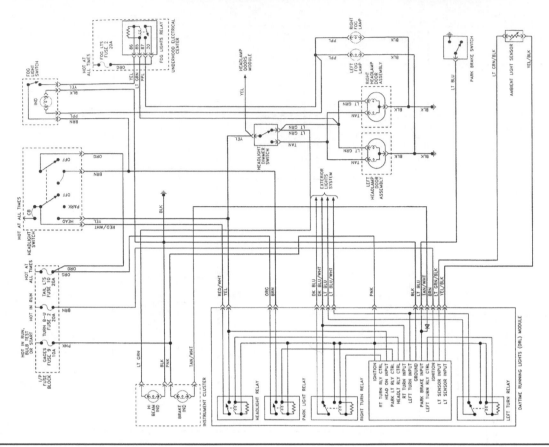

Typical Firebird headlight and fog light system 1995 and earlier models

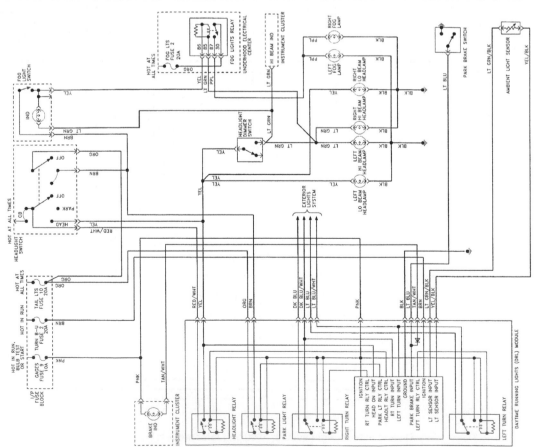

Typical Camaro headlight and fog light system 1995 and earlier models

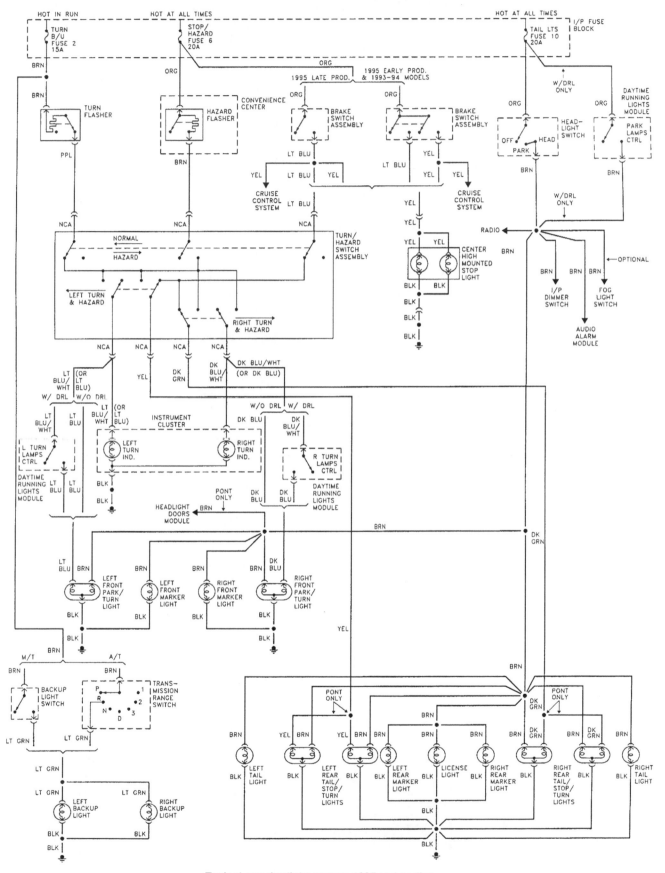

Typical exterior light system 1995 and earlier

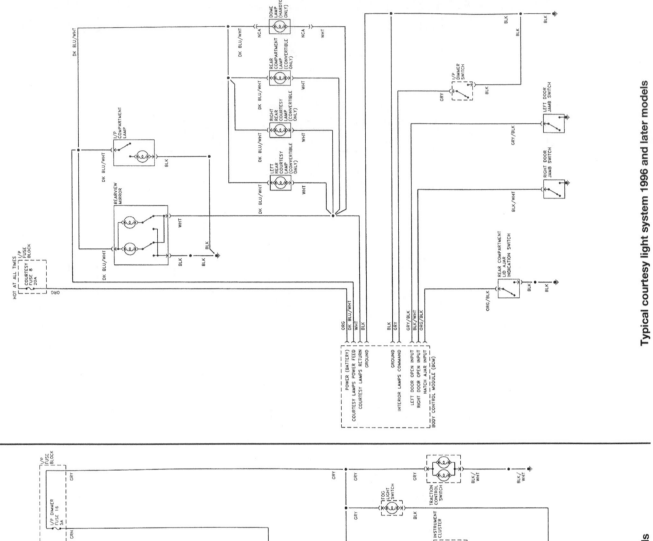

Typical courtesy light system 1996 and later models

Typical illumination system 1996 and later models

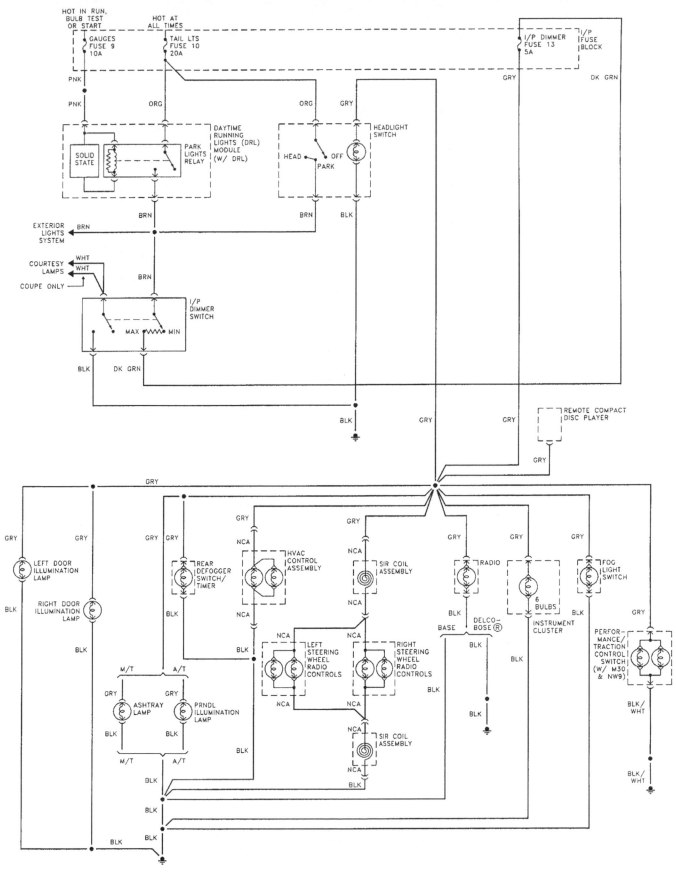

Typical illumination system 1995 and earlier models

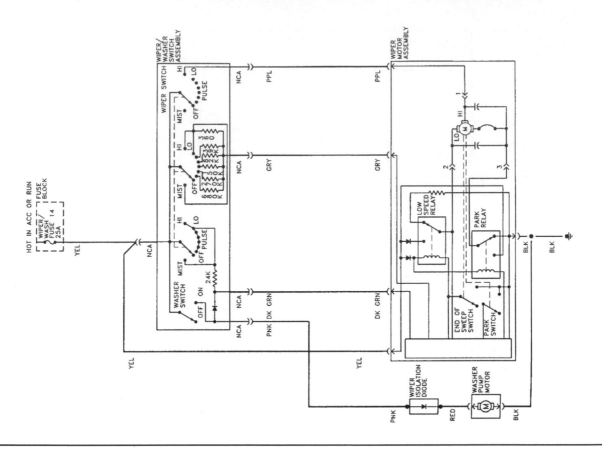

Typical windshield wiper and washer system

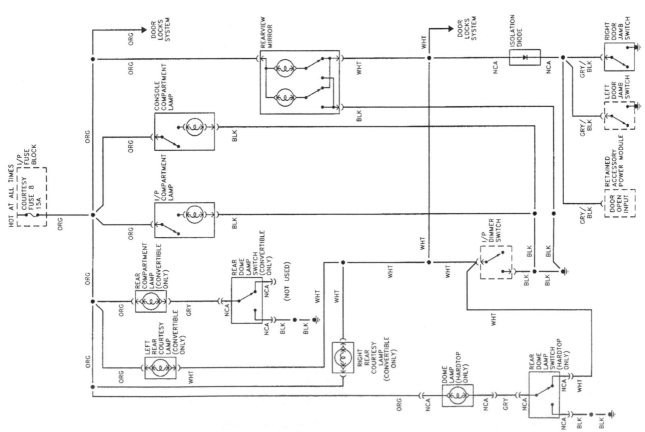

Typical courtesy light system 1995 and earlier models

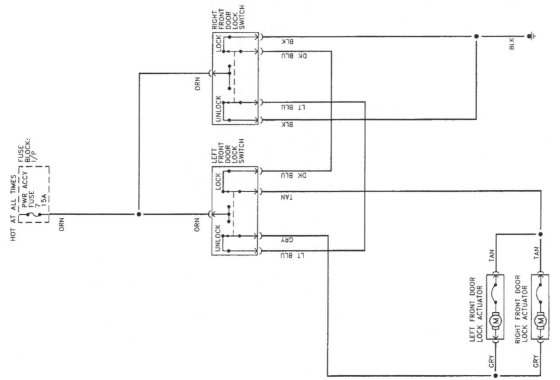

Typical power door lock system 1995 and earlier models

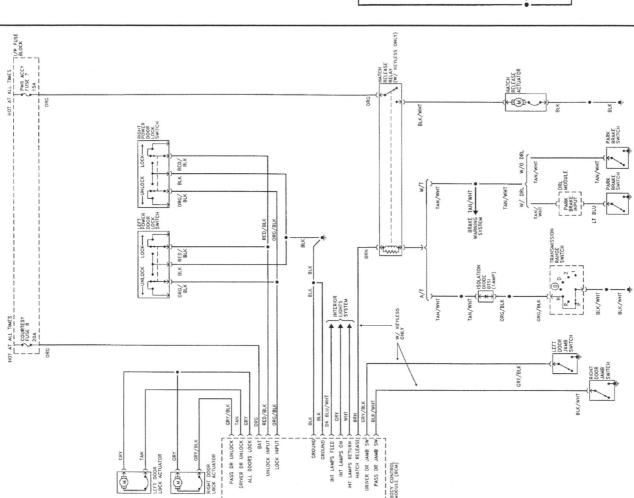

Typical power door lock system 1996 and later models (including remote keyless entry)

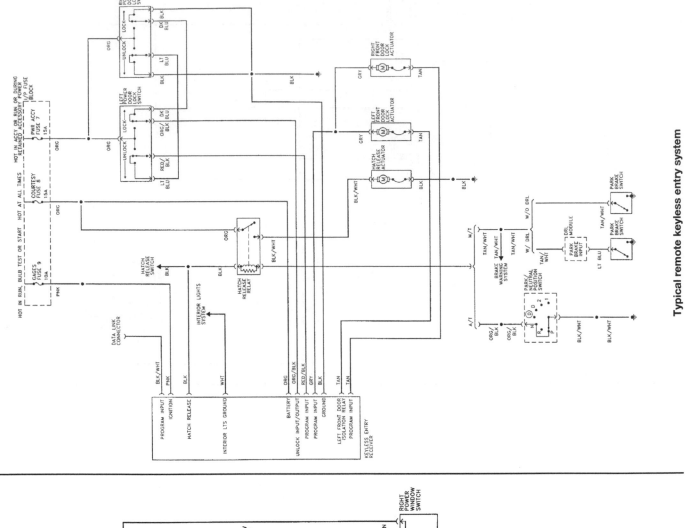

Typical remote keyless entry system

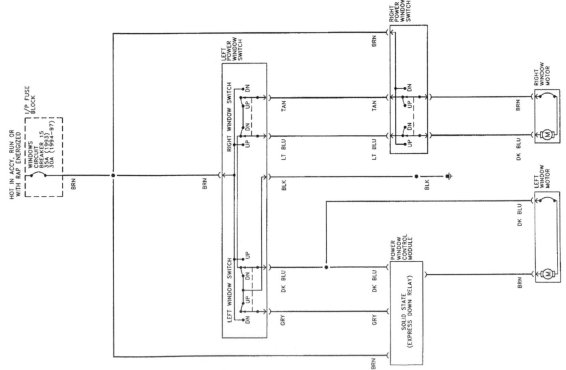

Typical power window system

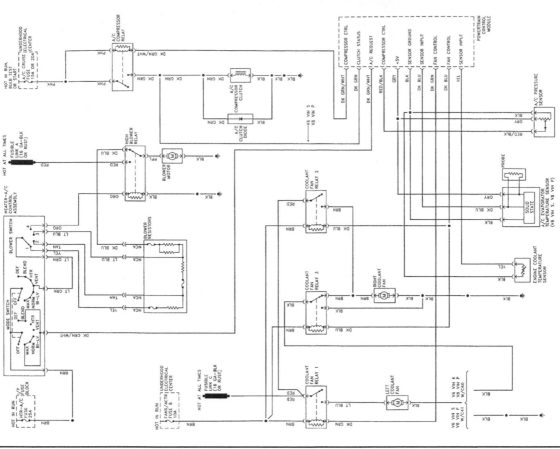

Typical heating and air conditioning sytem 1995 and earlier models

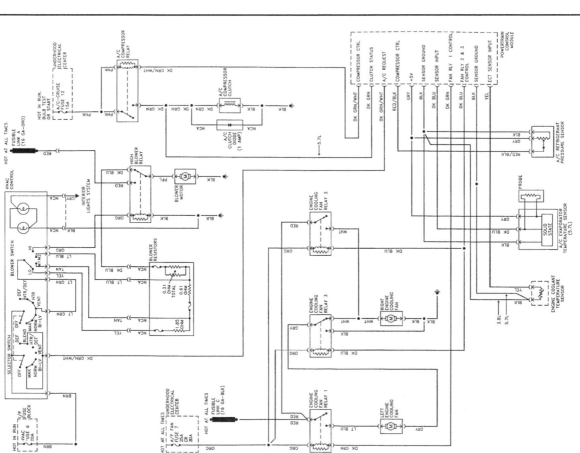

Typical heating and air conditioning sytem 1996 and later models

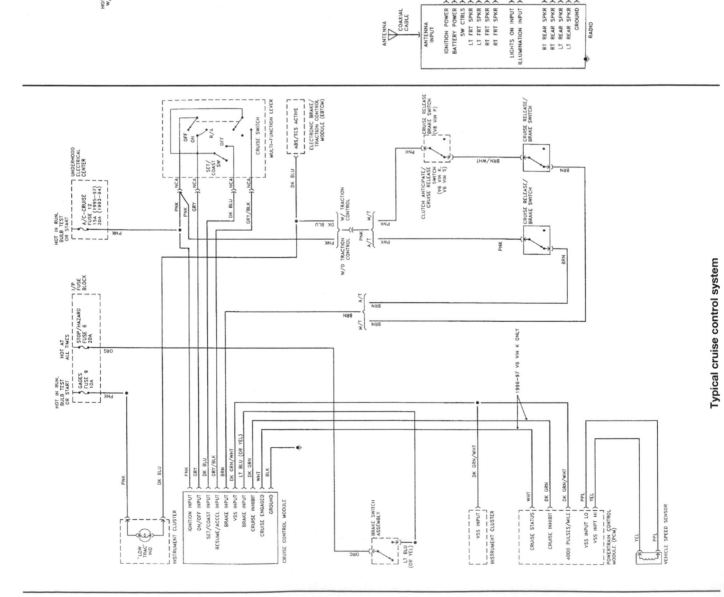

Typical audio system

Typical cruise control system

Index

F

Haynes Automotive Manuals

ACURA
- **12020** Integra '86 thru '89 & Legend '86 thru '90
- **12021** Integra '90 thru '93 & Legend '91 thru '95
- Integra '94 thru '00 - *see HONDA Civic (42025)*
- MDX '01 thru '07 - *see HONDA Pilot (42037)*
- **12050** Acura TL all models '99 thru '08

AMC
- **14020** Mid-size models '70 thru '83
- **14025** (Renault) Alliance & Encore '83 thru '87

AUDI
- **15020** 4000 all models '80 thru '87
- **15025** 5000 all models '77 thru '83
- **15026** 5000 all models '84 thru '88
- Audi A4 '96 thru '01 - *see VW Passat (96023)*
- **15030** Audi A4 '02 thru '08

AUSTIN-HEALEY
- Sprite - *see MG Midget (66015)*

BMW
- **18020** 3/5 Series '82 thru '92
- **18021** 3-Series incl. Z3 models '92 thru '98
- **18022** 3-Series incl. Z4 models '99 thru '05
- **18023** 3-Series '06 thru '14
- **18025** 320i all 4-cylinder models '75 thru '83
- **18050** 1500 thru 2002 except Turbo '59 thru '77

BUICK
- **19010** Buick Century '97 thru '05
- Century (front-wheel drive) - *see GM (38005)*
- **19020** Buick, Oldsmobile & Pontiac Full-size (Front-wheel drive) '85 thru '05
- Buick Electra, LeSabre and Park Avenue; Oldsmobile Delta 88 Royale, Ninety Eight and Regency; Pontiac Bonneville
- **19025** Buick, Oldsmobile & Pontiac Full-size (Rear wheel drive) '70 thru '90
- Buick Estate, Electra, LeSabre, Limited, Oldsmobile Custom Cruiser, Delta 88, Ninety-eight, Pontiac Bonneville, Catalina, Grandville, Parisienne
- **19027** Buick LaCrosse '05 thru '13
- Enclave - *see GENERAL MOTORS (38001)*
- Rainier - *see CHEVROLET (24072)*
- Regal - *see GENERAL MOTORS (38010)*
- Riviera - *see GENERAL MOTORS (38030, 38031)*
- Roadmaster - *see CHEVROLET (24046)*
- Skyhawk - *see GENERAL MOTORS (38015)*
- Skylark - *see GENERAL MOTORS (38020, 38025)*
- Somerset - *see GENERAL MOTORS (38025)*

CADILLAC
- **21015** CTS & CTS-V '03 thru '14
- **21030** Cadillac Rear Wheel Drive '70 thru '93
- Cimarron - *see GENERAL MOTORS (38015)*
- DeVille - *see GENERAL MOTORS (38031 & 38032)*
- Eldorado - *see GENERAL MOTORS (38030)*
- Fleetwood - *see GENERAL MOTORS (38031)*
- Seville - *see GM (38030, 38031 & 38032)*

CHEVROLET
- **10305** Chevrolet Engine Overhaul Manual
- **24010** Astro & GMC Safari Mini-vans '85 thru '05
- **24013** Aveo '04 thru '11
- **24015** Camaro V8 all models '70 thru '81
- **24016** Camaro all models '82 thru '92
- **24017** Camaro & Firebird '93 thru '02
- Cavalier - *see GENERAL MOTORS (38016)*
- Celebrity - *see GENERAL MOTORS (38005)*
- **24018** Camaro '10 thru '15
- **24020** Chevelle, Malibu & El Camino '69 thru '87
- Cobalt - *see GENERAL MOTORS (38017)*
- **24024** Chevette & Pontiac T1000 '76 thru '87
- Citation - *see GENERAL MOTORS (38020)*
- **24027** Colorado & GMC Canyon '04 thru '12
- **24032** Corsica & Beretta all models '87 thru '96
- **24040** Corvette all V8 models '68 thru '82
- **24041** Corvette all models '84 thru '96
- **24042** Corvette all models '97 thru '13
- **24044** Cruze '11 thru '19
- **24045** Full-size Sedans Caprice, Impala, Biscayne, Bel Air & Wagons '69 thru '90
- **24046** Impala SS & Caprice and Buick Roadmaster '91 thru '96
- Impala '00 thru '05 - *see LUMINA (24048)*
- **24047** Impala & Monte Carlo all models '06 thru '11
- Lumina '90 thru '94 - *see GM (38010)*
- **24048** Lumina & Monte Carlo '95 thru '05
- Lumina APV - *see GM (38035)*
- **24050** Luv Pick-up all 2WD & 4WD '72 thru '82
- **24051** Malibu '13 thru '19
- **24055** Monte Carlo all models '70 thru '88
- Monte Carlo '95 thru '01 - *see LUMINA (24048)*
- **24059** Nova all V8 models '69 thru '79
- **24060** Nova and Geo Prizm '85 thru '92
- **24064** Pick-ups '67 thru '87 - Chevrolet & GMC
- **24065** Pick-ups '88 thru '98 - Chevrolet & GMC
- **24066** Pick-ups '99 thru '06 - Chevrolet & GMC
- **24067** Chevrolet Silverado & GMC Sierra '07 thru '14
- **24068** Chevrolet Silverado & GMC Sierra '14 thru '19
- **24070** S-10 & S-15 Pick-ups '82 thru '93, Blazer & Jimmy '83 thru '94,
- **24071** S-10 & Sonoma Pick-ups '94 thru '04, including Blazer, Jimmy & Hombre
- **24072** Chevrolet TrailBlazer, GMC Envoy & Oldsmobile Bravada '02 thru '09
- **24075** Sprint '85 thru '88 & Geo Metro '89 thru '01
- **24080** Vans - Chevrolet & GMC '68 thru '96
- **24081** Chevrolet Express & GMC Savana Full-size Vans '96 thru '19

CHRYSLER
- **10310** Chrysler Engine Overhaul Manual
- **25015** Chrysler Cirrus, Dodge Stratus, Plymouth Breeze '95 thru '00
- **25020** Full-size Front-Wheel Drive '88 thru '93
- K-Cars - *see DODGE Aries (30008)*
- Laser - *see DODGE Daytona (30030)*
- **25025** Chrysler LHS, Concorde, New Yorker, Dodge Intrepid, Eagle Vision, '93 thru '97
- **25026** Chrysler LHS, Concorde, 300M, Dodge Intrepid, '98 thru '04
- **25027** Chrysler 300 '05 thru '18, Dodge Charger '06 thru '18, Magnum '05 thru '08 & Challenger '08 thru '18
- **25030** Chrysler & Plymouth Mid-size front wheel drive '82 thru '95
- Rear-wheel Drive - *see Dodge (30050)*
- **25035** PT Cruiser all models '01 thru '10
- **25040** Chrysler Sebring '95 thru '06, Dodge Stratus '01 thru '06 & Dodge Avenger '95 thru '00
- **25041** Chrysler Sebring '07 thru '10, 200 '11 thru '17 Dodge Avenger '08 thru '14

DATSUN
- **28005** 200SX all models '80 thru '83
- **28012** 240Z, 260Z & 280Z Coupe '70 thru '78
- **28014** 280ZX Coupe & 2+2 '79 thru '83
- 300ZX - *see NISSAN (72010)*
- **28018** 510 & PL521 Pick-up '68 thru '73
- **28020** 510 all models '78 thru '81
- **28022** 620 Series Pick-up all models '73 thru '79
- 720 Series Pick-up - *see NISSAN (72030)*

DODGE
- 400 & 600 - *see CHRYSLER (25030)*
- **30008** Aries & Plymouth Reliant '81 thru '89
- **30010** Caravan & Plymouth Voyager '84 thru '95
- **30011** Caravan & Plymouth Voyager '96 thru '02
- **30012** Challenger & Plymouth Sapporo '78 thru '83
- **30013** Caravan, Chrysler Voyager & Town & Country '03 thru '07
- **30014** Grand Caravan & Chrysler Town & Country '08 thru '18
- **30016** Colt & Plymouth Champ '78 thru '87
- **30020** Dakota Pick-ups all models '87 thru '96
- **30021** Durango '98 & '99 & Dakota '97 thru '99
- **30022** Durango '00 thru '03 & Dakota '00 thru '04
- **30023** Durango '04 thru '09 & Dakota '05 thru '11
- **30025** Dart, Demon, Plymouth Barracuda, Duster & Valiant 6-cylinder models '67 thru '76
- **30030** Daytona & Chrysler Laser '84 thru '89
- Intrepid - *see CHRYSLER (25025, 25026)*
- **30034** Neon all models '95 thru '99
- **30035** Omni & Plymouth Horizon '78 thru '90
- **30036** Dodge & Plymouth Neon '00 thru '05
- **30040** Pick-ups full-size models '74 thru '93
- **30042** Pick-ups full-size models '94 thru '08
- **30043** Pick-ups full-size models '09 thru '18
- **30045** Ram 50/D50 Pick-ups & Raider and Plymouth Arrow Pick-ups '79 thru '93
- **30050** Dodge/Plymouth/Chrysler RWD '71 thru '89
- **30055** Shadow & Plymouth Sundance '87 thru '94
- **30060** Spirit & Plymouth Acclaim '89 thru '95
- **30065** Vans - Dodge & Plymouth '71 thru '03

EAGLE
- Talon - *see MITSUBISHI (68030, 68031)*
- Vision - *see CHRYSLER (25025)*

FIAT
- **34010** 124 Sport Coupe & Spider '68 thru '78
- **34025** X1/9 all models '74 thru '80

FORD
- **10320** Ford Engine Overhaul Manual
- **10355** Ford Automatic Transmission Overhaul
- **11500** Mustang '64-1/2 thru '70 Restoration Guide
- **36004** Aerostar Mini-vans all models '86 thru '97
- **36006** Contour & Mercury Mystique '95 thru '00
- **36008** Courier Pick-up all models '72 thru '82

- **36012** Crown Victoria & Mercury Grand Marquis '88 thru '11
- **36014** Edge '07 thru '19 & Lincoln MKX '07 thru '18
- **36016** Escort & Mercury Lynx all models '81 thru '90
- **36020** Escort & Mercury Tracer '91 thru '02
- **36022** Escape '01 thru '17, Mazda Tribute '01 thru '11, & Mercury Mariner '05 thru '11
- **36024** Explorer & Mazda Navajo '91 thru '01
- **36025** Explorer & Mercury Mountaineer '02 thru '10
- **36026** Explorer '11 thru '17
- **36028** Fairmont & Mercury Zephyr '78 thru '83
- **36030** Festiva & Aspire '88 thru '97
- **36032** Fiesta all models '77 thru '80
- **36034** Focus all models '00 thru '11
- **36035** Focus '12 thru '14
- **36045** Fusion '06 thru '14 & Mercury Milan '06 thru '11
- **36048** Mustang V8 all models '64-1/2 thru '73
- **36049** Mustang II 4-cylinder, V6 & V8 models '74 thru '78
- **36050** Mustang & Mercury Capri '79 thru '93
- **36051** Mustang all models '94 thru '04
- **36052** Mustang '05 thru '14
- **36054** Pick-ups & Bronco '73 thru '79
- **36058** Pick-ups & Bronco '80 thru '96
- **36059** F-150 '97 thru '03, Expedition '97 thru '17, F-250 '97 thru '99, F-150 Heritage '04 & Lincoln Navigator '98 thru '17
- **36060** Super Duty Pick-ups & Excursion '99 thru '10
- **36061** F-150 full-size '04 thru '14
- **36062** Pinto & Mercury Bobcat '75 thru '80
- **36063** F-150 full-size '15 thru '17
- **36064** Super Duty Pick-ups '11 thru '16
- **36066** Probe all models '89 thru '92
- Probe '93 thru '97 - *see MAZDA 626 (61042)*
- **36070** Ranger & Bronco II gas models '83 thru '92
- **36071** Ranger '93 thru '11 & Mazda Pick-ups '94 thru '09
- **36074** Taurus & Mercury Sable '86 thru '95
- **36075** Taurus & Mercury Sable '96 thru '07
- **36076** Taurus '08 thru '14, Five Hundred '05 thru '07, Mercury Montego '05 thru '07 & Sable '08 thru '09
- **36078** Tempo & Mercury Topaz '84 thru '94
- **36082** Thunderbird & Mercury Cougar '83 thru '88
- **36086** Thunderbird & Mercury Cougar '89 thru '97
- **36090** Vans all V8 Econoline models '69 thru '91
- **36094** Vans full size '92 thru '14
- **36097** Windstar '95 thru '03, Freestar & Mercury Monterey Mini-van '04 thru '07

GENERAL MOTORS
- **10360** GM Automatic Transmission Overhaul
- **38001** GMC Acadia '07 thru '16, Buick Enclave '08 thru '17, Saturn Outlook '07 thru '10 & Chevrolet Traverse '09 thru '17
- **38005** Buick Century, Chevrolet Celebrity, Oldsmobile Cutlass Ciera & Pontiac 6000 all models '82 thru '96
- **38010** Buick Regal '88 thru '04, Chevrolet Lumina '88 thru '04, Oldsmobile Cutlass Supreme '88 thru '97 & Pontiac Grand Prix '88 thru '07
- **38015** Buick Skyhawk, Cadillac Cimarron, Chevrolet Cavalier, Oldsmobile Firenza, Pontiac J-2000 & Sunbird '82 thru '94
- **38016** Chevrolet Cavalier & Pontiac Sunfire '95 thru '05
- **38017** Chevrolet Cobalt '05 thru '10, HHR '06 thru '11, Pontiac G5 '07 thru '09, Pursuit '05 thru '06 & Saturn ION '03 thru '07
- **38020** Buick Skylark, Chevrolet Citation, Oldsmobile Omega, Pontiac Phoenix '80 thru '85
- **38025** Buick Skylark '86 thru '98, Somerset '85 thru '87, Oldsmobile Achieva '92 thru '98, Calais '85 thru '91, & Pontiac Grand Am all models '85 thru '98
- **38026** Chevrolet Malibu '97 thru '03, Classic '04 thru '05, Oldsmobile Alero '99 thru '03, Cutlass '97 thru '00, & Pontiac Grand Am '99 thru '03
- **38027** Chevrolet Malibu '04 thru '12, Pontiac G6 '05 thru '10 & Saturn Aura '07 thru '10
- **38030** Cadillac Eldorado, Seville, Oldsmobile Toronado & Buick Riviera '71 thru '85
- **38031** Cadillac Eldorado, Seville, DeVille, Fleetwood, Oldsmobile Toronado & Buick Riviera '86 thru '93
- **38032** Cadillac DeVille '94 thru '05, Seville '92 thru '04 & Cadillac DTS '06 thru '10
- **38035** Chevrolet Lumina APV, Oldsmobile Silhouette & Pontiac Trans Sport all models '90 thru '96
- **38036** Chevrolet Venture '97 thru '05, Oldsmobile Silhouette '97 thru '04, Pontiac Trans Sport '97 thru '98 & Montana '99 thru '05
- **38040** Chevrolet Equinox '05 thru '17, GMC Terrain '10 thru '17 & Pontiac Torrent '06 thru '09

GEO
- Metro - *see CHEVROLET Sprint (24075)*
- Prizm - '85 thru '92 see CHEVY (24060), '93 thru '02 see TOYOTA Corolla (92036)
- **40030** Storm all models '90 thru '93
- Tracker - *see SUZUKI Samurai (90010)*

(Continued on other side)

Haynes Automotive Manuals (continued)

NOTE: If you do not see a listing for your vehicle, please visit **haynes.com** for the latest product information and check out our **Online Manuals!**

GMC

Acadia - *see GENERAL MOTORS (38001)*
Pick-ups - *see CHEVROLET (24027, 24068)*
Vans - *see CHEVROLET (24081)*

HONDA

42010 **Accord CVCC** all models '76 thru '83
42011 **Accord** all models '84 thru '89
42012 **Accord** all models '90 thru '93
42013 **Accord** all models '94 thru '97
42014 **Accord** all models '98 thru '02
42015 **Accord** '03 thru '12 **& Crosstour** '10 thru '14
42016 **Accord** '13 thru '17
42020 **Civic 1200** all models '73 thru '79
42021 **Civic 1300 & 1500 CVCC** '80 thru '83
42022 **Civic 1500 CVCC** all models '75 thru '79
42023 **Civic** all models '84 thru '91
42024 **Civic & del Sol** '92 thru '95
42025 **Civic** '96 thru '00, **CR-V** '97 thru '01
 & Acura Integra '94 thru '00
42026 **Civic** '01 thru '11 **& CR-V** '02 thru '11
42027 **Civic** '12 thru '15 **& CR-V** '12 thru '16
42030 **Fit** '07 thru '13
42035 **Odyssey** all models '99 thru '10
 Passport - *see ISUZU Rodeo (47017)*
42037 **Honda Pilot** '03 thru '08, **Ridgeline** '06 thru '14
 & Acura MDX '01 thru '07
42040 **Prelude CVCC** all models '79 thru '89

HYUNDAI

43010 **Elantra** all models '96 thru '19
43015 **Excel & Accent** all models '86 thru '13
43050 **Santa Fe** all models '01 thru '12
43055 **Sonata** all models '99 thru '14

INFINITI

G35 '03 thru '08 - *see NISSAN 350Z (72011)*

ISUZU

Hombre - *see CHEVROLET S-10 (24071)*
47017 **Rodeo** '91 thru '02, **Amigo** '89 thru '94 & '98 thru '02
 & Honda Passport '95 thru '02
47020 **Trooper** '84 thru '91 **& Pick-up** '81 thru '93

JAGUAR

49010 **XJ6** all 6-cylinder models '68 thru '86
49011 **XJ6** all models '88 thru '94
49015 **XJ12 & XJS** all 12-cylinder models '72 thru '85

JEEP

50010 **Cherokee, Comanche & Wagoneer Limited**
 all models '84 thru '01
50011 **Cherokee** '14 thru '19
50020 **CJ** all models '49 thru '86
50025 **Grand Cherokee** all models '93 thru '04
50026 **Grand Cherokee** '05 thru '19
 & Dodge Durango '11 thru '19
50029 **Grand Wagoneer & Pick-up** '72 thru '91
 Grand Wagoneer '84 thru '91, Cherokee &
 Wagoneer '72 thru '83, Pick-up '72 thru '88
50030 **Wrangler** all models '87 thru '17
50035 **Liberty** '02 thru '12 **& Dodge Nitro** '07 thru '11
50050 **Patriot & Compass** '07 thru '17

KIA

54050 **Optima** '01 thru '10
54060 **Sedona** '02 thru '14
54070 **Sephia** '94 thru '01, **Spectra** '00 thru '09,
 Sportage '05 thru '20
54077 **Sorento** '03 thru '13

LEXUS

ES 300/330 - *see TOYOTA Camry (92007, 92008)*
ES 350 - *see TOYOTA Camry (92009)*
RX 300/330/350 - *see TOYOTA Highlander (92095)*

LINCOLN

MKX - *see FORD (36014)*
Navigator - *see FORD Pick-up (36059)*
59010 **Rear-Wheel Drive Continental** '70 thru '87,
 Mark Series '70 thru '92 **& Town Car** '81 thru '10

MAZDA

61010 **GLC (rear-wheel drive)** '77 thru '83
61011 **GLC (front-wheel drive)** '81 thru '85
61012 **Mazda3** '04 thru '11
61015 **323 & Protegé** '90 thru '03
61016 **MX-5 Miata** '90 thru '14
61020 **MPV** all models '89 thru '98
 Navajo - *see Ford Explorer (36024)*
61030 **Pick-ups** '72 thru '93
 Pick-ups '94 thru '09 - *see Ford Ranger (36071)*
61035 **RX-7** all models '79 thru '85
61036 **RX-7** all models '86 thru '91
61040 **626 (rear-wheel drive)** all models '79 thru '82
61041 **626 & MX-6 (front-wheel drive)** '83 thru '92
61042 **626** '93 thru '01 **& MX-6/Ford Probe** '93 thru '02
61043 **Mazda6** '03 thru '13

MERCEDES-BENZ

63012 **123 Series Diesel** '76 thru '85
63015 **190 Series** 4-cylinder gas models '84 thru '88
63020 **230/250/280** 6-cylinder SOHC models '68 thru '72
63025 **280 123 Series** gas models '77 thru '81
63030 **350 & 450** all models '71 thru '80
63040 **C-Class:** C230/C240/C280/C320/C350 '01 thru '07

MERCURY

64200 **Villager & Nissan Quest** '93 thru '01
 All other titles, see FORD Listing.

MG

66010 **MGB** Roadster & GT Coupe '62 thru '80
66015 **MG Midget, Austin Healey Sprite** '58 thru '80

MINI

67020 **Mini** '02 thru '13

MITSUBISHI

68020 **Cordia, Tredia, Galant, Precis & Mirage** '83 thru '93
68030 **Eclipse, Eagle Talon & Plymouth Laser** '90 thru '94
68031 **Eclipse** '95 thru '05 **& Eagle Talon** '95 thru '98
68035 **Galant** '94 thru '12
68040 **Pick-up** '83 thru '96 **& Montero** '83 thru '93

NISSAN

72010 **300ZX** all models including Turbo '84 thru '89
72011 **350Z & Infiniti G35** all models '03 thru '08
72015 **Altima** all models '93 thru '06
72016 **Altima** '07 thru '12
72020 **Maxima** all models '85 thru '92
72021 **Maxima** all models '93 thru '08
72025 **Murano** '03 thru '14
72030 **Pick-ups** '80 thru '97 **& Pathfinder** '87 thru '95
72031 **Frontier** '98 thru '04, **Xterra** '00 thru '04,
 & Pathfinder '96 thru '04
72032 **Frontier & Xterra** '05 thru '14
72037 **Pathfinder** '05 thru '14
72040 **Pulsar** all models '83 thru '86
72042 **Roque** all models '08 thru '20
72050 **Sentra** all models '82 thru '94
72051 **Sentra & 200SX** all models '95 thru '06
72060 **Stanza** all models '82 thru '90
72070 **Titan pick-ups** '04 thru '10, **Armada** '05 thru '10
 & Pathfinder Armada '04
72080 **Versa** all models '07 thru '19

OLDSMOBILE

73015 **Cutlass** V6 & V8 gas models '74 thru '88
 For other OLDSMOBILE titles, see BUICK,
 CHEVROLET or GENERAL MOTORS listings.

PLYMOUTH

For PLYMOUTH titles, see DODGE listing.

PONTIAC

79008 **Fiero** all models '84 thru '88
79018 **Firebird** V8 models except Turbo '70 thru '81
79019 **Firebird** all models '82 thru '92
79025 **G6** all models '05 thru '09
79040 **Mid-size Rear-wheel Drive** '70 thru '87
 Vibe '03 thru '10 - *see TOYOTA Corolla (92037)*
 For other PONTIAC titles, see BUICK,
 CHEVROLET or GENERAL MOTORS listings.

PORSCHE

80020 **911** Coupe & Targa models '65 thru '89
80025 **914** all 4-cylinder models '69 thru '76
80030 **924** all models including Turbo '76 thru '82
80035 **944** all models including Turbo '83 thru '89

RENAULT

Alliance & Encore - *see AMC (14025)*

SAAB

84010 **900** all models including Turbo '79 thru '88

SATURN

87010 **Saturn** all S-series models '91 thru '02
 Saturn Ion '03 thru '07- *see GM (38017)*
 Saturn Outlook - *see GM (38001)*
87020 **Saturn L-series** all models '00 thru '04
87040 **Saturn VUE** '02 thru '09

SUBARU

89002 **1100, 1300, 1400 & 1600** '71 thru '79
89003 **1600 & 1800** 2WD & 4WD '80 thru '94
89080 **Impreza** '02 thru '11, **WRX** '02 thru '14,
 & WRX STI '04 thru '14
89100 **Legacy** all models '90 thru '99
89101 **Legacy & Forester** '00 thru '09
89102 **Legacy** '10 thru '16 **& Forester** '12 thru '16

SUZUKI

90010 **Samurai/Sidekick & Geo Tracker** '86 thru '01

TOYOTA

92005 **Camry** all models '83 thru '91
92006 **Camry** '92 thru '96 **& Avalon** '95 thru '96
92007 **Camry, Avalon, Solara, Lexus ES 300** '97 thru '01

92008 **Camry, Avalon, Lexus ES 300/330** '02 thru '06
 & Solara '02 thru '08
92009 **Camry, Avalon & Lexus ES 350** '07 thru '17
92015 **Celica Rear-wheel Drive** '71 thru '85
92020 **Celica Front-wheel Drive** '86 thru '99
92025 **Celica Supra** all models '79 thru '92
92030 **Corolla** all models '75 thru '79
92032 **Corolla** all rear-wheel drive models '80 thru '87
92035 **Corolla** all front-wheel drive models '84 thru '92
92036 **Corolla & Geo/Chevrolet Prizm** '93 thru '02
92037 **Corolla** '03 thru '19, **Matrix** '03 thru '14,
 & Pontiac Vibe '03 thru '10
92040 **Corolla Tercel** all models '80 thru '82
92045 **Corona** all models '74 thru '82
92050 **Cressida** all models '78 thru '82
92055 **Land Cruiser FJ40, 43, 45, 55** '68 thru '82
92056 **Land Cruiser FJ60, 62, 80, FZJ80** '80 thru '96
92060 **Matrix** '03 thru '11 **& Pontiac Vibe** '03 thru '10
92065 **MR2** all models '85 thru '87
92070 **Pick-up** all models '69 thru '78
92075 **Pick-up** all models '79 thru '95
92076 **Tacoma** '95 thru '04, **4Runner** '96 thru '02
 & T100 '93 thru '08
92077 **Tacoma** all models '05 thru '18
92078 **Tundra** '00 thru '06 **& Sequoia** '01 thru '07
92079 **4Runner** all models '03 thru '09
92080 **Previa** all models '91 thru '95
92081 **Prius** all models '01 thru '12
92082 **RAV4** all models '96 thru '12
92085 **Tercel** all models '87 thru '94
92090 **Sienna** all models '98 thru '10
92095 **Highlander** '01 thru '19
 & Lexus RX330/330/350 '99 thru '19
92179 **Tundra** '07 thru '19 **& Sequoia** '08 thru '19

TRIUMPH

94007 **Spitfire** all models '62 thru '81
94010 **TR7** all models '75 thru '81

VW

96008 **Beetle & Karmann Ghia** '54 thru '79
96009 **New Beetle** '98 thru '10
96016 **Rabbit, Jetta, Scirocco & Pick-up**
 gas models '75 thru '92 **& Convertible** '80 thru '92
96017 **Golf, GTI & Jetta** '93 thru '98, **Cabrio** '95 thru '02
96018 **Golf, GTI, Jetta** '99 thru '05
96019 **Jetta, Rabbit, GLI, GTI & Golf** '05 thru '11
96020 **Rabbit, Jetta & Pick-up** diesel '77 thru '84
96021 **Jetta** '11 thru '18 **& Golf** '15 thru '19
96023 **Passat** '98 thru '05 **& Audi A4** '96 thru '01
96030 **Transporter 1600** all models '68 thru '79
96035 **Transporter 1700, 1800 & 2000** '72 thru '79
96040 **Type 3 1500 & 1600** all models '63 thru '73
96045 **Vanagon Air-Cooled** all models '80 thru '83

VOLVO

97010 **120, 130 Series & 1800 Sports** '61 thru '73
97015 **140 Series** all models '66 thru '74
97020 **240 Series** all models '76 thru '93
97040 **740 & 760 Series** all models '82 thru '88
97050 **850 Series** all models '93 thru '97

TECHBOOK MANUALS

10205 **Automotive Computer Codes**
10206 **OBD-II & Electronic Engine Management**
10210 **Automotive Emissions Control Manual**
10215 **Fuel Injection Manual** '78 thru '85
10225 **Holley Carburetor Manual**
10230 **Rochester Carburetor Manual**
10305 **Chevrolet Engine Overhaul Manual**
10320 **Ford Engine Overhaul Manual**
10330 **GM and Ford Diesel Engine Repair Manual**
10331 **Duramax Diesel Engines** '01 thru '19
10332 **Cummins Diesel Engine Performance Manual**
10333 **GM, Ford & Chrysler Engine Performance Manual**
10334 **GM Engine Performance Manual**
10340 **Small Engine Repair Manual,** 5 HP & Less
10341 **Small Engine Repair Manual,** 5.5 thru 20 HP
10345 **Suspension, Steering & Driveline Manual**
10355 **Ford Automatic Transmission Overhaul**
10360 **GM Automatic Transmission Overhaul**
10405 **Automotive Body Repair & Painting**
10410 **Automotive Brake Manual**
10411 **Automotive Anti-lock Brake (ABS) Systems**
10420 **Automotive Electrical Manual**
10425 **Automotive Heating & Air Conditioning**
10435 **Automotive Tools Manual**
10445 **Welding Manual**
10450 **ATV Basics**

Over a 100 Haynes
motorcycle manuals
also available

10/22